# The International Library of Bioethics



Volume 108

The *International Library of Bioethics* comprises international and interdisciplinary volumes on foundational and applied issues in bioethics. The series focuses on perennial topics of aging, mental health, preventive medicine, medical research issues, end of life, biolaw, and other areas of bioethics, whilst expanding into other current and future questions involving the living world. The *International Library of Bioethics* examines the role of bioethics against the background of increasing globalization and interdependency of the world's cultures and government.

The series explores meaningful ways to answer the questions of how to direct biotechnology to human and other living things' ends, how to deal with changed values in the areas of religion, society, and culture, and how to formulate a new way of thinking. That is, to create a new, practical bioethics.

The international Library of Bioethics welcomes book proposals representing the broad interest of the series' focus. We especially encourage proposals addressing aspects of changes in biological and medical research and clinical health care, health policy, medical and biotechnology, and other applied ethical areas involving living things, with an emphasis on those interventions and alterations that force us to re-examine foundational issues.

Andria Bianchi · Janet A. Vogt
Editors

# Intellectual Disabilities and Autism: Ethics and Practice

Springer

*Editors*
Andria Bianchi
Centre for Clinical Ethics
Unity Health Toronto
Toronto, ON, Canada

Dalla Lana School of Public Health
University of Toronto
Toronto, ON, Canada

Janet A. Vogt
Surrey Place
Toronto, ON, Canada

ISSN 2662-9186 ISSN 2662-9194 (electronic)
The International Library of Bioethics
ISBN 978-3-031-61564-1 ISBN 978-3-031-61565-8 (eBook)
https://doi.org/10.1007/978-3-031-61565-8



This Springer imprint is published by the registered company Springer Nature Switzerland AG
The registered company address is: Gewerbestrasse 11, 6330 Cham, Switzerland

*The editors would like to dedicate this book to the people we serve, their families, and all those who support and advocate for them.*

# Foreword

Our societies have perhaps never expressed such a clear and strong desire to foster inclusion and participation. The last decades have seen numerous strides in advancing the rights of people belonging to visible minorities, of people living with disabilities, of people with different sexual and gender orientations. There has been progress. More, obviously, needs to be done to foster human practices which allow each person, regardless of their skin color, their physical or intellectual abilities, or other differences to belong, to participate, and to enact who they are.

The ability to enact who we are and to live a life we aspire to live is part of the very meaning of ethics and the cornerstone idea of human flourishing. Human flourishing, historically, was referred to by Greek thinkers as "eudaimonia". Eudaimonia comes from the Greek "eu" (good) and "daimon" (a spirit or deity). It designated a state and a life where a person was accompanied by a good spirit such that the person made good and wise choices following the advice of this spirit in the form of an inner voice. Socrates famously claimed to be accompanied by a good daimon which he listened to attentively. By extension, eudaimonia came to be considered as a focus of ethical and political thinking. There was an intense and active search to identify what the flourishing life was and consisted of. Accordingly, different thinkers and their followers claimed to find a eudaimonic life in various sets of teachings about, for example, the nature of pleasure (e.g., Epicurus and the Epicureans) or about nature (e.g., Zeno and the Stoics) (Hadot 2002). Aristotle's teachings on ethics and flourishing are some of the best preserved and best known. Although ethics was much about thinking and enacting a flourishing life during Greek and Roman antiquity, philosophy shied away from these more entrenched considerations as it integrated theology in medieval times. After the Renaissance, ethics soon modeled itself to the emerging new science of Bacon, Descartes Galileo, and Newton such that it tried to identify universal ethical laws and principles. It was also influenced by liberalism and movements toward democracy which made ethics relevant to all, but often at the price of being general and allergic to considering differences as it idealized a universal prototypic individual. Ethics has sporadically come back to the deeper considerations about the flourishing life. This is starting to change (Racine 2024) following the impetus of contemporary psychology and welfare economics and work led by figures such as Nussbaum and Sen (Nussbaum 1997; Sen 2002).

The imagination of what a flourishing life is has historically been highly constrained, reflecting gender biases, sexual orientation biases, classism, and ableism. This can be seen in much of the Greek and Roman scholarly traditions which were developed by men of the elite and reflected their social position. It is also true of modern academics who reflect some of their own biases and preferences in their theories (e.g., in various instances of colonialism, sexism, and ableism). Only of late have our imaginative capabilities been questioned—mostly by the force of activism and of disability rights movements across the world—because of how excluded groups have called for inclusion and participation. These groups have contributed to expanding our imagination of what a flourishing life is. They have also made everyone realize how our own well-being and flourishing is connected to that of others. To make headway in thinking and enacting a more inclusive society, there needs to be thoughtful consideration given to the various challenges to inclusion such that constrained imaginations are enriched based on the acknowledgement of their limits. And there obviously needs to be resources dedicated to fostering different models of health and social services that afford for different views about what flourishing is and what a flourishing person is.

Such is how I understand the work and contribution of Andria Bianchi and Janet A. Vogt, the editors of *Intellectual Disabilities and Autism: Ethics and Practice.* They invite us to envision in theory and enact in practice how we can create and share opportunities for growth. It is a wonderful effort to carry forth ethics as a quest for human flourishing and human development. The editors are deeply involved in this task on a daily basis. Andria Bianchi is a clinical ethicist at Unity Health Toronto, Surrey Place, and University of Toronto, who specializes on considerations related to sex and sexual consent for people with cognitive challenges. Janet A. Vogt is a manager and senior research associate at Surrey Place, a not-for-profit organization at the service of people with developmental disabilities of all ages for the populous region of Toronto and also of the more remote Northwestern Ontario. In undertaking their volume, they wanted to sensitize, inform, and support dialogue. They need to be commended as well as all their colleagues (numbering over 40) who have delivered a truly impressive and remarkable edited volume of 31 chapters. They have managed to recruit contributors from a wide range of backgrounds, including political science, law, medicine, self-advocacy, nursing, and bioethics. Readers will learn about and reflect on the nature of disability and its moral significance, consent issues, and practices in a variety of contexts such as intimate relationships and minimal-risk research, empowerment strategies based on compelling examples and practices, issues related to health care as well as a strong focus on romantic relationships and sexual education. The feat needs to be underscored since the ability to convene so much talent and diversity implies not only undertaking an impressive amount of work but also a capacity to lead so many people in a common and powerful direction. In many respects, to flourish is to learn about others and be on a learning trajectory. In this respect and many others, this volume is a major advance in the discussion about ethical aspects of intellectual disability. It is destined to remain a reference for a long time.

Montréal, Canada Eric Racine, Ph.D., MCAHS

## References

Hadot, Pierre. 2002. *La philosophie comme manière de vivre*. Paris: Albin-Michel.
Nussbaum, Martha C. 1997. Capabilities and human rights. *Fordham Law Review* 66 (2): 273–300.
Racine, Eric. 2024. *The theory of deliberative wisdom*. Cambridge, MA: MIT Press.
Sen, Amartya. 2002. Why health equity? *Health Economics* 11 (8): 659–666.

# Contents

# Editors and Contributors

## About the Editors

**Andria Bianchi** works as a clinical ethicist at the Centre for Clinical Ethics at Unity Health Toronto. As part of her role, she provides ethics support to Surrey Place. Surrey Place is the lead agency for Developmental Services Ontario, Toronto Region, providing support for people with intellectual disabilities and those on the autism spectrum. Given her experiences in the developmental sector, she is often consulted at hospitals within Toronto when ethically complex circumstances arise regarding the provision of care for people with intellectual and developmental disabilities, as well as autistic people. She is also an assistant professor (status-only) at the University of Toronto in the Dalla Lana School of Public Health and an Affiliate Scientist at the KITE Research Institute—University Health Network. She focuses her teaching and research on various ethics challenges related to persons with cognitive impairments and disability. Prior to her career in ethics, she completed her Ph.D. in Philosophy where she focused on sexual decision-making and people with dementia. She has published several peer-reviewed publications, non-peer-reviewed publications, and book chapters on ethical complexities related to persons with cognitive impairment. She frequently presents on topics related to ethics and intellectual disabilities in local, national, and international settings.

**Janet A. Vogt** is a senior research associate and manager at Surrey Place, with experience on their Research Ethics Board, as a co-chair of their Clinical Ethics Committee, and a founder of their Ethics Education working group. She completed her Ph.D. in Nutritional Sciences and M.H.Sc. in Bioethics at the University of Toronto and enjoys mentoring medical students, engaging them in projects to raise their awareness of bioethics issues arising in the delivery of services for people with developmental disabilities. She is on the Board of Directors for Toronto's Community Ethics Network and has led the development of an organizational ethics framework and toolkit for them. She has presented at local, national, and international conferences;

taught university courses in nutrition and research methods; and given guest lectures in bioethics at the graduate level. Her publications include peer-reviewed articles, book chapters, and tools to support health care access for people with developmental disabilities. Her education in developmental disabilities began in childhood as the sister to an older sibling with a severe developmental disability.

## Contributors

**Julie S. Armin** The University of Arizona, Tucson, AZ, USA

**Marjorie Aunos** Department of Applied Disability Studies, Brock University, St. Catharines, ON, Canada

**Jonas-Sébastien Beaudry** Department of Equity, Ethics, and Policy, Faculty of Law, McGill University, Montreal, QC, Canada

**Andria Bianchi** Centre for Clinical Ethics, Unity Health Toronto, Toronto, ON, Canada;
KITE Research Institute, Toronto, Canada;
Dalla Lana School of Public Health, University of Toronto, Toronto, ON, Canada

**Laura R. Bowman** Holland Bloorview Kids Rehabilitation Hospital, Toronto, Canada

**Elspeth Bradley** University of Toronto, Toronto, ON, Canada

**Hilary K. Brown** University of Toronto Scarborough and University of Toronto, Toronto, ON, Canada

**Daniel Z. Buchman** Education, Centre for Addiction and Mental Health, Toronto, ON, Canada;
Bioethics, Centre for Addiction and Mental Health; Clinical Public Health, Dalla Lana School of Public Health, University of Toronto, Toronto, ON, Canada;
Joint Centre for Bioethics, University of Toronto, Toronto, ON, Canada

**Meghan M. Burke** Vanderbilt University, Nashville, TN, USA

**M. Ariel Cascio** Michigan State University, East Lansing, USA

**Shay Dawson** Central Michigan University, Mount Pleasant, USA

**Maurice Feldman** Department of Applied Disability Studies, Brock University, St. Catharines, ON, Canada

**Parker Gamble** Wabash College, Crawfordsville, USA

**Abraham Graber** Nisonger Center Affiliate Faculty, Division of Bioethics, Department of Biomedical Education and Anatomy, College of Medicine, The Ohio State University, Columbus, OH, USA

**Marina Heifetz** York University, Toronto, Canada

**Joy F. Johnson** Spectrum Support, LLC, Timonium, USA

**Amanda Johnston** Vanderbilt University, Nashville, TN, USA

**Chris Kaposy** Faculty of Medicine, Centre for Bioethics, Memorial University, St. John's, NL, Canada

**Trudo Lemmens** Faculty of Law, Dalla Lana School of Public Health, University of Toronto, Toronto, Canada

**Chak Li** Vanderbilt University, Nashville, TN, USA

**Nicole Luke** Brock University, St. Catharines, Canada

**Brooke Magel** Education, Centre for Addiction and Mental Health, Toronto, ON, Canada

**Lorraine Krall McCrary** Political Science, Wabash College, Crawfordsville, USA

**Carolyn McDougall** Holland Bloorview Kids Rehabilitation Hospital, Toronto, Canada

**Dolly Menna-Dack** Holland Bloorview Kids Rehabilitation Hospital, Toronto, Canada

**Laura Pacheco** School of Social Work, Memorial University, St. John's, NL, Canada;
School of Social Work, McGill University, Montreal, Quebec, Canada

**Robyn M. Powell** University of Oklahoma College of Law, Norman, OK, USA;
Lurie Institute for Disability Policy, Brandeis University, Waltham, MA, USA

**Neli Ragina** Central Michigan University, Mount Pleasant, USA

**Dora M. Raymaker** Regional Research Institute for Human Services, School of Social Work & Complex Systems Program, School of Earth, Environment, and Society, Portland State University, Portland, USA

**Gabriel M. Ronen** Department of Pediatrics and CanChild Centre for Childhood Disability Research, McMaster University, Hamilton, ON, Canada

**Peter L. Rosenbaum** Department of Pediatrics and CanChild Centre for Childhood Disability Research, McMaster University, Hamilton, ON, Canada

**Jacob A. Sadavoy** CB Consultants, Sag Harbour, USA

**Elizabeth K. Schmidt** Department of Occupational Therapy, Lincoln Memorial University, Knoxville, USA

**Jane Seale** The Open University, Milton Keynes, UK

**Tess Sheldon** Faculty of Law, University of Windsor, Windsor, Canada

**Stacy Clifford Simplican** Department of Gender and Sexuality Studies, Vanderbilt University, Nashville, TN, USA

**Tim Stainton** Canadian Institute for Inclusion and Citizenship, University of British Columbia, Vancouver, BC, Canada

**Peter Sturmey** The Graduate Center and Department of Psychology, Queens College, City University of New York, New York, USA

**Valerie Temple** Surrey Place, Toronto, ON, Canada

**Adriana Kaori Terol** University of Illinois at Urbana-Champaign, Champaign, IL, USA

**Kiruthiha Vimalakanthan** Surrey Place, Toronto, ON, Canada

**Janet A. Vogt** Surrey Place, Toronto, ON, Canada;
Surrey Place, Scarborough, Canada

**Meredith Wiles** Central Michigan University, Mount Pleasant, USA

**Heather J. Williamson** Northern Arizona University, Flagstaff, AZ, USA

# Abbreviations

| | |
|---|---|
| AASPIRE | Academic Autism Spectrum Partnership in Research and Education |
| ABA | Applied Behavior Analysis |
| ABAI | Applied Behavior Analysis International |
| ADA | Americans with Disabilities Act |
| ADAPT | American Disabled for Attendant Programs Today |
| ASAN | Autistic Self Advocacy Network |
| AT | Assistive Technology |
| AU | Alcohol Use |
| AUD | Alcohol Use Disorder |
| BACB® | Behavior Analyst Certification Board |
| CAHPS | Consumer Assessment of Healthcare Providers and Systems |
| CDS | Critical Disability Studies |
| CPR | Cardiopulmonary Resuscitation |
| CRPD | Convention on the Rights of Persons with Disabilities |
| CW | Child Welfare |
| DMC | Decision Making Capacity |
| EPP | Employment Participation Pathway |
| FASD | Fetal Alcohol Spectrum Disorder |
| FBA | Functional Behavioral Assessment |
| FMS | Full Moral Status |
| HEEDS | Healthcare Education Engages Disability Studies |
| HIV | Human Immunodeficiency Virus |
| HPV | Human Papillomavirus |
| ICE | Initial Clinical Experience |
| ICF | International Classification of Functioning Disability and Health |
| ICIDH | International Classification of Impairments, Disabilities and Handicaps |
| ICT | Information and Communication Technologies |
| ID | Intellectual Disability(ies) |
| IDD | Intellectual and Developmental Disability(ies) |
| IQ | Intelligence Quotient |

| | |
|---|---|
| IT | Information Technologies |
| IUD | Intrauterine Device |
| LGBTQ+ | Lesbian, Gay, Bisexual, Transgender, Queer or Questioning, Intersex, Asexual, and more |
| MEES | Model of Embedded Employment Supports |
| OPZ | Openbaar Psychiatrisch Zorgcentrum |
| OSCE | Objective Standardized Clinical Examination |
| PAE | Prenatal Alcohol Exposure |
| PCA | Parenting Capacity Assessment |
| PECS | Picture Exchange Communication System |
| PLAN | Planned Lifetime Advocacy Network |
| PRN | Pro Re Nata |
| PSID | People with Severe Intellectual Disabilities |
| PWD | People with Disabilities |
| RDSP | Registered Disability Savings Plan |
| SIB | Self-Injurious Behavior |
| SP | Standardized Patient |
| STIs | Sexually Transmitted Infections |
| UN | United Nations |
| WHO | World Health Organization |
| YAC | Youth Advisory Council |
| YES | Youth Engagement Strategy |

# Chapter 1
# Introduction

**Andria Bianchi and Janet A. Vogt**

Have you ever considered that people with intellectual disabilities have a right to become parents? Have you ever thought about the ways in which healthcare services tend to "treat" autistic people? How do you feel about selective termination of pregnancy based on a gene linked to a condition conferring intellectual impairment? To what extent might your feelings about these—and other questions regarding neurodiverse populations—be based on widespread ableism that pervades our organizations, systems and society? The purpose of this volume is to engage with ethically complex questions relevant to people with intellectual and developmental disabilities (IDD) and/or autistic people, as well as those caring for and working with neurodiverse populations. As editors of this collection, we bring our perspectives as a clinical ethicist and a program evaluation specialist (and sibling of a brother with a severe intellectual disability) who work together in a large developmental service agency in Toronto, Canada. The agency serves individuals of all ages, including autistic people, and those with developmental and sensory-related concerns. As the organization's ethicist and ethics resource lead, respectively, we hear first-hand and through ethics needs assessments, about the ethical issues our clinicians and leaders encounter in their daily work. Together, we lead the Ethics Education working group in the provision of online events and resources to support the ethics education of staff.

The idea for this book came about when we discussed our respective experiences completing graduate degrees in ethics. Both of us found a paucity of examples and discussion regarding ethical issues arising from the treatment of neurodiverse

A. Bianchi (✉)
Centre for Clinical Ethics, Unity Health Toronto, Toronto, ON, Canada
e-mail: andria.bianchi@utoronto.ca

Dalla Lana School of Public Health, University of Toronto, Toronto, ON, Canada

J. A. Vogt
Surrey Place, Toronto, ON, Canada
e-mail: j.vogt@alumni.utoronto.ca; janet.vogt@surreyplace.ca

 
A. Bianchi and J. A. Vogt (eds.), *Intellectual Disabilities and Autism: Ethics and Practice*, The International Library of Bioethics 108,
https://doi.org/10.1007/978-3-031-61565-8_1

populations and felt the ethics programs with which we were familiar paid insufficient attention to these populations. Relatedly, as an ethicist working in hospital settings, Andria has found that ethical conundrums often arise amongst clinical teams when it comes to working with autistic patients and/or patients with IDD who may present or communicate differently than people considered neurotypical. Thus, the primary reason for developing this collected volume is to give due and deserved attention to ethical issues affecting those we serve—those who are influenced by ableist stereotypes.

Ableism assumes the existence of a perfect kind of human self and body to which "disabled" people are compared unfavorably (Campbell 2001). Consequently, the life of a person with a disability may be deemed as less valuable than the life of a person without a disability (Morris 2023). Ableism involves "… stereotyping, prejudice, discrimination, and social oppression toward people with disabilities" (Bogart and Dunn 2019). Conversely, we take the view that: "… disability is an evolving concept and that disability results from the interaction between persons with impairments and attitudinal and environmental barriers that hinders their full and effective participation in society on an equal bases [sic] with others, …" (UN 2006, Preamble). Ableism contributes to the barriers to social inclusion experienced by people with IDD and many autistic people. In fact, such barriers can pervade their lives; some examples being barriers to equitable health care, sexual autonomy, meaningful employment, the necessary technology and the skills to use it, and participation in research and politics.

Starting with a framework encompassing six key areas of ethical tension that have practical implications: moral status (Chaps. 2–5); capacity and consent (Chaps. 6–9); inclusion and empowerment (Chaps. 10–18); sexuality, intimacy, romance, and parenting (Chaps. 19–22); health care (Chaps. 23–26); and ethical responses to 'behaviours' (Chaps. 27–31), we approached a range of early-career to seasoned authors who had published recent work in these areas. We encouraged each author to propose a specific chapter topic related to one of the aforementioned areas of tension; this process contributed to the many issues identified and perspectives taken throughout the volume. Our assertion is that in order to effectively understand and analyze ethical issues arising in our work with these populations, we must be aware of, and informed by, the work of our colleagues in other disciplines (Cox 2012). Although principles and values will always be fundamental to the process of ethical deliberation, the perspectives contributed by other disciplines can enrich our understanding of the clinical context in important ways.

In order to capture the multiple perspectives of many disciplines involved in supporting and advocating for neurodiverse populations, we invited not only ethicists, but also scholars and practitioners from various fields (psychology, philosophy/ethics, applied behavior analysis, disability studies, political science/policy, occupational therapy, social work, law, education, rehabilitation sciences, medical anthropology, epidemiology, developmental pediatrics, neurology, genetics, psychiatry, gender and sexuality, public health, and system science) to contribute chapters. Many chapters are written by clinicians and researchers whose work focuses on areas that are fraught with ethical challenges, and who offer practical ways to respond to such challenges. Other authors turn to philosophical assistance in grappling with ethically complex questions, considering what it means to be a person, to have agency, and to make autonomous decisions. Some authors respond to relatively narrow questions (e.g., what constitutes an ethical response to alcohol use and people with IDD in Chap. 28), whereas others take up more general perennial issues (e.g., consent and capacity in Chaps. 2–5). This blend of theory and practice is intended to empower our readers to: (a) reflect on complex ethical challenges relevant to the developmental sector, each of which stem from ableist structures in our general society; and (b) learn about what people, organizations, and/or disciplines are doing in response to such ethical complexities *in practice*. Overall, each contributor to this volume critically and meaningfully examines how we ought to include and respond to autistic people and individuals with IDD in our ever-changing and ethically complex world; a world that needs to better understand and embrace neurodiversity.

Prior to collating chapters, we hosted online meetings with contributors to discuss our motive for developing the volume and explore overarching topics that relate to it. During these meetings, two particularly impactful discussions occurred. One revolved around the use of person-first language and identity-first language.[1] Several contributors to this volume have expertise in and focus their work on autistic people. As such, it was brought to our attention that neither we—as editors of this volume—nor others were necessarily nor consistently using the language endorsed by the autistic community on a large scale; namely, that of identity-first language (e.g., "Autistic person"). In response to this discussion, authors were encouraged to consider highlighting their rationale for using person- or identity-first language, specifically if their chapters focused exclusively on autistic people. The resulting volume offers some chapters that use identity-first language, with others using person-first language (e.g., people with IDD) and, as such, is fairly representative of the evolving use of language. However, it is worth noting that in our meetings and across the volume, there exists a consensus that it should be up to the person themselves whether they choose to use person- or identity-first language, and others ought to respect their preference.

The second discussion questioned the inclusion of autism spectrum disorder and intellectual (and developmental) disabilities in the same volume and under the same umbrella, as it were. The simple response is that we, as co-editors, work for an organization that supports both groups and consequently, the ethical tensions we identified are applicable, to a greater or lesser extent, to both populations. One reason that the ethical tensions apply to both groups is, at least in part, because we live in an ableist society (Nario-Redmond et al. 2019)—a society that continues not to support neurodiversity in the right kinds of ways. As a consequence, both groups may experience similar kinds of challenges in health care environments, stigma in relation to parenting, questions about their moral worthiness, and require similar accommodations in order to participate meaningfully in day-to-day activities. In addition, recent data suggests that approximately 40 percent of autistic individuals

[1] For a brief overview of person-first and identity-first languages, and the importance of a person-centered approach, refer to Chap. 23, Sect. 3.3.

are likely to have a co-occurring intellectual impairment (Christensen and Zubler 2020). Consequently, readers will find that the focus of individual chapters may be solely on autism, on IDD, or inclusive of people with IDD and/or on the autism spectrum; however, the theme of ableism and the way in which its prevalence pervades the lives of, and workings with, those who are neurodiverse is embedded across all.

With a desire to identify and respond to the many ways that ableism infiltrates our society, subjecting people with IDD and/or those on the autism spectrum to unjust treatment, this book aims to influence bioethicists, ethics students, social justice scholars, and those interested in autism and IDD (e.g., clinicians, service providers, researchers, and policy makers) to learn about and meaningfully reflect upon difficult, timely, and debated ethics questions. Although there exists a range of disciplinary perspectives and approaches to ethics questions across the volume, many of the arguments and discussions put forward in this book resonate with various articles of the United Nations (UN) Convention on the Rights of Persons with Disabilities (CRPD) (UN 2006). The purpose of the CRPD is to help protect the rights of people with disabilities worldwide, as guided by eight general principles: respect for the individual's inherent dignity and autonomy; non-discrimination; full and effective inclusion and participation in society; respect for difference and acceptance of persons with disabilities as part of human diversity and humanity; equality of opportunity; accessibility; gender equality; and respect for the evolving capacities of children with disabilities (CRPD, Article 3). In spite of the broad range of rights promoted and protected in the CRPD, people with IDD still encounter barriers to accessing these rights and the people who support them may experience significant moral distress as a result (Fisher et al. 2020).

The primary themes explored in this book are what separate each section, though there exists some overlap across sections. By working through this volume, we hope that readers will be better positioned to think differently about and/or respond differently in practice to ethically complex circumstances that arise in their day-to-day work and lives in relation to those who are neurodiverse.

## 1.1 Moral Status

The first section explores how one might reconceptualize the way we determine the moral status of people with intellectual disabilities. This broad topic is explored in different ways across four chapters, each of which ultimately argues in support of the moral status of people with IDD. In Chap. 2, Jonas-Sébastien Beaudry examines shortcomings of the social contract tradition when considering the moral status of people with severe IDD. He introduces strategies that others have employed to try to include people with disabilities in contractual justice theories. Ultimately, Beaudry demonstrates that none of the strategies can ground a robust moral status for people with severe IDD or, if they do, then their status is merely derivative or contingent. The chapter succinctly presents some key arguments that can be found in Beaudry's book (2021). In Chap. 3, Abraham Graber approaches the debate regarding moral status and individuals with IDD in an entirely different way. More specifically, Graber introduces a particular quality of love that is sufficient for its object to have full moral status. In the end, Graber's proposal demonstrates that people with IDD have full moral status based on their being loved in this particular kind of way—a way that would not necessarily apply to non-human animals. Similar to Beaudry's opening chapter, Chap. 4 written by Stacy Clifford Simplican, explores the moral status of people with IDD through the lens of the social contract tradition. However, instead of making arguments in relation to the social contract itself, Simplican uses a lens of social contract theory to analyze the 2019 film, *The Peanut Butter Falcon.* Based on her analysis, Simplican argues that the film is important not only because it focuses on a person with an intellectual disability, but also because of the ways that it envisions ethical futures for vulnerable adults in a landscape of neoliberal neglect. Chapter 5 concludes this section of the book, where Chris Kaposy discusses moral progress around the neonatal care of infants born with Down Syndrome during the period from Baby Doe to the present day. In this chapter, Kaposy discusses morally relevant similarities that exist between selective abortion and infanticide/not treating infants with Down Syndrome who have curable conditions, followed by how the Baby Doe case helped make moral progress for people with Down Syndrome. Although Kaposy concludes by saying that moral progress is tangible, he notes that widespread ableism contributes to its fragility. This chapter resonates with Article 10 of the CRPD (right to life).

While the approaches to exploring moral status differ, each chapter in this section acknowledges challenges that exist in thinking about moral status in certain kinds of ways in relation to people with IDD. Furthermore, all authors put forward some possible way to reconceive moral status such that people with IDD would be recognized as moral agents. And insofar as people with IDD have moral status, then they ought to be treated in ways that are exemplified by the CRPD, which are not necessarily reflected in our ableist society.

## 1.2 Capacity and Consent

The chapters in this section of this volume focus on the perennial topics of consent and capacity. In Chap. 6, Abraham Graber discusses Britney Spears's conservatorship to describe the tension that may exist between the moral obligation to respect a person's autonomy and the legal authority often granted to others who oversee the affairs of a person with IDD. Graber distinguishes between capacity (a clinical property) and competence (a legal property), where it is presumed that Spears had *capacity* to consent to certain medical interventions but lacked *competence* due to her conservatorship. In similar kinds of ways, many people with IDD may be capable of consenting to certain health interventions yet have a guardian in place who has legal authorization to make all decisions on their behalf. Through his analysis, Graber encourages readers to explore the moral obligations that may exist to respect a person's capable preferences even if they lack competence. In Chap. 7, Marjorie Aunos, Maurice Feldman, and Laura Pacheco present challenges associated with parenting capacity assessments that are frequently used within the child welfare decision-making process for parents with IDD. This area of the developmental sector is fraught with ableist biases regarding who should versus should not parent. In response, the authors present a parenting model specific to the needs of parents with IDD, which may be a more ethically defensible approach to child protection intervention cases involving these parents, particularly insofar as the right to parent is protected by the CRPD. In Chap. 8, Andria Bianchi discusses the value-laden and taboo topic of sexual consent and people with IDD. Bianchi highlights that traditional conceptions of sexual consent may be unhelpful for some people with IDD, which may lead members of our society to inaccurately conclude that they cannot consent in ways that would make the relevant act ethically licit. However, insofar as some people with IDD may want and have a right to engage in relationships that are sexual in nature (as per Article 23 of the CRPD), then it is important to explore different consent mechanisms. Bianchi offers one possible model of consent for consideration, though stops short of necessarily endorsing it. In Chap. 9, Janet A. Vogt examines the informed consent process that is traditionally used for research participation in relation to people with IDD. Vogt argues that there exists an ethical imperative to provide accommodations during research consent processes to reduce the context-specific vulnerability of adults with IDD. Her argument aligns with one of the main messages of the CRPD; namely the importance of reasonable accommodation(s) to support "… the recognition, enjoyment or exercise, on an equal basis with others, of all human rights and fundamental freedoms in the political, economic, social, cultural, civil or any other field" (UN 2006).

Ethical complexities associated with consent and capacity are prevalent in our ableist society. In general, there exists a widespread assumption that people with IDD are incapable of consenting to, or even making rational decisions about important aspects related to their lives. This assumption is inaccurate. The four chapters included in this section speak to a handful of the many areas within which our assumptions about consent and capacity in relation to people with IDD can cause harm and prevent certain rights from being realized. As co-editors, we recognize that the challenges surrounding consent and capacity could fill an entire volume. For example, we recommend Khemka and Hickson's excellent book, *Decision Making by Individuals with Intellectual and Developmental Disabilities: Integrating Research into Practice* for discussion of such topics as supported decision making (Khemka and Hickson 2021). However, we elected to include just four primary topics in this volume to serve as an introduction to this important area.

## 1.3 Inclusion and Empowerment: Society, Systems, and Ethics

This section includes chapters that focus on inclusion and empowerment within communities, systems, and society. In Chap. 10, Tess Sheldon examines the distance that exists between law and justice for people labelled with intellectual disabilities and calls for a disability-inclusive approach to understanding access to justice. Citing CRPD Articles 12 (equal recognition before the law) and 13 (access to justice), Sheldon critiques the use of clear language judgments as a strategy to improve the understanding of legal information. In Chap. 11, Dora M. Raymaker introduces four ethical tensions in the landscape of autism research (models of disability, definitions of normality, politics of inclusion, and research priorities) and helps us understand the knowledge-power dynamics that contribute to such tensions by applying a critical-systems-thinking lens. Raymaker then describes how Meadow's framework for leverage can be used to intentionally work toward more inclusive and empowered futures. In Chap. 12, Joy F. Johnson and Jacob A. Sadavoy discuss the value of social validity in relation to the field of behaviour analysis. Additionally, they argue that systemic ableism limits a practitioner's ability to provide care that is both meaningful and individualized. In Chap. 13, Parker Gamble and Lorraine Krall McCrary argue that the ability for people with disabilities to participate in local associations and local government is a matter of justice; work that resonates with CRPD Article 29 (participation in political and public life). With this in mind, the authors introduce the principle of subsidiarity as one that would support the empowerment of people with disabilities to participate in local governance while also protecting them against abuse.

The four chapters in this section discuss different parts of our society (e.g., law, research, health care, politics) wherein autistic people and/or people with IDD are neither recognized nor included in meaningful and ethically defensible ways. This failure to treat those who are neurodiverse in ways that are inclusive is, in large part, due to the prevalence of ableist norms. In response, each author presents arguments in support of inclusion and empowerment and offers suggestions of how our society can enhance our inclusion of neurodiverse populations.

## 1.4 Inclusion and Empowerment: Practical Approaches

While the chapters in Sect. 1.3 discuss inclusion and empowerment from a primarily theoretical perspective, those in this section each provide a robust practical approach to inclusion and empowerment. In Chap. 14, Gabriel M. Ronen and Peter L. Rosenbaum argue for the use of the World Health Organization's International Classification of Functioning, Disability and Health (the ICF) as an ethical framework for working with people with IDD (WHO 2001). Referencing Beauchamp and Childress's Four Principles approach (2013), they argue that the ICF offers opportunities to contextualize a person with any health "impairment" in terms of their functioning, personal values and preferences, and environments. In Chap. 15, Jane Seale discusses conflicting risk discourses that have influenced the support practices that enable adults with learning disabilities to access and use technologies. In a chapter that aligns with CRPD Article 9 (accessibility), Seale examines the potential of ethical frameworks to offer a way forward, concludes that the ethics literature supports a possibility-focused approach, and argues that ethicists should collaborate with people with learning disabilities and researchers to develop ethically-informed digitally inclusive support practices for adults with IDD. In Chap. 16, Meghan M. Burke, Chak Li, Amanda Johnston, and Adriana Kaori Terol examine parent advocacy and self-advocacy, focusing on the need to balance the two, especially in cases where their respective goals may differ. The authors argue that person-family interdependent planning ought to be used in order to harmonize these advocacy efforts and ensure positive outcomes for the individuals concerned. In Chap. 17, Laura R. Bowman, Carolyn McDougall, and Dolly Menna-Dack consider how employment settings, policies, community supports, and other contextual factors might need to change to meet the needs of individuals with IDD and/or autism. Referencing CRPD Articles 26 (habilitation and rehabilitation) and 27 (work and employment), they advocate for ecologically-based employment pathways to facilitate a more just and inclusive employment context and describe an example of one such approach. In the chapter, a practical example from the authors' workplace is provided, namely that of the Employment Participation Pathway model, which demonstrates a best practice approach to employment engagement. In Chap. 18, Dolly Menna-Dack and Laura R. Bowman argue for the importance and benefits of empowering youth with disabilities to develop self-advocacy skills for themselves and on behalf of their community. The authors describe their Youth Engagement Strategy which provides opportunities for youth to develop advocacy skills while contributing to the creation of new programs and policies within their organization.

This section of the book offers concrete examples of how we can include, empower, and respond to autistic people and/or those with IDD in our society in the right kinds of ways, i.e., in ways that respond to ableist norms and structures. Although each author focuses on a different area of practice, each practical approach is ethically defensible by virtue of arguments offered within and outside of this section. This section of the book provides various tables and diagrams that may be particularly helpful for readers working on the ground (e.g., clinical staff, program development specialists, etc.).

## 1.5 Sexuality, Intimacy, Romance, and Parenting

These chapters consider the challenges people with IDD and/or autism may face when trying to express their sexuality, engage in intimate relationships, or have children. Many of these challenges resonate with specific CRPD articles such as the rights: to access education, including sex education (Article 24); to access the same range of health programs provided to others, including reproductive health services (Article 25); to marry and have children (Article 23); and to be afforded equal recognition before the law (Article 12). In Chap. 19, Robyn M. Powell reviews the profound impacts of ableism on the sexual, reproductive, and parenting rights of people with IDD. Powell offers recommendations to confront and dismantle these impacts by transforming laws, policies, and practices. In Chap. 20, Marina Heifetz provides an overview of how people with IDD conceptualize and experience romantic relationships, discussing the implications of sheltering these individuals from open and informative discussions of romantic relationships and sexuality. Ultimately, Heifetz poses recommendations to mitigate the risks associated with unplanned pregnancies, sexually transmitted infection, and abuse. In Chap. 21, Elizabeth K. Schmidt argues for accessible, inclusive, and comprehensive sexual health education to support the sexuality-related rights of people with IDD. This can be provided by parents, caregivers, educators, health providers, and peers and may utilize a variety of different approaches. Done properly, such education can improve sexual self-advocacy and support the sexuality-related rights articulated in the CRPD. In Chap. 22, Laura Pacheco examines the ways in which dominant and community cultures shape the life story of a parent with IDD. She introduces us to Maria, a resilient woman with IDD who is determined to raise her children in a loving environment. Pacheco explores the implications for socially just ethical practice.

The chapters in this section each discuss a unique aspect related to sex, romance, parenting, and/or intimacy in relation to people with IDD. These topics are value-laden and ethically controversial. As a society, we seldom discuss sex, intimacy, parenting, and romance *in general*, and certainly not regarding those who are neurodiverse. At least part of the reason that we do not give sufficient attention to these matters when thinking about people with IDD is because of ableist stereotypes, which lead members of our society to believe that people with IDD are not sexual, not fit for parenthood, and not interested in relationships. Each chapter in this section responds to the widespread ableist stereotypes by discussing these taboo topics in a manner that accords with the CRPD.

## 1.6 Health Care

This section includes chapters related to health care, broadly speaking. People with IDD experience a wide range of health inequities (WHO 2022). In Chap. 23, Shay Dawson, M. Ariel Cascio, Meredith Wiles, and Neli Ragina tackle the challenge of diversifying healthcare students' knowledge of disabled patient populations and fostering compassionate practice techniques. To this end, the authors provide an overview of their innovative approach to medical curriculum change—HEED. In Chap. 24, Hilary K. Brown presents data that highlights the many inequities in women's reproductive health, as experienced by women with IDD; data which flies in the face of common assumptions regarding the reproductive status of individuals with IDD. Citing CRPD Articles 23 (respect for home and the family) and 25 (health), Brown advocates for significant changes to health care provision, including reasonable accommodations for people with IDD and specific training for healthcare professionals. In Chap. 25, Julie S. Armin and Heather J. Williamson examine case reports of cancer treatment for people with IDD, including examples of their not receiving standard-of-cancer-care; their exclusion from cancer research; and their limited involvement in decision making about their care. Their chapter concludes with recommendations for ethical treatment practices. In Chap. 26, Tim Stainton and Trudo Lemmens discuss euthanasia and people with IDD, elucidating some key areas of ethical concern, and posing a critical question. With the rapid expansion of euthanasia and assisted suicide, are we heading into an era of quiet, more "subtle" eugenics?

The chapters in this section focus on different areas of health care but take a similar approach. Each chapter acknowledges that gaps exist in our traditional health care system. Our health care system is not built *with* nor *for* autistic people or people with IDD, making it such that those who are neurodiverse can easily fall through the cracks. Each author in this section encourages readers to think hard about how health care is structured, the ways that neurodiverse individuals are/are not treated, and particular kinds of challenges that may exist for those who are neurodiverse given the context within which they live and receive care (i.e., one that is ableist).

## 1.7 Ethical Responses to 'Behaviours'

The chapters in this section explore ethical responses to behaviours that may be perceived as "difficult". In Chap. 27, Elspeth Bradley and Janet A. Vogt argue that organizations ought to broaden their understanding of "patient" safety and embrace practices that contribute to the person's *felt sense* of being safe. Drawing on Porges's polyvagal theory and Ramsden's cultural safety theory, the authors argue that we have an ethical imperative to engender a feeling of safety in service recipients, thereby preventing emotional distress and behaviours that challenge services. In Chap. 28, Brooke Magel and Daniel Z. Buchman introduce the topic of harm reduction as an approach for people with IDD who use alcohol and/or may have alcohol use disorder, highlighting the alignment between the principles of harm reduction and concepts such as dignity of risk, choice, and autonomy. Ultimately, they argue that not only is harm reduction an ethically justifiable approach; people with IDD ought to lead harm reduction strategies to ensure alignment with the community's needs. In Chap. 29, Peter Sturmey provides an overview of the reduction of restraint and restrictive behaviour management practices. In his chapter, Sturmey points to their widespread use in educational and residential settings for people with disabilities and argues that the use of restraint raises numerous ethical questions for human services. In Chap. 30, Kiruthiha Vimalakanthan and Valerie Temple discuss the many unique ethical challenges for professionals arising in the assessment, treatment, and support of individuals and families living with Fetal Alcohol Spectrum Disorder (FASD), stemming from its link with maternal alcohol consumption. Their chapter closes with recommendations for improving our approach and support for these families. Finally, in the closing chapter, Nicole Luke and Janet A. Vogt propose an approach to training behaviour analysts in ethical deliberation, informed by the principles of reflective practice, and demonstrated using a case involving behaviours that challenge services. As they point out, this approach can be used to train learners and practitioners in any of the many disciplines who support autistic people and those with IDD.

Although various actions expressed by autistic people and/or people with IDD may be considered "behaviours" of some kind, the chapters in this section each focus on behaviours that may be considered particularly challenging or problematic from clinical perspectives. Value judgments may be placed on many, if not all, of the behaviours discussed in this section (e.g., "bad", "poor", "challenging"), and these judgments may influence those on the ground to respond in ways that are ethically concerning (e.g., inappropriate use of restraints). Each of the chapters in this final section encourage readers to question their assumptions about behaviours, which may be rooted in ableist traditions, and offer proportional responses that can be ethically defended.

In conclusion, as editors of this volume, we have aimed to offer a broad-ranging look at some of the ethical challenges experienced by, or in relation to, autistic people and persons with IDD. Although the contributors come from a multitude of backgrounds, each of them shares a common interest in raising awareness of and providing ethical responses to the needs of people with IDD and those on the autism spectrum. The reader will note that the boundaries between sections of the book are not inflexible and similar ideas and arguments will be encountered in chapters residing in different sections. The reader is therefore encouraged to explore the whole book to gain an appreciation of some of the many viewpoints currently occupying this space. That said, we do not claim to have encompassed all important topics related to ethics and IDD and/or autism. There is room for many more volumes of this kind.

We are grateful to have worked with such accomplished and dedicated authors in developing this volume. This book is intended to appeal to bioethicists (and bioethics students) as well as people with an interest in IDD and/or autism, such as researchers, clinicians, service providers, and policy makers. The volume strives to contain materials to challenge readers from undergraduate to post-doctoral levels. Our overall hope is that with more awareness of the issues and challenges, and some exposure to a positive framing of these, the volume may contribute to a better understanding of, and broader advocacy for, the rights and ethical treatment of autistic people and individuals with IDD.

## References

Beauchamp, Tom L., and James F. Childress. 2013. *Principles of biomedical ethics*, 7th ed. Oxford: Oxford University Press.

Beaudry, Jonas-Sébastien. 2021. *The disabled contract: Severe intellectual disability, justice and morality*. Cambridge: Cambridge University Press.

Bogart, Kathleen R., and Dana S. Dunn. 2019. Ableism special issue introduction. *Journal of Social Issues* 75 (3): 650–664.

Campbell, Fiona A.K. 2001. Inciting legal fictions: 'Disability's date with ontology and the ableist body of the law.' *Griffith Law Review* 10 (1): 42–62.

Christensen, Deborah, and Jennifer Zubler. 2020. CE: From the CDC: Understanding autism spectrum disorder. *AJN The American Journal of Nursing* 120 (10): 30–37.

Cox, David J. 2012. From interdisciplinary to integrated care of the child with autism: The essential role for a code of ethics. *Journal of Autism and Developmental Disorders* 42: 2729–2738.

Fisher, Kathleen, Catherine Robichaux, Jeanie Sauerland, and Felicia Stokes. 2020. A nurses' ethical commitment to people with intellectual and developmental disabilities. *Nursing Ethics* 27 (4): 1066–1076. https://doi.org/10.1177/0969733019900310.

Khemka, Ishita, and Linda Hickson. 2021. *Decision making by individuals with intellectual and developmental disabilities*. Springer International Publishing.

Morris, Megan A. 2023. Death by ableism. *New England Journal of Medicine* 388 (1): 5–7. https://doi.org/10.1056/NEJMp2212109.

Nario-Redmond, Michelle R., Alexia A. Kemerling, and Arielle Silverman. 2019. Hostile, benevolent, and ambivalent ableism: Contemporary manifestations. *Journal of Social Issues* 75: 726–756.

UN (The United Nations). 2006. *(CRPD) Convention on the Rights of Persons with Disabilities*. Treaty Series, 2515, 3, G.A. Res. 61/106, 76th plen. Mtg., U.N. Doc A/Res/61/106. https://social.desa.un.org/issues/disability/crpd/convention-on-the-rights-of-persons-with-disabilities-crpd. Accessed May 26, 2024.

WHO (World Health Organization). 2001. *International classification of functioning, disability and health*. Geneva: WHO Press.

WHO (World Health Organization). 2022. *Global report on health equity for persons with disabilities*. Switzerland. License: CC BY-NC-SA 3.0 IGO.

# Part I
# Moral Status

# Chapter 2
# The Disabled Contract: A Précis

**Jonas-Sébastien Beaudry**

**Abstract** This chapter presents key arguments made in my book, *The Disabled Contract: Severe Intellectual Disability, Justice and Morality*. It examines how people with severe intellectual disabilities (PSID) fare within the social contract tradition. More specifically, it contends that even strategies that attempt to integrate disability within the realm of contractual justice and morality are not entirely successful. These strategies cannot ground a robust moral status for PSID; or, if they do so, it is at the cost of making this status merely derivative or contingent. The failure of social contract theory to bring PSID within its purview is significant. At best, it reveals a gap that should impel moral and political theorists to give fiduciary and caring ideals due weight next to contractual ideals. At worst, the tradition is not only incomplete, but necessarily creates and oppresses the "disabled subject."

**Keywords** Social contract · Contractualism · Contractarianism · Rawls · Disability · Intellectual disability

## 2.1 Introduction

A primary challenge that people with disabilities (PWD) face is social exclusion. Instead of adjusting social structures and institutions to be universally welcoming, societies have traditionally marginalized individuals with atypical bodies or minds. Political, economic, and cultural settings create barriers for PWD to exercise abilities, express talents, and formulate a conception of the good life. This chapter does not examine particular instances of exclusion but focuses instead on fundamental reasons that explain why the exclusion of PWD has often been justified and accepted as a natural fact rather than condemned as an injustice.

J.-S. Beaudry (✉)
Department of Equity, Ethics, and Policy, Faculty of Law, McGill University, Montreal, QC, Canada
e-mail: jonas-sebastien.beaudry@mcgill.ca

 
A. Bianchi and J. A. Vogt (eds.), *Intellectual Disabilities and Autism: Ethics and Practice*, The International Library of Bioethics 108,
https://doi.org/10.1007/978-3-031-61565-8_2

More specifically, this chapter examines how people with severe intellectual disabilities (PSID) fare within the social contract tradition, which has had an all-pervasive impact on our political ethos. It contends that even recent strategies that attempt to integrate disability within contractual justice and morality are not entirely successful. These strategies cannot ground a robust moral status for PSID; or, if they do so, it is at the cost of making this status merely derivative or contingent. The failure of social contract theory to bring PSID within its purview should not be seen as a marginal theoretical defect affecting only a small segment of human populations. At best, it reveals a gap that should impel moral and political theorists to give fiduciary and caring ideals their due weight next to contractual ideals. At worst, the social contract tradition is not only incomplete, but necessarily creates and oppresses the "disabled subject."

This chapter is a revised précis of my book, *The Disabled Contract: Severe Intellectual Disability, Justice and Morality* (Beaudry 2021a). It introduces readers to some conclusions I reach in an accessible format. The arguments are therefore illustrative rather than exhaustive.[1]

## 2.2 The Ubiquity of Social Contract Theory

Social contract theories predicate the authority of rules that govern society on the idea that these rules are the product of a contractual agreement struck between members of society. Theories of the social contract assume the importance of certain contractual features (e.g., equality and reciprocity between contractors), and take for granted that certain characteristics are held by the parties (e.g., mutual disinterest and rationality) (Rawls 1999). Social contract theories employ a normative or descriptive narrative about the origins of society, presenting parties to the contract as endorsing certain rules and principles when they set up basic social institutions. Contractual procedures may also be used to assess the morality of smaller scale interactions (e.g., a relationship that is not mutually beneficial) (Dimock 2008).

Social contract theory relies on reason, rather than on "an order antecedent and given to us" (Rawls 1980, 519) (e.g., scriptures) to justify and delineate moral and political obligations. Rather than a fixed set of moral norms, the social contract is an intellectual procedure, which capitalizes on people's capacities to think autonomously and recognize whether particular normative principles are reasonable (Ibid.).

Social contract theory has been divided in two variants: *contractarianism* and *contractualism* (Darwall 2002).[2] Contractarianism lays out a prudential approach to morality and politics, according to which people are self-interested bargainers who

---

[1] Most of this chapter was previously published in the Canadian Journal of Disability Studies (Beaudry 2021b), which was itself a précis of my book (Beaudry 2021a). I would like to thank Ms. Genny Plumptre and the editors of this collection for their excellent editorial work on this piece.

[2] These traditions have been described in various ways: as Hobbesian and Kantian (Hampton 2006); as "subject-centered justice" and "justice as reciprocity" (Buchanan 1990); as theories of justice

see an advantage in negotiating mutually beneficial terms of coexistence with others. Contractualist theories, by contrast, focus on personal autonomy. They rationalize authority by setting up social institutions and imposing norms that are reasonable and justifiable to the parties bound by them, thus respecting citizens' capacity to self-legislate.

The influence of contractual thinking in public affairs is undeniable. Both strands of the social contract tradition are appealing precisely because they rely on widely shared individual traits that also happen to be values of the liberal ethos, namely rationality, reasonability, prudence, self-interest, autonomy, as well as shared intuitions that fair social arrangements require reciprocity and impartiality.

## 2.3 The Oppressive Potential of Contractual Theory

If contractual intuitions are, indeed, structuring our social and political arrangements, we have cause for concern if they fail in important ways.[3] Notably, contractual thought may not suffice in making moral and political sense of the human experience, and the ideals it conveys may become a palatable ideological veneer for oppressive social structures. For instance, some have criticized social contract theories for neglecting injustice within family institutions, as well as the unequal treatment of people (mostly women) caring for dependents (Okin 1989; Kittay 1999). By contrast, an ethics of care suggests that properly responding to vulnerability and interdependency is a more compelling ethical foundation for moral and political theories, as opposed to ideals of reciprocity between independent bargainers (Noddings 1984; Tronto 1993).

The social contract tradition is especially problematic for PSID[4] primarily because they may lack the capacities that enable one to be a contractor (Nussbaum 2006). From a contractarian, "scratch my back, I'll scratch yours" perspective, PSID do not seem to be a threat, and collaborating with them does not seem beneficial from a mutually disinterested point of view. In fact, a political community conceived as a joint venture would only welcome "fully cooperating members of society over a complete life" (Rawls 2001, 18–19). From a contractualist perspective, PSID seem to lack the cognitive capacities required to be rational, reasonable, and autonomous. From both those contractual perspectives, PSID would fall short of being full moral and political subjects who imperatively deserve respect. People unable to reciprocate, cooperate, or devise a conception of the good life may not even be "persons" in this

"founded on the notions of impartiality and individual well-being" and as theories of justice "founded on the notions of self-interest and mutual advantage" (Becker 2005, 9).

[3] As is well known, John Rawls's theory of justice revived the social contract tradition in the seventies, but also garnered an abundance of critiques (Mulhall and Swift 1996).

[4] "Intellectual disability" refers to the neurodevelopmental disorder as defined in the American Psychiatric Association's *Diagnostic and Statistical Manual of Mental Disorders* (*DSM-5*) (American Psychiatric Association 2013, 31). Severely and profoundly intellectually disabled persons require constant support and are more limited in their ability to participate in their communities.

cooperative political sense, nor in the moral sense of being able to exercise freedom (Rawls 2001).

The exclusion of PSID is detrimental to social contract theories because the adequacy of theories of justice—as Becker (2005, 9) notes—"is now measured partly by [their] success in dealing with justice for the disabled." Some innovative contractual thinkers have therefore put forward arguments to ensure that PWDs are protected within social contract theories. It is important to assess how compelling these "integrationist"[5] arguments are, and whether their conclusions can justify the inclusion of PSID. Understanding how these theories fail teaches us about the nature of a *Disabled Contract*, that is, the ableist ideology underlying contractual thought that parties to the contract are trained not to even detect.

## 2.4 Contractual Strategies of Integration

### *2.4.1 Inclusive Contractarianism: PSID Within a Society of Self-interested Contractors*

The most obvious contractarian argument to bring PSID into our political community is by showing that it is advantageous for other community members to include them in their *cooperative venture*. This inclusion can be done either directly—by thinking of PSID's social input differently—or indirectly—by showing that integrating PSID would indirectly benefit others, even if PSID were mostly passive beneficiaries. However, the first kind of (direct) justification seems empirically dubious, and the latter offers a problematically derivative or contingent status to PSID.

Consider the direct justification: following the social model of disability's key insight that disability is socially constructed,[6] one could argue that PSID are merely "unprofitable" contractors because of current social arrangements that prevent them from being productive and cooperative. After all, many organizations offer vocational activities and training to people with intellectual disabilities. However, if profitability is to be the guiding principle, rather than ensuring that people with intellectual disability have a flourishing life, some disabled people would presumably still not be profitable enough.

We might then point to other ways of contributing to society. For instance, we could say, first, that PSID help to create jobs (e.g., caretaking professions). Second, we could say that vulnerable people help their community to develop more comprehensive healthcare institutions that may benefit many other community members. Third, we could say that the vulnerability of PSID teaches us about our own vulnerability, which is a "profitable" insight.

---

[5] I call them "integrationist" because they propose to integrate severe disability within the domain of contractual justice and morality.

[6] On the nature and scope of the social model of disability, see e.g., Oliver (1996) and Beaudry (2016).

However, each suggestion assumes what needs to be proven. The first argument assumes that we ought to subsidize care work for PSID rather than considering the bleak alternative of not doing so, or of letting caretaking costs be assumed by those personally attached to PSID. It may also assume that developing a caring self is more valuable than pursuing other preferences. At worst, it treats PSID like any other source of work, implying that care work for PSID could be replaced by another kind of work that would achieve similar social results.

The same goes for the second argument. If contractarian society found a way to develop scientific knowledge and caring institutions without PSID, then society could do without these disabled people. Thus, their inclusion relies on an empirical gamble.

The same objection can be made in relation to the third hypothesis, which does not assume a concern for PSID, but only for our (non-severely intellectually disabled) selves. However, the way in which it voices this concern—acknowledging our vulnerable nature and the importance of this fact on our happiness—implies a social theory of justice that goes beyond core contractarianism assumptions. More precisely, the realization that vulnerability and dependence matter would, because of the implications of these rich concepts, imply a normative relational stance. According to such a stance, the goods required to deal with our dependency-related needs cannot be properly socially pursued within a contractarian framework that is founded on self-interest (Kittay 1999).

### 2.4.2 The Capacity to Trust as a Contractual Basis for Robust Moral Status

In my view, Silvers and Francis (2005) make the best case to include PSID among the rank of contributing contractors. They insist on abandoning the misleading metaphor of the social contract as a tit-for-tat exchange. Instead, they argue that society is more accurately conceived of as a complex net of transactions that requires a climate of trust in which any being able to trust or be trustworthy can participate. PSID's contribution to fostering a climate of trust can be said to occur not in spite of their disability and its corollary vulnerability, but partly because of it. However, trust is a complex phenomenon with cognitive, affective, and conative dimensions (Baier 1991), and can give rise to different meanings. For instance, I *trust* that a purse would be taken if I left it at a busy store, because I *trust* in the predictability of people's untrustworthiness. Silvers and Francis (2005) use the concept in two different contexts, whereby trust could be reasonably expected to imply different features: the trust that members of a society want to foster for cooperative endeavors to take place, and the trust that people with intellectual disabilities can have or receive. It is essential that these two notions overlap—or are connected in a causal way—if Silvers and Francis's argument that fostering the latter kind of trust is desirable because fostering the former kind of trust is desirable, is to succeed.

The kind of trust necessary to stabilize social cooperation may have attitudinal and behavioral elements that are different from the ones taking place in relations with PSID. For instance, if trust is taken to mean an expectation that people will behave a certain way, all that should be needed to secure stable cooperation would be inductive confidence in the ability to predict someone's behavior. This would be a different understanding of trust from the kind that a PSID has towards her caregiver—namely, that the caregiver has her best interests at heart.

One could insist that a trust-infused society would benefit from a more other-regarding understanding of trust. However, we may then wonder if the reasons for interpreting the social contract tradition through this lens are necessarily (or even best) expressed through contractual thought. A more *prudential* conception of trust would better correspond to a mutually beneficial bargaining paradigm and a culture of promoting self-interestedness. Revisiting contractarianism, detaching it from a bargaining paradigm, or infusing it with other-regarding attitudes may be driven by moral or political pulls that perhaps precede or subordinate contractual ones. Those ethical pulls (like concern for the well-being of vulnerable people) are certainly worth pursuing. My point is only that theorizing them as "contractual" may fail to give them their due importance when they clash with traditional contractual pulls.

These tensions tend to confirm that contractarian theories have difficulty genuinely accommodating PSID. Let me now turn to two categories of contractualist revisionist strategies.

### *2.4.3 PSID as Active Citizens*

Contractualist thinkers have introduced at least five strategies to conceptualize PSID as active participants of the social contract. These focus on: (1) PSID's talents; (2) their capacity to have a conception of the good; (3) their ability to engage with others or play a part in society; (4) their potential to develop (further) contractual capacities; and (5) their need for assistance by *collaborators* or *cognitive prostheses* in the exercising of these capacities.

These strategies attempt to "normalize" PSID by modifying benchmark requirements for counting as a contractor, or by arguing that PSID do meet these requirements. For instance, Silvers (1998, 2009) proposes a dichotomy between PWD as vulnerable, dependent and burdensome and PWD as potential social participants. The operational goals of justice associated with the former conception are distribution, separation, special treatments, and accommodations, whereas the goals associated with the latter conception are integration, similar treatment, and neutrality.

The conceptual shift that Silvers (1998, 2009) makes is part of her contractualist-integrationist strategy of conceptualizing PWD as active contractors. On her account, focusing on redistribution at a later stage of the social contract procedures has inherently exclusivist consequences. Comparatively, by emphasizing social integration instead of accommodation or compensation, she characterizes disabled people as active participants, bringing them closer to the model of active contractors.

Although Silvers (1998, 2009) criticizes cases of wrongful normalization, such as capping PWD's social support relative to the level of "average talents", her own characterization has normalizing features. In some respects, forcing PWD into the traditional liberal model of the person (or contractualist model of the social member) risks denying the fact that central aspects of certain PWD's lives—and of the relationships they have with non-disabled people—sometimes have as much, if not more, to do with caring for them as dependents as with enabling their participation and expression of talents to contribute to the common good. If it does not broaden its reach, Silvers's (2009, 184) "justice for talent" risks excluding a category of severely disabled individuals for whom available accommodation measures will not suffice.

A response to this objection may be to limit the reach of integrationist contractual theories to people with less severe disabilities, and make PSID a special case. This response would, however, concede the theory's inability to justify the robust moral status that some of the most vulnerable humans would have within our communities.

It remains that many arguments offered to include people with less severe intellectual disabilities amongst the ranks of contractors are quite convincing. These arguments lower the cognitive standards required to count as "autonomous" or to cooperate with others, or suggest ways of supporting PSID in developing and pursuing a conception of the good life. They also debunk the myth that autonomy implies independence (Francis and Silvers 2007, 2010; Francis 2009; Southwood 2010; Silvers 1998, 2009; Hartley 2009).

These integrationist strategies are meritorious. Yet, they share an unspoken moral background. They assume that PSID are members of society, even though they all rely on PSID's supported abilities or assets as requirements for membership as a contractor or active citizen. This assumed membership seems to be grounded in a kind of concern for PSID which cannot be explained on contractual terms. If we analyze contractualist and contractarian strategies without taking this concern into account, they begin to look like conceptual stretches or dubious empirical wagers. No matter how severe a person's disability, revisionist contractualist authors assume that we must give her some concern as a person—that is, as a member of our society to whom robust duties are owed. In this sense, contractualist and contractarian thinkers may find it natural to frame PSID not only as persons, but also as contractors. Otherwise, they must concede that the normative concept of "person" is not exhausted by that of a contractor. Doing so without abandoning a contractual stance is part of the next strategy.

### 2.4.4 *People with Severe Intellectual Disabilities as Passive Citizens*

If PSID cannot be active contractors, the next best thing integrationist contractual thinkers can do is to argue that one need not be a contractor to benefit from the social contract.

This is not as obvious as it seems, as parties to, and beneficiaries of, the social contract are often conflated. Under the contractualist tradition, the conflation is best understood by considering Kant's (2012) theory that only moral legislators—those able to enact the moral law—are owed non-derivative respect. To put this another way, only respecters need be respected. A similar idea is found in Rawls's (1999, 446) theory of justice when he says that only those able to "give justice are owed justice." The contractualist logic behind this conflation is that the only beings worthy of intrinsic respect are persons *qua* enactors of the rational law. What is worth respecting in these persons is their embodiment of the moral law: their rational, autonomous self. Respect of autonomy is so central that nothing else warrants granting a strong moral status.[7]

Nonetheless, there are a few strategies to secure PSID's status within a community in indirect ways. The most promising assimilate PSID to full contractors. In other words, PSID do not have what it takes to participate in the social contract as active subjects, but they are similar enough to active subjects so as to share their moral and political status.

A Kantian version of this strategy connects autonomous beings with non-autonomous beings by suggesting that autonomous beings would legislate over human needs—needs that are shared by PWD, and even some animals. As a result, contractors would issue rules that would apply to any entity. What is more, contractors, as *valuers* and *value-givers*, could grant intrinsic value to beings possessing those needs.

This approach explains why PSID would be contractually given a robust kind of value, but does not delineate their moral and political status and the set of rights they would enjoy, unless we assume that all beings sharing our *natural good* will have such status and rights, provide an account of the relative normative stringency of different moral statuses, or explain why and how intrinsically valuable beings make demands upon certain moral agents. Although I can imagine some non-contractual criteria to do so, I cannot think of a criterion that would delineate what is owed to PSID and distinguish it from what is owed to other living entities. This argument nonetheless succeeds in showing that Kantians can endorse the idea that beings partaking of human goods are intrinsically valuable. This would make PSID important *like us* instead of important *for us*.

Other integrationist arguments for the passive inclusion of PWD on the contractualist side have been offered within a Rawlsian framework. The most obvious argument would be to extend Rawls's (1999) "veil of ignorance"[8] so that contractors would not know whether they might themselves be PSID and therefore want to make provisions for the benefit of PSID. However, on a Rawlsian account, the veil is justified because of the moral claim that certain facts about people should not matter when setting

---

[7] Nussbaum (2006, 13) talks of a "Kantian split" between Kantian reasonable and rational persons and "everything else in nature."

[8] Which keeps parties to the social contract ignorant of their distinctive personal and social traits and talents, like gender and intelligence, and of their specific conceptions of the good (Rawls 1999, 11, 118–123).

up a system of distributive justice—their skin color or religion, for instance. While parties behind the veil of ignorance can make provisions for the worst that could happen to the people they represent, these represented people are defined as fully cooperating throughout a lifetime, which presumably excludes PSID. To circumvent this constraint without abandoning Rawlsian commitments is no small challenge, considering that the main idea behind justice as fairness is to secure conditions for a "mutually advantageous cooperative venture" (Rawls 1999, 96).

Most theorists tinkering with, or interpreting, a Rawlsian framework to integrate PSID assume some kind of bond between traditional contractors and fellow PSID. To offer just one example: Cureton (2008) suggests that Rawls's theory of distributive justice is only a "limited project," but that this limited project rests on a Rawlsian framework that can be extended to the disabled. He suggests asking parties to choose principles of justice that would apply to "a society that also includes radically and moderately cognitively impaired people" (Ibid., §7). Actual human societies do, after all, comprise many permanently dependent and non-autonomous people. That said, it is not clear why the fact that PSID are born in every society would warrant that they would come within the purview of contractual justice.

Attempts to explain the robust moral status PSID have as passive citizens not only rely on non-Rawlsian or non-contractualist values. They also risk being too indeterminate to provide PSID and other PWD with the care they actually need. This is partly because their needs risk being assimilated to the needs of non-disabled contractors.

## 2.5 Beyond Contractual Relations

I have briefly surveyed some strategies that social contract theorists could adopt to include PSID specifically (and PWD more generally) in the category of full subjects of justice, or to confer a robust moral status upon them. However, each of the integrationist strategies have two main shortcomings: (1) they cannot convincingly ground a robust moral status for PSID; and (2) they attempt to do so at the cost of making this status merely derivative or contingent.

Let me conclude by examining the implications of contractual theories' failures to satisfactorily justify the robust moral status of PSID.

A possible implication is that PSID should be excluded from our moral and political communities. Widespread intuitions about the wrongness of eliminating vulnerable, disabled people would turn this implication into a *reductio ad absurdum* counting against contractual theory. A minority of revisionist theorists would endorse such a suggestion (McMahan 2008; Singer 2011), but responding to their arguments must wait for another time. Most contractual thinkers, like Rawls (2005, 21), are uncomfortable with this implication, and concede that disability poses a problem that contractual thought must eventually deal with.

I suggest that contractual justice is far from a broken idea; it is simply a narrow one intended to deal with limited interpersonal relations. Contractual justice is so intuitively compelling, so traditionally important in philosophy, and such a natural fit for a capitalist bargaining ethos, that it has unsurprisingly overreached and yet fallen short of meeting the demands of disability policy. Other moral and political theories could better guide our never-ending social endeavour to conceptualize bodies and minds that defy norms commonly used by scholars and policymakers.

My conclusions do not encourage me to reject social contract theory altogether, nor suggest that an alternative theory to explain our duties toward PSID should be expanded to the whole of society. On the contrary, my findings only urge different moral theories to stay within the boundaries of the relationships, people, and circumstances to which they apply, and to refrain from claiming that they cover all morality.

The argument that this claim points to would require: (1) defending a relational or role-based theory of morality; and (2) defending the position that the moral relationships that our community has with PSID are not primarily contractual, but are instead (at first glance) fiduciary or caring relationships. Such a pluralist account of what is owed to human beings with varying degrees of autonomy, dependency, and needs would also have to (3) provide a principled way to adjudicate between competing claims or values that derive from these relationships.

To be sure, a pluralist relational account would face objections. First, if drives to endorse a role would be partially explained by emotions toward others, it may grant undue moral weight to psychological phenomena (e.g., affective dispositions). Second, even if we allow that the psychological facts central to fiduciary relations are morally relevant, they may not be relevant to justice. Third, pluralist accounts make it hard to adjudicate between competing moral pulls experienced by a moral agent wearing different "moral hats." For this reason, some may explain the value of these relations in the kind of reductionist way that a pluralist approach tries to resist, for instance by reducing their value to "quality of life" assessments.

Yet, I contend that such conceptual challenges are worth addressing and the complexity they raise is part and parcel of morality. Reductionist approaches do not solve morality's messiness, but rather sweep it under the rug.

Importantly, the failure of social contract theory should not be interpreted as a failure affecting only a small segment of human populations. To cite Geuss's (2008, 94) critique of other shortcomings of contractual theory, overlooking the phenomenal reality of PSID "is not a tiny mole that serves as a beauty spot to set off the radiance of the rest of the face, but the epidermal sign of a lethal tumour." People with the most severe disabilities may contribute to our understanding of human vulnerability and of the importance of caring relations in human life. Reducing the morality of human relations to their contractual aspects is a mistake, though it is a promising political strategy of procedural reconciliation between disagreeing strangers.

A bolder conclusion, and a less friendly one to the social contract tradition, would take its cue from disability studies' cultural critiques by arguing that the social contract tradition is not only incomplete, but also inevitably part of the problem.

While I do not reject the social contract tradition wholesale, I agree that the myth of the autonomous, independent, fully-cooperating subject who lies at the heart of the social contract fails to take into account the reality of the human condition. The social contract tradition is also inevitably complicit in the construction of the "disabled subject". The philosophical discourses asking what is owed to PSID are simultaneously narratives that construct PSID and assert specific characteristics (e.g., a sense of justice) and relations (e.g., mutually beneficial bargaining) as starting points. If we took a different starting point—for instance, the social model of disability—we would not ask, "What is owed to PWD as a matter of fairness?" but rather, "What are the elements that socially disable them in the first place?" A systemic turn heralded by political theorists like Young (1990, 2011) would seek to redress the roots, not only the symptoms, of injustice.

Yet, the same concern with oppressive uses of theoretical discourses might warrant that we preserve integrationist contractual arguments. This is because contractual values like reciprocity, impartiality, equality, etc. are politically powerful within liberal rhetoric. Telling a story according to which we owe, as a matter of justice and morality, important duties to PWD may be more convincing if we manage to show how "the disabled" are, in fact, a category of differently embodied people that were conceptualized from the get-go to be justifiably disempowered and dispossessed. This is so because repairing society's fault will likely seem to be a more pressing, less optional, and more morally and politically stringent duty than helping those in need. These are practical and cultural considerations that disability theorists must consider. A political framework to manage "disability" may be more morally accurate in the abstract, but not as practically efficient as an imperfect fiction. The social model of disability, still immensely politically important in spite of its theoretical imperfections, is a worthy illustration of this.

## References


American Psychiatric Association. 2013. *Diagnostic and statistical manual of mental disorders (DSM-5)*, 5th ed. Arlington: American Psychiatric Association.

Baier, Annette C. 1991. Trust. In *Tanner lectures on human values*, March 6–8. https://tannerlectures.utah.edu/_resources/documents/a-to-z/b/baier92.pdf. Accessed June 24, 2023.

Beaudry, Jonas-Sébastien. 2016. Beyond (models of) disability? *The Journal of Medicine and Philosophy* 41: 210–229. https://doi.org/10.1093/jmp/jhv063.

Beaudry, Jonas-Sébastien. 2021a. *The disabled contract: Severe intellectual disability, justice and morality*. Cambridge: Cambridge University Press.

Beaudry, Jonas-Sébastien. 2021b. Disability and contractual expectations. *Canadian Journal of Disability Studies* 10: 207–237. https://doi.org/10.15353/cjds.v10i1.736.

Becker, Lawrence C. 2005. Reciprocity, justice, and disability. *Ethics* 116: 9–39. https://doi.org/10.1086/453150.

Buchanan, Allen. 1990. Justice as reciprocity versus subject-centered justice. *Philosophy and Public Affairs* 19: 227–252.

Cureton, Adam. 2008. A Rawlsian perspective on justice for the disabled. *Essays in Philosophy* 9: 55–76. https://doi.org/10.5840/eip20089119.

Darwall, Stephen, ed. 2002. *Contractarianism/contractualism*. Malden: Blackwell.

Dimock, Susan. 2008. Why all feminists should be contractarians. *Dialogue: Canadian Philosophical Review* 47: 273–290.

Francis, Leslie P. 2009. Understanding autonomy in light of intellectual disability. In *Disability and disadvantage*, ed. Kimberley Brownlee and Adam Cureton, 200–215. Oxford: Oxford University Press.

Francis, Leslie Pickering, and Anita Silvers. 2007. Liberalism and individually scripted ideas of the good: Meeting the challenge of dependent agency. *Social Theory and Practice* 33: 311–334.

Francis, Leslie Pickering, and Anita Silvers. 2010. Thinking about the good: Reconfiguring liberal metaphysics (or not) for people with cognitive disabilities. In *Cognitive disability and its challenge to moral philosophy*, ed. Eva Feder Kittay and Licia Carlson, 237–259. Malden, MA: Wiley-Blackwell.

Geuss, Raymond. 2008. *Philosophy and real politics*. Princeton: Princeton University Press.

Hampton, Jean. 2006. *The intrinsic worth of persons: Contractarianism in moral and political philosophy*. Edited by Daniel Farnham. New York: Cambridge University Press.

Hartley, Christie. 2009. Justice for the disabled: A contractualist approach. *Journal of Social Philosophy* 40: 17–36. https://doi.org/10.1111/j.1467-9833.2009.01436.x.

Kant, Immanuel. 2012. *Groundwork of the metaphysics of morals*. Edited by Mary Gregor and Jens Timmermann. Cambridge: Cambridge University Press.

Kittay, Eva Feder. 1999. *Love's labor: Essays on women, equality and dependency*. New York: Routledge.

McMahan, Jeff. 2008. Challenges to human equality. *Journal of Ethics* 12: 81–104. https://doi.org/10.1007/s10892-007-9020-9.

Mulhall, Stephen, and Adam Swift. 1996. *Liberals and communitarians*, 2nd ed. Oxford: Wiley-Blackwell.

Noddings, Nel. 1984. *Caring: A feminine approach to ethics and moral education*. Berkeley: University of California Press.

Nussbaum, Martha C. 2006. *Frontiers of justice: Disability, nationality, species membership*. Cambridge: Harvard University Press.

Okin, Susan Moller. 1989. *Justice, gender, and the family*. New York: Basic Books.

Oliver, Michael. 1996. *Understanding disability: From theory to practice*. New York: Palgrave Macmillan.

Rawls, John. 1980. Kantian constructivism in moral theory. *Journal of Philosophy* 77: 515–572. https://doi.org/10.2307/2025790.

Rawls, John. 1999. *A theory of justice*. Revised. Cambridge: Harvard University Press.

Rawls, John. 2001. *Justice as fairness: A restatement*. Edited by Erin Kelly. Cambridge: Harvard University Press.

Rawls, John. 2005. *Political liberalism*. Expanded. New York: Columbia University Press.

Silvers, Anita. 1998. Formal justice. In *Disability, difference, discrimination: Perspectives on justice in bioethics and public policy*, ed. Anita Silvers, David Wasserman, and Mary B. Mahowald, 13–145. Lanham: Rowman and Littlefield.

Silvers, Anita. 2009. No talent? Beyond the worst off! A diverse theory of justice for disability. In *Disability and disadvantage*, ed. Kimberley Brownlee and Adam Cureton, 163–199. Oxford: Oxford University Press.

Silvers, Anita, and Leslie Pickering Francis. 2005. Justice through trust: Disability and the "outlier problem" in social contract theory. *Ethics* 116: 40–76. https://doi.org/10.1086/454368.

Singer, Peter. 2011. *Practical ethics*, 3rd ed. New York: Cambridge University Press.

Southwood, Nicholas. 2010. *Contractualism and the foundations of morality*. Oxford: Oxford University Press.

Tronto, Joan. 1993. *Moral boundaries: A political argument for an ethic of care*. New York: Routledge.

Young, Iris Marion. 1990. *Justice and the politics of difference*. Princeton: Princeton University Press.

Young, Iris Marion. 2011. *Responsibility for justice*. New York: Oxford University Press.

# Chapter 3
# All You Need Is Love (for Full Moral Status)

**Abraham Graber**

**Abstract** The nature of full moral status remains one of the most hotly contested issues in normative ethics. At stake are fundamental issues about how we should treat non-human animals and individuals with severe or profound intellectual disability. Those accounts of full moral status that count out non-human animals also count out individuals with severe or profound intellectual disability. Relatedly, those accounts of full moral status that hold that those with severe or profound intellectual disability have full moral status generally also give full moral status to at least some non-human animals. Rather than attempting to provide an account of full moral status that navigates this dilemma, this manuscript aims to put substantive constraints on the desiderata for a theory of full moral status. In brief, there is a particular quality of love that is sufficient for its object to have full moral status. Individuals with severe or profound intellectual disability are loved in this way. It follows that individuals with severe or profound intellectual disability have full moral status and, concomitantly, that any theory that entails otherwise is false.

**Keywords** Full moral status · Intellectual disability · Love · Friendship · Non-human animals

## 3.1 Introduction

Something has moral status insofar as "it matters (to some degree) from the moral point of view for its own sake" (Jaworska and Tannenbaum 2023, §1). Rocks, motes of dust, and soccer balls all lack moral status; they do not matter, at all, for their own sakes. By contrast, humans and many non-human animals possess moral status; your tìo Xavier matters for his own sake, as does my neighbor's cat, Sally.

A. Graber (✉)
Nisonger Center Affiliate Faculty, Division of Bioethics, Department of Biomedical Education and Anatomy, College of Medicine, The Ohio State University, Columbus, OH, USA
e-mail: Abraham.Graber@osumc.edu

 
A. Bianchi and J. A. Vogt (eds.), *Intellectual Disabilities and Autism: Ethics and Practice*, The International Library of Bioethics 108,
https://doi.org/10.1007/978-3-031-61565-8_3

Yet though Xavier and Sally both matter for their own sakes, they do not have the same moral standing. The idea of treating Xavier as if he matters no more than a domestic cat is morally repugnant. Thus, while many things have moral status, they do not all have the same degree of moral status.

Moral status is thus a spectrum. At one end of the spectrum are rocks, dust, and soccer balls. They have no moral status. Worms, snails, and bugs are (plausibly) higher on the spectrum. They matter for their own sakes; there are ways of treating these animals (e.g., torture) that is morally wrong. Even higher on the spectrum, one finds Sally the cat. Higher still is tìo Xavier.

The spectrum of moral status does not, however, continue indefinitely. Full moral status (henceforth, "FMS") constitutes the apex of the spectrum. Full moral status is widely taken to be a "range property" (McMahan 2002; Wasserman et al. 2017). While there are gradations of moral status beneath the threshold of FMS, there are no gradations in moral status above the threshold. Everything that has FMS has it to the same degree.

Full moral status is central to our moral lives. The (putative) fact that humans have, and non-human animals lack, FMS can explain why it is permissible to hunt animals but not similarly permissible to hunt humans. The fact that my dog lacks FMS can explain why it is permissible to lock him in the house while I am at work, while it would be horrifying for my employer to similarly restrict my freedom during the workday. The fact that our legal system offers you and me, but not songbirds, due process is justified by the (putative) fact that we have, and songbirds lack, FMS. Full moral status is thus central to both normative theorizing and to how we choose to live.

Though FMS is a central moral category, the correct account of the grounds of FMS and the extension of FMS are contested. By way of illustrating the difficulty of giving a satisfactory account of the grounds of FMS, consider McMahan's account: "How a being ought to be treated depends, to some significant extent, on its intrinsic properties—in particular, its psychological properties and capacities" (McMahan 1996, 32; for related views see Tooley 1983; Harris 2006; Singer 1993). On McMahan's view (and relevantly similar ones), "With respect to this dimension of morality, there is nothing to distinguish the cognitively impaired from comparably endowed nonhuman animals" (McMahan 1996). McMahan's (2002) account of FMS thus has the deeply counter-intuitive consequence that people with severe or profound intellectual disability (henceforth, "ID") lack FMS. Insofar as McMahan's view represents the dominant approach to theorizing about FMS, the most popular accounts of FMS condemn people with ID to the same level of moral consideration as non-human animals (Wasserman et al. 2017).

This deeply problematic consequence can be avoided by lowering the bar regarding the capacities necessary for FMS. Consider, e.g., Wood who argues that something has FMS just in case it has "fragments" of rationality (Wood 2003, for a related account see Regan 2004). On Wood's view, people with ID have FMS. Unfortunately, so do, e.g., dogs and cats.

A satisfactory account of the grounds of FMS will hold (i) that nearly all humans, including those with ID, have FMS and (ii) that run-of-the-mill non-human animals lack FMS. Yet it is very difficult to identify some property that (i) nearly all humans, including those with ID, have and (ii) that run-of-the-mill non-human animals lack. Call this *the problem of FMS*.

Several theories have successfully threaded the needle. For example, Dworkin (1993) has argued that FMS is grounded in membership in the human species. Yet this type of species-membership view faces its own challenges. Neither Superman, Thor, nor HAL 9000 are members of the human species. Yet they all seem to possess FMS. Similarly, there is growing biomedical promise in the implantation of non-human animal parts into human bodies. The resultant human-animal chimeras may not have any species membership (Piotrowska 2014), yet clearly have FMS.

Kittay (2021) avoids these counterexamples by holding that membership in the human species is sufficient, but not necessary, for FMS. She thus gets the result that all humans, including those with ID, have FMS while potentially avoiding the counterintuitive consequence that many run-of-the-mill non-human animals have FMS.

Yet even her view faces its own problems. First, species membership does not seem like the kind of thing that carries moral weight. Species are defined variously, including via genetic similarity and mating habits (Robert and Francoise 2003). Genetic similarity is nothing more than similarity of the chemicals in DNA. Yet the arrangement of chemicals seems like a paradigmatically non-moral property. It is thus unclear how it could serve as grounds for FMS. Similarly, the type of thing with which one reproduces is a morally inert property, incapable of serving as the grounds for FMS.

Furthermore, there are good reasons to think that species are merely valuable taxonomic units that fail to correspond with anything of biological importance (Ereshefsky 2022). It is deeply implausible that whether something has FMS is dependent on a historical accident regarding how biologists carved up their taxonomies to best afford pragmatic success.

Giving an account of FMS is notoriously difficult. In making the case that individuals with ID have FMS, one can attempt to develop one's own theory of FMS. Yet any such theory is likely to have counter-intuitive consequences. This in no way sets it apart from theories of FMS that deny that people with ID have FMS; all extant theories of FMS have substantive shortcomings. But given that many philosophers seem willing to accept shortcomings of theories that deny FMS to those with ID, offering competing theories with shortcomings with which philosophers have not made peace is unlikely to be successful, at least as a rhetorical strategy.

However, if one's primary interest in the debate over the grounds of FMS is less in arriving at the correct theory of FMS and is rather in the extension of FMS—who can rightly be said to possess FMS—one need not give a full-fledged theory of FMS. In this vein, Graber (2021) recently argued that, in theorizing about FMS, one should prioritize the intuitions of those in close personal relationships with individuals with ID. As those in close relationships with individuals with ID tend to have the intuition that those with ID have FMS, Graber's argument aims to make holding that people with ID have FMS a core desideratum for any theory of FMS.

This paper will embrace a similar strategy. In what follows, I will argue that individuals with ID have FMS. Though I will not attempt to develop a theory of FMS, if sound, the arguments in this paper entail the falsity of any theory of FMS that denies FMS to individuals with ID.

## 3.2 A Related Problem: Animals, Love, and Friendship

### *3.2.1 On Types of Friendship*

There is an interesting symmetry between debates about FMS and seemingly unrelated debates about love and friendship. In this section I will explore this symmetry. I will then leverage the symmetry to argue that (at least some) people with ID have FMS. Finally, I will consider and respond to objections.

Since at least Aristotle, it is common to distinguish between three different types of love (Helm 2021). The first is *agape*. Agape "is a universal good will, best characterized by the love of the Christian [G-d] for humankind" (Smuts 2014, 508). Agape is impersonal—the kind of love that one may hold for a stranger.

Unlike agape, both *eros* and *philia* are personal; except in odd and likely defective cases, one does not feel either eros or philia for a stranger. While eros is romantic love, "philia is … the kind of love one feels for friends and family" (Smuts 2014, 508). Romantic love is tangential to the focus of this project. Thus, though much of the following applies *mutatis mutandis* to eros, I will nonetheless exclude it from the discussion.

Though rarely a central point for debate, a key question remains: toward what kind of thing can one have philia? For example, while Helm (2021) holds that any philia one might have for non-human animals is "a kind of deficient mode of the sort of love we typically reserve for persons," Smuts rejects this view: "Although many philosophers are comfortable denying that we can love animals, this seems preposterous to all but those sheltering some pet theory of love [pun intended?]" (Smuts 2014, 509).

Yet even amongst those who hold that we can have philia toward non-human animals, one finds agreement that there are important differences between friendships that hold between humans and those that hold between a human and a non-human animal. For example, while Jordan argues for vegetarianism on the grounds that humans and non-human animals can be friends, he nonetheless notes, "it is true that a human and, say, a dog cannot be friends in the same way that one [human] is friends with another human …" (2001, 315). Similarly, while Gheaus argues that human love for non-human animals can serve as a foundation for animal ethics, she nonetheless highlights important differences between the friendships we have with other humans and the friendships we can have with non-human animals:

> The fact that some animals can meet the need—more important for some people than for others—for affectionate relationships with beings who are never judgmental, offensive, or

> disrespectful adds to the ethical character of our friendships with them. It qualifies such friendship both positively and negatively: contributing to its special preciousness, and, at the same time, distinguishing it from close, emotional, needy relationships with people. (2012, 589)

Other authors go so far as to make the case that friendships with animals are of lesser value than are (many) friendships between humans. Aristotle identifies and ranks three distinct grounds for philia. For Aristotle, philia may be based (non-exclusively) on (i) mere mutual advantage, (ii) mutual pleasure, or (iii) mutual admiration (Aristotle 2002). Building on Aristotle's three-fold analysis of types of philia, Frööding and Peterson argue that humans and non-human animals can be friends by way of mutual advantage, while explicitly noting that "friendship based on mutual advantage is not as valuable as friendship based on mutual admiration" (2011, 64).

Some authors hold that philia is unique to human relationships. Others disagree. But this surface level dispute hides an undercurrent of agreement: philia between two humans often has a unique quality not found in philia between a human and a non-human animal. In what follows, I will use "LOVE" to refer to the type of philia characteristic of human–human friendships. This type of philia is of particular interest for the debate over FMS.

### 3.2.2 LOVE and Full Moral Status

One may wonder: in virtue of what is the object of LOVE restricted to humans? The most straightforward answer is that *being human* is necessary for an organism to be the target of LOVE. Yet familiar counterexamples arise. Imagine a BELOVED friend that has multiple organs replaced with substitutes from non-human animals. In virtue of being a chimera, your friend is no longer a member of the human species. But it is deeply implausible to suppose that, while you LOVED this friend before the operation, they are LOVEABLE no more.

Accounts of LOVE thus face a now familiar challenge. They must identify the grounds in virtue of which some things are LOVEABLE, while others are not. Just as the intuitive extension of FMS closely—but not exactly—tracks the boundaries around humanity, the boundaries between the LOVEABLE and the non-LOVEABLE closely—but not exactly—track the boundaries around humanity. Just as there is a philosophical puzzle regarding the grounds of FMS, there appears to be an extensionally equivalent puzzle regarding the grounds of LOVEABILITY.

Yet the puzzle about LOVE has a solution. Why do all and only things with FMS appear to be LOVEABLE? The answer is that *having FMS is necessary for being LOVED*. Or, put the other way, *being LOVED is sufficient for having FMS*. This is, in fact, a starting place for many philosophers who theorize about philia. For example, Helm takes philia to be the kind of love that can only be felt toward something with FMS (Helm 2021; cf. Jaworska and Wonderly 2020). Even Smuts, who finds the idea that we cannot love animals to be "preposterous" nonetheless notes that "love of persons" is central to his understanding of love (Smuts 2014).

The idea that being LOVED is sufficient for FMS not only both reflects a key theme in the philosophical literature on love and offers to elegantly explain the boundaries of LOVE, it is supported by intuition. I have a particularly close relationship with my dog, Nelson. We have spent hundreds of hours together and he kept me company through several years where the vicissitudes of the academic job market often meant I had to be far from my wife. While I love Nelson, I do not LOVE him. What changes in Nelson would it take for me to LOVE him? When I think about the things that would have to be different about Nelson in order for me to feel the deep, nuanced, and rich love definitive of LOVE, the animal I find myself imagining is increasingly one with FMS.

### 3.2.3 Two Ways LOVE Could Be Sufficient for Full Moral Status

Though there are many kinds of love, there appears to be a specific kind of philia characteristic of deep friendships between persons. The nature of this specific kind of philia—which we are calling "LOVE"—is that being LOVED is sufficient for FMS. In a moment I will draw on this key claim in arguing that (some) individuals with ID have FMS. But first it is worth taking a moment to consider two very different ways in which LOVE might be sufficient for FMS.

Much of the philosophical literature on love is devoted to the question, "what reason(s) do we have for loving the beloved?" Two answers are of particular interest here. One family of answers, called "appraisal views," hold that our love is a reaction to the value of the beloved (Stump 2006). Appraisal views are a particularly good fit for explaining why LOVE is sufficient for FMS. In addition to LOVE being a reaction to particular features of particular persons, one can hold that LOVE is a reaction to the value only possessed by those with FMS. Where such value is lacking, e.g., in the case of my dog, LOVE is impossible.

A different family of answers starts with the same core thought: that the beloved is of unique and special value. But rather than holding that the value of the beloved is antecedent to the love, so called "volitional" views hold that the beloved is valuable *because* they are loved (Stump 2006). On such views, the value of the beloved is not inherent, but is rather conferred.

Volitional views of love generally take *being loved* to confer only agent-relative value (Frankfurt 2006). Thus, on such views, X loving Y is sufficient for Y to be valuable for X, but has no implications for whether Y is valuable for Z. Yet, unlike special obligations, FMS is supposed to generate reasons for all agents (McMahan 2002). Extant volitional views of love may thus not be immediately well positioned to explain why being LOVED is sufficient for FMS.

Consideration of volitional and appraisal options may raise concern regarding the informativeness of the claim that *being LOVED is sufficient for FMS*. On an appraisal approach, FMS is a necessary condition for something being LOVED. As such, our understanding of LOVE will remain incomplete until we have a satisfactory account of FMS. It is thus unclear how, on an appraisal view of love, noting that *being LOVED is sufficient for FMS* could be of value in guiding our theorizing about FMS.

In a similar vein, a volitional approach may explain how *being LOVED is sufficient for FMS* but fails to offer any insight into why we can, e.g., LOVE other humans but cannot LOVE a dog. As such, a volitional approach merely seems to push the question back a level. We can explain why certain kinds of beings have FMS—they are LOVED—but are left without any explanation why only certain kinds of beings can be LOVED.

Nonetheless, because "love has an epistemic role" (Gheaus 2012, 594), we can use LOVE as a FMS detector. Once we have identified that some organism is LOVED, we know that any adequate theory of FMS must entail that this organism has FMS. This is nothing more than the familiar method of cases ubiquitous in theorizing about FMS (Graber 2021). Noting that *LOVE is sufficient for FMS* thus allows us to better identify cases of organisms with FMS.

## 3.3 LOVE and ID

At the start of the COVID-19 pandemic, my daughter was about two-and-a-half years old. She is typically developing and, concomitant with her age at the time, was talking quite a bit. Her vocabulary was still limited, and she spoke in the way of young children, where only their parents can decipher what they are saying. Like many children, our daughter was not in childcare during the pandemic. Instead, she spent her days with me while my wife worked, and spent her evenings with my wife, while I worked. Despite the challenges of the pandemic, I will never regret the opportunity the pandemic afforded; without the pandemic, I never would have gotten to spend so much time with my daughter at such a formative moment in her life.

I have, of course, always loved my daughter. However, at some point in the year-and-a-half I spent at home with her—my best guess is sometime around her third birthday—I noticed a qualitative change in my love. As she aged, and as I grew to know her better, my love for her became richer, deeper, and more nuanced. At the start of the pandemic, my love was perhaps best described as *natural philia*—a kind of love grounded in biological relationships (Torres 2021). But soon, I LOVED my daughter. My appreciation for her quirks and eccentricities, for her love of adventure, for her unique sense of humor, was on par with the similar appreciation I have for the adults that I LOVE.

We can now complete the syllogism. Being LOVED is sufficient for FMS. At the age of three, my daughter was LOVED. Therefore, at the age of three, my daughter had FMS.

This, by itself, tells us nothing about the moral status of individuals with ID; my daughter is typically developing. But it is nonetheless a small inferential step toward substantive constraints on accounts of FMS. An adult with similar cognitive and emotional capacities would have severe or profound intellectual disability. Yet if having these cognitive and emotional capacities did not disqualify my three year old daughter from FMS, it should similarly not disqualify someone with ID.

Though I personally do not have an intimate relationship with anyone with ID, in a number of places, Kittay describes her love for her daughter, Sesha. I will, here, offer but a tiny sampling of excerpts, all from (Kittay 2019):

> Knowing her [Sesha's] capabilities, one gets a glimpse into the richness of her life and the remarkable quality of her very being. Nonetheless, the limitations shape her life and the life of her family, so we all must address them if we are to realize her possibility of flourishing. At the same time, it is only by considering Sesha in the fullness of her joys and capacities that we can view her impairments in light of her life, her interests, her happiness … (Ibid., 6)

> Our love and appreciation of Sesha *as she was* grew and became more expansive. Sesha has taught me that a profound love of another can change one's values and priorities. (Ibid., 9)

> It is through my daughter's bodily limitations and her mental incapacities, not despite them, that I experience my daughter as a person—and, in more mystified terms, as a soul. In fact, what I see in her by means of her body is a beautiful soul. When I say of my daughter that she is a lovely person, a beautiful soul, those who know her do not for a moment hesitate to assent—they do not have to stop to figure out what I mean by this. Sesha's soul is experienced more directly, not through what she does but by how she is. (Ibid., 249)

LOVE is rich and deep, vibrant and bold, yet also nuanced and subtle. While a handful of quotes cannot do justice to Kittay's descriptions of her love for Sesha, one cannot read Kittay's work without becoming convinced that Kittay LOVES her daughter.

We can, again, complete the syllogism. If Kittay LOVES Sesha, then Sesha has FMS. Kittay LOVES Sesha. Therefore, Sesha has FMS. Sesha, however, has ID. It follows, *a fortiori*, that someone with ID can have FMS and, concomitantly, that any theory of FMS that entails otherwise is false.

## 3.4 Objections and Responses

I have now developed the following argument and defended each of its two premises:

(1) LOVE is sufficient for FMS.
(2) (Some) individuals with ID are LOVED.
(3) Therefore, (some) individuals with ID have FMS (From 1 and 2).

The argument can then be expanded via the addition of (4):

(4) If some individuals with ID have FMS, then any account of FMS that entails that no individuals with ID have FMS is false.
(5) Any account of FMS that entails that no individuals with ID have FMS is false (From 3 and 4).

As promised, if sound, this argument substantively constrains available views on FMS. For example, if sound, this argument likely rules out the accounts of FMS such as those developed in McMahan (2002), Singer (1993), and Tooley (1983). This final portion of this paper will be dedicated to considering and responding to objections.

### *3.4.1 Is LOVE Sufficient for Full Moral Status?*

One may be tempted to push back on the claim that LOVE is sufficient for FMS. We have spent some time characterizing LOVE and distinguishing it from other types of love. The claim that *LOVE is sufficient for FMS* is not intended to be analytic. Rather, it is supposed to follow from an accurate characterization of a particular kind of love with which we are familiar. However, while it is true that a person is generally the object of LOVE, is this always the case?

The easiest way to show that LOVE is not sufficient for FMS would be to identify an instance of LOVE in which the object of the LOVE lacked FMS. By way of example, consider the following excerpt, involving quotations from Poole (1996), who spent much of her life in the company of elephants, and corresponding commentary provided by Wichert and Nussbaum (2021, 97):

> Poole … clearly thought of herself as like an elephant: "I went through labor and the birth of my baby as an elephant would, surrounded and assisted by my female companions … And when at last my child came into the world, she arrived like a baby elephant, amid great commotion and ceremony" (270). To Poole, mutual support and community are elephant traits, which she prefers to the common traits of human society, divisiveness, sexism, and selfishness. After an absence of two years she returned to the Amboseli elephants. They surrounded her car, "and, with their trunks outstretched, deafened us with a cacophony of rumbles, trumpets, and screams until our bodies vibrated with the sound. They pressed against one another, urinating and defecating, their faces streaming with the fresh black stain of temporal gland secretions" (276). As Poole knows, "this is a ceremony elephants typically reserve for family and group members who have been separated for a long time. It is also associated with the birth of a new baby elephant" (276).

Does an example like this suggest that elephants are LOVED? It is important to distinguish between *being LOVED* and *being in a loving relationship* (Smuts 2014). The nature of love is importantly distinct from the nature of loving relationships. The excerpts from Poole quoted by Wichert and Nussbaum offer evidence that Poole is friends with elephants, where this is understood as a claim about a *relationship*. As Jeske (2019) notes, special concern is a defining feature of friendship. The above excerpt offers good evidence that the elephants have special concern for Poole and thus that the relationship between Poole and the elephants can appropriately be described as a "friendship." But this is evidence about the nature of the *relationship* between Poole and the elephants, rather than evidence regarding whether Poole LOVES the elephants. What is needed for evidence that Poole LOVES the elephants is a description of the quality of Poole's love, something lacking from the excerpt.

Nonetheless, the sense remains that the above passage offers evidence of Poole's LOVE for the elephants. Poole's depiction of the elephants highlights a depth of both cognitive and emotional sophistication that one may have doubted elephants possess. Yet such cognitive and emotional sophistication is often associated with

FMS. Given that FMS is necessary for being LOVED, it is no surprise that evidence of elephants' cognitive and emotional sophistication offers some evidence of Poole's LOVE. But it follows that the excerpt does *not* offer evidence that elephants are LOVED *despite lacking FMS*. Rather, the very evidence that increases our confidence that the elephants are LOVED simultaneously increases our confidence that the elephants have FMS. Rather than undermining the claim that *being LOVED is sufficient for FMS*, the excerpt from Wichert and Nussbaum merely highlights the fact that there may be compelling evidence that some non-human animals have FMS.

But suppose one is nonetheless convinced that being LOVED is *not* sufficient for FMS. Even this does not skuttle the argument. The claim that *LOVE is sufficient for FMS* is not strictly required for a version of the argument to be effective. For, even if LOVE is not sufficient for FMS, it is certainly the case that most things that are LOVED have FMS. And this is, itself, enough to move forward:

(1) Most things that are LOVED have FMS.
(2) (Some) individuals with ID are LOVED.
(3) Therefore, (some) individuals with ID have FMS.

Someone skeptical of this statistical syllogism might point out that most things that are LOVED do not have ID. There is thus a relevant difference—namely having ID—between most of the things that are LOVED and individuals with ID, and the inference is thus putatively blocked.

This move is, however, too fast. Statistical syllogisms have the following form:

(1) Such-and-such percent of As are Bs.
(2) X is an A.
(3) Therefore, X is a B.

One can undermine a statistical syllogism by pointing out differences that hold between X and other As. But in employing this strategy it will not do to point out just any differences. Rather, the differences need to be relevant to whether X is a B. The proposed response to the above statistical syllogism is thus only effective if having ID is relevant to whether something has FMS. But that's what's up for grabs in the debate. The proposed response is thus straightforwardly question-begging and the central argument of this paper survives *even if* being LOVED is *not* sufficient for FMS.

### *3.4.2 Are Individuals with ID LOVED?*

One could also push back on the second premise and try to draw into doubt the claim that *individuals with ID are LOVED*. Given, e.g., Kittay's many descriptions of her love for Sesha, it's difficult to see this as a promising strategy. At a minimum, one would want to offer an error theoretic story: despite appearances, people who seem to LOVE an individual with ID don't actually because of such-and-such. For example, one might want to argue that, in virtue of their love for an individual with ID, we ought to be skeptical of the testimony of someone like Kittay. Perhaps Kittay's deep love for Sesha somehow undermines the epistemic import of her testimony.

Yet, in the context of testimony about love, such a move is not compelling. LOVE is constituted by exactly this kind of deep yet nuanced attachment! The very features of emotional attachment that are supposed to undermine the epistemic import of Kittay's testimony themselves provide evidence that (some) people with ID are LOVED after all.

## 3.5 Conclusion

There is a unique quality of love reserved for those with FMS. People with ID are the object of this type of love. It follows that (at least some) people with ID have FMS which, in turn, entails the falsity of any account of FMS that is incompatible with individuals with ID having FMS. The argument for this conclusion is surprisingly simple, requiring only two premises. Furthermore, there is compelling evidence in support of each of the premises and it is unclear how to make good on objections to either. The result is not a theory of the grounds of FMS but, rather, substantive constraints on how we approach such theorizing.

## References


Aristotle. 2002. *The nicomachean ethics*. Edited by Sarah Broadie and Christopher Rowe. Oxford University Press.

Dworkin, Ronald. 1993. *Life's dominion: An argument about abortion, euthanasia, and individual freedom*. New York: Vintage.

Ereshefsky, Marc. 2022. *Species*. Stanford Encylopedia of Philosophy. https://plato.stanford.edu/archives/sum2022/entries/species. Accessed April 9, 2024.

Frankfurt, Harry. 2006. *The reasons of love*. Princeton University Press.

Fröôding, Barbro, and Martin Peterson. 2011. Animal ethics based on friendship. *Journal of Animal Ethics* 1 (1): 58–69.

Gheaus, Anca. 2012. The role of love in animal ethics. *Hypatia* 27 (3): 583–600.

Graber, Abraham. 2021. A parent's intuition is always right: Weighing intuitions in the debate over the nature of full moral status. *Metaphilosophy* 52 (5): 570–582.

Harris, John. 2006. *The value of life: An introduction to medical ethics*. Routledge.

Helm, Bennett. 2021. *Love*. Stanford Encyclopedia of Philosophy. https://plato.stanford.edu/archives/fall2021/entries/love/. Accessed April 9, 2024.

Jaworska, Agnieszka, and Julie Tannenbaum. 2023. *The grounds of moral status*. Stanford Encyclopedia of Philosophy. https://plato.stanford.edu/archives/spr2023/entries/grounds-moral-status. Accessed April 9, 2024.

Jaworska, Agnieszka, and Monique Wonderly. 2020. Love and caring. In *The Oxford handbook of philosophy of love*, ed. Christopher Grau and Aaron Smuts. Oxford University Press.

Jeske, Diane. 2019. *Friendship and social media: A philosophical exploration*. Routledge.

Jordan, Jeff. 2001. Why friends shouldn't let friends be eaten: An argument for vegetarianism. *Social Theory and Practice* 27 (2): 309–322.

Kittay, Eva Feder. 2019. *Learning from my daughter: The value and care of disabled minds*. Oxford University Press.

Kittay, Eva Feder. 2021. Why human difference is critical to a conception of moral standing: An argument for the sufficiency of being human for full moral status. *The Journal of Philosophy of Disability* 1: 79–103.

McMahan, Jeff. 1996. Cognitive disability, misfortunate, and justice. *Philosophy and Public Affairs* 25 (1): 3–35.

McMahan, Jeff. 2002. *The ethics of killing: Problems at the margins of life*. Oxford University Press.

Piotrowska, Monika. 2014. Transferring morality to human-nonhuman chimeras. *The American Journal of Bioethics* 14 (2): 4–12.

Poole, Joyce. 1996. *Coming of age with elephants: A memoir*. New York: Hyperion.

Regan, Tom. 2004. *The case for animal rights*. Berkeley: University of California Press.

Robert, Jason Scott, and Baylis Francoise. 2003. Crossing species boundaries. *American Journal of Bioethics* 3 (3): 1–13.

Singer, Peter. 1993. *Practical ethics*. Cambridge University Press.

Smuts, Aaron. 2014. Normative reasons for love, part 1. *Philosophy Compass* 8 (9): 507–517.

Stump, Eleonore. 2006. Love, by all accounts. *Proceedings and Addresses of the American Philosophical Association* 80 (2): 25–43.

Tooley, Michael. 1983. *Abortion and infanticide*. Oxford University Press.

Torres, Jorge. 2021. Philia: The biological foundations of Aristotle's ethics. *History and Philosophy of the Life Sciences* 43 (4): 119.

Wasserman, David, Asch Adrienne, Jeffrey Blustein, and Daniel Putnam. 2017. *Cognitive disability and moral status*. Fall. Accessed May 12, 2023. https://plato.stanford.edu/archives/fall2017/entries/cognitive-disability/. Accessed April 9, 2024.

Wichert, Rachel Nussbaum, and Martha Nussbaum. 2021. Can there be friendship between human beings and wild animals? *Journal of Human Development and Capabilities* 22 (1): 87–107.

Wood, Allen. 1998. Kant on duties regarding nonrational nature. *The Aristotelian Society Supplementary Volume* 72 (1): 189–210.

# Chapter 4
# *The Peanut Butter Falcon*, Social Contract Theory, and the Dream of Mutual Aid

**Stacy Clifford Simplican**

**Abstract** This chapter analyzes the 2019 film *The Peanut Butter Falcon* through an analysis of social contract theory. The film focuses on two men: Zak, a man with Down syndrome who has escaped from a nursing home, and Tyler, a young man running from the law and his brother's death. Fugitives of the state, they forge a friendship against a deserted and coastal backdrop, conjuring a modern-day state of nature. With the help of Zak's nursing home employee Eleanor, they reconstitute a new kind of family to provide mutual care for one another. The film is important not only because it stars a man with an intellectual disability, but also because of the ways that it envisions ethical futures for vulnerable adults in a landscape of neoliberal neglect.

**Keywords** Social contract theory · Social inclusion · Family · Intellectual disability · State of nature · Masculinity

## 4.1 Introduction

The 2019 film *The Peanut Butter Falcon* is a must-watch movie. The film stars Zack[1] Gottsagen, a man with Down Syndrome: his performance proves that actors with intellectual disabilities can play complex and leading roles in Hollywood films. Additionally, the film offers a rare glimpse into United States' cultural understandings of the lives and dreams of adults with intellectual disability. Too often, popular films use disabled characters as a kind of "narrative prosthetic": they are used to give meaning to the plot of nondisabled characters but have little autonomous significance of their own (Mitchell and Snyder 2001). Landmark (2021) criticizes *The Peanut*

[1] The actor's first name is Zack; the character in the film has the same name but is spelled Zak. I use "Zack" or Gottsagen to refer to the actor, and Zak to refer to the character he plays.

S. C. Simplican (✉)
Department of Gender and Sexuality Studies, Vanderbilt University, Nashville, TN, USA
e-mail: stacy.c.simplican@vanderbilt.edu



A. Bianchi and J. A. Vogt (eds.), *Intellectual Disabilities and Autism: Ethics and Practice*, The International Library of Bioethics 108,
https://doi.org/10.1007/978-3-031-61565-8_4

*Butter Falcon* for its patronizing treatment of Zak as well as casting him within a familiar frame of the supercrip—as Zak is given superhuman strength. Though Landmark's analysis is accurate, the film cannot be reduced to familiar ableist tropes. Instead, it raises vital questions for our futures: What is required for meaningful social inclusion? How should we promote autonomous decision making? How should we think about the necessary long-term supports and services that many adults with intellectual disability will need throughout their lives? And who do we envision providing these supports?

*The Peanut Butter Falcon* embraces the mythology of American folklore and also captures elements of social contract theory. Described on its website as a "modern Mark Twain style adventure story," the film follows the journey of two white men, Zak and Tyler (Shia LaBeouf), who journey south on a homemade raft, pursued by violent outlaws (*The Peanut Butter Falcon* 2019). The film's location on the outer banks of the North Carolina coast and the film's imagery—lone gas stations, empty fields, secluded cabins, and amateur fighting rings—conjures a modern-day state of nature, adrift from law and order. When Zak escapes from his nursing home, he sets out to find his wrestling idol, The Salt Water Redneck (Thomas Haden Church). Eleanor (Dakota Johnson), who works at the nursing home, is charged with finding Zak and, along the way, cannot resist Tyler's charm. Against a backdrop of rural neglect, Zak, Eleanor, and Tyler forge a kind of social contract among themselves, not because they are equal in capacities, but because they recognize each other's mutual vulnerability (Simplican 2015).

I draw on the concept of the capacity contract to analyze the film—a concept that builds on critical contract theory pioneered by Pateman (1988) and Mills (1997). We can think of two types of capacity contract. The more traditional version of the capacity contract grants political membership only to those who pass a threshold level of cognitive capacity and thereby excludes anyone deemed inferior. We can also construct a more emancipatory capacity contract when we imagine political solidarity as the recognition of mutual vulnerability.

*The Peanut Butter Falcon* envisions a mutual aid society built from the recognition of shared mutual vulnerability in contemporary America. But it also harbors a deep suspicion of state-supported safety nets for vulnerable adults. The film's depiction of a reciprocal and nonfamilial caring network between disabled and nondisabled adults is particularly salient in the wake of COVID-19, as the pandemic provided new spotlight to the idea of mutual aid. Tolentino (2020) heralded mutual aid in *The New Yorker* as "a radical practice [that] is suddenly getting mainstream attention." Drawing on her interview with disability activist Mia Mingus, Tolentino described the idea of neighborhood "pods": groups of five to twenty people who commit to providing mutual aid to one another. Tolentino's article also acknowledged and responded to skepticism of whether mutual care networks can realistically provide actual solutions to entrenched societal problems. Citing political scientist Nancy L. Rosenblum, "when government and politics disappear from view as they do, we are left with the not-so-innocuous fantasy of ungoverned reciprocity as the best and fully adequate society" (Rosenblum, quoted by Tolentino 2020). As Rosenblum (2016) suggests, the fantasy of ungoverned reciprocity as a solution to political problems may be a beloved, yet dangerous, American myth.

## 4.2 Housing, Employment, and Social Inclusion

*The Peanut Butter Falcon* should be credited with bringing attention to the lives of adults with intellectual disabilities, many of whom experience a range of inequalities and vulnerabilities. People with intellectual disability are far less likely to be employed in comparison to nondisabled adults, as well as people with other types of disability (Siperstein et al. 2013).[2] In fact, "[t]heir unemployment rates exceed 70%, nearly ten times the rate of their nondisabled peers" (Sannicandro et al. 2018, 412). Relatedly, people with disabilities are more likely to live in poverty; research suggests that one in four children with disabilities live with families with earnings below the poverty level (Hughes and Avoke 2010). Furthermore, people with intellectual disabilities, specifically, are more likely to be victims of violence, including sexual violence (Harrell 2021; Tomsa et al. 2021). During the COVID-19 pandemic, people with intellectual disabilities experienced some of the highest mortality rates and were identified as having heightened risks in the summer of 2020 (Shapiro 2020). However, the Centers for Disease Control and Prevention (the United States' national public health agency) did not include people with intellectual disabilities on the list of those with medical conditions with an elevated risk of developing severe illness until February 2022.

Although poverty is undoubtedly one of the most significant issues experienced by people with intellectual disabilities, the political issues most salient to *The Peanut Butter Falcon* are housing and social inclusion, as the film begins with the lead character trying to escape from his residential nursing home. The Independent Living Movement has advanced deinstitutionalization for people with disabilities since the 1960s. But as Knight (2021) argues, the deinstitutionalization movement is unfinished, at least in part due to the health system and coverage that exists in the United States, and its gravitational pull toward institutional placements rather than smaller community-based options. According to Woodman et al. (2014), "As more [individuals] with intellectual disability age into adulthood, they are increasingly likely to move away from the parental home and require residential services for greater lengths of time. The demand for residential services is therefore on the rise" (496).

The pandemic, in combination with worker shortages, has contributed to fewer housing options available for people with intellectual disabilities. According to Weber and Miller (2022):

> A 2021 survey by the American Network of Community Options and Resources,[3] an advocacy group for providers that support those with intellectual and developmental disabilities, found that because of staffing shortages, more than three-quarters of service providers were

---

[2] For more information related to employment and people with ID, see Chap. 17 in this volume.

[3] https://www.ancor.org/resources/the-state-of-americas-direct-support-workforce-crisis-2022/.

turning away new referrals, more than half were discontinuing programs and services, and the vast majority were struggling with recruitment and retention of staff.

In *The Peanut Butter Falcon*, both housing *and* quality of life for people with ID are explored, where housing is a contributing, though not exclusive, factor. One's quality of life encompasses various factors, including the ability to pursue one's passions, the right to make one's own decisions, and the quality of relationships one has with other people, especially people who are not paid staff or one's parents. According to Merrells et al. (2019, 13), "Spending time in valued activities, having meaningful relationships and feeling included are three core aspects of social inclusion." In their study, they found that adults with ID who had access to disability support programs still felt excluded, like outcasts in their communities, were frequently bored, and had difficulty maintaining peer relationships.

What can we infer from this research? First, many people with intellectual disabilities will require support to find housing options, taking into consideration that many will be unable to live alone and meet their own needs. Unfortunately, most places in the United States do not dedicate sufficient resources to meet the needs of people with intellectual disability, thus relying on unpaid families for support or on institutional supports and services. Second, people with intellectual disability may benefit from having opportunities to cultivate relationships with community members, i.e., not exclusively with parents and paid staff. As neoliberal ideologies continue to pull support away from governmental solutions, can mutual aid societies step in to fill these gaps? *The Peanut Butter Falcon* offers one way that we can think about implementing long-term supports and services in a neoliberal landscape of twenty-first century America.

## 4.3 Making the Film: Nothing About Us Without Us

The story behind the making of the film is just as important as the film itself, particularly when it comes to thinking about the importance of meaningful inclusion and developing relationships with community members. As it turns out, the movie was written specifically *for* Zack Gottsagen, a person with intellectual disability and the lead actor in the movie. Film directors Michael Schwartz and Tyler Nilson met Zack at an acting camp for disabled actors. Zack shared with them his dream to be in a Hollywood movie. Schwartz and Nilson, however, were skeptical. For Nilson, "Michael and I didn't sugar-coat it. We explained to him that though he was talented and had been studying acting for years, there just weren't roles written in Hollywood for actors with Down syndrome (or any disability) and there was a *very* small chance that he'd ever get an opportunity to play a major role" (Schwartz and Nilson 2019). According to Nilson, Zack smiled and said, "'Well … you guys make movies, why don't you write one and I can be in it?!'"

In response to Zack's request, Schwartz and Nilson wrote a film that was centered around important aspects of Zack's life, including his love for professional wrestling and swimming. It took five years to get a studio to agree to produce the film. Several wanted to make the film, but only if they could cast someone other than Zack in the lead role. Prejudice against people with intellectual disabilities—fears that they may be difficult to work with or might slow the production or harm the final product—led to hesitation within the industry (Norlian 2019). In fact, less than three percent of films made in the last decade have a disabled character, and ninety-five percent of those roles are played by nondisabled actors (Smith et al. 2017). Yet, the writers were adamant: they would only make the film with Zack in the lead role. After five years of trying, they finally found a studio that would greenlight the production.

Central to the film's plot is the relationship between Zak and Tyler, and media attention to the film focused on the off-screen relationship between the actors Gottsagen and LaBeouf. Both on and off-screen, Gottsagen and LaBeouf are portrayed as having a special bond in which Gottsagen can connect with LaBeouf's troubled soul. Press articles focused on the fact that LaBeouf was arrested for drunk driving in Georgia while filming the movie and the role Zack played in getting LaBeouf back on track (Harrison 2019). While arrested, LaBeouf was caught on camera using racist and sexist language toward a Black police officer (Ibid.). According to the *Independent* (Ibid.), LaBeouf credited Gottsagen for changing his behaviour. Gottsagen made LaBeouf realize that his own actions were "ruining" Gottsagen's chances of successfully pursuing this film and fulfilling his dream.

Across these stories told by the writers and actors of the film, is a thread about (some men's) mutual vulnerability. Within the movie, their mutual vulnerability serves as the reason that Zak and Tyler forge a new and non-ableist social contract. Outside of the film, it seems plausible that Gottsagen and LaBeouf develop a kind of social contract as well, irrespective of their being in different positions within society. Although these new, less ableist social contracts develop both within and outside of the film, it may be worth highlighting that these relationships occur between white men, as the outcome of the Black police officer who LaBeouf harassed is erased. The way that intersecting identities may influence whether and/or the extent to which social contracts develop is worth further consideration and is not addressed in the film.

## 4.4 Friendship and the Fraternal Social Contract

The film demonstrates both the lack of and the importance of sufficient supports for people with intellectual disability; it is only through having certain supports that some people with intellectual disability may have an opportunity to achieve their full potential and have a positive quality of life. Zak meets Tyler when he takes shelter in his boat on his escape from his nursing home. But Tyler abandons him once on land. It is only when Tyler witnesses Zak being bullied and physically attacked that he decides to let Zak accompany him. Even once together, however, the film emphasizes their isolation from one another, with the camera taking long shots, positioning Tyler and Zak on opposite sides of the screen.

The importance of supports is especially highlighted when Zak and Tyler must cross a river, but Zak can't swim. Upon recognizing Zak's vulnerability and the difference he can make, Tyler rigs a floatation device for Zak, and ropes their bodies together. Zak is scared. He asks, "Tyler … am I gonna die?" Tyler answers plainly: "Yeah, yer gonna die. It's a matter a' time. That ain't the question. The question is whether they're gonna have good stories to tell aboucha when you're gone. Now don't be a bitch, let's go." Though not elegantly stated, Tyler's words acknowledge the shared vulnerability of them both, while also making plain that it is not just their safety that is important—but also their quality of life. As Tyler swims, the camera again takes a wide shot from above, the men distant from one another, but now tied together. Suddenly, a horn blares: a large boat with huge nets dredging for fish is barreling down the river. Tyler swims frantically and scrambles to shallow water, now pulling Zak as fast as he can. When Zak survives, Tyler collapses to the shore. But Zak is ecstatic. "That will be a great story," laughs Zak, slapping Tyler on the shoulder.

This scene may be interpreted as one that represents the value of supported decision making, which is when a person with intellectual disability is supported in their ability to make their own decision(s). Tyler supports Zak in crossing the river, taking into consideration his individual needs and the level of accommodation required. Furthermore, Zak and Tyler's conversation reveals the film's investment in *risk* as a key part of a well-lived life, and as one that people with intellectual disability can pursue if desired. The "dignity of risk" is a concept introduced in the 1970s (Perske 1972) that connects the value of risk-taking with that of leading a dignified life for people with intellectual disability.

Alternatively, we can interpret this scene through the lens of social contract theory, as Tyler and Zak forge a capacity contract in the face of their mutual vulnerability. Importantly, this contract is forged through vulnerability—the threat of their imminent and mutual death by the oncoming fishing boat—*and* their shared masculinity. As Tyler warns Zak, "don't be a bitch." In the ensuing scene montage, Tyler teaches Zak to swim, to wrestle, and shoot a rifle. They eat watermelon and fashion the rinds into hats. Often in cut-off shorts and no-shirts, it is a distinctly masculine-tinged adolescent play and care that seals their fraternal capacity contract.

Zak and Tyler's fraternal contract strengthens, even as Eleanor joins them. Eleanor, who has continued to search for Zak, finds them after a night of drunken revelry: Zak is passed out on the beach clad only in underwear while Tyler pees in the ocean. Aghast, Eleanor scolds Tyler, "You have a young boy with Down syndrome in the middle of nowhere." Eleanor wants to return Zak to the nursing home, but Tyler refuses, stating, "I'm the one who's been taking care of him, because you can't do it unless you put him in a cage." Eleanor's response captures both her understanding of proper care and her perception that Tyler and Zak are living in a stage of masculine adolescence. She replies, "I'm sorry to break up your little primitive gang, or whatever's going on … – There's rules, regulations. He has medication. There's paperwork." Their argument challenges Eleanor and the nursing home as the site of appropriate care, but it is Zak who makes the final decision, as he tosses Eleanor's keys into the ocean, demonstrating both his preference to remain with Tyler and his capacity for autonomous decision making.

By referring to Zak as a boy and insisting on his return to institutionalized care, Eleanor demonstrates potential harms that can come from substitute (surrogate) decision making and failure to take into adequate consideration the preferences, beliefs, and values of a person with intellectual disability. In the film, Eleanor's presumption to decide what is best for Zak overrules his own autonomous preferences, which is a common experience of people with intellectual disability (Werner 2012). Moreover, Eleanor mistakes the bureaucratic regulation of Zak's life—the paperwork, rules, regulations, and medications—for care. Likewise, we can see Eleanor through a kind of domination capacity contract, in which she presumes that Zak cannot be an equal participant in his own life's decisions.

*The Peanut Butter Falcon* directly confronts the problems with paternalism and the resultant stigma towards people with intellectual disability. During one scene, Eleanor tries to mold Zak's behavior, telling him to sit down, eat fruit, apply sunblock—all in the name of care and concern. In response, Tyler says to Eleanor, "You'd better stop calling him a retard." Although Eleanor states that she "ha[s] never called anyone a…that word.", Tyler clarifies that her *not* saying the word is irrelevant since she is making him "feel retarded."

Tyler's argument goes beyond simple language choices and instead penetrates to the kinds of harms that people with intellectual disability experience because of stigma, substitute decision making and/or the types of well-intentioned supports provided to people with intellectual disability, and even philosophical approaches to political governance. In the film, Eleanor is forced to rethink her relationship with Zak and her understanding of what it means to care for someone. Rethinking what it means to care for someone with intellectual disability in an ethically defensible way is important in our current society. As Alberts (2020) argues, well-intentioned, though paternalistic and stigmatizing approaches to working with people with intellectual disability continue to exist, and different models of supporting and caring for people with intellectual disability ought to be explored. In relation to the film, it may be worth asking whether the fraternal contract that exists between Zak and Tyler can combat ableist stigma and provide the support and inclusion that Zak desires. If yes, then considering how similar models may translate to the real world is important.

## 4.5 Violence and the Importance of a Chosen Family

Much of the film takes place when characters Eleanor, Tyler, and Zak are on the way to Salt Water Redneck. Along the way, they confront threats to their lives and freedom. The importance of building and maintaining trust amongst people with intellectual disability is highlighted in certain scenes, particularly when these threats occur. For instance, in one scene, an outlaw gang sets Tyler, Eleanor, and Zak's raft afire. Though the three escape physically unharmed, it shatters the trust that Tyler had built with Eleanor and Zak. Trustworthiness is an essential characteristic to demonstrate when seeking to build relationships with and support people with intellectual disability.

In this way, the film adds to prior analysis around the emancipatory potential of the capacity contract, in which I argued (Simplican 2015) that participants must acknowledge and accept mutual vulnerability as a way to build and maintain solidarity. Acceptance of mutual vulnerability is not enough. Instead, participants need to trust one another—that the recognition of mutual vulnerability comes with the responsibility to care for that vulnerability. Eleanor and Zak feel betrayed. Tyler knowingly jeopardized their safety.

Parts of the film also highlight how power imbalances and those in positions of authority may influence the outcomes to which people with intellectual disability may be subjected. In this way, the film illustrates the limits of social contract theorizing, wherein participants are presumed equal and exist prior to state authority. For instance, in one scene, Eleanor's boss insists that Zak return to a new residential placement. Although Zak does not want to return to his residential care environment, and—at this point in the film—Eleanor disagrees with the bureaucratic caring approach to Zak, Eleanor is unwilling to resist the authority of state bureaucracy.

*The Peanut Butter Falcon* thus reminds us that people with intellectual disability (like all people) are inevitably influenced by the societies and systems within which they reside. If the people and spaces available to us do not value autonomy, independence, and the dignity of risk, then this may have a detrimental impact on quality of life outcomes. Furthermore, insofar as people with intellectual disability have the right to have their "inherent dignity, individual autonomy including the freedom to make one's own choices, and independence of persons" respected, as indicated in Article three of the United Nations Convention on the Rights of Persons with Disabilities (CRPD) (UN 2006), then overriding decisions based on one's own, or the system's, values and capacities is ethically indefensible.

What is the answer to these problems? For *The Peanut Butter Falcon*, it is not a new capacity contract. It's family and, most importantly, a chosen family in which members are bonded together not by biology, but by love, care, and commitment. In one part of the film, Zak suggests to Eleanor and Tyler that they "could be a family." Zak finds fulfillment and meaning in his relationship with Tyler and Eleanor. Though abandoned by his biological family, Zak still believes in the value of family and his journey reignites his desire to belong to one. Zak seems to perceive his family as those who provide support, as well as offer opportunities for him to engage in adventurous, risky, and joyful activities. The way that Eleanor and Tyler respond to Zak throughout the film allows him to maintain "full and effective participation and inclusion in society", which is an important principle as per Article three (c) of the CRPD (UN 2006). Ultimately, the film and Zak's desire to have a family encourages us to ask: What is family? Can you choose your family? What, if any, differences may exist when thinking about the concept and importance of family for people with intellectual disability, particularly if they live in a congregate living environment?

When the trio eventually arrives at the home of the now retired Salt Water Redneck, a.k.a. Clint, they find him initially reluctant to become involved in Zak's plan. However, as they make their way down the road leading away from his home, we hear an engine and see a hot rod speed toward them. Now, in full costume of blue face paint, a sleeveless cut off denim shirt, and a bandana, Clint has transformed into the Salt Water Redneck and he beckons Zak to join his wrestling school. Hope is restored—as they see in Clint the same values that brought them together: recognition of their shared vulnerability, a determination to contribute to Zak's joy and quality of life, and the willingness to abandon oneself to imaginative play. The trio return to Clint's home, now transformed with a homemade ring of rope under a shaded tree, where Salt Water teaches Zak wrestling moves.

Zak wants to learn the "Atomic Throw"—Salt Water's old signature move—but Clint explains that they "just made that up. It was smoke and mirrors." As Clint explains, we see an older man grunt and gristle in the background. Another blue-collar man, but he is unlike Clint, Zak, and Tyler—who share not only in their ability to express their own vulnerability—but also in a spirit of play and wonder. In contrast, Clint's pal Samson is coarsened, distant from anything that may warn of vulnerability. Again, *Peanut Butter Falcon* suggests the limits of an emancipatory capacity contract in which all participants recognize mutual vulnerability and entrust each other with the responsibility of mutual aid.

These tensions lead to the near-death experiences of both Zak and Tyler. On a euphoric high of Zak's joy and confidence, Tyler and Clint sign him up to compete in a "real match" with Samson. Meanwhile, vengeful anglers are still after Tyler. This all culminates in a match. Zak is frightened of the actual battle, so Tyler uses cardboard to create a fantastical wrestling persona of *The Peanut Butter Falcon* (an identity dreamed up during an earlier drunken night of revelry between the two men). Eleanor tries to crack Tyler's delusional thinking about the match: "We care about [Zak]. They don't." In using the language of "they"—and not just the singular Samson—Eleanor's warning extends far beyond the wrestling match, of a hostile and dangerous world. We watch Samson beat, harass, and ridicule Zak. But Zak stays in the ring. With Tyler coaching him, Zak draws deep upon his anger and miraculously (fantastically) lifts Samson above his head and tosses him far into the air in a true atomic throw. At the same time, the ringleader of the violent gang cracks Tyler on the head with a lead pipe. The scene goes black.

In the following scene, Zak and Eleanor are at the hospital, with Eleanor in the background talking to a doctor. Zak sits alone. It's here, I believe, that Zak processes fully how his dreams for his future have shifted. He escaped from the nursing home in search of adventure and to become a celebrated wrestler. But he has discovered that what he was in search of—all along—was love, trust, and connection. Love from the chosen family of himself, Eleanor, and Tyler. The movie does not disappoint. In the final scene, Eleanor drives south to Florida with Zak in the passenger seat, and a bruised and bandaged Tyler in the back seat. Together.

## 4.6 Conclusion

Throughout the film, important political themes are highlighted, including the everyday threats of violence and stigma that people with disabilities continue to face, the future of long-term services and supports for vulnerable populations, the importance of friendship, authentic connection, and the dignity of risk. It also demonstrates some limitations that may be more likely to occur in some institutional settings (or in any place wherein people with ID are not included, able to connect with others, etc.). Though the movie gives viewers many feel-good moments, as intended by the directors (Schwartz and Nilson 2019), I am left feeling uneasy, for three main reasons.

First, I worry about the dream of mutual aid, especially as an answer to long-term supports and living arrangements for people with intellectual disability. According to the Residential Information Systems Project (2023), a large majority of people with intellectual disability, sixty percent, live with a related family member. The next largest share, sixteen percent, resides in a group home. For the rest: eleven percent own or rent their own home; eight percent reside in an institution, and the smallest share, five percent live with a host or foster family. *The Peanut Butter Falcon* makes us feel good about a situation that, frankly, does not exist for most people with intellectual disability. In celebrating the unpaid labor of a chosen family, I worry that the film bolsters neoliberal rejections of state responsibility for vulnerable people under the familiar reliance on (unpaid) family care. The feel-good ending of Eleanor, Tyler, and Zak driving to Florida is *made* happy because it is set against Zak's escape from institutionalized and bad care. But should we really feel good about a film that celebrates the withering responsibility of state bureaucracies? Alternatively, we might imagine Eleanor returning to her vocation, demanding institutional changes, and working to build a stronger public safety net for Zak and others like him. Instead, she is called to return to the private realm, where she and Tyler are presumed to live happily ever after in their heteronormative romance, along with Zak. Whether Zak desires a romance of his own is left unaddressed—as are questions of Zak's future economics, healthcare, employment, or dreams.

Second, I worry that the film misrepresents the kinds of relationships that people with intellectual disability desire and are most likely to have. *The Peanut Butter Falcon* primarily tells the story of men, and particularly white men. Yet, we know that rates of intellectual disability occur among all people across differences of gender, race, and ethnicity. We need to diversify the stories we tell about disability. But that's not the only problem with the film's pull toward masculinity. In centering the relationships between men with and without intellectual disabilities, I worry that *The Peanut Butter Falcon* distorts gendered and raced realities in the intellectual disability landscape. For example, the direct support workforce serving people with intellectual disabilities is predominately made up of women and people of color. In fact, the percentage of white direct support professionals has decreased, whereas Hispanic women are the fasting growing segment of the workforce (The Case for Inclusion 2022, 18). In telling a feel-good story about the relationships between men with and without intellectual disability, I hope the film serves as a call to action among men and boys, to take up vocations, careers, and friendships across disability.

Finally, and theoretically, I worry about the emancipatory potential of a capacity contract forged across the recognition of mutual vulnerability. Yes, we need to acknowledge our mutual vulnerability, but *The Peanut Butter Falcon* shows us that we deserve and desire so much more from our political and ethical futures. Perhaps, what the film and social contract theory share most of all is their embrace of human imagination. Social contract theorists, like Thomas Hobbes, John Locke, and Jean-Jacques Rousseau, imagined wildly divergent states of nature and radical new ways to legitimize state governance. In the film, people discover a sense of belonging not just because they see in one another mutual vulnerability. But more so because they give themselves over to play, to imagination, and to fun. I hope that Hollywood can imagine and produce new stories in which more diverse casts with intellectual disability can embody these playful futures. More importantly, I hope we can imagine and help create these new futures for each other.

## References

Alberts, Lisa. 2020. Paternalism to trusteeship: Limitations of autonomy caring for persons with intellectual disabilities. *International Journal for Human Caring* 24 (3): 213–221. https://doi.org/10.20467/HumanCaring-D-19-0005.

Harrell, Erika. 2021. *Crime against persons with disabilities, 2009–2019—Statistical tables*. U.S. Department of Justice, Office of Justice Programs. Bureau of Justice Statistics. https://bjs.ojp.gov/content/pub/pdf/capd0919st.pdf. Accessed April 9, 2024.

Harrison, Ellie. 2019. Dakota Johnson recalls Shia LaBeouf's 'terrifying and difficult' arrest during *Peanut Butter Falcon* filming. *Independent*. https://www.independent.co.uk/arts-entertainment/films/news/dakota-johnson-shia-labeouf-arrest-peanut-butter-falcon-a9221246.html. Accessed April 9, 2024.

Hughes, Carolyn, and Selete K. Avoke. 2010. The elephant in the room: Poverty, disability, and employment. *Research and Practice for Persons with Severe Disabilities* 35 (1–2): 5–14. https://doi.org/10.2511/rpsd.35.1-2.5.

Knight, Amber. 2021. Unfinished business: Deinstitutionalization and Medicaid policy. *Politics, Groups & Identities* 9 (2): 395–408. https://doi.org/10.1080/21565503.2020.1854324.

Landmark, Shelby E. 2021. The patronized supercrip: A textual analysis of *The Peanut Butter Falcon*. In *Disability representation in film, TV, and print media*, ed. Michael S. Jeffress, 128–144. New York, NY: Routledge.

Merrells, Jessica, Angus Buchanan, and Rebecca Waters. 2019. 'We feel left out': Experiences of social inclusion from the perspective of young adults with intellectual disability. *Journal of Intellectual and Developmental Disability* 44 (1): 13–22. https://doi.org/10.3109/13668250.2017.1310822.

Mills, Charles W. 1997. *The racial contract*. Ithaca, NY: Cornell University Press.

Mitchell, David T., and S.L. Snyder. 2001. *Narrative prosthesis: Disability and the dependencies of discourse*. Ann Arbor, MI: University of Michigan Press.

Norlian, Alison. 2019. '*The Peanut Butter Falcon*' writers on how their movie went from a fun idea to making a positive impact. Final Craft—A Cast and Crew Company. https://blog.finaldraft.com/the-peanut-butter-falcon-writers-on-how-their-movie-went-from-a-fun-idea-to-making-a-positive-impact. Accessed April 8, 2024.

Pateman, Carole. 1988. *The sexual contract*. Oxford, UK: Polity Press.

Perske, Robert. 1972. The dignity of risk and the mentally retarded. *Mental Retardation* 10 (1): 24–27.

Residential Information Systems Project. 2023. *Living arrangements for people with IDD*. Minneapolis: University of Minnesota, RISP, Research and Training Center on Community Living, Institute on Community Integration. https://publications.ici.umn.edu/risp/infographics/where-do-people-with-intellectual-or-developmental-disabilities-live-when-they-dont-live-with-their-family. Accessed April 8, 2024.

Rosenblum, Nancy L. 2016. *Good neighbors: The democracy of everyday life in America*. Princeton, NJ: Princeton University Press.

Sannicandro, Thomas, Susan L. Parish, Steve Fournier, Monika Mitra, and Maria Paiewonsky. 2018. Employment, income, and SSI effects of postsecondary education for people with intellectual disability. *American Journal on Intellectual and Developmental Disabilities* 123 (5): 412–425. https://doi.org/10.1352/1944-7558-123.5.412.

Schwartz, Michael, and Tyler Nilson. 2019. '*Peanut Butter Falcon*' writers crafted the film for actor friend who has Down syndrome. *Los Angeles Times*. https://www.latimes.com/entertainment-arts/movies/story/2019-12-30/peanut-butter-falcon-writers-crafted-film-for-down-syndrome-actor-friend. Accessed April 9, 2024.

Shapiro, Joseph. 2020. COVID-19 infections and deaths are higher among those with intellectual disabilities. *Morning Edition*. https://www.npr.org/2020/06/09/872401607/covid-19-infections-and-deaths-are-higher-among-those-with-intellectual-disabili?fbclid=IwAR33qWHWXVrXok9qpnkhicIXRLmZB22qCIhrpFC6o_R4d4vCsITxe7wudi8. Accessed April 9, 2024.

Simplican, Stacy Clifford. 2015. *The capacity contract: Intellectual disability and the question of citizenship*. Minneapolis, MN: University of Minnesota Press.

Siperstein, Gary N., Robert C. Parker, and Max Drascher. 2013. National snapshot of adults with intellectual disabilities in the labor force. *Journal of Vocational Rehabilitation* 39 (3): 157–165. https://doi.org/10.3233/JVR-130658.

Smith, Stacy L., Marc Choueiti, and Katherine Pieper. 2017. Inequality in 900 popular films: Examining portrayals of gender, race/ethnicity, LGBT, and disability from 2007–2016. *Annenberg Foundation and USC Annenberg*. https://annenberg.usc.edu/sites/default/files/Dr_Stacy_L_Smith-Inequality_in_900_Popular_Films.pdf. Accessed April 9, 2024.

*The Case for Inclusion 2022: Blazing trails to sustainability for community disability services*. 2022. ANCOR Foundation, United Cerebral Palsy. https://caseforinclusion.org/application/files/1716/4658/7147/Case_for_Inclusion_2022_Blazing_Trails_to_Sustainability_for_Community_Disability_Services.pdf. Accessed April 9, 2024.

*The Peanut Butter Falcon*. 2019. Directed by Michael Schwartz and Tyler Nilson. Armory Films.

Tolentino, Jia. 2020. What mutual aid can do during a pandemic. *The New Yorker*. https://www.newyorker.com/magazine/2020/05/18/what-mutual-aid-can-do-during-a-pandemic. Accessed April 9, 2024.

Tomsa, Raluca, Smaranda Gutu, Daniel Cojocaru, Belén Gutiérrez-Bermejo, Noelia Flores, and Cristina Jenaro. 2021. Prevalence of sexual abuse in adults with intellectual disability: Systematic review and meta-analysis. *International Journal of Environmental Research and Public Health* 18 (4): 1980. https://doi.org/10.3390/ijerph18041980.

UN (The United Nations). 2006. *(CRPD) Convention on the Rights of Persons with Disabilities*. Treaty Series, 2515, 3, G.A. Res. 61/106, 76th plen. Mtg., U.N. Doc A/Res/61/106. https://social.desa.un.org/issues/disability/crpd/convention-on-the-rights-of-persons-with-disabilities-crpd. Accessed July 9, 2023.

Weber, Lauren, and Andy Miller. 2022. *Disability Scoop*. https://www.disabilityscoop.com/2022/03/09/programs-for-people-with-disabilities-still-not-seeing-federal-funds/29746/. Accessed April 9, 2024.

Werner, Shirli. 2012. Individuals with intellectual disabilities: A review of the literature on decision-making since the convention on the rights of people with disabilities (CRPD). *Public Health Reviews* 34 (2): 1–27. https://doi.org/10.1007/BF03391682.

Woodman, Ashley C., Marsha R. Mailick, Kristy A. Anderson, and Anna J. Esbensen. 2014. Residential transitions among adults with intellectual disability across 20 years. *American Journal on Intellectual and Developmental Disabilities* 119 (6): 496–515. https://doi.org/10.1352/1944-7558-119.6.496.

# Chapter 5
# From Baby Doe to Selective Termination for Down Syndrome

**Chris Kaposy**

**Abstract** From the early 1980s to today, much has changed in the neonatal care of infants born with Down syndrome. Leading up to the Baby Doe case in the US in 1982, infants born with Down syndrome were often denied life-saving treatment solely because of their disability. These norms of non-treatment changed in the aftermath of Baby Doe. Around the same time, prenatal screening tests for Down syndrome became more widely available, leading to the prevalence of selective abortion for Down syndrome. This essay considers the question of moral progress during the era in which infants with Down syndrome were no longer denied life-saving treatment, but increasing numbers of pregnancies were terminated because of a prenatal diagnosis of Down syndrome. I look at three aspects of this history and of the debate in bioethics about Down syndrome in the perinatal context. First, I examine the comparison between the ethics of infanticide and abortion. Second, I discuss the historical contingency of the resolution provided by the Baby Doe case. Third, I present the view that ableism continues to underlie widespread selective abortion. I conclude that moral progress during this period from Baby Doe to the present day is tangible, but also fragile.

**Keywords** Abortion · Bioethics · Down syndrome · Infanticide

## 5.1 Introduction

The story of Down syndrome over the past hundred years is, in many ways, a story of progress. People with Down syndrome are now living much longer than they did in the 1920s (Wright 2011, 185). In the Western world, children with Down syndrome go to public schools like most other children, and many adults with Down syndrome have jobs and live independently. This essay examines the idea of "moral progress" in the ethics of medical care during the perinatal period when fetuses or

C. Kaposy (✉)
Faculty of Medicine, Centre for Bioethics, Memorial University, St. John's, NL, Canada
e-mail: ckaposy@mun.ca

 
A. Bianchi and J. A. Vogt (eds.), *Intellectual Disabilities and Autism: Ethics and Practice*, The International Library of Bioethics 108,
https://doi.org/10.1007/978-3-031-61565-8_5

infants are diagnosed with Down syndrome. I examine a span of time in history from roughly the mid-twentieth century up until the present day. One major turning point during this period was the death of an infant with Down syndrome in 1982 who was given the name "Baby Doe". Baby Doe died because he was denied surgery that would have allowed him to live. Prior to Baby Doe, infants with Down syndrome were often denied life-saving treatment, and were left to die, simply because of their disability. In the years immediately following Baby Doe's death in the early- to mid-1980s, this norm of the acceptability of denying life-saving treatment to infants with Down syndrome began to change. Around the same time, prenatal screening tests designed to reveal the likelihood that fetuses had Down syndrome or a few other disabilities, became increasingly available. The availability of prenatal screening for Down syndrome made selective abortion a common occurrence.

In the context of this progression from the death of Baby Doe in the 1980s to the use of selective abortion in our era, I focus on the question of whether this history signals moral progress. I examine three aspects of this history for the purposes of my analysis. First, I examine the question of comparing abortion to infanticide—more specifically the comparison between (1) the denial of treatment to infants like Baby Doe, and (2) selective abortion. I consider the use of this comparison for the purpose of offering an ethical justification for selective abortion. I argue that abortion is not morally questionable in the way that infanticide clearly is.

Second, I examine the changes brought about in the care of infants with Down syndrome around the time of the Baby Doe controversy to suggest that these changes were a contingent occurrence that were not inevitable. The first two points lead up to my third point, and the overall conclusion of my analysis: that moral progress during this period from Baby Doe to selective termination in the present day is incomplete, and that the progressive gains for people with Down syndrome are fragile.

Throughout this chapter, I often refer to the denial of life-saving treatment to infants with Down syndrome as "infanticide". Some may wonder whether this terminological choice begs the conceptual question of whether the denial of treatment is the same as infanticide. I am unable to fully address this conceptual question. But I find it convincing that if an infant is in the care of a doctor or a parent, and the caregiver denies the infant care or treatment necessary to save his or her life, or necessary for his or her survival, then this denial is a form of killing. For this reason, I use the term "infanticide". Those who disagree with this stipulation can nonetheless regard my use of the term "infanticide" merely as a stylistic shorthand for the more cumbersome term "the denial of life-saving treatment".

## 5.2 Background

Baby Doe was born in 1982 in the US state of Indiana. He had Down syndrome, along with an esophageal atresia, which is an interruption in the tract leading from the mouth to the stomach, preventing food and water from passing into his stomach. Baby Doe's parents decided not to have the atresia surgically corrected, even though he would die without the intervention, and even though surgery was likely to be successful. The parents apparently agreed with the opinion of their obstetrician who thought that, because of his Down syndrome, Baby Doe would not enjoy a "minimally adequate quality of life" (Lyon 1985, 31). A judge agreed that Baby Doe's parents could deny him surgery. A complicated series of legal appeals ensued, but these were unsuccessful and a few days after his birth Baby Doe died from being denied food, water, and potentially life-saving surgery (Lyon 1985).

Baby Doe was not a unique case. He was in fact among the last in a line of infants who were denied medical care and who were left to die because of their disabilities. And as we will see, Baby Doe's death was actually a major reason this tradition of medical infanticide came to an end. There were other infants with Down syndrome who came before him. In 1963, a baby with Down syndrome was born at Johns Hopkins Hospital with a duodenal atresia—a discontinuity in the duodenum. The infant's parents similarly refused surgical intervention, and this decision was not legally challenged by the hospital. The infant died, apparently of starvation, after eleven days (Gustafson 1973). A second infant was born with a duodenal atresia and Down syndrome at Johns Hopkins in 1971. This infant was also denied potentially life-saving medical care and died (Antommaria 2006). Together, the deaths of these two infants at Johns Hopkins were the inspiration for a documentary titled "Who Should Survive?: One of the Choices on Our Conscience", which was screened at a bioethics conference in 1971 and funded by the Joseph P. Kennedy Jr. Foundation (Antommaria 2006). The film is available for viewing on YouTube (Joseph P. Kennedy Jr. Foundation, and Johns Hopkins Hospital 1971).

In between the infants born at Johns Hopkins hospital in the 1960s and 70s and Baby Doe in 1982 there was also a well-known case in the UK involving Dr. Leonard Arthur. In 1980 a baby with Down syndrome was born in the city of Derby. The baby's parents did not want the infant to survive. Unlike Baby Doe, or the infants at Johns Hopkins hospital, this baby did not require any surgery in order to survive. There is some dispute about whether the baby had underlying health issues that ultimately led to his death, but he was not diagnosed with any condition other than Down syndrome at birth. In accordance with the request of the parents, Dr. Arthur made orders for "nursing care only" and for periodic treatment with a morphine-like drug to manage any pain. The infant died after roughly three days without food or water (Brahams and Brahams 1983). Dr. Arthur was charged with murder but was ultimately acquitted at trial (Brahams and Brahams 1983).

During the era from the 1950s to the 1980s, different surgical disciplines routinely denied potentially life-saving care to infants with Down syndrome. In 1974, surgeons at Toronto's Hospital for Sick Children published a retrospective analysis of cases of infants born with duodenal obstruction dating back two decades. The analysis shows that during that time, 27 infants born with Down syndrome and duodenal obstruction were denied potentially life-saving surgery with the consent of their parents (Girvan and Stephens 1974). Similarly, between the 1970s and mid-1980s, pediatric cardiac surgeons often decided not to repair atrioventricular septal defects in infants with Down syndrome, typically leading to the death of these infants at a young age, even when they would repair the same heart issue in almost all infants born without Down syndrome (Champagne et al. 2014).

If we broaden the perspective to include infants with other disabilities, during this period it was also common for physicians and parents to deny potentially life-saving treatment to infants born with myelomeningocele or spina bifida (Lorber 1971). So Down syndrome was certainly not the only disability singled out for fatal non-treatment.

Today, newborns with Down syndrome are no longer left to die when they have conditions like an esophageal or duodenal atresia or a cardiac malformation. In the current context of clinical care, it would be virtually unthinkable to refuse life-saving care to an infant with Down syndrome. Looking back on the prior fifty years of ethics in pediatric care, bioethicist Lynn Gillam (2015, 9) noted that, "Non-treatment of readily treatable conditions in babies with Down syndrome would not be accepted today." Several other authors have made the same point (Lantos and Meadow 2006; Dresser 2005; Robertson 2004). In pediatric care, there is now a better understanding of disability, a greater awareness of how ableism can influence treatment decisions, and a determination to avoid the influence of ableism. To give an example, in the UK the Ethics and Law Advisory Committee of the Royal College of Paediatrics and Child Health formulated an extensive framework on decisions to withhold or withdraw treatment for life-limiting and life-threatening conditions in children. According to this framework:

> the provision of care to those with disability should not be reduced and there must always be a commitment to the provision of high-quality care for them and … decisions to limit LST (life-sustaining treatment) in children with disabilities should be made on the same basis as in non-disabled children. (Larcher et al. 2015, sec. 12)

In the case of Baby Doe in 1982, and others like him, life-saving treatment was refused simply on the basis of disability when similar treatment would have been provided to an infant without a disability. The contemporary UK framework would prohibit such discrimination.

Another part of this story is the increased availability of prenatal testing since the 1980s. Prior to the early 1980s, the only prenatal testing modality available was amniocentesis, an invasive procedure that involves a small risk of miscarriage. Beginning in the mid-1980s, noninvasive screening programs for conditions like Down syndrome began to be developed (Löwy 2014; Russo and Blakemore 2014). These first screening tests relied on biochemical markers from maternal blood that indicate an increased likelihood that the fetus has Down syndrome or a few other conditions. The difference between screening tests and a diagnostic test like amniocentesis is accuracy. Typically, screening tests need to be followed by a more accurate diagnostic test like amniocentesis in order to provide an actionable diagnosis. Even today with the availability of more accurate cell-free fetal DNA screening tests, prenatal screening tests require follow-up with diagnostic testing in order to confirm the diagnosis of a genetic condition like Down syndrome (Mayo Clinic 2022). The greater availability of prenatal screening for Down syndrome since the 1980s has resulted in high rates of selective termination for Down syndrome (Natoli et al.

2012) and a diminished number of births of children with the condition (de Graaf et al. 2015). Currently, the majority of pregnancies in which a fetus is diagnosed prenatally with Down syndrome end in termination (Natoli et al. 2012).

While changes in the norms of pediatric and neonatal care have resulted in the unacceptability of denying life-saving treatment for Down syndrome, any parent or couple determined not to have a child with Down syndrome would today rarely have to face such treatment decisions. A contemporary couple who wishes to avoid having a child with Down syndrome (similar to the parents of Baby Doe) would today be able to access prenatal testing and selective abortion (where it is legal) rather than contemplating infanticide. It is probably an oversimplification to claim that selective abortion has superseded Down syndrome-based infanticide. The two are not ethically equivalent, and selective abortion is likely much more common than infanticide ever was. Yet it is nonetheless true that technological changes in prenatal care have changed the landscape of parental and clinical decision making in the postnatal period with regard to Down syndrome.

## 5.3 The Cruelty of the Past: Comparing Infanticide and Abortion

I would like to discuss three aspects of this history of the progression from Baby Doe in the 1980s to the prevalence of selective abortion for Down syndrome in our era. I will address one aspect in each of the next three sections. The first issue is about the way in which today's norm of selective abortion can be compared to the refusal of life-saving treatment in the past, for the purposes of advancing an ethical justification for selective abortion.

The refusal of treatment for infants with Down syndrome was a way to avoid becoming a parent of a child with Down syndrome. The use of prenatal testing and selective abortion is also a way to avoid becoming such a parent. There are clear ethical differences between the two, but in this regard they can both be used to realize the same goal. One of the ethical differences between refusing treatment and selective abortion is cruelty. The termination of a pregnancy in the first, second, or into the third trimester does not cause pain or suffering to a fetus (American College of Obstetricians and Gynecologists 2021). In contrast, the denial of life-saving treatment to infants with Down syndrome was often cruel. The reports of such cases usually include descriptions of pain management, sometimes unsuccessful, or suffering of the infant (Lyon 1985; Brahams and Brahams 1983). Baby Doe in particular is said to have endured a gruesome death in which his stomach acid ate through his organs, corroded his lungs, and caused a fatal pneumonia while he periodically cried out in hunger (Lyon 1985).

Historically, another way to avoid parenting a child with Down syndrome was institutionalization. Prior to the closure of institutions like the Willowbrook State School in New York in the 1970s and 1980s, many parents with children who had intellectual disabilities like Down syndrome would confine their children to these institutional environments. Again, the parallel goal of avoiding the parenting of a child with Down syndrome does not make institutionalization ethically equivalent to selective abortion. One of the reasons for the closure of institutions like Willowbrook was cruelty. Many children and adults with intellectual disabilities died or sustained injuries in Willowbrook because of illness, violence, or neglect (Rothman and Rothman 1984).

When it comes to this theme of avoiding parenting of children with Down syndrome, the cruelty of the past has been used to justify practices in the historical present. The denial of life-saving treatment to infants with disabilities (while itself cruel) was often justified by the consequent reduction of the number of children who would be consigned to abusive institutions (Rothman and Rothman 1984, 196). For example, in a book published in 1979, law professor Joseph Goldstein, along with therapist Anna Freud, and researcher Albert J. Solnit proposed a set of guidelines for state intervention in the decision making of parents in order to protect the well-being of their children (Goldstein et al. 1979). These model guidelines included discussion of the refusal to authorize life-saving medical care for infants. The authors recommended respecting parental autonomy in cases involving infants with "highly predictable mental retardation" (Ibid., 96)—meaning that parents should be able to choose the denial of treatment. One of the reasons stated for this position was that "as long as the state offers institutions that provide little more than storage space and 'hay, oats, and water' for these children, the law must err on the side of the strong presumption in favor of parental autonomy" (Ibid., 98). Notice the implicit endorsement of treating children with intellectual disabilities like animals. Rather than criticize the negligence of the state, Goldstein, Freud, and Solnit presupposed this negligence as an unquestioned "given". The authors referenced the abuse of children at Willowbrook in a footnote to this passage. Their model guidelines would allow parents to deny life-saving treatment to their infants with intellectual disabilities because such decisions would prevent the infants from experiencing the cruelty of institutionalization, which was taken to be an unchangeable fact of life for such children.

With regard to the progression from Baby Doe-like denial of treatment in the 1980s to selective abortion for Down syndrome thereafter, we might assume a similar argument could be made justifying selective abortion by virtue of the consequent reduction in the cruel withholding of treatment. As a short form, call this the "avoidance of cruelty" argument for the ethics of prenatal testing and selective abortion. According to the avoidance of cruelty argument, prenatal testing and selective abortion are ethically justified because they are a better alternative than infanticide, which often involved inflicting pain and suffering on infants with disabilities during the pre-Baby Doe era. The case for prenatal testing and selective abortion is sometimes made on the basis of comparison with the denial of treatment. For example, prenatal testing and selective abortion are sometimes described as less controversial than infanticide (Kuhse and Singer 1985, 173). But by-and-large, the "avoidance of cruelty argument" is not commonly used as a justification for selective abortion by philosophers and bioethicists.

Our current ethical norms in neonatal care preclude the cruelty of denying life-saving treatment to infants with Down syndrome. If this state of affairs arose in part because of the increasing usage of prenatal testing and selective abortion beginning in the 1980s, then I regard this historical development as a form of moral progress. My claim here is not that selective abortion for disability is morally innocuous. Abortion is ethically justified for reasons that go beyond the "avoidance of cruelty" argument (Kaposy 2012). But since ableist bias is prevalent in our culture, it is inevitable that much prenatal decision making will be premised on biased motives (Kaposy 2018). However, the fact that we no longer deny life-saving treatment to infants with Down syndrome is progress, nonetheless.

The next section discusses the second related aspect of the progression from the pre-Baby Doe era into the present: the issue of the contingency of moral progress in the ethical treatment of infants with Down syndrome and other disabilities.

## 5.4 Contingency

My second point is about the historical contingency of the ultimate resolution of the ethical problem exemplified by the Baby Doe case. After protracted legal battles, the Baby Doe case resulted in improvements in pediatric care and a repudiation of the position that infants with disabilities have a diminished claim to life-saving treatment solely because of their disability. But these outcomes were not guaranteed and could have gone in an opposite direction.

When the Reagan administration became aware of the Baby Doe case in 1982, the US Department of Health and Human Services put regulations in place that would prevent further cases of denial of treatment on the basis of disability. These regulations were the so-called "Baby Doe rules" (Resnik 2011). The Department set up a hotline so that people across the country could report suspicions that a baby was being denied life-saving treatment. The American Academy of Pediatrics successfully blocked these regulations in court (American Academy of Pediatrics v. Heckler 1983). Subsequent reformulations of the rules were also struck down. Hospital and physicians' groups, including the American Hospital Association and the American Medical Association, consistently opposed any effort by the government to regulate medical discrimination of this kind (Lantos and Meadow 2006). To be sure, it makes sense that these groups would want to avoid intrusion by the federal government. Professionals prefer to practice autonomously. But by resisting regulation, hospitals and physicians left the unavoidable impression that they did not want to prevent the possibility of infanticide.

Eventually, legislation provided a resolution. The federal government made amendments to the *Child Abuse Prevention and Treatment Act* in October 1984 (Robertson 2004). Legislation was less vulnerable to court challenge than the previous regulatory strategy. In the amendments, the government stipulated three scenarios in which life-saving treatment could justifiably be withheld from an infant.

1. The infant is chronically and irreversibly comatose.
2. The provision of such treatment would merely prolong dying, not be effective in ameliorating or correcting all of the infant's life-threatening conditions, or otherwise be futile in terms of the survival of the infant.
3. The provision of such treatment would be virtually futile in terms of the survival of the infant and the treatment itself under such circumstances would be inhumane (Lantos and Meadow 2006, 73).

Baby Doe could not have been denied surgical treatment under any of these scenarios. These amendments to the *Child Abuse Prevention and Treatment Act* effectively ended the longstanding practice of denying treatment to infants based on disability alone. According to the legal scholar John Robertson, "many doctors, hospital administrators, and even lawyers perceived its passage as creating a legal presumption in favor of treating children likely to have disabilities" (Robertson 2004, 34).

Physicians and hospitals fell into line. Whereas infants with Down syndrome had been routinely denied cardiac surgery prior to Baby Doe, by 1994 the American Academy of Pediatrics "recommended routine cardiac screening of all infants with Down syndrome, suggesting that offering cardiac repairs to children with Down syndrome had become standard of care" (Champagne et al. 2014, 1049).

But prior to the Baby Doe controversy, none of these developments were preordained. Prior to Baby Doe, many doctors, lawyers, bioethicists, and advocacy groups were advocating for legal recognition of the justifiability of denying life-saving treatment on the basis of disability. One example is a "proposal for legislation" made by legal scholars Diana Brahams and Malcolm Brahams in the aftermath of the Leonard Arthur case—the doctor in Derby, UK charged with the murder of an infant with Down syndrome (Brahams and Brahams 1983). According to their "Draft Limitation of Treatment Bill", registered medical practitioners would be permitted to withhold or withdraw from an infant treatment necessary to preserve or prolong life (Brahams and Brahams 1983, 14). The "Draft" envisioned roughly three conditions under which this form of infanticide would be permitted. The denial of such treatment would be allowed when the parents or guardians consent, when the infant has a "severe physical and or mental disability" that could not be cured, and when the disability would cause the infant to "enjoy no worthwhile quality of life" (Brahams and Brahams 1983, 14–15). Since the draft legislation was meant to respond to the Leonard Arthur case, the authors clearly regarded Down syndrome as a condition that causes the infant to have no worthwhile "quality of life". This belief reflects a lack of understanding about Down syndrome and a tendency common in that era to lump together all disabling conditions—mild-to-moderate cognitive disabilities alongside conditions that are highly painful and life-threatening. The article, written in response to a well-publicized case of infanticide involving a baby with Down syndrome, and so close in time to the killing of Baby Doe, sought to codify infanticide in the law, rather than to progress beyond such cruelty and discrimination.

Similarly, I have mentioned already Goldstein, Freud, and Solnit, authors of *Before the Best Interests of the Child* (1979). Their book argues that "only *after* the family fails in its function should the child's interests become a matter for state intrusion" (Goldstein et al. 1979, iii). One major exception to this precept is the question of whether a "congenitally malformed newborn with an ascertainable neurological deficiency and highly predictable mental retardation should be provided with treatment which might avoid death" (Ibid., 96). They answer in the negative if the parents refuse treatment and the alternative is institutionalization in a place like Willowbrook, while giving "the state" a pass for the concentration-camp-like conditions of such institutions (Ibid., 98). Again, at a crucial moment in the history of the inclusion of people with disabilities in our communities, child welfare experts sought to provide legal justification for ableist infanticide, rather than taking steps to reform the system and ideologies so hostile to people with disabilities.

A third example comes from the advocacy group "Prospect". According to Helga Kuhse and Peter Singer, this group was made up of parents, some of whom had children with disabilities (Kuhse and Singer 1985, 173). From what I can tell, the group no longer exists, and has left few traces behind. In 1983, Prospect lobbied the UK government to pass a "Limitation of Treatment Bill" that allowed for the denial of treatment to infants with disabilities within 28 days after birth, with the consent of the parents (Ibid.). The lobbying was unsuccessful, but again in the example we can see that during the late 1970s and early 1980s various experts and advocates were seeking to further entrench ableist infanticide as a social reality.

Insofar as the Baby Doe controversy of 1982 ultimately led to a reversal of these practices, this reversal was not a *fait accompli*. In fact, the repudiation of infanticide as a part of neonatal care was a contingent development that might not have happened had the voices of prominent bioethicists, physicians, many lawyers, and some advocacy groups prevailed. The greater inclusion of people with Down syndrome and other disabilities in our social lives is not something to be taken for granted.

## 5.5 Implicit Ableism and Moral Progress

On the one hand, as I have argued, it is a form of moral progress if infants with Down syndrome are no longer denied life-saving treatment soon after birth. We may have moved past the era of selective non-treatment of infants with disabilities because of the increasing reliance on prenatal testing beginning in the 1980s. Though we may have supplanted occasional cases of infanticide with an even greater incidence of abortion, I regard this as progress. For one thing, fetuses are not infants. And as I have argued, selective abortion does not involve cruelty in the way that the refusal of life-saving treatment often resulted in cruelty.

On the other hand, there are reasons to be suspicious of this claim of moral progress, or to regard this progress as only a partial victory. Ableist bias directed at people with Down syndrome is still prevalent in our culture (Kaposy 2018). The presence of this bias is one reason to be circumspect about widespread prenatal testing and selective abortion for Down syndrome. Research into implicit bias reveals that people with cognitive disabilities are commonly subject to this form of bias (Wilson and Scior 2015; Proctor 2011). There is at least one study showing that negative implicit attitudes towards people with Down syndrome in particular are very common (Enea-Drapeau et al. 2012). Implicit bias psychological research is designed to tap into the beliefs and attitudes held by members of the general public—beliefs and attitudes that research participants might not admit explicitly to researchers or even to themselves.

To some extent, we all probably hold implicit biases toward one group or another. But the prevalence of implicit ableism suggests that while doctors and parents no longer deny life-saving treatment to infants with Down syndrome, ableism still influences the decisions we make in our lives, including prenatal decision making. People do not shed their implicit biases when they step into the prenatal clinic. Outside of the clinic and in their daily lives, people with Down syndrome still live in an atmosphere of implicit bias directed at them. Some aspects of life have gotten noticeably better for people with Down syndrome since the mid-twentieth century—the right to education has improved, as have employment opportunities, and human rights legislation protects people with disabilities in many countries. But these gains are fragile as long as ableism remains.

On the one hand, the fact that Baby Does are no longer killed in hospitals anymore in the Western world is progress. On the other hand, this progress was never guaranteed. Even in the aftermath of the Baby Doe case, many influential people wanted to continue to allow infanticide, and even wanted to strengthen law and policy to support the autonomy of parents to kill their infants with Down syndrome. Furthermore, the underlying cause of the death of Baby Doe and the many other infants like him—ableism—remains with us today. So, the progression from Baby Doe to selective abortion for Down syndrome is a story of moral progress, but the story is not over.

## References

American Academy of Pediatrics v. Heckler 561 F. Supp. 395 (D.D.C. 1983).

American College of Obstetricians and Gynecologists. 2021. Facts are important: Gestational development and capacity for pain. https://www.acog.org/advocacy/facts-are-important/gestational-development-capacity-for-pain. Accessed April 26, 2023.

Antommaria, Armand M. 2006. "Who should survive?: One of the choices of our conscience": Mental retardation and the history of contemporary bioethics. *Kennedy Institute of Ethics Journal* 16 (3): 205–224.

Brahams, Diana, and Malcolm Brahams. 1983. The Arthur case—A proposal for legislation. *Journal of Medical Ethics* 9: 12–15.

Champagne, Chantelle R., Melanie Lewis, and Dawna M. Gilchrist. 2014. Should we mend their broken hearts? The history of cardiac repairs in children with Down syndrome. *Pediatrics* 134 (6): 1048–1050.

de Graaf, Gert, Frank Buckley, and Brian G. Skotko. 2015. Estimates of the live births, natural losses, and elective terminations with Down syndrome in the United States. *American Journal of Medical Genetics Part A* 167 (4): 756–767.

Dresser, Rebecca. 2005. Schiavo's legacy: The need for an objective standard. *Hastings Center Report* 35 (3): 20–22.

Enea-Drapeau, Claire, Michèle Carlier, and Pascal Huguet. 2012. Tracking subtle stereotypes of children with Trisomy 21: From facial-feature-based to implicit stereotyping. *PLoS ONE*. https://doi.org/10.1371/journal.pone.0034369.

Gillam, Lynn. 2015. Fifty years of paediatric ethics. *Journal of Paediatrics and Child Health* 51: 8–11.

Girvan, David P., and Clinton A. Stephens. 1974. Congenital intrinsic duodenal obstruction: A twenty-year review of its surgical management and consequences. *Journal of Pediatric Surgery* 9 (6): 833–839.

Goldstein, Joseph, Anna Freud, and Albert J. Solnit. 1979. *Before the best interests of the child*. New York: The Free Press.

Gustafson, James M. 1973. Mongolism, parental desires, and the right to life. *Perspectives in Biology and Medicine* 16 (4): 529–557.

Joseph P. Kennedy Jr. Foundation, and Johns Hopkins Hospital. 1971. Who should survive?: One of the choices of our conscience. https://www.youtube.com/watch?v=fHTs9D_OEqE. Accessed April 9, 2024.

Kaposy, Chris. 2012. Two stalemates in the philosophical debate about abortion and why they cannot be resolved using analogical arguments. *Bioethics* 26 (2): 84–92.

Kaposy, Chris. 2018. *Choosing Down syndrome: Ethics and new prenatal testing technologies*. Cambridge, MA: MIT Press.

Kuhse, Helga, and Peter Singer. 1985. *Should the baby live? The problem of handicapped infants*. New York: Oxford University Press.

Lantos, John D., and William L. Meadow. 2006. *Neonatal bioethics: The moral challenges of medical innovation*. Baltimore: Johns Hopkins University Press.

Larcher, Vic, Finella Craig, Kiran Bhogal, et al. 2015. Making decisions to limit treatment in life-limiting and life-threatening conditions in children: A framework for practice. *Archives of Disease in Childhood* 100 (Suppl 2): s1–s23.

Lorber, John. 1971. Results of treatment of myelomeningocele: An analysis of 524 unselected cases, with special reference to possible selection for treatment. *Developmental Medicine and Child Neurology* 13 (3): 279–303.

Löwy, Ilana. 2014. Prenatal diagnosis: The irresistible rise of the "visible" fetus. *Studies in the History and Philosophy of the Biological and Biomedical Sciences* 47: 290–299.

Lyon, Jeff. 1985. *Playing God in the nursery*. New York: W.W. Norton.

Mayo Clinic. 2022. Prenatal testing: Is it right for you? https://www.mayoclinic.org/healthy-lifestyle/pregnancy-week-by-week/in-depth/prenatal-testing/art-20045177. Accessed April 25, 2023.

Natoli, Jaime, Deborah L. Ackerman, Suzanne McDermott, et al. 2012. Prenatal diagnosis of Down syndrome: A systematic review of termination rates (1995–2011). *Prenatal Diagnosis* 32 (2): 142–153.

Proctor, Stephon N. 2011. Implicit bias, attributions, and emotions in decisions about parents with intellectual disabilities by child protection workers. Ph.D. dissertation, Pennsylvania State University.

Resnik, Jack. 2011. *The Baby Doe rules (1984)*. The Embryo Project Encyclopedia. https://embryo.asu.edu/pages/baby-doe-rules-1984. Accessed April 26, 2023.

Robertson, John A. 2004. Extreme prematurity and parental rights after Baby Doe. *Hastings Center Report* 34 (4): 32–39.

Rothman, David J., and Sheila M. Rothman. 1984. *The Willowbrook wars*. New York: Harper and Row.

Russo, Melissa L., and Karin J. Blakemore. 2014. A historical and practical review of first trimester aneuploidy screening. *Seminars in Fetal and Neonatal Medicine* 19 (3): 183–187.

Wilson, Michelle C., and Katrina Scior. 2015. Implicit attitudes towards people with intellectual disabilities: Their relationship with explicit attitudes, social distance, emotions and contact. *PLoS ONE*. https://doi.org/10.1371/journal.pone.0137902.

Wright, David. 2011. *Downs: The history of a disability*. New York: Oxford University Press.

# Part II
# Capacity and Consent

# Chapter 6
# Free Britney! Capacity, Competence, and Consent for Those with Diminished Decision-Making Abilities

**Abraham Graber**

**Abstract** It is commonplace for people with intellectual and developmental disabilities to have a guardian who is legally authorized to make decisions on their behalf. Such guardianships are often comprehensive, providing the guardian with legal authorization to make all decisions on behalf of a person with intellectual and developmental disabilities. This comprehensive legal authority clashes with the understanding of patient autonomy dominant in clinical ethics whereby nearly every patient has the right to make some of their own decisions. Drawing lessons from Britney Spears's conservatorship as a case study of tension between moral obligations and legal authority, this manuscript explores an identical tension ubiquitous in the lives of those with intellectual and developmental disabilities.

## 6.1 Introduction

Questions of capacity, competence, and consent are ubiquitous in the lives of individuals with intellectual and developmental disabilities (IDD). On one hand, because many people with IDD have a guardian tasked with making decisions on their behalf, people with IDD can be left with little effective control over their lives. On the other hand, because many people with IDD lack the capacity required to make quotidian decisions, failure to appropriately intervene in decision-making can have disastrous implications for their health and well-being. This difficult moral landscape is further complicated by tension between the demands of morality and relevant legal codes. Drawing on the conservatorship of Britney Spears, I will highlight a key difference between capacity, a moral concept, and competence, capacity's legal analog. As a

A. Graber (✉)
Nisonger Center Affiliate Faculty, Division of Bioethics, Department of Biomedical Education and Anatomy, College of Medicine, The Ohio State University, Columbus, USA
e-mail: Abraham.Graber@osumc.edu

 
A. Bianchi and J. A. Vogt (eds.), *Intellectual Disabilities and Autism: Ethics and Practice*, The International Library of Bioethics 108,
https://doi.org/10.1007/978-3-031-61565-8_6

result of this key difference, two notions of consent—one moral and one legal—are often in tension. This tension is ubiquitous in the lives of many with IDD. Recommendations for healthcare providers will be provided.

## 6.2 Free Britney! On the Distinction Between Capacity and Competence

Contemporary American bioethics is largely a response to the violation of patient autonomy central to historical medical atrocities: Nazi experimentation, the Tuskegee Syphilis Study, the involuntary sterilization of hundreds of thousands of poor and non-White women (Cascais 1997). In large part as a response to historical wrongdoing, respect for autonomy has taken center stage in contemporary biomedical ethics. Respect for patient autonomy is operationalized, in both research and clinical contexts, by the demand that patients provide informed consent. In order to count as "informed consent," consent must meet three criteria (Eyal 2019). First, it must be informed; the patient must have access to all the information they need to make their decision. Second, it must be voluntary; patients cannot be subject to coercion or undue influence. Finally, a patient must have decisional capacity/competence.

Though distinct, capacity and competence are tightly connected. Capacity is a clinical property; decisions about whether a patient has decisional capacity are made by healthcare providers (Ganzini et al. 2004). By contrast, competence is a legal property; decisions about whether a patient has competence are made by a judge (Ibid.) In order to have capacity/competence, a patient must meet four criteria (Appelbaum 2007). First, a patient must be able to understand relevant information. Second, a patient must be able to reason through relevant consequences. Third, a patient must be able to appreciate their circumstances. Finally, a patient must be able to communicate a choice.

Capacity and competence are often used interchangeably and bioethicists rarely recognize the philosophical or ethical distinction between the two (Ibid.). Nonetheless, there are at least two key theoretical differences between capacity and competence. Unlike competence, capacity is both (i) fluid and (ii) decision relative (Larkin et al. 2001, 282).

Capacity is fluid in that it can change from moment-to-moment. A drunk patient may lack capacity to make some particular decision at 3:00 a.m. yet have capacity to make that same decision at 8:00 a.m. Similarly, a patient with dementia may lack capacity to accept a medical intervention one day but may have capacity to accept that same intervention the following day. Competence is not similarly fluid. Once a court determines that a person lacks competence, only a subsequent court order can restore competence.

Capacity is also decision-relative. Capacity depends on both a person's intellectual and emotional abilities and the difficulty of the question at hand. For example, deciding whether to undergo a prophylactic mastectomy involves balancing guaranteed but non-fatal harm against the mere potential for future, but potentially devastating, cancer. By contrast, the decision to take ibuprofen for a headache has a clear benefit, minimal downside, and few implications for the trajectory of one's life. A patient may thus have capacity to decide whether to take ibuprofen yet lack capacity to decide whether to undergo a prophylactic mastectomy. The decision-relative nature of capacity entails that nearly every patient has the capacity to make at least *some* decisions.

By contrast, competence is not decision relative (Ganzini et al. 2004). A court ruling that a patient lacks competence to make their own healthcare decisions constitutes an umbrella ruling in which a proxy-decision maker is given broad authority to make treatment decisions on behalf of the patient. Because capacity is decision relative, but competence is not, the two often come apart.

In 2008, Britney Spears was placed under a conservatorship (Tsioulcas 2021).[1] This rare legal arrangement ceded Spears's legal rights to make financial and bodily decisions. Under the conservatorship, Spears lacked the legal authority to, e.g., decide to go for a walk or choose to have a phone. During the thirteen year period of her conservatorship, Spears "released four albums, headlined a global tour that grossed a hundred and thirty-one million dollars, and performed for four years in a hit Las Vegas residency" (Farrow and Tolentino 2021). Spears's case provides a clear example of someone with the moral right to make their own decisions but who is legally barred from doing so.

During Spears's conservatorship, "… a team led by her father, who is her conservator, prevented her from having her IUD removed because the team did not want her to have more children" (Hoffman 2021). Consider Spears's wish to have her IUD removed. Was she in a position to give informed consent for the procedure? The question is ambiguous. Spears could give *informed consent*$_{capacity}$ just in case: (i) Spears was adequately informed, (ii) Spears was making the decision voluntarily, and (iii) Spears had *decisional capacity*. There is every reason to think that Spears could *consent*$_{capacity}$.

One might, alternatively, be asking if Spears could give *informed consent*$_{competence}$. Spears could give *informed consent*$_{competence}$ just in case: (i) Spears was adequately informed, (ii) Spears was making the decision voluntarily, and (iii) Spears had *decisional competence*. Spears could not *consent*$_{competence}$; a court had removed from Spears the legal authority to make her own healthcare choices.

[1] While there are differences between a conservatorship and a guardianship the two are, for the purposes of this paper, effectively synonymous. I will thus use the terms interchangeably.

## 6.3 What Should Britney's Doctors Have Done

Spears's unfulfilled desire to have her IUD removed appears to constitute a clear case of a capacitated patient who nonetheless lacks competence. Furthermore, her legal guardian overrode her capacitated decision to have her IUD removed. In cases like this, what should healthcare providers do?

Autonomy is often taken to be a negative obligation (Beauchamp and Childress 2013). A *positive obligation* is an obligation to do, or provide, something. The healthcare provider's obligation to beneficence—the obligation to help—is an example of a positive obligation. In order to fulfill this obligation, the healthcare provider must provide (when appropriate) treatment or intervention. By contrast, a *negative obligation* is an obligation to refrain from acting. Because autonomy is a negative obligation, respect for patient autonomy requires that a healthcare provider refrain from providing any intervention without first receiving appropriate consent. It does not, however, require that the healthcare provider do whatever the patient asks.

Cases like Spears's request to have her IUD removed seem to put pressure on autonomy's status as a purely negative obligation. In cases like these, it seems a patient's wishes generate at least some moral presumption in favor of the requested intervention. Yet on the orthodox understanding of respect for patient autonomy, Spears's request to have her IUD removed does not bestow a moral obligation on the healthcare provider. Consequently, while it is morally permissible for a healthcare provider to remove Spears's IUD, it may not be obligatory. Furthermore, because Spears's conservator did not consent to the removal of the IUD, it was legally impermissible to provide the intervention. Importantly, though Spears's guardian's refusal to allow the removal of the IUD makes the intervention legally impermissible, her guardian's wishes are *morally* impotent. Spears has the capacity to make her own decision about IUD removal; her guardian's preferences are, at best, peripherally morally relevant.

Given the potentially serious legal and professional consequences that a physician could face if they were to remove Spears's IUD without receiving her guardian's consent, providing the intervention is likely morally supererogatory. A physician who thus decided to risk their career in order to respect Spears's capacitated decisions should be lauded for their unflinching dedication to their patient. Yet such dedication is plausibly not morally required of a physician. A physician could thus recognize that Spears has decisional capacity regarding the choice to have her IUD removed and still act in a way that is ethically permissible by refusing to move forward with the intervention until guardian approval has been secured.

While it may be morally permissible for a physician to refuse to remove Spears's IUD, the case nonetheless remains deeply troubling. Spears has every moral right to make her own decisions. Yet she is prevented from doing so by her conservator whose opinion on the matter, while legally binding, is morally irrelevant. While a healthcare provider may not be obligated to remove the IUD, they nonetheless cannot (ethically) sit by and do nothing.

While some models of informed consent assume that patients arrive with clearly articulated values and goals, it is unlikely that many patients have this antecedent understanding of their life projects (Emanuel and Emanuel 1992). More plausible views of informed consent involve the physician helping patients to understand and precisify their values and goals in the context of the treatment decision at hand. While it may be permissible for the healthcare provider to refuse to remove Spears's IUD when guardian consent is lacking, informed consent requires that the healthcare provider (i) push the proxy-decision maker to understand that Spears has the moral authority to make this kind of decision for herself, and (ii) help the proxy-decision maker to understand the magnitude of the harm involved in denying a decisionally-capacitated person the ability to make her own healthcare decisions.

Where the proxy-decision maker refuses to engage in such conversations, or continues to insist on overriding Spears's decision to have her IUD removed, the treating physician should engage a clinical ethics consultation service or clinical ethics committee. Mediation is a core component of the work done by clinical ethics consultants and/or committees (Fiester 2015). When a physician has unaddressed concerns regarding a proxy-decision maker's understanding of the harm caused by their decision, mediation may help address those concerns.

Finally, in those cases where a proxy-decision maker either is not concerned enough about a patient's well-being to take the time to understand the harm caused by overriding a capacitated patient's decision, or when the guardian recognizes the harm yet remains unmoved, it may be appropriate to contact adult protective services. It is unclear if either attitude toward the harm caused by refusing Spears her request for IUD removal is compatible with a guardian's obligation to act in Spears's best interest.

Let us now consider a closely related case. Suppose that Spears's guardian wanted Spears to have an IUD implanted, but Spears did not consent. What are the obligations of the healthcare provider in this instance? This is a particularly important question because this situation is more likely to mirror quotidian tension between *informed consent*$_{competence}$ and *informed consent*$_{capacity}$, as cases in which providers are asked to remove medical devices embedded in patients are comparatively rare.

In this hypothetical case, Spears has capacity to decide whether to have an IUD implanted yet lacks competence. Spears thus does not provide *consent*$_{capacity}$; however, her guardian does *consent*$_{competence}$ on her behalf. Both the ethical analysis and the legal analysis are straightforward.

It is not morally permissible to move forward with a medical intervention with a capacitated patient unless the patient has provided informed consent. Because Spears has the capacity to make the relevant decision and because she has not *consented*$_{capacity}$, it is morally impermissible to implant the IUD. Ethically, this case is no different from any other instance of a medical intervention being forced on an unwilling and fully autonomous adult. Under such conditions, implanting an IUD constitutes a clear violation of Spears's right to bodily autonomy and is thus unambiguously wrong.

The legal analysis is similarly straightforward. In the hypothetical situation, Spears does not have competence to make her own decision regarding IUD implantation and, as such, *consent*$_{competence}$ must be provided by her guardian. Spears's guardian has provided *consent*$_{competence}$ in her stead and it is thus legally permissible to move forward with the implantation of the IUD.

What should a provider do? The fact that a proxy has provided *consent*$_{competence}$ makes implantation legally permissible; however, there is no legal obligation to perform the procedure. By contrast, because Spears is decisionally capacitated and because she has not provided *consent*$_{capacity}$, the provider is morally obligated not to move forward with implantation of an IUD. There is thus no legal obligation to implant the IUD but a clear moral obligation not to.

## 6.4 Beyond Britney: On the Ubiquity of a Disconnect Between Consent$_{capacity}$ and Consent$_{competence}$ in the Lives of Individuals with IDD

Taken at face value, Spears's conservatorship may appear to have little to do with the lives of individuals with IDD. This apparent distance is a key component of what makes the case instructive. Spears's conservatorship pushes us to engage with questions about how healthcare providers should proceed in cases where there is conflict between *consent*$_{capacity}$ and *consent*$_{competence}$, without the threat of ableist bias infecting our analysis. The next step is to demonstrate that tension between *consent*$_{capacity}$ and *consent*$_{competence}$ is ubiquitous in the lives of individuals with IDD. We can then generalize from our analysis of Spears's conservatorship to lessons about how healthcare providers should navigate analogous cases in the care of those with IDD.

Capacity is decision-relative. Competence is not. If a person with IDD is judged to lack decisional competence to make their own healthcare decisions, a proxy must provide *consent*$_{competence}$ for all healthcare procedures. But because capacity is decision relative, a person with IDD will have capacity to make at least *some* healthcare decisions—even someone with severe or profound IDD likely has capacity to decide, e.g., what flavor vitamin or antibiotic to choose. It follows from the decision-relative nature of capacity and the blanket nature of competence that, in cases where someone with IDD lacks decisional competence, they have capacity to make some decisions. Tension between *consent*$_{competence}$ and *consent*$_{capacity}$ is thus likely to be ubiquitous in the lives of individuals with IDD and guardianships.

Two features of capacity assessment further highlight this point. First, a patient need never demonstrate that they have the capacity to make some decision (Ganzini et al. 2004). By contrast, we start with the standing assumption that a patient has capacity and only withdraw this assumption when there is countervailing evidence to the contrary. Second, setting aside extreme cases such as those involving patients in persistent vegetative states, a diagnosis is never adequate to determine that a patient lacks capacity to make some decision (Ganzini et al. 2004). These two tenets of capacity assessment combine to establish a presumption in favor of the decisional capacity of a person with IDD. This presumption can, of course, be overridden. But a finding that a patient lacks capacity must be made with the gravity and care appropriate for a decision with the potential to deprive a person of their autonomy.

## 6.5 Competence Versus Capacity in the Lives of Individuals with IDD

**Case Study 6.1: Living Alone**

Mario is a 37-year-old male with moderate intellectual disability. He currently lives in a group home but has a strong preference for living on his own. Mario's care team believes that, given a mix of remote support services and limited in-person supports, Mario could successfully transition to living alone. Mario's legal guardian is not, however, confident in Mario's ability to live on his own, even with the suggested supports, and thus is not willing to approve the change.

This case is analogous to Britney Spears's request to have her IUD removed. In both instances, the client is asking the care team to make a change, yet the legal guardian refuses consent. However, while it is clear that Spears had capacity regarding the decision to remove her IUD, is Mario similarly decisionally capacitated?

Consider decisions that Mario clearly has capacity to make. Suppose Mario wants a snack and is deciding between eating an apple or a pear. Similarly, suppose Mario wants to watch TV and is choosing between sitting on the couch or sitting in a recliner. It should be uncontroversial that Mario has capacity to make both of these decisions. What features of these decisions make this judgment uncontroversial?

There are potential long-term consequences to each decision that Mario may, or may not, understand. For example, variation in lumbar support across the couch and recliner could have potential implications for Mario developing backpain or even have (minute) implications for the probability that Mario will develop a slipped disk. Even if Mario cannot grasp these implications, he remains capacitated. One need not be able to understand all of the implications of one's decision in order to have decisional capacity. Much more modestly, one need only understand the information we take to be relevant to making the choice.

In order to be capacitated regarding some decision, Mario must be able to understand and reason through all of the information regarding his decision that a reasonable (i.e., average) person would want to know in making that choice (Murray 2012). Miniscule risks and high risk of miniscule effects are not things that the average person generally takes to be relevant to decision making. Thus, though there may be potential implications that Mario cannot reason through regarding his choice to eat a pear rather than an apple, or the choice to sit on the couch rather than in the recliner, this inability does not have implications for his decisional capacity.

What implications of these decisions are relevant to Mario's capacity? Outlandish complications aside, the relevant impacts are limited to preference fulfillment and happiness. Eating the pear will fulfill Mario's preferences better than eating the apple. Sitting in the recliner will make Mario happier than sitting on the couch.

These are implications of his decision that Mario can clearly understand. As such, he has capacity to make both decisions.

Something very similar holds regarding the decision to transition to living alone versus continuing to stay in the group home. The ins and outs of this decision are certainly more complicated than the ins and outs of deciding what to eat for a snack or where to sit. But the key piece of information is that Mario can, with appropriate supports, successfully transition to living on his own. Mario can thus be confident that the consequences for his health and safety come out a wash—living in a group home or on his own, his fundamental needs will be met.

It may well be that Mario is not capable of following the various evidence that leads to this conclusion; however, just as capacity does not require that we understand the chemical mechanisms or Bayesian analysis underlying a treatment or diagnostic recommendation, capacity does not require that Mario understands the mechanisms or evidence that lead to the conclusion that he can successfully transition to living on his own. All he needs to understand is that, with supports, he will be as safe and well living on his own as he is living in the group home.

What's left, other than these considerations regarding health and safety? Mario's preferences! While some people with moderate intellectual disability may struggle to understand how living on their own will impact their happiness and desire satisfaction, many will not. It is thus quite realistic to suppose that, in our hypothetical case study, Mario has capacity to decide whether to live on his own.

Given that Mario has provided $consent_{capacity}$, transitioning Mario to living on his own is morally permissible. Without the consent of Mario's guardian it is, however, likely legally impermissible. Both because autonomy is a negative obligation and because there are likely to be serious legal repercussions for moving Mario without guardian consent, transitioning Mario to living on his own is not morally obligatory. Much as it was morally permissible to refuse Britney Spears's request to have her IUD removed, it is morally permissible to refuse Mario's request to move until guardian consent is received.

Nonetheless, even if it is permissible not to transition Mario to living on his own, Mario's care team likely has undischarged moral obligations to Mario. There is, morally, no difference between ignoring Mario's capacitated decisions and the capacitated decisions of any typically developed adult. Thus, even if Mario's care team is not obligated to act on his specific wish to transition to living alone, they nonetheless owe Mario an unfulfilled obligation to respond to the injustice of Mario being denied the right to make fundamental decisions about the course of his life.

Mario's care team can take steps toward resolving this obligation by advocating for Mario's transition to living alone and by working with Mario's guardian to develop a plan for moving toward that goal that would alleviate the guardian's concerns. Mario's care team may or may not have access to a hospital ethics committee or to a clinical ethics consultation service. If Mario's guardian continues to resist Mario's move to living on his own, the care team should consider contacting an ethics committee or clinical ethics consultation service. Where neither are readily available, human rights committees or community-based disability advocates may be able to fill the void.

Regarding her guardian's refusal to $consent_{competence}$ to the removal of Britney Spears's IUD, I recommended referral to adult protective services. I am much more hesitant to make a similar recommendation with regard to Mario's situation. There are pragmatic considerations in Mario's case that are not relevant in Spears's case, e.g., if the organization that employs Mario's care team gains a reputation for escalating disagreements with guardians to adult protective services, they are likely to quickly find themselves without any clients as guardians look elsewhere for services. Under such circumstances, the goal of better advocating for Mario would have backfired, undermining efforts both to advocate for Mario and to advocate for other clients. That said, this style of justification for remaining passive in the face of injustice can easily overgeneralize, which raises the concern that it only seems compelling in this context because of ableist bias. While there is a lot to be unpacked here, there is unfortunately not room in this paper to do that work.

Let us now turn to a second hypothetical case study in which there is tension between $consent_{capacity}$ and $consent_{competence}$:

**Case Study 6.2: Hand-Flapping**

Felicia is a non-verbal 29-year-old with profound intellectual disability. She often engages in repetitive hand flapping. Her guardian is uncomfortable with her hand flapping and has requested that her care team implement a behavioral program aimed at eliminating Felicia's hand flapping.

A number of authors, including, e.g., Graber and Graber (2023), Ne'eman (2021), and Kirkham (2017), have argued that there are independent reasons to have ethical concerns about the requested intervention.[2] For the purposes of this project, let us focus specifically on questions related to capacity. The first question is: does Felicia have decisional capacity to engage in hand flapping?

As always, much will depend on the specifics of the case. Stereotypic behavior like hand flapping can be self-injurious (Hagopian et al. 2020). Furthermore, it is unlikely that Felicia can understand and reason through the implications of treating self-injurious hand flapping. Under such a circumstance, Felicia likely does not have capacity to decide to hand flap. Let us stipulate that, in our hypothetical case study, Felicia's hand flapping is not causing her any physical harm.

As Ne'eman (2021) discusses, because the behavior is stigmatized, hand flapping or other behavioral differences often associated with autism can present a barrier to social interaction. Given that Felicia has profound intellectual disability, it is quite likely that she does not have the capacity to understand and reason through the pros and cons of her hand flapping for her social interactions.

This is not, however, enough to conclude that Felicia lacks decisional capacity regarding hand flapping. We all make decisions about when and how we would like to engage in social interactions. Reading a book on the bus or putting headphones on in a coffee shop are common ways of signaling that one is uninterested in social

[2] For more about ethical considerations related to applied behavioral analysis, see Chaps. 11, 12, and 31 of this volume.

interaction. Felicia's hand flapping may, similarly, be a behavior she engages in when she's not otherwise interested in interpersonal interaction.

One cannot ask Felicia about her preferences regarding interpersonal interactions, nor can one ask about how those preferences intersect with her hand flapping. There are, however, a wealth of empirically validated preference assessments that can be used to assess Felicia's preferences and similarly a wealth of functional behavioral assessments (FBAs) that can be used to determine the function of Felicia's behavior (Cooper et al. 2020). Of particular interest in this case is the temporal relationship between Felicia's hand flapping and her preference for social interaction. If the two often overlap, and if hand flapping isn't giving rise to a preferred kind of social interaction, then it's likely that Felicia does not have capacity to decide to hand flap—she likely cannot understand or reason through the impact of hand flapping on her preferences for interpersonal interaction.

But let us suppose, as is quite plausible, that Felicia's hand flapping is not interfering with her preference for social interaction. Maybe Felicia generally has a preference for hand flapping over social interaction. Or perhaps hand flapping actually serves as a way for Felicia to escape social interactions. Granting these assumptions, Felicia likely has decisional capacity to choose to engage in hand flapping; this is no different than Mario choosing between an apple and a pear. The only relevant implications involve Felicia's happiness and desire satisfaction and there's every reason to suppose that Felicia grasps the connection between hand flapping and her happiness and the satisfaction of her desires.

It is worthwhile taking a moment to reflect on this result. One might have thought that someone with profound intellectual disability could not be capacitated to make their own healthcare decisions. Yet here we see the opposite. It is true that some intervention—in the form of preference assessments and FBAs—was required for us to determine that Felicia was capacitated. But that does not seem any different from the use of interpreters in the case of a language barrier or the use of eye-tracking technology in the case of locked-in syndrome (Bernat 2020). Even profound intellectual disability does not prevent one from having capacity to make at least some healthcare decisions.

Assuming that Felicia has capacity to engage in hand flapping, this case looks closely analogous to the hypothetical case in which Spears did not consent to her conservator's request that she get an IUD. When a patient has capacity, it is morally impermissible to move forward with a medical intervention without the patient's consent. Just as it would be morally impermissible to implant an IUD without Spears's consent, it would be similarly morally impermissible to implement an intervention aimed at eliminating Felicia's hand flapping without her consent. Unlike Spears, Felicia is non-verbal and cannot express opposition to the proposed intervention; however, while a capacitated patient expressing their refusal to undergo a proposed intervention is surely sufficient to make it wrong to move forward with an intervention, such express resistance is not necessary for the intervention to be impermissible. Where a capacitated patient has not given consent, the intervention is impermissible. And Felicia has not given consent.

Furthermore, while consent of Felicia's guardian makes it legally permissible to move forward with the proposed intervention, the guardian's consent does not make implementation of the intervention legally obligatory. A healthcare provider who refuses to intervene to eliminate Felicia's hand flapping will consequently not face adverse legal consequences for refusing to implement the intervention. Thus, while morality demands that one not move forward with the intervention, the law merely provides permission to implement the intervention. It follows that, all things considered, the healthcare provider should not move forward with an intervention aimed at eliminating Felicia's hand flapping.

## 6.6 Conclusion

We have now covered significant ground. What are the most important takeaways? There are a number. First, our concept of informed consent is not, contrary to the way we regularly use it, univocal. Rather, there are two distinct concepts in play: $consent_{competence}$ and $consent_{capacity}$. Thinking clearly about consent in the clinical context, especially when caring for patients with IDD, requires that we be careful to distinguish between the two different notions of informed consent.

Second, and relatedly: there is often a tendency to slide from recognition that someone lacks competence to the presumption that they lack capacity as well. A case like Felicia's illustrates this well, as do my own experiences parenting. When someone lacks competence it is easy to fall into the habit of making all of their decisions for them. But just because we can, does not mean that we should. It is thus important not only to recognize the distinction between competence and capacity, but further to resist sliding from the quotidian decision making that can characterize being a guardian into unjustifiable violations of autonomy.

Finally, it is important to recognize that nearly every person has the capacity to make at least some (medical) decisions. A person that, like Mario, has moderate intellectual disability will have capacity to make a wide range of decisions. A person like Felicia, with profound intellectual disability, may be capacitated to make a narrower range of decisions than Mario, but nonetheless has capacity to make any number of her own decisions, medical decisions included.

There is no moral difference between overriding the capacitated decisions of a typically developed adult, like Britney Spears, and overriding the capacitated decisions of someone with IDD, like Mario or Felicia. Guardians may not always appropriately respect the autonomy of their charges. Because a guardian's failure to respect the autonomy of a person with IDD is likely often more a result of ignorance than malice, healthcare providers can serve as an invaluable educational resource. But where a guardian cannot see, or refuses to see, the harm in their behavior, healthcare providers must not be party to overriding the autonomy of individuals with IDD.

## References

Appelbaum, Paul S. 2007. Assessment of patients' competence to consent to treatment. *New England Journal of Medicine* 357 (18): 1834–1840.

Beauchamp, Tom, and James Childress. 2013. *Principles of biomedical ethics*, 8th ed. New York: Oxford University Press.

Bernat, James L. 2020. Medical decision making by patients in the locked-in syndrome. *Neuroethics* 13 (2): 229–238.

Cascais, A.F. 1997. Bioethics: History, scope, object. *Global Bioethics* 10 (1–4): 9–24.

Cooper, John O., Timothy E. Heron, and William L. Heward. 2020. *Applied behavior analysis*, 3rd ed. Pearson.

Emanuel, Ezekiel, and Linda Emanuel. 1992. Four models of the physician-patient relationship. *JAMA* 267 (16): 2221–2226.

Eyal, Nir. 2019. Informed consent. *Stanford Encyclopedia of Philosophy*. https://plato.stanford.edu/entries/informed-consent/. Accessed 9 April 2024.

Farrow, Ronan, and Jia Tolentino. 2021. How Britney Spears got free and what comes next. November 13. https://www.newyorker.com/news/news-desk/how-britney-spears-got-free-and-what-comes-next. Accessed 9 April 2024.

Fiester, Autumn. 2015. Neglected ends: Clinical ethics consultation and the prospects for closure. *The American Journal of Bioethics* 15 (1): 29–36.

Ganzini, Linda, Ladislav Volicer, William A. Nelson, Ellen Fox, and Arthur R. Derse. 2004. Ten myths about decision-making capacity. *Journal of the American Medical Directors Association* 5 (4): 263–267.

Graber, Abraham, and Jessica Graber. 2023. Applied behavior analysis and the abolitionist neurodiversity critique: An ethical analysis. *Behavior Analysis in Practice* 16 (4): 921–937.

Hagopian, Louis P., Michelle A. Frank-Crawford, Noor Javed, Alyssa B. Fisher, Christopher M. Dillon, Jennifer R. Zarcone, and Griffin W. Rooker. 2020. Initial outcomes of an augmented competing stimulus assessment. *Journal of Applied Behavior Analysis* 53 (4): 2172–2185.

Hoffman, Jan. 2021. Is the forced contraception alleged by Britney Spears legal? https://www.nytimes.com/2021/06/24/health/britney-spears-forced-IUD.html. Accessed 7 July 2024.

Kirkham, Patrick. 2017. 'The line between intervention and abuse'—Autism and applied behavior analysis. *History of the Human Sciences* 30 (2): 107–126.

Larkin, G.L., C.A. Marco, and J.T. Abbott. 2001. Emergency determination of decision-making capacity: Balancing autonomy and beneficence in the emergency department. *Academic Emergency Medicine* 8 (3): 282–284.

Murray, Bryan. 2012. Informed consent: What must a physician disclose to a patient? *AMA Journal of Ethics* 14 (7): 563–566.

Ne'eman, Ari. 2021. When disability is defined by behavior, outcome measures should not promote "passing." *AMA Journal of Ethics* 23 (7): E569.

Tsioulcas, Anastasia. 2021. Britney Spears' conservatorship has finally ended. https://www.npr.org/2021/11/12/1054860726/britney-spears-conservato. Accessed 7 July 2024.

# Chapter 7
# Capacity-Based Competency Assessment and Its Impact on Child Custody Cases Amongst Parents with Intellectual Disabilities

**Marjorie Aunos, Maurice Feldman, and Laura Pacheco**

**Abstract** Parents with intellectual disabilities are often subject to child protection interventions. The research shows that cases involving a parent with an intellectual disability will most likely end up in child welfare court. For many of these families, parenting capacity assessments are often requested to make decisions involving the best interest of the child/ren. Given the weight of parenting capacity assessments within the child welfare decision-making process, it is imperative to use a parenting model specific to the needs of parents with intellectual disabilities. The following chapter explores Feldman's contextual parenting model that outlines the personal and environmental variables that impact parenting and provides a protocol for evidence-based assessment and intervention.

## 7.1 Introduction

In 1939, a study was published about how increased stimulation can have a positive impact on children's development (Skeels and Dye 1939). Unbeknown to many, this study is one of the first in the field of parenting with an intellectual disability (ID).

M. Aunos (✉)
Department of Applied Disability Studies, Brock University, St. Catharines, ON, Canada
e-mail: marjorieaunos@gmail.com

M. Feldman
Department of Applied Disability Studies, Brock University, St. Catharines, ON, Canada
e-mail: mfeldman@brocku.ca

L. Pacheco
School of Social Work, Memorial University, St. John's, NL, Canada

School of Social Work, McGill University, Montreal, Quebec, Canada

L. Pacheco
e-mail: lpacheco@mun.ca

 
A. Bianchi and J. A. Vogt (eds.), *Intellectual Disabilities and Autism: Ethics and Practice*, The International Library of Bioethics 108,
https://doi.org/10.1007/978-3-031-61565-8_7

Women who were said to be “feebleminded” were asked to provide stimulation to orphan children with developmental delays. After two years of experimentation, the development of those children who had received stimulation by these women had drastically improved, showing that the cognitive limitations of the young women did not prevent them from offering a stimulating environment. Yet, at that time, several US states and several Canadian provinces had passed laws to systematically sterilise women with ID as means to prevent their procreation and the potential “threat” of passing on their disability.

Decades later, principles of the eugenics movement continue to influence practices questioning the ability of women with ID to raise children and practices that have gone from systemic to disguised sterilisation of women and girls perceived as “unfit mothers” (McConnell and Phelan 2022).

## 7.2 The Relationship Between Intelligence and Parenting Capacity

As soon as we had tools to measure how a person might think differently, they were used politically and socially to legitimise the marginalisation of subgroups of the population (Clarizio 1979). The tools helped concretise the divide between those who demonstrated cognitive limitations in comparison to those who demonstrated “normal intelligence”. To this day, issues around the purpose and the use of those tests remain.

When a parent is known to have a cognitive disability, related to ID or other conditions (e.g., acquired brain injury, learning disability), the capacity of the parent is called into question. This is when an adult intelligence test might be used (sometimes inappropriately) to determine parenting capacity, even though such tests were not designed to measure parenting per se (Callow et al. 2017). Unfortunately, for many, receiving a low IQ score or a diagnosis of ID will often be misinterpreted as an inherent and invariant lack of competence and inability to benefit from interventions designed to improve parenting skills (Pacheco et al. 2022; Sigurjónsdóttir and Rice 2017). The simple mention of ID in a case vignette led child welfare (CW) professionals to show more pity toward the parents and an increased willingness to help, but also made them think that their children were at a greater risk (Proctor and Azar 2012). Consequently, parents with ID are over-represented in child custody proceedings and more likely to lose custody of their children than any other group of parents (McConnell et al. 2011); an over-representation that cannot be explained by a lack of skill or by the parents’ disability per se (Lightfoot et al. 2010). In addition, their CW cases stay open longer, more often end up in court actions, and services aimed at reunification are less likely to be offered.

## 7.3 Best Practices in Parenting Capacity Assessments

Parenting Capacity Assessments (PCAs) are used to determine if a parent is able to provide "good-enough" parenting—that is, to provide sufficient care to keep a child healthy; safe; and promote secure attachment, development, and prosocial behaviors (Feldman and Aunos 2010). In the context of parents with ID, the issue of capacity is usually related to a lack of knowledge and skills that put the child at risk for neglectful care (McConnell et al. 2011).

Yet, the focus on intellectual capacity to determine parenting capacity ignores the myriad of other factors that are known to impact parenting and child outcomes. Family life has long been viewed through an ecological-contextual lens in which child outcomes are influenced not just by heredity, but also by the environments to which the child is exposed (Bronfenbrenner 1979; Bronfenbrenner and Ceci 1994). Belsky (1993) expanded this model to child maltreatment and Feldman (2002) adapted the model for families where parents have learning difficulties.

## 7.4 Feldman's Contextual Model

Borrowing from Bronfenbrenner and Belsky's work, Feldman identified key variables that likely influence parenting abilities and child and family outcomes in families headed by parents with ID. These variables include the parent's history, health, life circumstances, and social support. Child characteristics also will play an important role.

Due in part to access to large national cohort studies that include parent educational and disability variables, empirical support for the application of the contextual model to families with parents who have learning difficulties is growing (Feldman and Aunos 2020). Parents with learning difficulties and their children have multiple adversities, including economic, health, psychosocial, and systemic disadvantages (McConnell et al. 2011). Studies show that these disadvantages are stronger predictors of parenting failure and negative child outcomes and placement than parental IQ scores (Feldman et al. 2012; Granqvist et al. 2014). Furthermore, when these disadvantages are considered, differences in CW decisions and child outcomes between parents with and without ID are significantly reduced or eliminated (Hindmarsh et al. 2017; McConnell et al. 2011).

Parental history plays an important role. Many parents with ID have experienced trauma (e.g., physical and sexual abuse), and grown up in poverty, in a chaotic home environment, with poor parental role models, and a lack of social support (Feldman and Aunos 2010). In support of Feldman's model, Granqvist et al. (2014) found that parents' childhood trauma was a stronger predictor of limited sensitivity from mothers with ID and attachment issues with their children, than was the parent's intelligence.

Social variables can also impact parenting. Feldman et al. (2002) showed that social support was not only associated with lower parental stress, but also related to more positive interactions between parents with learning difficulties and their children. Feldman (2002) also highlights how society's views on disability can impact family outcomes. For instance, a person with ID may learn to disguise their disability, putting on what Edgerton (1993) called a "cloak of competence" to avoid stigmatisation, oppression, discrimination, and bullying. This strategy may back-fire when the person is to become or is a parent. They may decline supports and parenting services when they could actually benefit from them. By refusing such supports, the child may be endangered and CW authorities may feel that it is unsafe for them to be in the care of the parent.

Feldman's contextual model provides the basis of a comprehensive PCA protocol for parents with learning difficulties that includes observations of the parents caring for their child(ren) at home (Feldman and Aunos 2010). These observations make use of validated checklists from the Step-by-Step Parenting Program © (Feldman 1998; Feldman and Tahir 2016) and help pinpoint the parent's strengths and skills in need of training. In addition, the comprehensive approach relies on multiple sources of information, including review of records; interviews with the parents, workers, and significant others (including the children if old enough); and standardised questionnaires to gather contextual, health, and psychosocial information from the parents. This information is summarized in terms of strengths, current supports, and barriers to successful parenting. As it is commonly assumed that parents with learning difficulties cannot benefit from training, the third component is a mini training session to determine capacity to learn when provided with evidence-based parenting training specifically tailored for parents with learning difficulties (Feldman and Tahir 2016). The information collected is used to make recommendations on the services and supports the family needs to increase the chances of the children staying, or being returned home, no longer in need of surveillance and protection.

## 7.5 Systemic Discrimination and Ethical Issues

PCAs are meant to be used in litigious cases or when the position of CW is in opposition with the position of either or both parents. It is at least the case when parents do *not* have an ID. Since PCAs are at the center of decision making within CW courts, a comprehensive competence-based assessment framework needs to be universally used with parents with ID (Feldman 2002; Feldman and Aunos 2010). Yet, PCAs are seldom used and when they are presented in court, they often do not include Feldman's recommendations in terms of content and structure (Aunos and Pacheco 2020). Most documents presented are just plain psychological assessments that report on the results of an IQ evaluation.

## 7.6 Preconceived Notions and Ableism

Implicit and explicit biases seep through the assessment process of parents with ID as most PCAs showcase hegemonic, Eurocentric, and ableist ideals (i.e., white, able-bodied, middle-class standards) that further disadvantage marginalized families (Albert and Powell 2022; Aunos et al. 2022; Norlin and Randell 2022). Universal standards and benchmarks for "good enough parenting" fail to address the unique needs, realities, and structural forces that impact the lives of families facing intersecting levels of oppression, including but not limited to parents with ID (Aunos and Pacheco 2020; Callow et al. 2017). These parents are constructed within the CW system as being "disabled" and therefore "unfit" to care for and raise their children (Pacheco et al. 2022). Another example of institutional disablism is that parents with ID are subject to increased notifications to the CW system (given their social position and presumed incapacity) and the number of notifications holds more weight than the legitimacy of the claim (Sigurjónsdóttir and Rice 2017).

A non-disabled professional cannot fully understand the disabled person and how they navigate their everyday life, including parenting. We can be disabled and have impairments yet be quite capable of doing many things. And we cannot disregard the positive "reciprocal adaptation to disability" that children of parents with disabilities and disabled parents showcase (Kirshbaum and Olkin 2002).

## 7.7 Expert Knowledge and Training

Courts prefer having Ph.D. psychologists, with an expertise and previous experiences testifying in court (Aunos and Pacheco 2020), and most professionals conducting PCAs are psychologists or social workers. The assessments conducted by specialised clinics, compared to those conducted by CW professionals more often include best practices (Aunos and Pacheco 2020). As expert testimonial carries so much weight, professionals should be queried regularly by their licensing board to ensure continued expertise (Cox et al. 2015; Kollinsky et al. 2012), and to determine whom could supervise early career professionals. American Psychological Association Guidelines do exist but additional knowledge of best practices in this field, including the knowledge of validated tools for this specific population, is required.

Professionals know they would benefit from more training and supervision (McGaw et al. 2010), and a lack of knowledge make them more prone to confirmation biases (Cox et al. 2015). But even with up-to-date training, economic and time constraints prevent assessors from performing PCAs that would be considered evidence-based (McGaw et al. 2010), which results in evaluations being done quickly and poorly. Additionally, assessments are usually done for the purpose of the courts, and as such, many do not give their conclusions and recommendations a clear direction for intervention.

## 7.8 Defining Parenting and Identifying "Best Interests"

Systemic discrimination toward parents with ID is evident, as they seldom receive the services that a family without cognitive limitations would receive to keep their family intact, including services around family reunification (McConnell et al. 2017). Variables brought forward in assessments are often ill-defined and theoretical models that inform the assessment are absent (Aunos and Pacheco 2020). Most also try to measure the parent's capacity outside of the support received from their partner, family, or services, and they never question the services offered or determine if those services were adapted to the needs of the parent (Alexius and Hollander 2014).

In the coparenting literature, parenting is conceptualised as a series of tasks delivered by a group of adults who offer love and support to one child (McHale and Sirotkin 2019). Each adult is involved in their own way and all adults can navigate and negotiate their roles and responsibilities to ensure an effective collaboration for the child's benefit. A coparent's role can also be seen as a "Mother behind a Mother", thus offering support based on a continuum where they would either be doing with or for the other parent, or as a safety net or a companion (Traustadóttir and Sigurjónsdóttir 2008). These arrangements can even be established between biological parents and foster parents (McHale and Sirotkin 2019). Yet having support, or a coparent that is not one of the two biological parents, is often frowned upon for parents with ID. Parents with ID are often expected to raise their children on their own, or at least demonstrate their capacity to do so. We argue that the "best interests" of the child could include the benefits of having several loving adults involved in their lives.

If services were organised differently, PCAs would not be required. Yet the fear is that other types of reports would be used instead. In an analysis of 119 court decisions concerning families where one or both parents had ID, one third used some report as evidence.[1] When we looked closely, those reports were either written by the children's teachers or by CW professionals. Only ten percent were written by a psychologist. And half of those were a PCA performed by a professional that is independent from CW.

As long as the courts require data and information upon which to base their judgment, PCAs will be necessary. And ensuring that those evaluations are performed in the best way possible then becomes essential as "experts" positions are more often taken into consideration than the parents' voices (Pacheco et al. 2022). We have never seen courts question the independence or objectivity of the assessor, yet the overall conclusion around the (in)capacity of the parent to parent with or without help is often systematically linked to whomever mandated the evaluation (Aunos and Pacheco 2020). Furthermore, even when the best practices and guidelines are applied, the assessment can still be mis-represented, mis-quoted, mis-used, or just plain ignored.

---

[1] Aunos, Marjorie, Rahel More, Rachelle Rose, and Laura Pacheco. In preparation. *Parents with ID in Québec Child Welfare tribunals: An explorative analysis of court reports.*

## 7.9 (Mis)Understanding of Parents and Building Rapport

Many parents with ID have had CW involvement in their own lives as children (McConnell et al. 2017, 2021). Given that many of these parents have had previous and sometimes negative relationships with professionals, trauma-informed practices remind us of the importance of building rapport within the context of the assessment process (Pacheco et al. 2021). Parents might associate the assessor or process with great stress since the outcome might lead to either child removal and/or reunification (Aunos and Pacheco 2020; Norlin and Randell 2022).

Assessors need to reflect and identify how their own social position may impact and/or influence the process. Social position(s) can include intersecting variables related to one's identity along the lines of race, gender, socio-economic status, dis/ability, and religion that can influence our worldview and thus the assessment (Hailes et al. 2021). Assessors also need to acknowledge the power and privilege inherent within their professional role and intersecting social identities and ensure that they use their authority responsibly (Cameron 2014; Pacheco et al. 2022). Ensuring objectivity and neutrality, by being transparent concerning the nature, function, and limitations of the assessment as well as by acknowledging who has mandated and paid for the PCA is also primordial. A PCA should include what it intends to measure, what the process entails, the nature of the meetings (i.e., location and frequency), and what will happen upon completion (Feldman and Aunos 2010; Norlin and Randell 2022). Building rapport should include exploring the family's understanding of the PCA, what is important to them, and exploring their hopes and fears related to the conclusion (Meppelder et al. 2014). Assessors need to develop an understanding of the parents' learning and communication style (Tarleton and Turney 2020). In addition, they need to restrain from pathologizing the parent (i.e., identifying them as anxious rather than acknowledging that the process produces anxiety) within the evaluation (Aunos and Pacheco 2020).

## 7.10 Understanding the Parents' Values

Culture shapes our understanding of what is normal and acceptable within the field of disability and beyond (Campos and Kim 2017; Elkhateeb and Peter 2019; Waldschmidt 2017). Within a cultural model of disability framework, there are multiple layers of culture that can impact (and be impacted by) lived experience (Garland-Thomson 2017). This includes family culture, cultural community, and macro-level culture.

Family culture applies to the way in which the family is structured (i.e., single parent family, intergenerational family), roles, rules, and dynamics within the family system, that must go beyond hegemonic benchmarks of "normality" (Campos and Kim 2017). Choate and Tortorelli (2022, 10) insist that part of the assessment should include exploring "how collectivistic cultures accomplish attachment and provide for long-term stability and security". Therefore, assessors need to move beyond Euro-centric ideologies around parenting and explore the strengths and assets within the community and extended family network (Aunos et al. 2022). Family resiliency and routine, embedded within the fabric of the family, helps illustrate how the family has worked together to achieve their goals (McConnell et al. 2017). "Culture" deriving from an ethnocultural group can shape the ways in which families identify social roles, childrearing practices, caregiving, support, and interpretations of disability (Elkhateeb and Peter 2019).

For families headed by parents with ID that identify as Indigenous or as part of a cultural community (including first- and second-generation migrants), it is important to explore the way in which disability and social roles (e.g., ideals of mothering) are defined and reinforced within their community. In countries where colonization brought incredible disparities and abuse toward Indigenous and First Nations peoples, policies and practices need to be completely revisited (Powell 2022; Collings et al. 2018). Moreover, the social consequences for not measuring up to cultural ideals should be examined (Pacheco and McConnell 2017).

Macro-level culture refers to ideologies within broader society, that construct people into social categories creating unequal power relations between institutions and marginalized communities. Within the lives of mothers with ID, this includes the overarching narrative that persons with ID are incapable of "good enough parenting" and/or are unable to improve their parenting skills (Pacheco et al. 2022; More 2022). Exploring the experience(s) of oppression and discrimination within and across intersecting identities (including biased attitudes and gaps in services) and the impact this has on family life, is another contextual factor that needs examination (Feldman and Aunos 2010).

## 7.11 The Missing Voices and Acts of Resistance

The PCA is often focused on the assessor's evaluation and thus, often excludes or minimizes the parents' and/or families' perspectives (Aunos and Pacheco 2020). The family's perspective of their own needs should be woven throughout the report, as the objectives and/or issues, resources and strengths, perspectives of CW involvement, identification of helpful services and supports, and reactions to the recommendations, yet they are not.

Women with ID have been subject to reproductive coercion and violation of their parenting rights for decades (McConnell and Phelan 2022). There is a consistent trail of evidence that identifies how professionals have been posited as having "authoritative knowledge", often dismissing the experiential knowledge of mothers with ID (MacIntyre et al. 2019). The "professional knows best" rhetoric is a substantive theme within the discourse analysis of court reports where professionals' perspectives were identified as "valuable" and demonstrated "expertise" (sometimes with little evidence to substantiate their claims), while mothers' testimonies were often framed as "unreliable" and "lacking insight" in telling their story (Pacheco et al. 2022).

A small but growing body of research explores the ways that parents with ID challenge and exercise agency in the face of oppression. These parents used overt and covert strategies of resistance that included concealing their pregnancy (as they wanted to raise their children); refusing to accept services (as they were aware of biases and/or had a negative past experience); pretending to conform (in order to reduce the risk of child removal); joining forces with empowering allies; collaborating in research projects; taking concrete action against discrimination; and staying in the public eye, as they were aware that resisting or hiding could increase the risk of child removal (Franklin et al. 2021; Pacheco and McConnell 2017; Traustadóttir and Sigurjónsdóttir 2008). Another way of demonstrating resistance is by reclaiming their mothering identity and holding onto what is most dear to them, and their children (Pacheco and McConnell 2017). The love for their children is often, however, not deemed enough given that these parents are rendered as "nonetheless disabled" within and across diverse social contexts (Sigurjónsdóttir and Rice 2017).

## 7.12 Conclusion

Parents with ID are more likely to enter the CW system (Lightfoot et al. 2020). And parents who enter the CW system are more likely to lose custody of their children, even when they do everything they are asked to do to remediate the "risky" situation (Lightfoot et al. 2017). Once these files get to court, judges ask for predictive recommendations that professionals are technically and ethically not in a position to give, as assessments can only report on what the parents are capable of doing that day (Bow et al. 2011). The anxiety that parents often experience within this context may influence their "performance". This phenomenon is often not explored within the assessment, instead, parents' "anxiety" can be seen as another risk factor that could negatively impact their parenting (Aunos and Pacheco 2020; Norlin and Randell 2022).

The role of PCA in the justification for removal, or the lack of evidence-based PCA that could demonstrate their capacity, is a result of systemic ableism. As such, PCAs that do not have references or a clear conceptualisation of parenting, or without justifications of the tools used and their validity for this specific population, should not be received by the courts, as they offer a biased perception of the parents' capacity. In addition, the responsibilities of fathers and other adults involved should also be considered, as more often the weight of the "capacity" solely relies on the mothers.[2]

PCAs rarely, if ever, bring forward the voice of the children, even when they are of age to have their opinions heard. Naming a lawyer to represent the children in court is not enough. These lawyers actually have to bring forward questions and solutions so that the perspectives and best interests of the children are considered

[2] Aunos et al., in preparation for publication (same work cited in footnote one).

during deliberations. This is essential given that most decisions within CW courts center on what is in the best interest of the child (McConnell et al. 2011; Pacheco et al. 2022).

There continues to exist a notion or perception that persons with disabilities don't, or should not become parents, despite a solid trail of evidence to the contrary. We know that these ideologies underpin much of the discrimination these parents face within their parenting role. The voices and perspectives of parents with ID and their children are missing or devalued within the broader discourse. There is a need for more representation, particularly of successful parents, so that society can see "how it is done"—differently, but still done with care, love, and determination. Representation also needs to come from the children who are being raised (or have been raised) by parents with ID, as the narrative will otherwise always focus on the risks and negative outcomes on their development. Instead, we must showcase the families that have done it successfully (Bachrach 2022). When proper supports are provided, children do not feel an increased sense of responsibility nor experience a role reversal, even though they may have to contribute to the family's functioning at times (Bachrach 2022; Traustadóttir and Sigurjónsdóttir 2008). How we define family, support, and functioning across cultures, within and outside of social and health care institutions, also needs to shift to deliver inclusive, equitable, and individualized supports to families headed by parents with intellectual disabilities.

In truth, the best PCAs are the ones that are not performed, as everyone involved (including formal and informal support systems) would have worked together to offer the necessary supports and adaptations, based on the needs of each child.

## References

Albert, Sasha M., and Robyn M. Powell. 2022. Ableism in the child welfare system: Findings from a qualitative study. *Social Work Research* 46 (2): 141–152. https://doi.org/10.1093/swr/svac005.

Alexius, Katarina, and Anna Hollander. 2014. Care assessments concerning involuntary removal of children from intellectually disabled parents. *Journal of Social Welfare and Family Law* 36 (3): 295–310.

Aunos, Marjorie, and Laura Pacheco. 2020. Able or unable: How do professionals determine the parenting capacity of mothers with intellectual disabilities. *Journal of Public Child Welfare* 15 (3): 357–383. https://doi.org/10.1080/15548732.2020.1729923.

Aunos, Marjorie, Margaret Spencer, Laura Pacheco, and Evelina Pituch. 2024. This changes everything: A critical reflection on the impact of internalized ableist constructs on becoming a disabled mother. *Disability & Society* 39 (5): 079-1101.https://doi.org/10.1080/09687599.2022.2137392.

Bachrach, Tammy. 2022. Reflections of adult children of mothers with intellectual disabilities. *Journal of Intellectual Disabilities* 27 (3): 671–688. https://doi.org/10.1177/17446295221096930.

Belsky, Jay. 1993. Etiology of child maltreatment: A developmental-ecological analysis. *Psychological Bulletin* 114 (3): 413–434. https://doi.org/10.1037/0033-2909.114.3.413.

Bow, James N., Michael C. Gottlieb, and Dianna Gould-Saltman. 2011. Attorneys beliefs and opinions about child custody evaluations. *Family Court Review* 49 (2): 301–312.

Bronfenbrenner, Urie. 1979. *The ecology of human development*. Harvard University Press.

Bronfenbrenner, Urie, and Stephen J. Ceci. 1994. Nature-nurture reconceptualised in developmental perspective: A bioecological model. *Psychological Review* 101 (4): 568–586. https://doi.org/10.1037/0033-295x.101.4.568.

Callow, Ella, Munazza Tahir, and Maurice Feldman. 2017. Judicial reliance on parental IQ in appellate level child welfare cases involving parents with intellectual and developmental disabilities. *Journal of Applied Research in Intellectual Disabilities* 30 (3): 553–562. https://doi.org/10.1111/jar.12296.

Cameron, Colin. 2014. Turning experience into theory: The affirmation model as a tool for critical Praxis. *Social Work & Social Sciences Review* 17 (3): 108–121.

Campos, Belinda, and Heejung Kim. 2017. Incorporating the cultural diversity of family and close relationships into the study of health. *American Psychologist* 72 (6): 543–554.

Choate, Peter, and Christina Tortorelli. 2022. Attachment theory: A barrier for indigenous children involved with child protection. *International Journal of Environmental Research and Public Health* 19 (14): 8754. https://doi.org/10.3390/ijerph19148754.

Clarizio, Harvey F. 1979. In defense of the IQ test. *School Psychology Review* 8 (1): 79–88. https://doi.org/10.1080/02796015.1979.12086472.

Collings, Susan, Angela Dew, Tess Gordon, Margaret Spencer, and Leanne Dowse. 2018. Intersectional disadvantage: Exploring differences between aboriginal and non-aboriginal parents with intellectual disability in the New South Wales child protection system. *Journal of Public Child Welfare* 12 (2): 170–189.

Cox, Rachel, Biza Stenfert-Kroese, and Roger Evans. 2015. Solicitors' experiences of representing parents with intellectual disabilities in care proceedings: Attitudes, influence and legal processes. *Disability & Society* 30 (2): 284–298.

Edgerton, Robert B. 1993. *The cloak of competence.* Revised and Updated ed. Berkeley CA: University of California Press.

Elkhateeb, Itaf, and Dimity Peter. 2019. Negotiation of the right of women with disabilities in Palestine to marry: Cultural considerations for disability. *Sexuality and Disability* 37 (4): 559–570.

Feldman, Maurice. 1998. Preventing child neglect: Child-care training for parents with intellectual disabilities. *Infants & Young Children* 11 (2): 1–11.

Feldman, Maurice. 2002. Parents with intellectual disabilities and their children: Impediments and supports. In *Ethical dilemmas: Sexuality and developmental disability,* ed. Dorothy Griffiths, Debbie Richards, and Paul Federoff, 255–292. Kingston: NADD Press.

Feldman, Maurice, and Marjorie Aunos. 2010. *Comprehensive competence-based parenting capacity assessment for parents with learning difficulties.* Kingston: NADD Press.

Feldman, Maurice, and Marjorie Aunos. 2020. Recent trends and future directions in research regarding parents with intellectual and developmental disabilities. *Current Developmental Disorders Reports* 7 (3): 173–181. https://doi.org/10.1007/s40474-020-00204-y.

Feldman, Maurice, and Munazza Tahir. 2016. Skills training for parents with intellectual disabilities. In *Handbook of evidence-based practices for individuals with intellectual disabilities,* ed. Nirbhay N. Singh, 613–631. Springer Cham.

Feldman, Maurice, David McConnell, and Marjorie Aunos. 2012. Parental cognitive impairment, mental health and child outcomes in a child protection population. *Journal of Mental Health Research in Intellectual Disabilities* 5 (1): 66–90. https://www.tandfonline.com/doi/abs/10.1080/19315864.2011.587632.

Feldman, Maurice, Jean Varghese, Jennifer Ramsay, and Danuta Rajska. 2002. Relationship between social support, stress and mother-child interactions in mothers with intellectual disabilities. *Journal of Applied Research in Intellectual Disabilities* 15 (4): 314–323.

Franklin, Laura, Kate Theodore, Daniel Foulds, Mary Cooper, Lisa Mallaghan, Paul Wilshaw, Alison Colborne, Emily Flower, Deborah Dickinson, and Joyce Nga Yu Lee. 2021. "They don't think I can cope, because I have got a learning disability": Experiences of stigma in the lives of parents with intellectual disability. *Journal of Applied Research in Intellectual Disability* 35 (4): 935–947.

Garland-Thomson, Rosemarie. 2017. Building a world with disability in it. In *Culture-theory-disability, encounters between disability studies and cultural studies*, ed. Anne Waldschmidt, Hanjo Berressem, and Moritz Ingwersen, 51–63. Transcript Disability Studies.

Granqvist, Pehr, Tommie Forslund, Mari Fransson, Lydia Springer, and Lene Lindberg. 2014. Mothers with intellectual disability, their experiences of maltreatment, and their children's attachment representations: A small-group matched comparison study. *Attachment & Human Development* 16 (5): 417–436. https://www.tandfonline.com/doi/abs/10.1080/14616734.2014.926946.

Hailes, Helen P., Christopher J. Ceccolini, Ellen Gutowski, and Belle Liang. 2021. Ethical guidelines for social justice in psychology. *Professional Psychology: Research and Practice* 52 (1): 1–11. https://doi.org/10.1037/pro0000291.

Hindmarsh, Gabrielle, Gwynnyth Llewellyn, and Eric Emerson. 2017. The social-emotional well-being of children of mothers with intellectual impairment: A population-based analysis. *Journal of Applied Research in Intellectual Disabilities* 30 (3): 469–481. https://onlinelibrary.wiley.com/doi/10.1111/jar.12306.

Kirshbaum, Megan, and Rhoda Olkin. 2002. Parents with physical, systemic or visual disabilities. *Sexuality and Disability* 20 (1): 65–80.

Kollinsky, Laura, Laura Simonds, and Julie Nixon. 2012. Qualitative exploration of the views and experiences of family court magistrates making decisions in care proceedings involving parents with LD. *British Journal of Learning Disabilities* 41 (2): 86–93. https://onlinelibrary.wiley.com/doi/abs/10.1111/j.1468-3156.2012.00726.x.

Lightfoot, Elizabeth, Katharine Hill, and Traci Laliberte. 2010. The inclusion of disability as a condition for termination of parental rights. *Child Abuse & Neglect* 34 (12): 927–934.

Lightfoot, Elizabeth, Traci LaLiberte, and Minhae Cho. 2017. A case record review of termination of parental rights cases involving parents with a disability. *Children and Youth Services Review* 79: 399–407. https://www.sciencedirect.com/science/article/abs/pii/S01907409173.

Lightfoot, Elizabeth, Minyang Zheng, and Sharyn DeZelar. 2020. Substantiation of child maltreatment among parents with disabilities in the United States. *Journal of Public Child Welfare* 15 (5): 583–596. https://www.tandfonline.com/doi/full/10.1080/15548732.2020.1773369.

MacIntyre, Gillian, Alisa Stewart, and Sharon McGregor. 2019. The double-edged sword of vulnerability: Explaining the persistent challenges for practitioners in supporting parents with intellectual disabilities. *Journal of Applied Research in Intellectual Disabilities* 32 (6): 1523–1534. https://doi.org/10.1111/jar.12647.

McConnell, David, and Shannon Phelan. 2022. The devolution of eugenic practices: Sexual and reproductive health and oppression of people with intellectual disability. *Social Science Medicine* 298: 114877. https://doi.org/10.1016/j.socscimed.2022.114877.

McConnell, David, Marjorie Aunos, Maurice Feldman, and Narasimha Prasad. 2011. Parental cognitive impairment and child maltreatment in Canada. *Child Abuse & Neglect* 35 (8): 621–632.

McConnell, David, Maurice Feldman, and Marjorie Aunos. 2017. Parents and parenting with intellectual disabilities: An expanding field of research. *Journal of Applied Research in Intellectual Disabilities* 30 (3): 419–422.

McConnell, David, Marjorie Aunos, Laura Pacheco, and Maurice Feldman. 2021. Child maltreatment investigations in Canada: Main and moderating effects of primary caregiver cognitive impairment. *Child Maltreatment* 26 (1): 115–125. https://doi.org/10.1177/1077559520910806.

McGaw, Sue, Tamara Scully, and Colin Pritchard. 2010. Predicting the unpredictable? Identifying high-risk versus low-risk parents with intellectual disabilities. *Child Abuse and Neglect* 34 (9): 699–710.

McHale, James P., and Yana Segal Sirotkin. 2019. Coparenting in diverse family systems. In *Handbook of parenting*, ed. Marc H. Bornstein, 3rd ed., 137–166. Routledge.

Meppelder, Marieke, Marja Hodes, Sabina Kef, and Carlo Schuengel. 2014. Parents with intellectual disabilities seeking professional parenting support: The role of working alliance, stress and informal support. *Child Abuse & Neglect* 38 (9): 1478–1486.

More, Rahel. 2022. Meanings of parenting and dis/ability for mothers and fathers with intellectual disabilities in the context of social work in Austria: Potentials of deconstruction for shifts in meaning. *Qualitative Social Work* 22 (5): 859–878. https://doi.org/10.1177/14733250221109765.

Norlin, Jerry, and Eva Randell. 2022. Social workers' perceptions of assessing the parental capacity of parents with intellectual disabilities in child protection investigations. *Journal of Intellectual Disabilities* 27 (4): 976–993. https://doi.org/10.1177/17446295221113717.

Pacheco, Laura, and David McConnell. 2017. Love and resistance of mothers with intellectual disability from ethnocultural communities in Canada. *Journal of Applied Research in Intellectual Disabilities* 30 (3): 501–510. https://doi.org/10.1111/jar.12342.

Pacheco, Laura, Marjorie Aunos, Maurice Feldman, and David McConnell. 2021. Reasonable efforts? Child maltreatment investigations and service referrals of parents with ascribed cognitive impairments in Canada. *Child Maltreatment* 27 (3): 501–510. https://journals.sagepub.com/doi/10.1177/10775595211001109.

Pacheco, Laura, Rahel More, Marjorie Aunos, and Rachelle Rose. 2024. Unfit and cast aside: Portrayal of mothering with intellectual disability in Québec court reports. *Critical Discourse Studies* 21 (3): 322–340. https://doi.org/10.1080/17405904.2022.2152063.

Powell, Robyn. 2022. Achieving justice for disabled parents and their children: An abolitionist approach. *Yale Journal of Law and Feminism* 33 (2): 37–109.

Proctor, Stephon N., and S.T. Azar. 2012. The effect of parental intellectual disability status on child protection service worker decision making. *Journal of Intellectual Disability Research* 57 (12): 1104–1116. https://onlinelibrary.wiley.com/doi/abs/10.1111/j.1365-2788.2012.01623.x.

Sigurjónsdóttir, Hanna Björg, and James Rice. 2017. 'Framed': Terminating the parenting rights of parents with intellectual disability in Iceland. *Journal of Applied Research in Intellectual Disabilities* 30 (3): 543–552. https://onlinelibrary.wiley.com/doi/abs/10.1111/jar.12301.

Skeels, Harold M., and Harold B. Dye. 1939. A study of the effects of differential stimulation on mentally retarded children. *Proceedings. American Association on Mental Deficiency* 44: 114–136.

Tarleton, Beth, and Danielle Turney. 2020. Understanding 'successful practice/s' with parents with learning difficulties when there are concerns about child neglect: The contribution of social practice theory. *Child Indicators Research* 13 (2): 387–402. https://doi.org/10.1007/s12187-019-09682-y.

Traustadóttir, Rannvieg, and Hanna Björg Sigurjónsdóttir. 2008. The "mother" behind the mother: Three generations of mothers with intellectual disabilities and their family support networks. *Journal of Applied Research in Intellectual Disabilities* 21 (4): 331–340. https://doi.org/10.1111/j.1468-3148.2008.00450.x.

Waldschmidt, Anne. 2017. Disability goes cultural: The cultural model of disability as an analytical tool. In *Culture-theory-disability, encounters between disability studies and cultural studies*, ed. Anne Waldschmidt, Hanjo Berressem, and Moritz Ingwersen, 19–29. Transcript Disability Studies.

# Chapter 8
# Sexual Consent and People with Intellectual and Developmental Disability

**Andria Bianchi**

**Abstract** People with intellectual and developmental disability (IDD) are sexual beings. However, some may be unable to consent to sexual acts in particular ways due to characteristics associated with their disability. An inability to consent in ways that are traditionally considered acceptable poses a challenge insofar as sexual consent is legally and morally significant. Additionally, as per the United Nations Convention on the Rights of Persons with Disabilities, people with disabilities, including IDD, have sexual rights. Given (1) the importance of consent, (2) limitations of some sexual consent models for people with IDD, and (3) the fact that people with IDD have sexual rights, it is important to reflect upon whether and what kinds of sexual consent models may be helpful for some people with IDD who may be unable to consent in traditional ways. The purpose of this chapter is to offer reflections and possible future directions so that people with IDD can engage in consensual sexual acts in a manner regarded as ethically licit.

**Keywords** Sexual consent · Sexuality · Intellectual disability · Consent · Assent · Supported decision making

---

A. Bianchi (✉)
Centre for Clinical Ethics, Unity Health Toronto, Toronto, Canada
e-mail: andria.bianchi@utoronto.ca

KITE Research Institute, Toronto, Canada

Dalla Lana School of Public Health, University of Toronto, Toronto, Canada

 
A. Bianchi and J. A. Vogt (eds.), *Intellectual Disabilities and Autism: Ethics and Practice*, The International Library of Bioethics 108,
https://doi.org/10.1007/978-3-031-61565-8_8

## 8.1 Introduction

People with intellectual and developmental disability (IDD)[1] are sexual beings. However, some may be unable to consent to sexual acts in particular kinds of ways due to characteristics associated with their disability. An inability to consent to sex in ways that are traditionally considered acceptable poses a challenge insofar as consent is legally and morally significant.

The purpose of this chapter is to challenge the perspective that people with IDD cannot consent to sexual acts in ways that would make the act *ethically* licit. To achieve this goal, the first part of this chapter describes the importance of autonomy and consent. The next section introduces sexual consent and offers a brief overview of some sexual consent models. Following this, the chapter describes reasons that some people with IDD may be unable to consent to sex in accordance with certain models. Additionally, this section contests the belief that people with IDD are asexual. The final part of this chapter brings forward the idea that different ways of consenting to sex may be ethically licit. One potential new model, which I refer to as "supported sexual decision making", is introduced as an idea that may be worth future consideration, though the chapter stops short of necessarily endorsing it.

Throughout this chapter, I use the term "sexual act" to apply to any kind of sexually intimate activity that requires consent, including, but not exclusive to, penetrative intercourse. The purpose of this chapter is *not* to question the importance of consent in and of itself. Rather, the purpose is to reflect upon and bring forward potential limitations regarding various models of consent as applied to some people with IDD. I have written about limitations of sexual consent models in previous work, which primarily focused on people with dementia (Bianchi 2018, 2022a, 2022b). Dementia is an importantly distinct diagnosis from IDD, though I refer to some previous work in this chapter because of similar challenges that may exist when thinking about models of consent.

Before considering traditional sexual consent models and implications for some people with IDD, it is important to acknowledge that many people with IDD are in positions of vulnerability when it comes to sexual acts. People with IDD have a high prevalence of sexual abuse in comparison to those without IDD (Tomsa et al. 2021; Fisher et al. 2016) and are less likely to report it (McGilloway et al. 2020). As a society, we must continue to advance appropriate sexual health education for people with IDD[2] and combat barriers that contribute to under-reporting. At the same time, maintaining an exclusive focus on protecting people with IDD from sexual harm may inadvertently result in a further harm, wherein the sexual rights of people with

---

[1] This chapter uses person-first language, i.e., people with intellectual and developmental disability. The author appreciates the constant evolution and degrees of acceptance of person-first and identity-first language. At the time of writing this chapter, the spaces and communities within which the author is a part tend to use person-first language in relation to intellectual and developmental disability. It may be the case that readers of this chapter have different preferences and/or experiences, or perhaps this chapter is being read at a time where general acceptance of person-first language has evolved.

[2] For more on sexual education and people with IDD, see Chaps. 19, 20 and 21 of this volume.

disabilities, including IDD, fail to be honoured. As Tomsa et al. (2021, 12) note, "… it is a challenge to find a balance between providing individuals with intellectual disability with the necessary protective supervision so that they are not victims of [sexual] abuse, while allowing them to enjoy their [sexual] rights and freedom." This chapter considers how people with IDD can fulfill their sexual rights in an ethically defensible way with full appreciation of this delicate balancing act.

## 8.2 Autonomy and Consent

Sexual consent is important. At least part of the reason that sexual consent (and other kinds of consent) is important is because of the principle of autonomy. The principle of autonomy encompasses the perspective that people should be permitted to make decisions for themselves, where respect for autonomy grounds the idea that individuals should respect others' autonomous decisions.

As I have written about elsewhere, the concept of autonomy was historically associated with self-governance and the idea of a "self-made man", making it such that women (and other "dependent" individuals) were considered non-autonomous (Stoljar 2000, 2022; Bianchi 2016). Feminist philosophers rejected this understanding and re-conceptualized the antiquated conception to that of *relational autonomy*. Relational autonomy is an umbrella term that describes various relational theories (i.e., accounts) of autonomy (e.g., procedural theories, substantive accounts). The commonality among each account is that individuals are recognized as agents who are inevitably influenced by their social contexts (Stoljar 2000). Furthermore, relational accounts take into consideration the way that relationships can inhibit or facilitate autonomy (Ibid.).

One way that autonomy is realized in practice is through the mechanism of consent. If a person consents for some act to occur, then pursuing the act is generally considered ethically permissible because it reflects the person's autonomous choice. The principle of autonomy is one ethical principle relevant to health care contexts (Beauchamp and Childress 2013), and consent is the way in which patients' autonomous preferences can translate into practice. For instance, if a patient consents to a proposed medical intervention, then it is typically permissible for clinicians to proceed accordingly. Conversely, if a patient refuses to consent, then it is generally impermissible for healthcare professionals to proceed. In short, consent (or refusal) provides a way for people to express their autonomous preferences, which is of ethical significance given the importance of autonomy.

## 8.3 Sexual Consent

Similar to consent for medical decisions, sexual consent is significant in assessing the legal and ethical permissibility of a sexual act. As noted by David Archard (1998, 1), "consent makes a difference to whether some sexual activity is seen as immoral or not." This sentiment is echoed by Shrage and Stewart (2015) who consider sexual consent as a minimum requirement of moral decency.

Although sexual consent is recognized as both legally and ethically significant at the time of writing this chapter, it was not always considered essential. Historically, the law in the United States determined the legality of a sexual act based on a man's sexual interests and conduct, i.e., not a woman's willingness to engage in sex (Schulhofer 1987).[3] During this time, it was impossible for a husband to sexually assault his wife (Ryan 1995; Tuchinsky 2016).

This heteronormative legal structure that only considered men's autonomous sexual preferences as relevant started to change in the nineteenth century (Ibid.). It is at this time that the women's rights movement advocated for self-ownership (Ryan 1995, 945). Such criticism continued into the twentieth century, arguing that all individuals should have equal rights, including married women.

Two major legal milestones occurred because of these movements, both of which contributed to the importance and evolution of sexual consent. The first milestone took place in 1971, when the United States Supreme Court permitted people both in and outside of marriage to have equal access to contraceptives (Ryan 1995). This meant that a structure was now in place so that women could have their autonomous sexual and reproductive preferences honoured irrespective of marital status. The second milestone occurred in 1984 when a man was convicted for raping his wife, thereby abolishing the marital rape exemption. Abolishing the marital rape exemption meant that sexual consent was now an essential part of any sexual act (Ibid.).

Sexual consent is the way that a person can express their autonomous preference about engaging in a sexual act(s). However, there exist different ideas about what consent ought to entail. I offer a brief description of each conception below, and a more substantial description in other work (Bianchi 2016, 2018, 2022a, 2022b).

### *8.3.1 "'No' Means No"*

The "'no' means no" model of consent is the idea that a person who says "no" to some sexual act indicates a refusal of consent (Estrich 1987). To engage in an act with someone who says "no" is considered unreasonable (Ibid.). This model of consent was developed primarily to protect women from unwanted sexual acts.

---

[3] When describing the history of sexual consent, the terms "man" and "woman" are used intentionally, as the law only allowed for, recognized, and spoke to heterosexual acts (Schulhofer 1987).

Several concerns and arguments have been put forward in opposition to the "'no' means no" model. One concern is that it does not take into adequate consideration the way that women have been socialized to acquiesce to others' requests. As a consequence of socialization patterns, some women may feel disempowered and unable to express their lack of consent in response to another's sexual requests. Relatedly, it seems reasonable that if a person fears their sexual partner due to implicit or explicit power imbalances or threats, then they may be frozen in fear and *not* explicitly refuse (Haag 1999; Shrage and Stewart 2015). Finally, a person who does not *say* "no" may not necessarily want to engage in sex, but the model could allow for this interpretation.

### 8.3.2 Affirmative Consent

Rather than saying "no" to indicate refusal of consent, affirmative consent puts forward the idea that sexual partners should positively affirm their willingness to engage in a sexual act.

Many proponents of affirmative consent do not believe that a person's "yes" is all that consent should entail. For instance, Kukla (2018) suggests that sexual partners' preferences should be negotiated using language. According to this account, a simple "yes" is insufficient. Michelle Anderson (2005) also endorses a negotiation model, though it applies exclusively to sexual penetration (Anderson 2005, 1423). Anderson's model allows for non-verbal consent to occur based on the assumption that it may be unrealistic for long-term consensual partners to vocalize "yes" every time they want to participate in a sexual act. Finally, another consent model is Lois Pineau's (1989), who refers to her model as "communicative sexuality". Pineau's model requires preferences to be communicated ongoingly, both in advance of and during a sexual encounter. Pineau notes that while both verbal and non-verbal cues can be used to indicate one's desire, the cues must be agreed upon in advance.

Kukla's, Anderson's, and Pineau's frameworks each offer an example of how affirmative sexual consent may translate into practice. Although this is not an exhaustive list, the primary point is that affirmative consent requires sexual partners to indicate "yes" to sexual acts, with some interpretations requiring additional, ongoing, and/or different types of communication.

## 8.4 Sexual Consent and IDD

As I have argued before (2018, 24), "proponents of the "'no' means no" and affirmative consent models do not think about their models as being helpful for only specific types of people." The above-mentioned models of sexual consent are intended to be universally applicable. The problem with this intention, however, is that most policies, procedures, and mechanisms like consent cannot be universally applied given the relevant differences and needs that exist across populations. Furthermore, it is presumably the case that those who create models or mechanisms of consent with a desire for universal applicability end up unintentionally excluding those whose perspectives are not heard or understood, which is often those who are already marginalized, in positions of vulnerability, and/or who have varying needs, including those who are disabled. As noted by Jeffress et al. (2023, 2) "disability is defined medically and socially by deviation from what normal embodiment, emotional response, and cognitive processing are considered to be (Garland Thomson 1997)." I argue that the above models of sexual consent tend to reflect, for the most part, "normal" embodiment and cognitive processing.

### 8.4.1 Sexual Rights and Desires

People with IDD represent one sub-population of our society whose needs may not be reflected in the above models of sexual consent. It may be the case that some are of the impression that people with IDD do not *need* to be represented because of the widespread perspective that people with IDD are asexual. As Michael Gill (2015, 10) highlights, "those with intellectual disabilities are often assumed to be incompetent and unable to make meaningful decisions in their lives, let alone make and execute sexually fulfilling choices." However, the perspective that all people with IDD are asexual has been combatted time and time again (Taylor Gomez 2012). As highlighted in other parts of this volume,[4] it is undoubtedly the case that people with IDD are sexual beings.

In accordance with the fact that people with IDD are sexual beings, there exist some relevant rights within the United Nations Convention on the Rights of Persons with Disabilities (CRPD) (UN 2006). For instance, Article 23 of the CRPD states that those who ascribe to it must "eliminate discrimination against persons with disabilities in all matters relating to marriage, family, parenthood and relationships, on an equal basis with others" so that people with disabilities can fulfill their marital, reproductive, familial, and relational preferences and desires. As a complement to the CRPD, the World Health Organization (2022) states that sexual health and well-being is a *human* right.

As alluded to above, alongside every right exists some corresponding duty. And insofar as people with disabilities have a right to make decisions about and engage in sexual acts, then we must ask ourselves: what duty or duties may exist regarding people with IDD's sexual rights? In response, it seems plausible that there may exist a duty to respect people with IDD's autonomous sexual decisions so that their rights can be honoured. In other words, if a person with IDD has a right to engage in a sexual act with another person(s), and if they (and their partner(s)) autonomously decide to

---

[4] See Chap. 19 in this volume for information about ableism and sexual, reproductive, and parenting rights. Additionally, chapter 19 of this volume contains information about dating and intimacy for people with IDD, and information about sexual education can be found in Chap. 21.

engage in the sexual act, then respecting their decision may be the corresponding duty. And, as mentioned earlier in this chapter, one way that a person's sexual preferences, interests, and decisions can be communicated such that any corresponding act is considered ethically licit is via consent. This means that if a person with IDD wants *and consents* to engage in a sexual act, then the act may be considered ethically licit and honouring it would allow their sexual rights to be realized. If a sexual act is ever considered ethically illicit due to it being non-consensual and/or not reflective of the person's autonomous preferences, then there presumably does not exist a duty to honour it. In short, autonomous sexual preferences communicated via consent are ethically licit and, as such, ought to be respected in accordance with individuals' sexual rights.

### 8.4.2 Current Models of Sexual Consent and IDD

The challenge with consent being an ethically relevant component of what it means to (and why we should) honour sexual rights is that some people with IDD may be unable to consent in accordance with traditional consent models. In other words, if consenting to sexual acts *in accordance with these models*[5] is a determining factor as to whether a person's sexual preferences ought to be respected, then people with IDD's preferences and rights may not be honoured (i.e., if a person with IDD cannot consent in a typical and accepted way, then any corresponding act(s) would be ethically illicit). The reason that some people with IDD may be unable to consent in accordance with accepted consent models may be due to characteristics related to their disability.

As a part of their disability (and as required for diagnosis), people with IDD will have limitations related to intellectual and adaptive functioning (Government of British Columbia 2021). Intellectual functioning refers to a person's ability to learn as assessed via standardized tests, whereas adaptive functioning reflects one's ability to navigate or perform activities and skills for everyday life (Ibid.). "Adaptive functioning" is an umbrella term that encompasses three distinct types of skills: (1) conceptual skills (e.g., reading, reasoning, memory), (2) social skills (e.g., interpersonal skills, social judgment), and (3) practical skills (e.g., financial management, personal care, organizing tasks). Importantly, the American Association on Intellectual and Developmental Disabilities (2023) highlights that people with IDD are a heterogeneous population; every person with IDD will have varying strengths, challenges, experiences, and presentations associated with their disability. Relatedly, the amount and degree of support that a person with IDD requires to navigate society and enhance their quality of life will vary.

---

[5] As noted in the Introduction, consent is important. The purpose of this section is not to contest the importance of consent in and of itself. Rather, the challenge being highlighted is that of current models of consent and some challenges that may exist when thinking about the models and people with IDD.

Although there exist differences amongst people with IDD, it seems plausible that certain intellectual and adaptive functioning challenges may influence the ability to consent to sexual acts *in accordance with traditional consent models.* For instance, if a person with IDD experiences challenges with respect to interpersonal communication, then they may be unable to negotiate their and their partner's sexual preferences effectively, which is relevant to some models of consent. Interpersonal communication challenges may also limit one's ability to communicate ongoing consent and/or to accurately interpret whether one's partner does or does not want to engage in the activity.[6] Similarly, if a person with IDD experiences challenges with their memory, then they may be unable to recall how they and their partner(s) agreed to communicate any preferences during sexual acts. Additionally, if a person uses non-verbal communication, then it would be impossible for them to articulate "Yes" or "No", which may be considered the most basic requirement of some interpretations of "'no' means no" and affirmative consent.

In short, it seems as though the strengths and challenges experienced by some people with IDD may influence their ability to consent to sexual acts in ways that make the corresponding act ethically licit. Saying "yes" or "no", negotiating preferences, determining the use of verbal and non-verbal communication during sex, and providing ongoing consent, reflect certain skills and abilities of most people *without* IDD. But what about people with IDD? If accepted consent models are used to determine whether a sexual act is ethically licit and if people with IDD cannot consent accordingly, then what does this mean?

Given that we live in an ableist society (Nario-Redmond 2020), some people may respond to the question above by suggesting that many people with IDD cannot engage in ethically licit sex because of limitations in their intellectual and/or adaptive functioning skills. However, insofar as people with IDD are sexual beings with sexual interests and rights, then it may be incumbent upon our society to consider what consent ought to look like so that their rights can be honoured. In order to determine what ethically licit sexual consent could look like for people with IDD, it may make sense to explore what types of decision-making and consent models are used for non-sexual decisions as a first step. For the purposes of this chapter, the model introduced below is that of supported decision making. Although various decision-making models exist, starting with one widely recognized model that is relevant to people with IDD may, at the very least, allow us to explore whether and, if so, how autonomous and ethically licit sexual decisions could be made. If, after reasonable exploration, supported decision making is determined to be unhelpful for sexual decisions, then other models can be explored.

[6] I appreciate that this point is relevant to the discussion of sexual offenders with IDD since some people with IDD may experience challenges with accurately interpreting another person's consent or lack thereof. The topic of sexual offenders with IDD is not a topic covered in this chapter, though it is an important and related topic that warrants attention.

## 8.5 Supported Decision Making

Supported decision making is a model that enables people with IDD to consent or meaningfully contribute to their own healthcare decisions. As described by Sullivan and Heng (2018), it is traditionally the case that people both with and without IDD are encouraged to make medical decisions as independent individuals. The intention behind this traditional, individualized decision-making method is to protect patients from undue pressure that may result from others' involvement; the model is intended to promote individual autonomy and serve as an indicator of one's capacity to make decisions.

As it turns out, the traditional model of medical consent may make it such that people with IDD are wrongly regarded as unable to consent. As Sullivan and Heng (Ibid.) argue, some people with IDD may benefit from embracing relational components of decision making and receiving support from trusted people (e.g., family or friends) who can help them navigate challenges associated with intellectual and/or adaptive functioning skills. Decision-making support from a trusted person(s) who is authorized "to have access to the patient's health records, facilitate the patient's understanding, and interpret and communicate the patient's goals or values" (Ibid., S34) may promote autonomy and optimize their capacity for medical decision making, both of which are ethically significant. In short, supported decision making acknowledges the importance and legitimacy of relationships as a way to enhance autonomy and communicate consent. Moreover, this decision-making model enables people with IDD to serve in the role of "decision maker", albeit in a non-traditional format that responds to their personal needs and skills.

### *8.5.1 Supported Sexual Decision Making*

Earlier in this chapter, it was argued that some people with IDD may be unable to consent to sexual acts in accordance with traditional consent models because of characteristics related to their disability. This is similar to the way that some people with IDD were/are considered unable to consent to medical decisions in a "traditional" manner. In response to the challenge in medicine, supported decision making serves as a helpful response. However, no equivalent model for sexual consent has been explicitly acknowledged as having the potential to contribute to ethically licit sexual acts for people with IDD.

Insofar as people with IDD are sexual beings with sexual rights, then it may be worth exploring whether a supported decision-making model could potentially translate to the sexual domain, i.e., supported *sexual* decision making. It seems plausible that the answer may be "yes", though some small, distinct differences may be required given unique factors associated with sexual acts. When thinking about supported sexual decision making for people with IDD, ensuring that a trusted person is available to "... facilitate the [person's] understanding, and interpret and communicate the [person's] goals or values" (Sullivan and Heng 2018, S34) specifically in relation to sexual acts would be essential. For some people with IDD, ongoing education from a support person may be needed, particularly if they have minimal knowledge, memory, and/or ability to independently distinguish between abusive versus consensual relations (Murphy and O'Callaghan 2004). Supportive sexual decision making would help the person understand these important differences and make an informed decision about how to proceed.

Additionally, the support person would be available to help the individual with IDD navigate complexities associated with sexual acts and to ensure that any "risky" decisions are informed and reflective of their goals and values. A support person should serve in a way that reflects the name of their role, namely, that of a support. The role of a support person for sexual decision making should *not* be to dissuade, judge, or alter a person with IDD's sexual values, beliefs, desires, or decisions; this may make it such that a person with IDD's typical support person is or is *not* ideal when it comes to sexual decisions. Sex is a value-laden, moralized, and taboo topic, and it may be the case that a person's typical support person has sexual values or beliefs that conflict with the person with IDD. If this is the case, then the support person should step aside so that the person is supported in a responsible, autonomy-enhancing manner that allows their sexual rights to be honoured.

In thinking about who may serve as an appropriate support for supported sexual decision making, some may argue that a person's sexual partner(s) themselves should *not* be permitted to serve in the role. This argument would presumably be based on the fact that the sexual partner has a uniquely personal vested interest in the outcome of the person with IDD's decision(s). Although I will not provide a concrete argument in support of or in opposition to a sexual partner(s) serving as a support, I will pose one consideration for further reflection. The consideration is as follows:

Although supported decision making for medical decisions is importantly distinct from sexual decisions, those who serve as a support person for medical decisions presumably have a vested, though different kind of vested, interest in the person with IDD's decision. For instance, if a person with IDD must decide whether they want to undergo a life-saving surgical intervention, then their support person may deeply and desperately want them to consent. In fact, insofar as they care about the person with IDD, then the support person may want to dissuade them from refusing. However, as a trusted person who agrees to serve in the role of a support, they are responsible for recognizing and developing strategies to manage their own values, beliefs, and perspectives so the person with IDD is supported in an autonomy-enhancing kind of way *even if they want the person with IDD to make a different decision.* Similarly, a support person for sexual decisions may *want* to dissuade the person with IDD from refusing to consent, but inherent in the role of a support person exists a responsibility to navigate one's own values, perspectives, and desires to ensure the person with IDD makes an autonomous choice. If a person's sexual partner *can* prioritize the person with IDD's preferences in light of their own (i.e., if they are, in fact, a trusted person in regard to sexual decisions), then it may be unreasonable to prevent them from taking on this role. However, if the sexual partner's own sexual interests are too strong and/or they find it challenging to navigate the person with IDD's autonomous preferences and interests, as well as their own, then they would not be considered a trusted person who should serve in this capacity. A support person's trustworthiness is essential in determining if they are the right kind of support. Some sexual partners may be able to facilitate supported sexual decision making, whereas others may not.

## 8.6 Conclusion

The CRPD highlights that people with disabilities, including people with IDD, have sexual rights. And when it comes to engaging in ethically licit sex, sexual consent is important. However, traditional sexual consent models fail to account for relevant intellectual and adaptive functioning differences that exist amongst some people with IDD.

In response to the challenges that exist with traditional sexual consent models and people with IDD, this chapter argues that society ought to explore ways in which people with IDD can be recognized as consenting to ethically licit sexual acts. Sexual consent models need to be more accessible so that people with IDD are not unfairly excluded. In order to do this, we ought to consider various decision-making models that are helpful in non-sexual domains as a first step, such as supported decision making. Upon exploration, it seems that a model like supported sexual decision making may be worth consideration when thinking about what ethically licit sexual consent may look like for people with IDD. However, additional research is required to ensure that people with IDD can consent to sex in the right kinds of ways, and in ways that respect their rights.[7]

## References

Anderson, Michelle. 2005. Negotiating sex. *Southern California Law Review* 78 (6): 1401–1438.

American Association on Intellectual and Developmental Disabilities. 2023. *FAQs on intellectual disability.* https://www.aaidd.org/intellectual-disability/faqs-on-intellectual-disability. Accessed 9 April 2024.

Beauchamp, Tom L., and James F. Childress. 2013. *Principles of biomedical ethics,* 7th ed. New York: Oxford University Press.

Bianchi, Andria. 2016. Autonomy, Sexuality, and Intellectual Disability. *Social Philosophy Today* 31: 107–121. https://doi.org/10.5840/socphiltoday20168232.

Bianchi, Andria. 2018. *Sex, dementia, and consent.* Ph.D. Dissertation. University of Waterloo. http://hdl.handle.net/10012/13520.

Bianchi, Andria. 2022a. Sex and aging populations. In *The Routledge handbook of the philosophy of sex and sexuality*, ed. Brian D. Earp, Clare Chambers, and Lori Watson, 286–296. New York: Routledge.

[7] By this, I am referring to both positive sexual rights and the right to be protected from undue sexual harm.

Bianchi, Andria. 2022b. Sex, dementia, and well-being. In *Sexual ethics handbook*, ed. David Boonin, 357–375. Cham: Springer International Publishing AG. https://doi.org/10.1007/978-3-030-87786-6.

Estrich, Susan. 1987. *Real rape.* Cambridge, Mass.: Harvard University Press.

Fisher, Marisa H., Jessie V. Baird, Andrew D. Currey, and Robert M. Hodapp. 2016. Victimisation and social vulnerability of adults with intellectual disability: A review of research extending beyond Wilson and Brewer. *Australian Psychologist* 51 (2): 114–127. https://doi.org/10.1111/ap.12180.

Garland-Thomson, Rosemarie. 1997. *Extraordinary bodies: Figuring physical disability in American culture and literature.* Columbia University Press.

Gill, Michael. 2015. *Already doing it: Intellectual disability and sexual agency.* Minneapolis: University of Minnesota Press.

Government of British Columbia. 2021. *Intellectual disabil*ities. https://www2.gov.bc.ca/gov/content/health/managing-your-health/child-behaviour-development/support-needs/intellectual-disabilities. Accessed 9 April 2024.

Haag, Pamela. 1999. *Consent: Sexual rights and the transformation of American liberalism.* Ithaca: Cornell University Press.

Jeffress, Michael S., Joy M. Cypher, Jim Ferris, and Julie-Ann. Scott-Pollock. 2023. *The Palgrave handbook of disability and communication.* Cham: Springer International Publishing. https://doi.org/10.1007/978-3-031-14447-9.

Kukla, Rebecca. 2018. That's what she said: The language of sexual negotiation. *Ethics* 129 (1): 70–97. https://doi.org/10.1086/698733.

McGilloway, Connie, David Smith, and Rose Galvin. 2020. Barriers faced by adults with intellectual disabilities who experience sexual assault: A systematic review and meta-synthesis. *Journal of Applied Research in Intellectual Disabilities* 33 (1): 51–66. https://doi.org/10.1111/jar.12445.

Murphy, Glynis H., and Ali O'Callaghan. 2004. Capacity of adults with intellectual disabilities to consent to sexual relationships. *Psychological Medicine* 34: 1347–1357.https://doi.org/10.1017/S0033291704001941.

Nario-Redmond, Michelle R. 2020. *Ableism: The causes and consequence of disability prejudice.* New Jersey: Wiley Blackwell.

Pineau, Lois. 1989. Date rape: A feminist analysis. *Law and Philosophy* 8 (2): 217–243.

Ryan, Rebecca M. 1995. The sex right: A legal history of the marital rape exemption. *Law & Social Inquiry* 20 (4): 941–1001.

Schulhofer, Stephen J. 1998. *Unwanted sex: The culture of intimidation and the failure of law.* Cambridge, Mass.: Harvard University Press.

Shrage, Laurie, and Robert Scott Stewart. 2015. *Philosophizing about Sex.* Peterborough, ON: Broadview Press.

Stoljar, Natalie. 2000. Autonomy and the feminist intuition. In *Relational autonomy: Feminist perspectives on autonomy, agency, and the social self*, ed. Catriona Mackenzie and Natalie Stoljar, 94–111. New York: Oxford University Press.

Stoljar, Natalie. 2022. Feminist perspectives on autonomy. In *The Stanford encyclopedia of philosophy*, ed. Edward N. Zalta and Uri Nodelman. https://plato.stanford.edu/archives/win2022/entries/feminism-autonomy/. Accessed 6 August 2023.

Sullivan, William F., and John Heng. 2018. Supporting adults with intellectual and developmental disabilites to participate in health care decision making. *Canadian Family Physician* 64: S32–S36.

Taylor Gomez, Miriam. 2012. The S words: Sexuality, sensuality, sexual expression and people with intellectual disability. *Sexuality and Disability*, 30 (2): 237–245. https://doi.org/10.1007/s11195-011-9250-4.

Tomsa, Raluca, Smaranda Gutu, Daniel Cojocaru, Belén Gutiérrez-Bermejo, Noelia Flores, and Cristina Jenaro. 2021. Prevalence of sexual abuse in adults with intellectual disability: Systematic review and meta-analysis. *International Journal of Environmental Research and Public Health* 18. https://doi.org/10.3390/ijerph18041980.

Tuchinsky, Adam. 2016. "Women and her needs": Elizabeth Oakes Smith and the divorce question. *Journal of Women's History* 28 (1): 38–59.

UN (The United Nations). 2006. (CRPD) *Convention on the rights of persons with disabilities*. Treaty Series, 2515, 3, G.A. Res. 61/106, 76th plen. Mtg., U.N. Doc A/Res/61/106. https://social.desa.un.org/issues/disability/crpd/convention-on-the-rights-of-persons-with-disabilities-crpd. Accessed 9 April 2024.

World Health Organization. 2022. Redefining sexual health for benefits throughout life. https://www.who.int/news/item/11-02-2022-redefining-sexual-health-for-benefits-throughout-life. Accessed 9 April 2024.

# Chapter 9
# Inclusion of Adults with Intellectual and Developmental Disabilities in Low-Risk Research: Consent and Accommodations

**Janet A. Vogt**

**Abstract** The ethical inclusion of adults with intellectual and developmental disabilities (IDD) in research entails practices that will enhance their autonomy without exposing them to undue risk. Although inclusive research involving people with IDD responds to many shortcomings of more traditional research approaches, this chapter focuses instead on how best to conduct traditional research. The author focuses on the process of informed consent and argues for the importance of accommodations to support adults with IDD. The topics discussed include: the development of research ethics in response to the exploitation of vulnerable populations; the ethical tensions associated with research involving people with IDD (including the concept of vulnerability, decision-making capacity, and informed consent); the ethical argument for providing accommodations; and well-informed strategies for accommodating adults with IDD throughout the research consent process. The purpose of the chapter is to convince researchers who use traditional research paradigms of the ethical imperative to provide accommodations that reduce the context-specific vulnerability of adults with IDD during the consent process.

## 9.1 Introduction

The principle of distributive justice argues for fair distribution of the benefits and burdens of research (Yan and Munir 2004). People with intellectual and developmental disabilities (IDD)[1] encompass a broad spectrum of cognitive abilities (Hart

[1] The author uses person-first language to refer to people with IDD in deference to the early self-advocates who advocated the empowerment of their community. In personal encounters, the author advocates a person-centered approach, respecting the right of any individual to choose how they wish to be referred to.

J. A. Vogt (✉)
Surrey Place, Toronto, ON, Canada
e-mail: j.vogt@alumni.utoronto.ca; janet.vogt@surreyplace.ca



A. Bianchi and J. A. Vogt (eds.), *Intellectual Disabilities and Autism: Ethics and Practice*, The International Library of Bioethics 108,
https://doi.org/10.1007/978-3-031-61565-8_9

et al. 2020). Finding the ideal balance between respect for their autonomy and rights, and the inclination to protect them from exploitation and harm due to their perceived vulnerability can be challenging (McDonald and Kidney 2012). Inclusive research approaches have become more common in research involving people with IDD and seek to actively involve community members in all stages of the research (Northway et al. 2015). Despite the clear advantages of inclusive research, much research involving people with IDD still employs traditional (i.e., less inclusive) approaches.

This chapter focuses on participation of adults with IDD in relatively low-risk self-report research that asks participants to report what they are experiencing, what they think of it, and how they feel about it (McDonald et al. 2015), typically through interviews and focus groups. The ethical tensions associated with such research (including the concept of vulnerability, decision-making capacity, and informed consent); the ethical argument for provision of accommodations; and well-informed strategies for accommodating an adult with IDD during the consent process are discussed. Researchers who use traditional research paradigms should recognize the ethical imperative to provide accommodations that reduce the context-specific vulnerability of adults with IDD during the consent process.

## 9.2 Ethical Tensions Arising in Research Involving People with IDD

To understand the ethical challenges of recruiting people with IDD as research participants, one must consider the principles of respect for the person's autonomy, beneficence, and distributive justice. In addition, although the concept of vulnerability has been important to the development of research ethics, its overemphasis can obscure the importance of considering ways to support capabilities. This tension can be felt in the consent process. Focussing on accommodations to address the context-specific vulnerabilities of the person aligns with respect for their autonomy, and provides them with access to the benefits of research.

### *9.2.1 Research Ethics and the Inclusion of People with IDD*

The development of human research ethics was propelled by historical examples of exploitation of participants, especially in research involving "vulnerable" populations; for example, hepatitis vaccine research in children with IDD (Mastroianni and Kahn 2001). In response to such abuses of trust, the international community developed regulations for the ethical conduct of research involving human subjects. Consequently, research ethics was initially focused on protecting human subjects, especially those deemed as vulnerable, from unacceptable or exploitive levels of research-associated risk (Ibid.). For people with IDD, this affected a shift from being regarded as subjects who could be used in research without their consent, to being considered vulnerable, prompting significant limitations to their research participation (Northway 2014). Intended to protect their interests, this shift actually increased the risk of making them vulnerable to *exclusion* from research (McDonald and Kidney 2012).

The Nuremberg Code established the first set of principles for ethical "human experimentation" which included the concept of voluntary consent and the right to withdraw consent (Nuremberg Code 1947). The Declaration of Helsinki, in 1964, introduced surrogate (or substitute) consent and supported "the rights of vulnerable populations to express their autonomy in the form of assent or dissent" (Hart et al. 2020, 157). It also stipulated that "vulnerable" groups be included only if the research was necessary to promote the health of the population and *could not be carried out with capable individuals* (World Medical Organization 1996). The Belmont Report proposed three principles for resolution of ethical problems arising in human research: respect for persons, beneficence, and justice (USDHHS 1979). The principle of *respect for persons* creates two moral requirements: (1) to acknowledge autonomy; and (2) to protect those with diminished autonomy. The report acknowledged that an individual's capacity for autonomy will vary in different situations and should be periodically re-assessed; two critically important points. A person with IDD is often assumed to lack the capacity to make their own decisions, including whether to participate in research (Longmore 1995). However, some are able to: (1) understand simple research protocols (e.g., an interview protocol); and (2) appreciate the consequences of their participation (Gates and Waight 2007).

The principle of *beneficence* aims to maximize benefits while minimizing risks of harm. Research-associated risks can be physical, psychological, social, legal, or economic (Levine 1988). The most salient risk for people with IDD participating in self-report research, and the most cited concern, is that of unanticipated psychological (or emotional) upset (Dresser 1996). Service providers and family members of people with IDD have also expressed hesitance to encourage the person's participation in research that might produce opportunities for confusion or misunderstandings (McDonald et al. 2015).

The benefits of participation for this group include opportunities to develop capabilities for self-realization or flourishing (Beauchamp and Childress 2013). These opportunities include: personal growth (e.g., learning about research, increased capacity to give informed consent (with implications for decisions in other areas), and participation in something viewed as making a valuable contribution to science and society) (McDonald et al. 2015). Self-advocates with IDD who have participated in self-report research have described it as an opportunity to speak for themselves and make a contribution on behalf of their community (Ibid).

The principle of *distributive justice* establishes the importance of inclusiveness in research and fair distribution of benefits and burdens (USDHHS 1979). The Belmont report argued for scientific justification for the inclusion *or exclusion* of any group and noted that the principle of respect for persons, if expressed by an inclination to exclude (and thus protect) "vulnerable" people from research, conflicts with the principle of distributive justice (Yan and Munir 2004). From the late 1980s, the application of justice in human subjects research began to shift from protection to access, with more emphasis on the fair distribution of research benefits (Mastroianni and Kahn 2001).

### 9.2.2 The Concept of Vulnerability as Applied to People with IDD in the Research Context

Vulnerability can be understood as "balancing the assurance of two equally important human rights, safety and being heard" (Hart et al. 2020, 156). In research ethics, the term *vulnerable* has been used varyingly to refer to individuals who lack decision-making capacity, and marginalized groups that have been subject to exploitation; both meanings being potentially applicable to people with IDD. However, although it is necessary to recognize vulnerable individuals in the research context, characterizing an entire community as "vulnerable" is problematic. It ignores differences in capabilities amongst community members; assuming instead that they are all vulnerable. This, in turn, can reinforce stigmatization and result in unnecessary denial of access to research that may benefit the individual and their community (Hart et al. 2020). Furthermore, it ignores the fact that some, and perhaps all, community members may see themselves as empowered and autonomous (Gustafson and Brunger 2014). Self-advocates with IDD do not necessarily see themselves as "vulnerable"; but rather, as engaged and self-directing individuals. Characterizing an entire community as "vulnerable" presents a significant perceptual barrier to the understanding of vulnerability as deriving from the research context (Sieber 2008; Lotan and Ells 2010), as opposed to a "deficit" inherent in the individual. It ignores problems inherent in the structure and implementation of research that can infringe on an individual's autonomy (Du Bois et al. 2012).

### 9.2.3 Decision-Making Capacity, Informed Consent and Individuals with IDD

In the Tri-Council Policy Statement: Ethical Conduct for Research Involving Humans[2] (TCPS2), discussion relevant to people with IDD centers around those with diminished decision-making capacity (TCPS2 2022, 55–57). The TCPS2 defines decision-making capacity as the determination of whether the potential participant sufficiently understands the nature of the proposed research project and its associated risks, consequences, and potential benefits. It also recognizes that decision-making capacity, rather than being a static state, is subject to change. Furthermore, a person

[2] The TCPS2 is Canada's guiding document for the ethical conduct of research involving human subjects.

can have diminished capacity in some respects but still be capable of deciding whether or not to participate in certain types of research (TCPS2 2022, 55). Doody (2018) suggests that a person's ability to engage in everyday decision making is an indicator of their ability to freely provide consent.

Ho et al. (2018) adapted Palmer et al.'s (2005) three-item capacity assessment questionnaire for use with people with IDD. Their tool contains three questions: (1) What is the purpose of the study?; (2) What are the risks?; and (3) What are the benefits? The participant's responses are scored as follows: incapable (zero points); questionable (one point); and capable (two points). A total score of four to six, and the demonstrated understanding of the purpose, benefits, and risks of the study, is taken as an indication of the participant's ability to provide consent independently (Ho et al. 2018, 94). However, Hamilton et al. (2017) argue that a checklist of questions is insufficient to "set out the constructs of capacity" (236).

Informed consent requires that the potential participant understands the information provided. Gates and Waight (2007) found that some people with IDD understand and appreciate the consequences and implications of research participation. Other studies have found that although participants demonstrated a basic understanding of the research and the time commitment required, they neither clearly understood and appreciated the consequences nor their right to withdraw at any time (Arscott et al. 1998). The latter study found those with higher receptive language skills seemed to better understand the information provided.

#### 9.2.3.1 Informed Consent

Informed consent can present one of the biggest barriers to respectful inclusion of people with IDD in research (McDonald and Kidney 2012) with some arguing that *strict adherence* to informed consent may be unattainable for many (Yan and Munir 2004). Conversely, potential participants may not need to grasp some of the more opaque elements (McCarthy 1998). These elements include: *all* reasonably foreseeable risks and potential benefits; and that although you may withdraw your consent and participation at any time, the researchers may not be able to remove your data from the results if these have already been disseminated.

Some have proposed psychological testing to assess an individual's ability to provide informed consent (Sturman 2005). Others have argued that requiring such testing could be sufficiently burdensome as to deter some researchers from working with this population (Iacono and Murray 2003; Becker et al. 2004). One alternative is the involvement of a substitute decision maker (SDM); a court-appointed person with the authority to give or refuse consent on the person's behalf. However, not all adults with IDD have a SDM and establishment of such an arrangement can be expensive. Thus, requiring involvement of a SDM could unjustly exclude many adults with IDD from research that might pose only minimal risk and convey benefit. Furthermore, such proxy consent denies people with IDD the dignity of risk and self-determination (McVilly and Dalton 2006). If a SDM is involved, the person with IDD should be included to as great an extent as possible in the consent process (TCPS2 2022, 56) and their assent should be sought (TCPS2 2022, 57).

#### 9.2.3.2 Supported Decision Making

Another alternative, *supported decision making*, is promoted by the United Nations Convention on the Rights of Persons with Disabilities (CRPD) (UN 2006). Article twelve of the CRPD requires a shift from the reliance on SDMs to a system that provides the necessary supports for a person to make their own choices, without being discriminated against on the basis of disability (Bach and Kerzner 2010). The CRPD has been interpreted as requiring that the necessary mental threshold for a person to make their own decisions be significantly lowered (Craigie 2015).

Supported decision making involves someone known to the potential participant (e.g., a trusted friend, family member, support professional) who helps explain the materials during the consent process so they can make their own decisions (Sullivan and Heng 2018). It is important to consider *who* the person with IDD *wants to involve* in their decision making (McDonald et al. 2013). Supported decision making in the research context has the potential to facilitate emerging self-determination by helping to develop the person's capacity for informed consent over repeated occurrences (McVilly and Dalton 2006). Craigie (2015) stresses the importance of helping the person develop the capacities of self-respect and self-trust necessary to perceive themselves as an agent, for the purpose of decision making; especially where their agency has heretofore been unrecognized. Similarly, Lotan and Ells (2010) promote conditions that empower the individual by increasing their self-confidence in expressing their point of view. One caveat to engaging a non-appointed person as a support is that, whereas an appointed guardian or SDM has a statutory obligation to make decisions in the *best interests* of the individual, depending on the jurisdiction, there may be no such obligation on the part of a non-appointed person acting in a supporting role (Craigie 2015).

#### 9.2.3.3 Reconsidering Some of the TCPS2 Consent Requirements

One challenging aspect of informed consent is the sheer amount of information the potential participant is required to comprehend. Article 3.2 of the TCPS2 lists *twelve* elements that researchers are generally required to disclose to potential participants (TCPS2 2022, 36–37); the inclusion of which can be overwhelming, perhaps especially for someone with IDD. Assuming the risk of adverse events is low, the inclusion of certain elements (e.g., disclosure of conflicts of interest; contact information for the principal investigator (PI) and research ethics board (REB) chair to answer questions regarding the study's scientific merit or ethical issues arising; a detailed description of how privacy will be protected and anticipated uses and handling of data) may not be necessary in order to protect the interests of the person. Insistence on their inclusion, and a demonstrated understanding of them, sets a very high threshold for capacity and almost certainly guarantees the inability of many people with IDD to meet these criteria for decision-making capacity. Simple changes could be made, such as a shortened project title and just the name of the PI (without titles and institutional names). Rather than identifying the PI and REB chair for questions, contact information for an institutional representative could be provided; someone with no investment in the specific research protocol who could act as a facilitator for participants (TCPS 2022, 38). For the purpose of our discussion, this person should have experience supporting people with IDD.

## 9.3 The Ethical Argument for Accommodation During Informed Consent

Yan and Munir (2004) assert that REBs share an obligation to help the researcher be as inclusive as possible in the informed consent process. Failure to do so constitutes a constraint on the freedom of the person with IDD (Morgan et al. 2014). When REB members and researchers in the field of IDD were asked how to conduct ethically strong research, their responses focused on manipulating the environment to facilitate inclusion and an informed consent process that reflects a relational ethic (McDonald and Patka 2012). According to Fisher (2003, 28), "reframing informed consent as a goodness-of-fit between person and consent context shifts responsibility for consent capacity away from the participant's mental status to examination and enhancement of those aspects of the consent setting that can reduce consent vulnerability." Lotan and Ells (2010) describe this as creating the right environmental conditions to foster autonomy. Thus, the accommodation is framed as an ethical imperative to support the individual's autonomy and their right to participate (McDonald and Patka 2012). Researchers have a morally binding obligation due to the "specific susceptibility to research risks" and the researcher's specific ability "to help alleviate these risks" (Fisher 2003, 29). This may entail reconceptualizing both the standard for competency to give informed consent and the role of others in consent decisions (Fisher (2003). Furthermore, when a group of adults with IDD who had experience as research participants were interviewed, most asserted that accommodations are critical to facilitating their equal access to research participation (McDonald 2012).

## 9.4 Strategies to Support the Person with IDD During the Consent Process[3]

One of the key concerns of adult research participants with IDD is the nature and tone of their interactions with researchers (McDonald 2012; McDonald et al. 2015). In particular, they want to feel valued and respected. Many researchers and practitioners have contributed to the discussion of well-informed strategies when seeking consent from people with IDD in research and healthcare settings. Drawing on these varied sources, the following sections provide strategies to foster and support the process of informed and voluntary consent for adults with IDD.

### *9.4.1 Adapting Consent Materials*

Employ research advisors with IDD to help design your tools, including consent forms and information letters (McDonald 2012). Create a lay language guide for the consent process which includes visuals and multiple everyday examples to describe, and enhance understanding of, the proposed research activities (McDonald et al. 2015). Use an appropriate reading level for any paperwork shared with the potential participant (Dalton and McVilly 2004). The writing should be short and concrete, avoiding long words (three or more syllables), long sentences (twenty or more words), using an active voice, and limiting ideas to one per sentence (Havlak 1998).

### *9.4.2 Approaching the Person*

Before the consent meeting, send an information package, including the adapted consent form, to the prospective participant and, if appropriate, their identified support person. Find out in advance what supports the person is likely to need (McDonald 2012). For example, these could be related to literacy level, visual or hearing impairment, level of receptive and expressive language, concern about anxiety, ability to manage change and stress, or sensory issues. Phone the person to remind them of the appointment. Ask if they have identified someone to support them during the consent process (McDonald et al. 2015).

[3] The advice and strategies shared in the following sections are appropriate for adults with mild or moderate IDD. Mild IDD assumes an IQ between fifty-five and seventy (± five) and an age equivalence of nine to twelve years. Someone with mild IDD typically uses simple to complex sentence types to communicate events, ideas, and opinions but may struggle to express feelings using words and/or to describe events in sequence. Moderate IDD assumes an IQ between forty and fifty (± five) and an age equivalence of six to nine years. Someone with moderate IDD typically uses phrases and simple sentences and can express preference, emotions, interests, and experiences. They can ask and answer questions involving concrete information (Edwards and Temple 2019).

### 9.4.3 Establishing a Comfortable Environment

The provision of individualized accommodations and a comfortable setting can contribute to feelings of safety (McDonald et al. 2015). Allow the person to choose the location if possible (Kidney and McDonald 2014) and reimburse their travel costs (McDonald 2012). Schedule the meeting at a preferred time for them (McNeil et al. 2019). Clarify your role early on (Doody 2018) and take time to build rapport with the person; getting to know them a bit to help you better understand their strengths and needs (McDonald 2012). Before beginning, ask the individual what would make them comfortable (McDonald et al. 2015). Pay attention to body language and try to identify, and modify, aspects of the consent setting that may be stress-provoking (Fisher 2003). Make eye contact (McDonald 2012) unless this makes the person uncomfortable. Allow ample time for the meeting and consider offering additional meetings to allow the person more time to process the information (Lotan and Ells 2010) or to make a decision about whether they want to participate (NIH 2009).

### 9.4.4 Facilitating Communication

The informed consent process must accommodate the communication and cognitive needs of the person with IDD (TCPS2 2022, 55). The onus is on research staff to modify their methods of communication and interaction in order to support the participant (Lotan and Ells 2010). Ho et al. (2018, 98) found that "… adapted communication strategies were the most important means of building rapport and subsequent engagement with participants." Communication difficulties should not be confused with incapacity (Government of Ontario 2005; Fisher 2003). There are many ways to support communication and the most relevant approaches will depend on the needs of the individual (Carlson 2013).

If the person with IDD requires highly individualized communication assistance or is more comfortable communicating with someone they trust, consider allowing that trusted person to explain the research, assisting them if necessary (Fisher 2003). Otherwise, explain the research directly to the person with IDD (McDonald 2012). When reviewing the consent form, you must be willing to invest time in effective means of communicating information (McDonald and Patka 2012). According to one self-advocate, you must be prepared to allow enough time for the person to integrate the information and to formulate a response (Elbard 2015). Review information a number of times (Fisher 2003) and ask the person to describe key elements in their own words (McDonald 2012). If their response is unclear, explain and clarify the question (Ho et al. 2018). Invite the person, *throughout the consent process*, to ask questions (McDonald et al. 2015).

Be attentive to deficits in the person's experience with both research and decision making and be prepared to educate them about research procedures and their rights as a research participant (Fisher 2003). The concepts of *public* and *private* information can help in explaining privacy and confidentiality concerns. Adapt your pace accordingly, pausing frequently to avoid overloading the person with information (McNeil et al. 2019). Use clear language and break information into discrete chunks (Kidney and McDonald 2014). Consider using multimodal presentation (e.g., pictures, videos, stories or gestures) to support the written content (Lotan and Ells 2010; McDonald et al. 2015). If unsure, offer visual aids but *respect the person's preference*; not all people with IDD need *or want* visual supports (Kidney and McDonald 2014). To accommodate short attention spans, keep questions brief and provide breaks (Ibid). Reiterate that participation is voluntary; assure the person that should they decide not to participate, this will not affect their access to services and future research opportunities (Fisher 2003). Stress that even if they decide they want to participate, they can change their mind at any point during the research (Doody 2018).

Above all, keep two things in mind. First, each person is unique; as is their capacity to provide informed and voluntary consent. Second, consent is an ongoing process. Check in with participants at various stages of the research to ask if they're still interested in continuing (Doody 2018). Consider allowing time between the initial consent decision and the research activities for the person to reflect on their decision (Kidney and McDonald 2014).

For researchers who have not yet embraced inclusive research models, adoption of the strategies presented in this section should help accommodate the needs of adults with IDD during the consent process for traditional research. The resources cited also provide advice regarding other aspects of research. One might start by adopting some of these strategies, being mindful of them throughout the research process, and eventually go on to co-design more inclusive types of research in collaboration with adults with IDD.

## 9.5 Other Key Challenges in Obtaining Informed Consent

This chapter did not discuss the related requirements of *voluntariness* and *undue influence.* Commonly utilized research practices call these into question, especially the role of support workers in decision making and the provision of compensation or incentives to participants (Becker et al. 2004; Gates and Waight 2007; McDonald et al. 2015). A lack of consensus exists amongst researchers and REBs regarding key issues such as: what constitutes undue influence, use of incentives, capacity to consent, and appropriate involvement of SDMs, amongst others (McDonald and Kidney 2012). These perennial challenges in research with adults with IDD certainly deserve consideration but exceed the scope of this chapter.

## 9.6 Conclusion

Many ethical dilemmas encountered when conducting research involving adults with IDD trace back to the enduring tension between protection, on the one hand, and self-determination, autonomy, and dignity, on the other (Fisher 2003). Several approaches can facilitate respectful inclusion of adults with IDD in self-report research. However, there is a need for systematic, empirical research to identify best ethical practices in obtaining informed and voluntary consent from people with IDD and to better understand actual, as opposed to perceived, risks (McDonald and Kidney 2012). Studies to address this should involve adults with IDD, trusted supporters of their choice, researchers, SDMs, bioethicists, and REB members, to probe the ethical tension between respect for autonomy and protection from harm and produce ethically-informed guidelines based on current well-informed practices. Such guidelines should support all key stakeholders to ensure that adults with IDD are included in research in meaningful, respectful, safe, and empowering ways.

## References

Arscott, Katy, Dave Dagnan, and Biza Stenfert Kroese. 1998. Consent to psychological research by people with an intellectual disability. *Journal of Applied Research in Intellectual Disabilities* 11 (1): 77–83. https://doi.org/10.1111/j.1468-3148.1998.tb00035.x.

Bach, Michael, and Lana Kerzner. 2010. A new paradigm for protecting autonomy and the right to legal capacity. *Law Commission of Ontario*, 1–196.

Beauchamp, Tom L., and James F. Childress. 2013. Justice. In *Principles of biomedical ethics*, 7th ed. New York: Oxford UP.

Becker, Heather, Greg Roberts, Janet Morrison, Julie Silver, and Sarah Taub. 2004. Recruiting people with disabilities as research participants: Challenges and strategies to address them. *Mental Retardation* 42 (6): 471–475.

Carlson, Licia. 2013. Research ethics and intellectual disability: Broadening the debates. *Yale Journal of Biology and Medicine* 86: 303–314.

Craigie, Jillian. 2015. A fine balance: Reconsidering patient autonomy in light of the UN convention on the rights of persons with disabilities. *Bioethics* 29 (6): 398–405. https://doi.org/10.1111/bioe.12133.

Dalton, Arthur J., and Keith R. McVilly. 2004. Ethics guidelines for international, multicenter research involving people with intellectual disabilities. *Journal of Policy and Practice in Intellectual Disabilities* 1 (2): 57–70.

Doody, Owen. 2018. Ethical challenges in intellectual disability research. *Mathews Journal of Nursing and Health Care* 1 (1): 5.

Dresser, Rebecca. 1996. Mentally disabled research subjects. *Journal of the American Medical Association* 276 (1): 67–72.

Du Bois, James M., Laura Beskow, Jean Campbell, Karen Dugosh, David Festinger, Sarah Hartz, Rosalina James, and Charles Lidz. 2012. Restoring balance: A consensus statement on the protection of vulnerable research participants. *American Journal of Public Health* 102 (12): 2220–2225. https://doi.org/10.2105/AJPH.2012.300757.

Edwards, Bruce, and Valerie Temple. 2019. Adaptive functioning and communication in adults with intellectual and developmental disabilities: Fact Sheet. Developmental Disabilities Primary Care

Program of Surrey Place, Toronto. https://ddprimarycare.surreyplace.ca/tools-2/general-health/adaptive-functioning/. Accessed 21 May 2024.

Elbard, Kareem. 2015. Inside out not outside in—A change of attitude for all: A self-advocate's vision for those with intellectual and developmental disabilities. *Journal of Religion & Society* Supplement 12: 8–12. https://cdr.creighton.edu/server/api/core/bitstreams/5636d3b3-04e4-4764-bb23-10c2fdd38cf2/content. Accessed 21 May 2024.

Fisher, Celia B. 2003. Goodness-of-fit ethic for informed consent to research involving adults with mental retardation and developmental disabilities. *Mental Retardation and Developmental Disabilities Research Reviews* 9 (1): 27–31.

Gates, Bob, and Mary Waight. 2007. Reflections on conducting focus groups with people with learning disabilities: Theoretical and practical issues. *Journal of Research in Nursing* 12 (2): 111–126. https://doi.org/10.1177/1744987106075617.

Government of Ontario, Ministry of the Attorney General, Capacity Assessment Office. 2005. Guidelines for conducting assessments of capacity. https://www.publications.gov.on.ca/guidelines-for-conducting-assessments-of-capacity. Accessed 21 May 2024.

Gustafson, Diana L., and Fern Brunger. 2014. Ethics, "vulnerability", and feminist participatory action research with a disability community. *Qualitative Health Research* 24 (7): 997–1005. https://doi.org/10.1177/1049732314538122.

Hamilton, Jahnese, Barry Ingham, Iain McKinnon, Jeremy R. Parr, Louise Yuen Chong Tam, and Ann Le Couteur. 2017. Mental capacity to consent to research? Experiences of consenting adults with intellectual disabilities and/or autism to research. *British Journal of Learning Disabilities* 45 (4): 230–237. https://doi.org/10.1111/bld.12198.

Hart, Sarah M., Mikayla Pascucci, Sheetal Sood, and Emily M. Barrett. 2020. Value, vulnerability and voice: An integrative review on research assent. *British Journal of Learning Disabilities* 48 (2): 154–161. https://doi.org/10.1111/bld.12309.

Havlak, Russ. 1998. *Consent for CDC research: A reference for developing consent forms and oral scripts*. Centers for Disease Control and Prevention. http://stacks.cdc.gov/view/cdc/12420 Accessed 21 May 2024.

Ho, Portia, Jenny Downs, Caroline Bulsara, Shane Patman, and Anne Marie Hill. 2018. Addressing challenges in gaining informed consent for a research study investigating falls in people with intellectual disability. *British Journal of Learning Disabilities* 46 (2): 92–100. https://doi.org/10.1111/bld.12217.

Iacono, Teresa, and Vanessa Murray. 2003. Issues of informed consent in conducting medical research involving people with intellectual disability. *Journal of Applied Research in Intellectual Disabilities* 16 (1): 41–51. https://doi.org/10.1046/j.1468-3148.2003.00141.x.

Kidney, Colleen A., and Katherine E. McDonald. 2014. A toolkit for accessible and respectful engagement in research. *Disability and Society* 29 (7): 1013–1030. https://doi.org/10.1080/09687599.2014.902357.

Levine, Robert J. 1988. Balance of harms and benefits. In *Ethics and regulation of clinical research*, 2nd ed., 37–65. New Haven and London: Yale UP.

Longmore, Paul K. 1995. Medical decision making and people with disabilities: A clash of cultures. *Journal of Law, Medicine & Ethics* 23 (1): 82–87.

Lotan, Gurit, and Carolyn Ells. 2010. Adults with intellectual and developmental disabilities and participation in decision making: Ethical considerations for professional-client practice. *Intellectual and Developmental Disabilities* 48 (2): 112–125. https://doi.org/10.1352/1934-9556-48.2.112.

Mastroianni, Anna, and Jeffrey Kahn. 2001. Swinging on the pendulum: Shifting views of justice in human subjects research. *Hastings Center Report* 31 (3): 21–28.

McCarthy, Michelle. 1998. Interviewing people with learning disabilities about sensitive topics: A discussion of ethical issues. *British Journal of Learning Disabilities* 26 (4): 140–145. https://doi.org/10.1111/j.1468-3156.1998.tb00070.x.

McDonald, Katherine E. 2012. We want respect: Adults with intellectual and developmental disabilities address respect in research. *American Journal on Intellectual and Developmental Disabilities* 117 (4): 263–274. https://doi.org/10.1352/1944-7558-117.4.263.

McDonald, Katherine, and Mazna Patka. 2012. "There is no black or white": Scientific community views on ethics in intellectual and developmental disability research. *Journal of Policy and Practice in Intellectual Disabilities* 9 (3): 206–214. https://doi.org/10.1111/j.1741-1130.2012.00348.x.

McDonald, Katherine E., Colleen A. Kidney, and Mazna Patka. 2013. 'You need to let your voice be heard': Research participants' views on research. *Journal of Intellectual Disability Research* 57 (3): 216-225. https://onlinelibrary.wiley.com/doi/10.1111/j.1365-2788.2011.01527.x.

McDonald, Katherine E., and Colleen A. Kidney. 2012. What is right? Ethics in intellectual disabilities research. *Journal of Policy and Practice in Intellectual Disabilities* 9 (1): 27–39. https://doi.org/10.1111/j.1741-1130.2011.00319.x.

McDonald, Katherine E., Nicole M. Schwartz, Colleen M. Gibbons, and Robert S. Olick. 2015. "You can't be cold and scientific": Community views on ethical issues in intellectual disability research. *Journal of Empirical Research on Human Research Ethics* 10 (2): 196–208. https://doi.org/10.1177/1556264615575512.

McNeil, Karen, Elizabeth Bell, Kerry Boyd, John Heng, William F. Sullivan, and Janet Vogt. 2019. *Communicate CARE: Guidance for person-centred care of adults with intellectual and developmental disabilities*. Toronto: Developmental Disabilities Primary Care Program of Surrey Place. https://ddprimarycare.surreyplace.ca/tools-2/general-health/communicating-effectively/. Accessed 21 May 2024.

McVilly, Keith R., and Arthur J. Dalton. 2006. Commentary on Iacono (2006): "Ethical challenges and complexities of including people with intellectual disability as participants in research." *Journal of Intellectual and Developmental Disability* 31 (3): 186–188. https://doi.org/10.1080/13668250600876426.

Morgan, Michelle F., Monica Cuskelly, and Karen B. Moni. 2014. Unanticipated ethical issues in a participatory research project with individuals with intellectual disability. *Disability and Society* 29 (8): 1305–1318. https://doi.org/10.1080/09687599.2014.934440.

NIH (National Institutes of Health). 2009. Research involving individuals with questionable capacity to consent: Points to consider. https://grants.nih.gov/grants/policy/questionablecapacity.htm#_ftn26. Accessed 21 May 2024.

Northway, Ruth. 2014. To include or not to include? That is the ethical question. *Journal of Intellectual Disabilities* 18 (3): 209–210. https://doi.org/10.1177/1744629514543863.

Northway, Ruth, Joyce Howarth, and Lynne Evans. 2015. Participatory research, people with intellectual disabilities and ethical approval: Making reasonable adjustments to enable participation. *Journal of Clinical Nursing* 24 (3–4): 573–581. https://doi.org/10.1111/jocn.12702.

Nuremberg Code. 1947. *Trials of war criminals before the Nuremberg Military Tribunals under Control Council Law* 10 (2): 181–182. Washington: US Government Printing Office. https://history.nih.gov/display/history/Nuremberg+Code. Accessed 21 May 2024.

Palmer, Barton W., Laura B. Dunn, Paul S. Appelbaum, Sunder Mudaliar, Leon Thal, Robert Henry, Shahrokh Golshan, and Dilip V. Jeste. 2005. Assessment of capacity to consent to research among older persons with schizophrenia, alzheimer disease, or diabetes mellitus: Comparison of a 3-item questionnaire with a comprehensive standardized capacity instrument. *Archives of General Psychiatry* 62 (7): 726–733.

Sieber, Joan E. 2008. Protecting the vulnerable: Who are they? *Journal of Empirical Research on Human Research Ethics* 3 (1): 1–2. https://doi.org/10.1525/jer.2008.3.1.1.

Sturman, Edward D. 2005. The capacity to consent to treatment and research: A review of standardized assessment tools. *Clinical Psychology Review* 25 (7): 954–974. https://doi.org/10.1016/j.cpr.2005.04.010.

Sullivan, William F., and John Heng. 2018. Supporting adults with intellectual and developmental disabilities to participate in health care decision making. *Canadian Family Physician* 64 (Suppl 2): S32–S36.

TCPS2: Canadian Institutes of Health Research, Natural Sciences and Engineering Research Council of Canada, and Social Sciences and Humanities Research Council of Canada. 2022. *Tri-council policy statement: Ethical conduct for research involving humans*. https://ethics.gc.ca/eng/documents/tcps2-2022-en.pdf. Accessed 21 May 2024.

UN (The United Nations). 2006. (CRPD) *Convention on the rights of persons with disabilities*. Treaty Series, 2515, 3, G.A. Res. 61/106, 76th plen. Mtg., U.N. Doc A/Res/61/106. https://social.desa.un.org/issues/disability/crpd/convention-on-the-rights-of-persons-with-disabilities-crpd. Accessed 21 May 2024.

USDHHS (US Department of Health and Human Services). 1979. The Belmont Report. *Office for Human Research Protections (OHRP)*. https://www.hhs.gov/ohrp/regulations-and-policy/belmont-report/. Accessed 21 May 2024.

World Medical Organization. 1996. Declaration of Helsinki. *British Medical Journal* 313 (7070): 1448–1449.

Yan, Eric G., and Kerim M. Munir. 2004. Regulatory and ethical principles in research involving children and individuals with developmental disabilities. *Ethics and Behavior* 14 (1): 31–49. https://doi.org/10.1207/s15327019eb1401_3.

# Part III
# Inclusion and Empowerment: Society, Systems, and Ethics

# Chapter 10
# Beyond Words: Access to Justice for Persons Labelled with Intellectual Disabilities and Clear Language Judgement Making

**Tess Sheldon**

**Abstract** This chapter is about the distance between law and justice, particularly for people labelled with intellectual disabilities. It calls for a disability-inclusive approach to understanding access to justice that shifts from a procedural justice paradigm to one that considers substantive conceptions of disability justice. The chapter also examines the critical perspectives of disability rights and disability justice, highlighting the gaps between idealized notions of rights and everyday experiences of rights violations. The chapter proposes that procedural reforms, alone, are insufficient and distract from the need for transformative change. In particular, the chapter discusses clear language judgments as a strategy to improve the comprehensibility of legal information. While clear language judgments have been praised, there are concerns that they may oversimplify complex legal concepts and fail to address underlying biases and injustices. This chapter warns against uncritical adoption and calls for broader transformative change toward collective liberation and elimination of disability injustice.

## 10.1 Introduction

There is a vast distance between law and justice, especially for people with disabilities. It is said that access to justice is in "crisis" (Stefanovich 2021). In Canada, access to justice has been described as the "greatest challenge to the rule of law in Canada today" (*Hryniak v. Mauldin*).[1] Legal documents are meaningless unless they

[1] *Hryniak v. Mauldin*, [2014] 1 S.C.R. 87, at ¶ 1.

T. Sheldon (✉)
Faculty of Law, University of Windsor, Windsor, Canada
e-mail: tess.sheldon@uwindsor.ca



A. Bianchi and J. A. Vogt (eds.), *Intellectual Disabilities and Autism: Ethics and Practice*, The International Library of Bioethics 108,
https://doi.org/10.1007/978-3-031-61565-8_10

are understood, and as found by the Supreme Court of Canada, "A person who does not understand his or her right cannot be expected to assert it" (*R. v. Evans*, at 871d).[2]

This chapter examines barriers mounted against disabled people in accessing that justice.[3] While access to justice is a human right, disabled individuals[4] often encounter barriers when navigating the legal system, impeding their ability to exercise their rights and seek redress.

The paper proposes a shift from a procedural justice paradigm to one that also accounts for substantive conceptions of disability justice. An impoverished version of access to justice focuses on the process of making a claim. Such an approach presumes that a fair process leads to just outcomes.

This chapter is inspired by a series of "family policing" decisions from the United Kingdom, written in clear language (Baobeid 2022). These decisions found in favour of the removal of children from their disabled parents. For clear language decisions to hold "genuine ethical import" (van Domselaar 2022, 111), they must address and confront systemic ableist and sanist barriers about, for example, disabled persons' capacity to parent.[5] Instead, clear language decisions divert attention from the violence embedded in such decisions.

This paper offers critical accounts of *accessibility* and *access to justice*, drawing on Mor's (2017) foundational scholarship. Accessibility for people with disabilities requires more than just formal or technical inclusion. Likewise, access to justice should not be limited to access to courts. While disability rights focus on equal treatment within existing systems, disability justice rejects a narrow focus on individual accommodations and emphasizes the importance of transformative change toward collective liberation.

This chapter has four parts. First, it draws on Critical Disability Studies (CDS) to illustrate the gap between idealized notions of rights and everyday experiences of violations. The second part highlights the evolution of the concept of *access to justice* beyond court-centric and lawyer-centric approaches. The third section introduces clear language strategies to improve the accessibility of legal information and offers examples of court decisions written in clear language. The final section warns of uncritical adoption of these strategies, where it oversimplifies legal concepts, masks power imbalances, and fails to address systemic injustices.

---

[2] *R. v. Evans*, [1991] 1 S.C.R. 869, at 871d.

[3] I am grateful to Joyce Dason, a law student at University of Windsor's Faculty of Law, for her invaluable editorial assistance and excellent research support.

[4] This chapter uses various terms including "disabled persons", "persons with disabilities" and "persons labelled with intellectual disabilities". These terms are politically and academically contested and evolving. "Person-first" language (i.e., persons with disabilities) is used sometimes and other times "identity-first" language (i.e., an autistic person) is adopted. This chapter consistently avoids over-emphasis on diagnostic labels, including from the *Diagnostic and Statistical Manual* of the American Psychiatric Association.

[5] For more about parenting and people with IDD see Chaps. 7, 19, 22, and 24 of this volume.

## 10.2 Critical Perspectives on Access to Justice: Distinguishing Disability Rights from Disability Justice

This section emphasizes critical perspectives on access to justice for disabled people. Critical Disability Studies offers a lens through which to examine the social, cultural, and political dimensions of disability. It understands disability, at least in part, as being constructed by barriers.

Ableism is evident in attitudes, practices, and policies that devalue and marginalize disabled persons, perpetuating stereotypes, barriers, and unequal treatment (Ontario Human Rights Commission 2016).[6] Sanism is a prejudice that portrays people with psychiatric disabilities as inferior, dangerous, or unworthy of respect and dignity (Perlin 1999, 2013). Sanism can lead to the denial of rights and opportunities (Poole et al. 2012) and is reflected in excessive reliance on institutionalization and forced treatment (Chamberlin 1990).

Ableism and sanism are persistent, at least in part because of the gap between the law on the books and the law in action (Bay 2006). CDS focuses on the "rights" of disabled people as well as how "rights" are expressed in an everyday sense. There are massive "gaps between idealized and abstract formulations of 'rights', and the everyday experience of people whose rights are violated" (Montenegro and Cornish 2015, 2). The enforcement of even carefully crafted legislation has a more profound effect than what is set out in its text, particularly for those with disabilities (Chouinard 2001).

By concentrating on its structural and ideological influences, CDS aims to get at the contexts and underpinnings of disability inequality, with an eye on meaningful change. On one hand, disability rights' primary focus is on achieving equal rights and opportunities for disabled persons *within* existing systems and structures (e.g., education).

On the other hand, disability justice works "beyond legal compliance" (Bennett and Hannah 2022, 328). While disability rights may focus on equal treatment within *existing* structures, disability justice refuses to narrowly focus on individual rights and accommodations. Instead, it recognizes that disability is not solely an individual issue but deeply intertwined with other forms of oppression (e.g., racism, sexism). Disability justice challenges interconnected systems of oppression while focusing on the experiences of disabled people, particularly from equity-denied communities (Piepzna-Samarasinha 2018). It aims to address the root causes of ableism and sanism, and the need for transformative change through collective liberation (Mingus 2018).

---

[6] For more information about ableism, see Chaps. 12, 19, and 26 of this volume.

## 10.3 Defining Access to Justice: The Risk of Rights Without Remedies

Access to justice is a "moving frontier" (Galanter 2005, 155). It has been said to mean "many things" (*BC v. CCD* 2022, 35),[7] and its definition has shifted as the meaning of "access" and "justice" have been "reinterpreted and redefined".[8]

Ensuring meaningful participation in the legal system has both substantive and procedural elements. Procedural access to justice refers to the availability of fair and effective legal procedures (e.g., legal representation, right to a fair trial).

Substantive access to justice, on the other hand, refers to the ability of individuals to exercise their legal rights effectively and obtain just outcomes. It emphasizes the realization of justice rather than merely the existence of legal remedies. Substantive justice is subjective and means different things to different people, including "having their "day in court," regardless of the risks of ultimately losing their case in the end" (Del Gobbo 2019, 289).

Procedural measures are "often a good thing in terms of making the legal system more efficient, user-friendly, and, overall, accessible" (Farrow 2014, 16). It is another question, though, whether "they are ultimately improving access to justice, or simply access to the tools, or processes, of law" (Farrow 2014, 16). Procedural reforms may overlook what is most important in our system of justice: *power* (Sealy-Harrington 2021).

The concept of access to justice initially emphasized the availability and affordability of services by legal *professionals*. Skeptics, though, balked at court-centric and lawyer-centric approaches that presumed the availability of justice through legal institutions (Cotterrell 2019).

In a recent publication, Mor (2017) classified three conceptions of access to justice. First, *access to courts* is the narrowest formulation. Second, *access to law* refers to the structural, spatial, mental, and other process-related barriers affecting one's ability to utilize the law. Finally, *access to outcomes* concerns biases in the structure and content of legal rules and judicial decisions, which are often infected by ableist and sanist biases (Sjoquist 2022).

Access to justice's substantive and procedural dimensions are interdependent. Even if a person receives a fair outcome from an objective standpoint, if the process does not enable them "to understand or believe that they got a fair shake, they aren't going to believe justice was served" (Glaves n.d.). While just outcomes are essential, individuals must feel they have experienced a fair process.

---

[7] *British Columbia (Attorney General) v. Council of Canadians with Disabilities*, [2022] S.C.C. 27, at ¶ 35.

[8] Vipond, Evan and Pierre Cloutier de Repentigny. Forthcoming. Searching for Justice: Moving Towards a Trans Inclusive Model of Access to Justice in Canada. *Dalhousie Law Journal* 47(1).

## 10.4 Procedural and Substance Barriers for Accessing Justice

This section enumerates the barriers that often prevent disabled litigants or prospective litigants from ensuring that their legal rights are respected. The Supreme Court in *British Columbia (Attorney General) v. Council of Canadians with Disabilities* (2022, ¶ 116) recently recognized the "significant personal and institutional hurdles" faced by individuals with mental health disabilities in pursuing and sustaining constitutional litigation.

Indeed, people with disabilities face complex barriers to commence and sustain legal claims. According to Lepofsky (2022, 152), people with disabilities "too often encounter recurring disability barriers when seeking legal services and/or when involved with courts, administrative tribunals, or regulatory regimes." To him, the legal system has operated on "the implicit, invisible, and pervasive premise that those they serve have no disabilities" (Ibid.).

The experience of poverty also impedes access to courts, which is relevant to many persons with disabilities (Ontario Human Rights Commission 2014, 2015). As noted by Hughes (2002, 580), poverty "is and always has been the key factor in the modern constitution of disability." Disabled litigants rarely have access to upstream resources to support their ability to file or continue a lawsuit, or to reach a just result (Lepofsky 1995).

The economic marginalization of people with disabilities is aggravated by physical, environmental, and systemic barriers they face in accessing and utilizing the justice system. Inaccessible courtrooms, police practices, and legal documents create barriers that limit effective participation in legal proceedings. According to one study from the UK, seventy-five percent of autistic respondents did not receive any accommodations during police questioning (Slavny-Cross et al. 2022).

These barriers result from power dynamics entrenched in legal processes, as well as unequal access to "the opportunity to shape the structure and content of legal rules" (Mor 2017, 634). Negative stereotypes, misconceptions, and paternalistic attitudes influence how legal professionals perceive and treat disabled litigants. For example, legal capacity and guardianship laws are a fundamental legal barrier facing disabled people (Law Commission of Ontario 2017). Justice professionals may presume the incapacity of individuals with disabilities, interfering with their credibility as witnesses or testimonial weight. Justice professionals may defer to others to speak for disabled persons. This practice disregards the rights of individuals with disabilities to speak for themselves and make their own decisions, which is a technique of disability exclusion.

In addition to legal barriers experienced by many disabled people, autistic litigants may fear disclosure of disability-related needs (Communications Disability Access Canada 2023; Slavny-Cross et al. 2022). They report "unfavorable experiences" about their interactions with justice professionals (Salerno and Schuller 2019). Autistic litigants with sensory sensitivities can also experience courtrooms and legal settings as overwhelming. According to Foster and Young (2021), if an autistic person becomes involved in the justice system, they face worse outcomes due to their social communication style (e.g., not directly looking at the judge) which may be perceived as non-compliant (Ibid.).[9]

People with communication disabilities, including autism, may experience particular barriers in legal settings, including not having ways to communicate with justice professionals using methods other than speech. Justice professionals are not trained on their responsibility to provide communication supports and may presume incapacity without providing accommodations. Established in 2021, Communication Access to Justice trains, supports, and publishes a national list of Communication Intermediaries, who assist people with communication, speech, and language disabilities to understand questions and to communicate answers effectively with justice professionals based on their communication needs (Communication Access to Justice n.d.).

These barriers are intersectional. Disabled persons face barriers to access justice, particularly when marginalized identities intersect. Sarrett and Ucar (2021) found that racialized persons labelled with intellectual disabilities have more interactions with the criminal system. Women labelled with intellectual disabilities face particular barriers reporting violence (Women Enabled International 2023). The "interplay of several social categories (identities) … produces multiple disadvantages" in legal settings (Lundberg and Simonsen 2017, 17).

## 10.5 A Strategy: Provision of Legal Information in Clear Language

Legal language is alienating. It is constituted by and constitutes "elite or professionalised power" (van Domselaar 2022, 110). Legal information is often written in a way that makes it impossible to know its content, given "excessively involuted sentence structure, the use of language other than in its plain or ordinary sense, and recursive drafting techniques that require the reader to interpret one section by reference to another" (Boyd 2015b). The complexity of legal language "has a significant potential impact on the extent to which Canada is a nation truly governed by the rule of law" (Ibid.). This complexity is a "serious barrier to justice" (Boyd 2015a, 2) since the law "must be understandable to those subject to it" (van Domselaar 2022, 96).

Clear language, also known as plain language (Canadian Centre for Occupational Health and Safety 2023), is one strategy to ensure accessible communication. Proponents extol its social benefits, including enabling people to make more informed decisions (Siegel 2010). Clear legal language enhances transparency, facilitates effective communication, and promotes trust in the legal system.

The Plain Language Movement has gained mainstream acceptance in common-law countries (Zödi 2019). For instance, the *Plain Writing Act* (2010) requires federal

[9] For more on ways that different communication styles and experiences of distress may be perceived, see Chaps. 12 and 27 of this volume.

agencies in the United States to use clear communication that the public can understand. In the European Union, the Movement has been key to the codification of the requirement for plain language in consumer contracts (Inclusion Europe 2017). In Canada, Pierre Poilievre, the leader of the Conservative Party of Canada at the time of writing this chapter, proposed a federal *Plain Language Act*[10] (Taylor 2022). Additionally, the Supreme Court of Canada prepares *plain-language* summaries of its judgments to explain "the context and judicial reasoning for its decisions" (Supreme Court of Canada 2023, 16). Finally, the Dutch Supreme Court committed to using "plain legal language" in its judgments by avoiding Latin terminology and adopting shorter sentences (Rechters voor Rechters 2019).

## 10.6 Examples of Clear Language Judgements

By adopting plain language, courts can facilitate more meaningful participation for individuals subject to its proceedings, including individuals with cognitive or communication-related disabilities. This section expands on the previous section that discussed clear legal information and offers examples of clear language judgements.

To offer an initial example, consider the case of Justice Nakatsuru of the Ontario Court of Justice, who released a clear language judgment in *R. v. Jesse Armitage (2015).*[11] He began the decision by noting the importance of plain language:

> [2] For Jesse Armitage, I have tried to say what I wanted to say in very plain language. I believe that this is very important for judges to do in every decision. However, judges often do not do a good job of this. I would describe myself as one of the worst sinners. As lawyers first and then judges, we get used to using words that are long and complicated. This only muddies the message we are trying to say…

Another example comes from a 2018 decision of Family Court in Swindon, England. In *A Council v. 'Jack's' Mother, 'Jack's' Dad and 'Jack'* (2018),[12] Judge Reed prepared the judgement with "simple words" (at ¶ seven) so that the father "can have it read and explained to [him]" (at ¶ 57). The father is a person labelled with intellectual disabilities who was supported by a Communication Intermediary to comprehend the proceedings. In this case, the decision was to remove Jack from his parents' home, since "… the problems that I have read about were so serious that I decided these orders had to be made" (at ¶ 52). Justice Reed continued: "'I would like Jack's mum and dad to know that I understand they love Jack very much" (at ¶ 53).

This decision has been "widely praised" with some saying, "All judgments should be written like this, including those for under eighteen-year-olds convicted of crimes"

---

[10] *Plain Writing Act*, Pub. L. No. 111–274 (2010).

[11] *R. v. Armitage*, [2015] ONCJ 64 (CanLII), https://www.canlii.org/en/on/oncj/doc/2015/2015oncj64/2015oncj64.html.

[12] *Jack (A Child: care and placement orders)*, [2018] EWFC B12.

(Fouzder 2018). Others have said that she has "rightly been lauded for the judgment" (basedrones 2018).

In addition to the decision in *A Council v. 'Jack's' Mother, 'Jack's' Dad, and 'Jack'*, there are other judgments from the United Kingdom written in clear and accessible language (listed in ascending chronological order).

- *Lancashire County Council v. M & Others*[13] (2016) involved four children removed from their home. The father of the younger two children refused to cooperate with social services, and the Court found that the mother could not create a stable home environment. The Court prepared the judgment so it was "as short as possible so that the mother and the older children can follow it" (at ¶ one).
- In *Re A: A Letter to a Young Person*[14] (2017), Justice Jackson presided over a dispute regarding the relocation of a child who was involved in multiple family court proceedings. Justice Jackson garnered praise for taking the "unusual step" (Fouzder 2017) of commencing the judgment with a letter addressed directly to the fourteen-year-old. This decision also incorporated the use of emojis, characterized as "refreshing" (Connelly 2016).
- In *Patrick v. Patrick*[15] (2017), Sheriff Anwar presided over a disagreement regarding the father's contact with three children and wrote a letter to the children. Ultimately, Sheriff Anwar decided that both parents would have to participate in courses. While the father did not receive custody, he was granted permission to maintain communication.
- The parents, in *A, B, C, and D*[16] (2021), faced barriers in caring for their children. Both parents had disability-related needs and the mother was autistic. Writing in clear language, the Family Court removed the children from their care.
- The decision of *A Borough Council v. N & Ors*[17] (2022) was prepared in clear language so the mother "can understand it with help" (at ¶ 61). The matter involved care and placement orders for two children. Their mother has "a learning disability which makes things difficult for her" (at ¶ four).

About *Patrick v. Patrick* (2017) and *A Letter to a Young Person* (2016), Barnes MacFarlane (2018, 104) asserts that clear language decisions accord with guidance offered by the United Nations Committee on the Rights of the Child (2009). General Comment No. 12 requires decision makers to "inform the child of the outcome of the process and explain how her or his views were considered" and that "the views of the child are not only heard as a formality but are taken seriously" (¶ 45).

Similarly, the United Nations Convention on the Rights of Persons with Disabilities (CRPD) supports a purposive interpretation of accessibility and access to justice (Ruskus 2023). Article nine ("Accessibility") protects the right to access all aspects of society "on an equal basis with others" and Article thirteen ("Access to Justice")

---

[13] *Lancashire County Council v. M and Others*, [2016] EWFC 9.

[14] *Re A (A Letter to a Young person)*, [2017] EWFC 48.

[15] *Patrick v. Patrick*, [2017] SC GLA 46.

[16] *A, B, C and D (care and placement orders)*, [2021] EWFC B47.

[17] *A Borough Council v. N & Ors*, [2022] EWFC 91.

requires "effective access to justice" in "all legal proceedings" (UN 2006). Equal access to justice is a core competent of the United Nations' work to protect the rule of law (United Nations Development Programme 2022).

Most of the cases reviewed in this section engage legal questions about family policing systems, sometimes called child protection (Baobeid 2022), including out-of-home placements of children or the termination of specific parental rights. It is worth noting that clear language judgements may be "particularly useful in family litigation" (Barnes Macfarlane 2018, 105), and in the Netherlands, many court rulings labeled as "plain language" pertain to care or placement orders.

At the same time, this area of law relies on ableist presumptions about the capacity to parent. Institutional ableism has historically manifested itself through practices such as forced sterilizations and restrictions on marriage, and "discriminatory child welfare practices that presume unfitness" (Bagenstos 2020, 287). This makes it such that the clear language used to communicate judgements is just, but the content included and/or reason for the judgement may be problematic. This, and other, challenges with clear language judgment making are described in the section below.

## 10.7 Navigating Complexity and Ambiguity: The Problem with Clear Language Judgement Making

The global movement toward clear language judgments is a step toward enhancing public trust in the justice system and ensuring equal access to justice. Language is important, and it "can structure or disrupt ideologies and societal "norms" and standards" (Jones and Williams 2017, 418).

The evolution toward clear language, however, requires a "critical eye" toward an "inclusive and social-justice stance" (Jones and Williams 2017, 413). Some are skeptical of the Plain Language Movement since it perpetuates a misconception that the law can directly communicate to those subjected to it "merely by simplifying its language" (Assy 2011, 378). Others challenge the push for plain language since precision in legal drafting often requires complexity. What may initially appear as legalese may be "indispensable" (Provincial Court of British Columbia 2015) to achieve clarity of intention and avoid ambiguity. In documents such as contracts, it is crucial to use language that unmistakably distinguishes between what is agreed upon and what is not. To Stephenson (2017, 89), "abandoning 'legalese' in favour of modern parlance may result in more legal wrangling" later.

Furthermore, clear language cannot address underlying biases and prejudices in judicial decision making. Even if the language is clear, the content and substance of the decision may still perpetuate systemic injustices or discriminatory practices. Those biases are "overlooked during the implementation of plain-language guidelines" (Jones and Williams 2017, 427). Clear language judgements, to Van Domselaar (2022, 110), obscure power imbalances: "The conversational style may hide from view that the relation in question is highly asymmetrical, in that the addressee will never be able to respond with a similar authority." In other words, procedural accommodations may reinforce power imbalances within the legal system by obscuring them.

For example, in R. v. Armitage, Justice Nakatsuru delivered a custodial sentence to the accused. While he intended to impose a non-custodial sentence, Armitage was re-arrested and requested to serve the remaining nine months of his sentence in custody to "be sure he had enough time in custody to fully make use of the help available" (70) in prison. Justice Nakatsuru, acknowledging Armitage's request, stated that this would be the longest period of incarceration to which Armitage has been subjected. While the words of the decision were clear, it was not a just result. It is fundamentally unjust that accused persons are unable to access necessary supports and services outside of correctional facilities.

## 10.8 Conclusion: Risking Access to Disability Injustice

This chapter considered the availability of *disability rights* and *disability justice* in conventional court-based processes. The chapter also discussed another dyad, namely, *accessibility* and *access to justice.* Both meanings of "access" are concerned with a society that fails disabled people.

The chapter introduced clear language and clear language judgements as important access to justice strategies that are ostensibly about "giving access to justice to average citizens" (van Domselaar 2022, 96). However, clear language only gets people with disabilities to "the table". It does not question who else is at the table, its key architectural features, or the system in which it participates. Only a "multilevel analysis" can achieve both equal access and just results (Mor 2017, 648).

In general, this chapter introduced various limitations of the traditional justice system and calls for attention to underlying systemic issues that perpetuate disability injustice. Other theoretical and practical strategies are necessary to collectively recognize and remedy disability violence. Without a comprehensive attention, procedural reforms—like clear language judgements—risk becoming "generic solutions"[18] that improve access to disability *injustice.*

[18] Vipond, Evan and Pierre Cloutier de Repentigny. Forthcoming. Searching for Justice: Moving Towards a Trans Inclusive Model of Access to Justice in Canada. *Dalhousie Law Journal* 47(1).

## References

Assy, Rabeea. 2011. Can the law speak directly to its subjects? The limitation of plain language. *Journal of Law and Society* 38: 376–404.

Bagenstos, Samuel. 2020. Disability and reproductive justice. *Harvard Law and Policy Review* 14: 273–293.

Baobeid, Iman. 2022. *The power of language: What do "family policing" and "child and family well-being" mean?* West Coast Legal Education and Action Fund. https://westcoastleaf.org/wp-content/uploads/2023/05/11-8-2022-Family-policing-resource-final.pdf. Accessed 13 April 2024.

Barnes Macfarlane, Lesley-Anne. 2018. Patrick v Patrick and re a letter to a young person: Judicial letters to children—An unannounced, but not an unwelcome, development. *Edinburgh Law Review* 22 (1): 101–107.

basedrones. 2018. A clear explanation of the case involving a Council, Jack's Mother, Jack's Dad and Jack. https://basedrones.wordpress.com/2018/04/13/a-clear-explanation-of-the-case-involving-a-council-jacks-mother-jacks-dad-and-jack/. Accessed 9 April 2024.

Bay, Michael. 2006. Making the law match the reality: Making the reality match the law. *Journal of Ethics in Mental Health* 1: 1–4.

Bennett, Kristin C., and Mark A. Hannah. 2022. Transforming the rights-based encounter: Disability rights, disability justice, and the ethics of access. *Journal of Business and Technical Communication* 36: 326–354.

Boyd, J.P. 2015a. A remarkable, plain language judgment from the Ontario Court of Justice. https://ablawg.ca/2015/03/06/a-remarkable-plain-language-judgment-from-the-ontario-court-of-justice/. Accessed 9 April 2024.

Boyd, J.P. 2015b. Legislative intelligibility and the rule of law. https://www.slaw.ca/2015/01/16/legislative-intelligibility-and-the-rule-of-law/. Accessed 9 April 2024.

Canadian Centre for Occupational Health and Safety. 2023. Health and safety programs. https://www.ccohs.ca/oshanswers/hsprograms/plain_language.html. Accessed 9 April 2024.

Chamberlin, Judi. 1990. The ex-patients' movement: Where we've been and where we're going. *The Journal of Mind and Behaviour* 11 (3/4): 323–326.

Chouinard, Vera. 2001. Legal peripheries: Struggles over disAbled Canadians' places in law, society and space. *Canadian Geographer* 45 (1): 187–192.

Committee on the Rights of the Child. "GENERAL COMMENT No. 12. (2009) The right of the child to be heard" presented at the 51st Committee on the Rights of the Child, Geneva, June 2009.

Communications Disability Access Canada. 2023. Barriers to justice. https://www.cdacanada.com/resources/access-to-justice-communication-intermediaries/about/barriers-to-justice-services/. Accessed 9 April 2024.

Connelly, Thomas. 2016. High court judge uses emojis to help children understand family law judgment. https://www.legalcheek.com/2016/09/high-court-judge-uses-emojis-to-help-children-understand-family-law-judgment/. Accessed 9 April 2024.

Cotterrell, Roger. 2019. Access to justice, moral distance and changing demands on law. *Windsor Yearbook Access to Justice* 36: 193–209.

Del Gobbo, Daniel. 2019. Queer dispute resolution. *Cardozo Journal of Conflict Resolution* 20: 283–327.

Farrow, Trevor C. 2014. What is access to justice. *Osgoode Hall Law Journal* 51: 957–988.

Foster, Tammie, and Robyn Young. 2021. Brief report: Sentencing outcomes for offenders on the autism spectrum. *Journal of Autism and Developmental Disorders* 52: 3314–3320.

Fouzder, Monidipa. 2017. 'Dear Sam': Judge writes to 14-year-old to explain custody ruling. *The Law Society Gazette*, July 28, 2017. https://www.lawgazette.co.uk/law/dear-sam-judge-writes-to-14-year-old-to-explain-custody-ruling/5062255.article. Accessed 9 April 2024.

Fouzder, Monidipa. 2018. Plaudits for judge's plain speaking in adoption case. *The Law Society Gazette*, April 11, 2018. https://www.lawgazette.co.uk/law/plaudits-for-judges-plain-speaking-in-adoption-case/5065607.article. Accessed 9 April 2024.

Galanter, Marc. 2005. Access to justice as a moving frontier. In *Access to justice for a new century: The way forward,* ed. Julia Bass, W.A. Bogart, and Frederick H. Zemens, 147–162. Toronto: Law Society of Upper Canada.

Glaves, Bob. n.d. What do we mean when we say justice? https://chicagobarfoundation.org/bobservations/what-do-we-mean-when-we-say-access-to-justice/. Accessed 9 April 2024.

Hughes, Bill. 2002. Bauman's strangers: Impairment and the invalidation of disabled people in modern and post-modern cultures. *Disability & Society* 17: 571–584.

Inclusion Europe. 2017. European standards for making information easy to read and understand. https://www.inclusion-europe.eu/wp-content/uploads/2017/06/EN_Information_for_all.pdf. Accessed 9 April 2024.

Jones, Natasha, and Miriam Williams. 2017. The social justice impact of plain language: A critical approach to plain-language analysis. *Institute of Electrical and Electronic Engineers* 60: 412–429.

Law Commission of Ontario. 2017. Legal capacity, decision-making and guardianship: Final report. http://www.lco-cdo.org/wp-content/uploads/2017/03/CG-Final-Report-EN-online.pdf. Accessed 9 April 2024.

Lepofsky, David. 1995. Equal access to Canada's judicial system for persons with disabilities—A time for reform. *National Journal of Constitutional Law* 5: 1–22.

Lepofsky, David. 2022. People with disabilities need lawyers too! A ready-to-use plan for law schools to educate law students to effectively serve the legal needs of clients with disabilities as well as clients without disabilities. *Windsor Yearbook of Access to Justice* 38: 148–191.

Lundberg, Camilla, and Eva Simonsen. 2017. Disability in court: Intersectionality and rule of law. *Scandinavian Journal of Disability Research* 17: 7–22.

Mingus, Mia. 2018. How our communities can move beyond access to wholeness. https://leavingevidence.wordpress.com/2011/02/12/changing-the-framework-disability-justice/. Accessed 9 April 2024.

Montenegro, C.R., and F. Cornish. 2015. 'It is not the State's fault that we have a person lie this': Relations institutions and the meaning of 'rights' to carers of people with psychosocial disabilities in Chile. *Global Mental Health* 2: E22. https://doi.org/10.1017/gmh.2015.20.

Mor, Sagit. 2014. Policy on preventing discrimination based on mental health disabilities and addictions. http://www.ohrc.on.ca/sites/default/files/Policy%20on%20Preventing%20discrimination%20basd%20on%20mental%20health%20disabilities%20and%20addictions_ENGLISH_accessible.pdf.

Mor, Sagit. 2015. By the numbers: A statistical profile of people with mental health and addiction disabilities in Ontario. http://www.ohrc.on.ca/en/numbers-statistical-profile-people-mental-health-and-addiction-disabilities-ontario. Accessed 9 April 2024.

Mor, Sagit. 2017. With access and justice for all. *Cardozo Law Review* 39: 611–647.

Ontario Human Rights Commission. 2014. Policy on preventing discrimination based on mental health disabilities and addictions. http://www.ohrc.on.ca/sites/default/files/Policy%20on%20Preventing%20discrimination%20based%20on%20mental%20health%20disabilities%20and%20addictions_ENGLISH_accessible.pdf.

Ontario Human Rights Commission. 2015. By the numbers: A statistical profile of people with mental health and addiction disabilities in Ontario. http://www.ohrc.on.ca/en/numbers-statistical- profile-people-mental-health-and-addiction-disabilities-ontario. Accessed 30 June 2023.

Ontario Human Rights Commission. 2016. Policy on ableism and discrimination based on disability. https://www.ohrc.on.ca/en/policy-ableism-and-discrimination-based-disability. Accessed 9 April 2024.

Perlin, Michael L. 1999. Half-wracked prejudice leaped forth: Sanism, pretextuality, and why and how mental disability law developed as it did. *Journal Contemporary Legal Issues* 10: 3–36.

Perlin, Michael L. 2013. Sanism and the law. *Medicine and Society* 10: 878–885.

Piepzna-Samarasinha, Leah. 2018. *Care work: Dreaming disability justice.* Vancouver: Arsenal Pulp Press.

Poole, Jennifer M., Tania Jivraj, Araxi Arslanian, Kristen Bellows, Sheila Chiasson, Husnia Hakimy, Jessica Pasini, and Jenna Reid. 2012. Sanism, mental health, and social work education: A review and call to action. *Intersectionalities: A Global Journal of Social Work Analysis Research, Polity, and Practice* 1: 20–23.

Provincial Court of British Columbia. 2015. Writing reasons for judgment simply is not easy. June 9. https://www.provincialcourt.bc.ca/enews/enews-09-06-2015. Accessed on 9 April 2024.

Rechters voor Rechters. 2019. *Matters of principle: Codes on the independence and impartiality of the judiciary.* Den Haag, Netherlands: Judges for Judges. https://www.rechtspraak.nl/sitecollectiondocuments/matters-of-principle.pdf. Accessed 9 April 2024.

Ruskus, Jonas. 2023. Transformative justice for elimination of barriers to access to justice for persons with psychosocial or intellectual disabilities. *Laws* 12 (3): 51. https://doi.org/10.3390/laws12030051.

Salerno, Alisha, and Regina Schuller. 2019. A mixed-methods study of police experiences of adults with autism spectrum disorder in Canada. *International Journal of Law and Psychiatry* 64: 18–25.

Sarrett, Jennifer, and Alexa Ucar. 2021. Beliefs about and perspectives of the criminal justice system of people with intellectual and developmental disabilities: A qualitative study. *Social Sciences & Humanities Open* 3: 1–8.

Sealy-Harrington, Joshua. 2021. Access to (in)justice: A critical race reflection. https://www.law360.ca/civillitigation/articles/25551. Accessed 9 April 2024.

Siegel, Alan. 2010. Let's simplify legal jargon! https://www.ted.com/talks/alan_siegel_let_s_simplify_legal_jargon. Accessed 9 April 2024.

Sjoquist, Geri C. 2022. Bias towards the disabled. https://www.americanbar.org/groups/family_law/publications/family-advocate/2022/winter/bias-towards-disabled/. Accessed 9 April 2024.

Slavny-Cross, Rachel, Carrie Allison, Sarah Griffiths, and Simon Baron-Cohen. 2022. Are autistic people disadvantaged in the justice system? *Autism* 27: 1173–1507.

Stefanovich, Olivia. 2021. *'We're in trouble': Advocates urge Ottawa to help close the access-to-justice gap.* Canadian Broadcasting Corporation. https://www.cbc.ca/news/politics/access-to-justice-federal-budget-2021-requests-1.5989872. Accessed 9 April 2024.

Stephenson, Michael. 2017. Harry Potter language? The Plain Language Movement and the case against abandoning legalese. *Northern Ireland Legal Quarterly* 68: 85–90.

Supreme Court of Canada. 2023. 2022 year in review: The work at the court. https://www.scc-csc.ca/review-revue/2022/Work-Travail-eng.html. Accessed 9 April 2024.

Taylor, Stephanie. 2022. *Pierre Poilievre promises new law against government jargon.* Canadian Broadcasting Corporation. https://www.cbc.ca/news/politics/pierre-poilievre-simple-language-1.6570129. Accessed 9 April 2024.

UN (The United Nations). 2006. (CRPD) *Convention on the Rights of Persons with Disabilities.* Treaty Series, 2515, 3, G.A. Res. 61/106, 76th plen. Mtg., U.N. Doc A/Res/61/106. https://social.desa.un.org/issues/disability/crpd/convention-on-the-rights-of-persons-with-disabilities-crpd. Accessed 9 April 2024.

United Nations Committee on the Rights of the Child. 2009. Convention on the Rights of the Child. https://www.ohchr.org/en/treaty-bodies/crc. Accessed 9 April 2024.

United Nations Development Programme. 2022. Human rights: The Global Programme for Strengthening the Rule of Law, Human Rights, Justice and Security for Sustainable Peace and Development. https://www.undp.org/sites/g/files/zskgke326/files/2022-11/UNDP%20GP_%20Justice.pdf. Accessed 9 April 2024.

van Domselaar, Iris. 2022. 'Plain' legal language by courts: Mere clarity, an expression of civic friendship or a masquerade of violence? *The Theory and Practice of Legislation* 10: 93–111.

Vipond, Evan, and Pierre Cloutier de Repentigny. Forthcoming. Searching for justice: Moving towards a trans inclusive model of access to justice in Canada. *Dalhousie Law Journal* 47 (1).

Women Enabled International. 2023. Fact sheet: Access to justice for women and girls with disabilities. https://womenenabled.org/reports/wei-fact-sheet-access-to-justice/. Accessed 9 April 2024.

Zödi, Zsolt. 2019. The limits of plain legal language: Understanding the comprehensible style in law. *International Journal of Law in Context* 15: 246–262.

# Chapter 11
# Disrupting Oppression: Critical Systems Thinking and Ethics in Autism Research

**Dora M. Raymaker**

**Abstract** In recent decades, autistic and other disabled people have fought for–and won–gains in research inclusion, both as participants and as researchers, and fueled shifts toward more ethical research with disability populations. Despite this, by their nature, large, complex systems like the social, cultural, and political systems driving bioethics shift very slowly. This chapter (1) presents four ethical tensions in the past and present landscape of autism research (models of disability, definitions of normality, politics of inclusion, priorities in research), (2) applies a critical systems thinking lens to understand the knowledge-power dynamics that generate these tensions, and (3) uses those dynamics in relationship to Meadow's leverage framework to consider points of impact for future change. Systems thinking teaches that, given the vast complexity of the world, we can neither control nor predict the future, but we can intentionally work toward the kinds of inclusive, empowered futures we would like to realize.

**Keywords** Critical disability · Critical systems thinking · Ethics · Autism · Participatory research · Inclusive research

## 11.1 Introduction

I entered academia in 2006 through a backdoor of neurodiversity and disability rights activism. As a then-graduate student in systems science, I was interested not in academic research, but in applying systems thinking toward a career in organizational management or building robot brains. In parallel, as an Autistic activist I was disrupting the systems of oppression threatening my ability to survive in a world hostile to neurodivergence, both for myself and for others in the neurodiversity community. Those two paths converged when I met Christina Nicolaidis, a

D. M. Raymaker (✉)
Regional Research Institute for Human Services, School of Social Work & Complex Systems Program, School of Earth, Environment, and Society, Portland State University, Portland, USA
e-mail: draymake@pdx.edu



A. Bianchi and J. A. Vogt (eds.), *Intellectual Disabilities and Autism: Ethics and Practice*, The International Library of Bioethics 108,
https://doi.org/10.1007/978-3-031-61565-8_11

health services physician-researcher, parent, and fellow neurodivergent who introduced me to community-based participatory research. Together, we founded the first documented (Jivraj et al. 2014) community-academic collaboration with the Autistic community, the Academic Autism Spectrum Partnership in Research and Education (AASPIRE).[1]

While our meeting may have been coincidence, our motivation to found AASPIRE was not. Ethical issues were rampant in autism and disability research yet ignored or unseen by the research community. Those issues were not just philosophical for us—unexamined questions related to the language of autism and disability, priorities in autism, exclusion of autistic and disabled people as research participants and researchers, and paradigmatic framing of autism and disability led to discrimination in our own lives, as well as for our families and our communities.

In 2009, fresh from my master's degree and learning about critical theories for the first time, I attended a conference to explore whether there might exist a new field of critical autism studies. My article from that meeting (in the book *Worlds of Autism*) describes a model for how community-based participatory research can act as a lever to generate empowerment and disrupt systems of oppression (Raymaker and Nicolaidis 2013, 169). In 2020, I wrote a follow-up chapter in *Autistic Community and the Neurodiversity Movement* reflecting on, 'well, we thought that might work ten years ago, but after living it, did we actually change anything?' (Raymaker 2020). While it's impossible to trace a complex outcome in the real world to any singular cause, there has been tremendous emergence of inclusive autism research, a wide-ranging reframing of autism and other neurodevelopmental conditions through a neurodiversity lens, a shift in the language of autism in academic literature toward greater cultural sensitivity, a development of a more formalized critical autism studies, and the hint of a shift in research priorities. Along with others working in this transformative space, it is likely AASPIRE's work has influenced these shifts both by breaking the original ground and by modeling what ethical autism research could look like. Examples of this include the *AASPIRE Inclusive Research Toolkit* (AASPIRE 2023) and the academic journal *Autism in Adulthood.*

However, by their nature, complex systems like culture and policy shift very slowly. While we've been fighting some ethical battles less lately (e.g., convincing scientific reviewers that autistic people are sufficiently human to participate in qualitative interviews (Anonymous 2010)), others remain; for example, ongoing work for recognition as a health disparities group (Raymaker and McDonald 2013). Other underlying ethical issues are coming to the fore as the more foundational issues to resolve; for example, questions related to whether normalizing therapy for the benefit of non-disabled people is worth the risk of long-term harm to mental health or increased mortality (Cassidy et al. 2020; Bradley et al. 2021).

The purpose of this chapter is to (1) present four ethical tensions in the past and present landscape of autism research, (2) take a critical systems thinking approach to understanding the power dynamics behind these tensions, and (3) use those dynamics to consider points of leverage for change. Systems thinking teaches us that, given

[1] https://aaspire.org/.

the vast complexity of the world, we can neither control nor predict the future, but we can work toward the kinds of inclusive, empowered futures we would prefer to have (Meadows 2002).

## 11.2 Ethical Tensions

The autistic and intellectual disability community share a history of systemic eugenics, incarceration, abuse, and oppression with other disability populations; however, we remain among the last to benefit from civil rights gains. While there exists a broad landscape of relevant ethical questions, I focus here on a few key illustrations of ethical tensions centering power dynamics in autism research.

### *11.2.1 Models of Disability*

Tensions arise when mental models—internal explanations for how things work, which we use for generalizing—do not align. In the recent history of the Western world the pathology or medical model of disability has dominated. In this model, disability results from injury, disease, or illness. To eliminate disability, one must eliminate the affliction. In this worldview, it is ethical to eradicate autism and other disabilities by prevention or cure. It is also ethical to remediate—or at least mask—symptoms of those suffering from the condition.

However, there are dozens of other models of disability (Richards 2023) including the social model. As disability rights activists have made gains in civil rights, they have increasingly brought the social model to dominance. In this model disability results from a mismatch between a person's functional abilities and the environment. After all, wouldn't someone who could hear be disabled in a world where no one else could, and thus little was done to moderate the volume of many sounds? Disability is a dynamic process, and in this world view it is ethical to make a more adaptive, universally accessible world. The neurodiversity paradigm is, in a sense, a special case of the social model; neurodiversity describes the type of marginalization experienced by people with neurodevelopmental differences (i.e., neurodivergence), and celebrates the identity, culture, and value of all bodyminds (Walker and Raymaker 2021).

I once attended a gathering of researchers at Oregon Health & Science University. An older, white, male researcher who did basic science related to Rhett's syndrome told us with sincerity how he loved his work because girls with Rhett's had no souls, and he might get them back. When several of us pressed him to interact with some girls with Rhett's before saying such things, he doubled down, relating how he'd been convinced by the father of a girl with Rhett's, and thus profoundly understood the tragedy. Those of us with more social model framing (some of us having been similarly dehumanized), were horrified by the ethical violation of the girls' humanity.

Which mental model one holds of disability can cause diametrically opposed views of what is ethical in biomedical research.

### *11.2.2 Definitions of Normality*

"Normal" is generally accepted to mean where most people fall relative to some dominant benchmark or statistical baseline. Outside of "normal" is "good" if it's something society values, as is the case with elite athletes or the possession of great material wealth. But outside of "normal" is "bad" if it's something society does not value, as is the case with autistic people who do not make eye contact in situations where non-autistic people would.

But–who decides what that benchmark of normality is? Baseline for whom, in what context, and why? Even when it seems possible to be objective—for example, to say, well, we can measure athletic ability in the general population, so we know that an elite athlete is outside of that range—there are ethical confounders to that definition. For example, take the case of elite athlete, double-amputee, and convicted murderer Oscar Pistorius who blurs questions of normativity and ethics in disability, technology, privilege, and aberrant behavior (Van Hilvoorde and Landeweerd 2010).

Autistic culture normalizes autistic behavior. Autreat was a large autism conference designed by Autistic people, for Autistic people (Sinclair 2005), and in that space, it was the neurotypical people who were abnormal. As a default we received interaction badges, a sensory-friendly and safe meeting location, and room in the conference schedule to process information and decompress. We stimmed and played with bubbles on the venue lawn between sessions, and no one called us "unprofessional."

### *11.2.3 Politics of Inclusion*

Situated in the section of this book on inclusion and empowerment, it is necessary to discuss ethical tensions around inclusion in autism research. Like "normal," the idea of "inclusion" seems simple and desirable on the surface. In practice within the autism and related communities, who is included, in what, and how is sharply contested territory. Key areas of tension are who speaks for whom, who is the expert, and what does inclusion in research mean in the first place.

#### **11.2.3.1 Who Speaks for Whom?**

As a marginalized population, people with disabilities have been systematically denied a voice in their own lives. This manifests on an individual level in ways such as forced incarceration, guardianship, and infantilism. It manifests on a communal level in lack of representation in activities that shape policy and practice from voting to political appointments to sitting on the board of foundations that provide funding for research and services. The "who speaks for whom" tension has a long history and spreads throughout disability communities.

In autism research, sharp divides are often made between "high functioning" autistic people who can speak for themselves (and therefore are either not autistic enough to speak for the "real" autistic people, or are too socially impaired by their autism to know what they are speaking about, for example (Brown 2011)) and "low functioning" or "profound" autistic people who cannot speak for themselves and therefore require proxies—non-autistic parents, caregivers, educators, clinicians, etc.—to speak for them. Those taking this perspective likely feel it is ethical to do so, giving voice to the voiceless or however they have defined "functioning" or "profundity," and ensuring that the population they care about isn't excluded.

Those of us in the Autistic community—including those who do not communicate with speech and/or who have co-occurring intellectual or learning disabilities, high support needs, difficulty controlling our emotions or behavior, or whatever else outsiders might mean by "low functioning" or "profound"—have been speaking for ourselves and advocating for our peers for decades. From our perspective, it is unethical to presume that a proxy can know a person's innermost thoughts (like a report, now lost to time, I read about autistic adults' perspectives of their intimate relationships that collected data solely from their parents). For us, the ethical stance is to provide supports to enable direct report, no matter what their functional abilities, and, if truly impossible, to be transparent about the limitations of proxy or caregiver reporting (e.g., designing measures that ask caregivers to report on their own perceptions rather than on the inner world of the autistic person, including positionality and potential biases of proxies or caregivers to contextualize findings, etc.).

#### 11.2.3.2 Who is the Expert?

Reviewing any arbitrary panel of autism researchers at conferences or in the press, it is highly likely that the majority—typically all—are not autistic. Similarly, in science reporting and popular media the "autism expert" is almost invariably a non-autistic person with a graduate degree in a related field. One facet of this rests with systemic oppressions that make it less likely for autistic people to receive education and support to become researchers, clinicians, or other "experts." However, it exposes another facet, which is that a certain type of expertise—academic knowledge created almost exclusively by non-disabled scientists—is privileged in autism research over the lived experience of autistic people themselves.

Science has an ethical obligation to rigor. Claims about findings could risk—or save—lives. Expertise must ground itself in ethical science. On the other hand, the type of knowledge generated by scientific research is only as ethical and accurate as the research designs and data sources that inform it, and but one type of expertise. It cannot substitute for experiential knowledge. I recall an online article from around 2008 about sighted flight attendants who donned blindfolds to test the accessibility of emergency procedures for blind passengers. Unable to navigate the procedures, they concluded blind passengers could not escape a damaged plane. However, when they asked blind people to test, the blind testers followed the emergency procedures without much difficulty due to their deep expertise in being blind. From an ethical standpoint, we may risk the very reason for scientifically grounding expertise if we do not accept the lived experience of disability as expertise of equivalent value to academic knowledge—and indeed, of greater value regarding the embodied experience of neurodivergence itself.

#### 11.2.3.3 What Does Inclusion in Research Mean in the First Place?

There are many ways to include autistic people in research as research participants, co-researchers in community-engaged projects, expert informants, academic or scientific researchers, research staff or consultants, or in intersecting roles. Ethically, it is important to be cognizant of these distinctions and deliberate about their application. Well-meaning autism researchers still ask me to take a community co-researcher role on their projects. This is an inappropriate ask of someone who could serve as a senior scientific consultant for their project—but more importantly, it is denying that position to someone who holds a valuable community perspective, uncontaminated by knowledge of survey psychometrics and the current state of the science. Similarly, if an autistic person is serving as student research assistant, is it ethical to ask them to provide the expertise of their lived experience without compensating them for it? There aren't easy answers for these types of tensions, as many overlap with additional power structures and areas of conflict such as between students and mentors, employees and supervisors, research subjects and researchers, and lay people and scientists.

Witnessing the emergence of dozens of participatory autism research projects from the nothing of twenty years ago is exciting, but more often I observe people implementing these projects without understanding the vast existing literature on inclusive research with disempowered and marginalized populations. Inclusive research stretches back to action research in the 1940s (Lewin 1946), and there are well-developed, formalized frameworks for implementing it [e.g., community-based participatory research (Israel 2005), patient and stakeholder engagement (Patient-Centered Outcomes Research Institute 2013)]. Many of these frameworks have well-developed means for being truly inclusive of people who have been systematically disempowered and include explicit structures to attend to building trust and balancing power. Rushing onto the participatory research bandwagon without first researching mature frameworks, choosing the best one for the project, and being transparent about the justifications behind these choices risks not just ethical violations but a loss of trust by the community in inclusive research itself. Tokenism, power imbalances, unmet expectations, and other pitfalls put well-meaning researchers at risk of alienating and further disempowering the very people they intended to center.

Tensions about inclusion also arise with respect to intersectionality [i.e., the experience of having more than one marginalized identity (Cho et al. 2013)]. Racial, ethnic, and gender disparities in the health system mean that the overwhelming number of individuals with an autism diagnosis are white males, often of higher socio-economic status (Shefcyk 2015; Aylward et al. 2021). Subsequently, much of autism research uses white male subjects, resulting in further marginalization of women and non-binary people, people of color, and others. As intersectional experiences are not additive (Cho et al. 2013), and the white male experience cannot speak for any intersectional experiences, intersectional segments of the population are unrepresented in much of autism research. This leads to ethical questions of whether autism research is reinforcing dominant narratives of autism by omission.

## 11.3 Priorities in Research

When AASPIRE started, we planned an intervention to enable autistic adults to access better quality healthcare. During early attempts to procure funding, we received the astonishing feedback that since there existed no evidence that autistic people experience difficulty accessing healthcare, intervention was not required. Thus, our first study was a survey to show that healthcare disparities existed. Despite the hundreds of millions of dollars going to autism research on a federal level alone in 2006, and an urgent plea to address healthcare access from both autistic adults and their families, no one had considered questions about healthcare disparities for autistic adults.

Since then, AASPIRE has consistently cut new ground and provided foundational literature in healthcare (Nicolaidis et al. 2013, 2015), instrumentation for use with autistic populations (Nicolaidis et al. 2020), inclusive autism research (Nicolaidis et al. 2011, 2019), autistic burnout (Raymaker et al. 2020), and autism and skilled employment (Raymaker et al. 2023). I don't think we've consistently broken this ground because we're special (though I am exceptionally proud of my team and do feel that they are all amazing and special!), but because as a community-based participatory research collaborative we ask autistic people what is important and then choose to research those things, even when what autistic people want researched is not the "experts'" priority.

Meanwhile, the majority of autism research remains concerned with causes and cures (see Sect. 11.2.1), remediation, and doing things that are largely for the comfort of non-autistic people (e.g., repressing hand flapping or other stims in public). In a 2017 report from the Interagency Autism Coordinating Committee (IACC), which directs federal autism funding policy, only six percent of tracked funding went to services and two percent to lifespan issues. The largest blocks of funding went to biology (thirty-nine percent); risk factors (presumably to prevent, twenty-one percent); and treatments (presumably to suppress or remediate, fifteen percent) (IACC 2017). The large research network, SPARK for Autism, that could have held much promise for including autistic people as research participants, focuses on genetic research which is ethically problematic for the Autistic community, where eugenics is an ongoing and active concern.

Despite some change in research priorities to reflect autistic people's priorities, and some efforts by the US government to add more autistic people to the IACC, the funding and priorities for autism research remain largely in the same domains as twenty years ago. See Sect. 11.2.3 on the Politics of Inclusion, and the dynamics of marginalization that occur when autistic voices are excluded from structures that generate social influence.

## 11.4 Applying Critical Systems Thinking

These four areas of tension in the ethics of autism research—models of disability, definitions of normality, politics of inclusion, and priorities in research—share a foundational dynamic of knowledge and power. Who has the power, who gets to use it to legitimize world views and sources of knowledge, and, thus, who gets to use it to shape reality radically alters the priorities and the outcomes of research. Is it the neurotypical majority or the neurodivergent minority? Is it those with academic expertise or those with lived experience? Is it mainstream culture or Autistic culture? Is it those who have the most privilege or the loudest voice, or is it those who are the most impacted by research, health policy, and bioethical choices? Questions of dynamic tensions in messy, intractable real-world settings are the domain of systems thinking.

Systems thinking is an anti-reductionist, or holistic, approach to inquiry and problem solving (Churchman 1979; Checkland 1999). It developed in the post-war 1940s from the need for a systematic means to better understand and intervene in complex socio-technical systems, non-mechanistic (i.e., unpredictable, non-physical) processes, and "wicked" problems which, by definition, have no clear formulation or single, testable solution (Rittel and Webber 1973); for example, most problems in public health. Systems thinking firstly acknowledges the "systemness" of the problem by seeing it as a collection of parts engaged in feedback with each other to create a whole in relation to an environment (i.e., applies the definition of a system to the problem) (Lendaris 1986). It then offers systemically focused methods and frameworks to analyze the phenomena without reducing it to component pieces (the reduction itself being undesirable because it destroys the phenomena as a whole). In essence, systems thinking provides tools to see the forest and the trees at the same time (Senge 1990), in part through the idea of generative causality—the structure of a system produces its behavior.

About fifty years later, in the 1990s, some systems thinkers began questioning the approach's considerations of power in its philosophy and methodology. Specifically, that seen through the lens of critical theory, frameworks such as Senge's learning organization (Caldwell 2012), or methods such as Checkland's Soft Systems Methodology did not pay adequate attention to power and emancipation (Jackson 1990). What good was a team visioning exercise, for example, if the employees are going to say what they think their managers want to hear for fear of retribution? Whose voice is excluded from systems practice, what is marginalized and centered? Critical

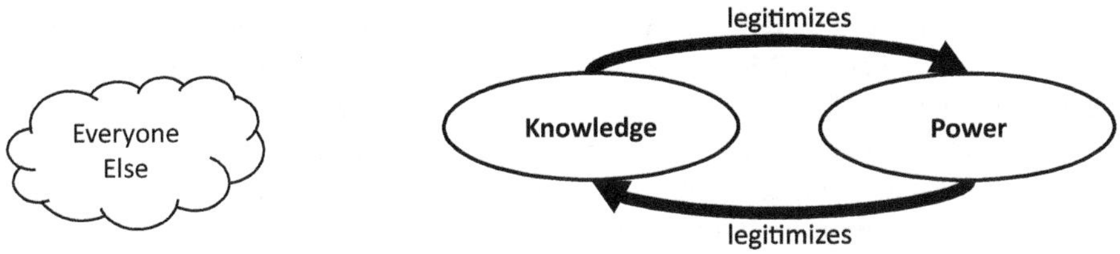


**Fig. 11.1** Knowledge/power reinforcing feedback dynamic

systems thinking is a reformulation of systems thinking, particularly through the philosophies of Marx, Habermas, and Foucault, to better consider the dynamics of knowledge and power (Jackson 1990).

Through a holistic, feedback-aware lens, the connection between knowledge and power (particularly in the sense described by Foucault (Foucault 1980)) is a reinforcing structure. Those with power legitimatize sources of knowledge; they then use these sources of knowledge to reinforce their own power. Anything falling outside of a legitimate knowledge source—for example the lived experience of autistic people—is disempowered. The disempowered are then unable to influence or change who holds power because the system excludes them; see Fig. 11.1.

This dynamic is illustrated, for example, by the large body of autism research focused on interventions to normalize autistic behavior. When autistic people start pushing back and asking for different intervention priorities based on their lived experience, they may be told it is their illness talking, dismissed because if they can speak out then they do not need intervention, told that's not what the research suggests, or simply ignored (e.g., Brown 2011).

Critical systems thinking is a useful approach for understanding and modeling dynamic feedback structures in coercive situations or contexts with power imbalances. Importantly, understanding how structure generates behavior, as with the knowledge-power feedback structure, provides insight into how the structures might change to influence the overall system behavior. If we want to truly include and empower neurodivergent people, we need to disrupt the closed, reinforcing knowledge-power dynamic.

## 11.5 Identifying Leverage and Solutions

Leverage is the place in a complex system where small changes to its parts can create large changes to the whole. Leverage works because it targets the structure generating the system's behavior. Leverage is powerful and elusive; for systems change it is a silver bullet. It can be hard to see, and even harder to implement, more so the more complex the system. But if one finds the right leverage, they can have immense impact.

Systems thinker Donella Meadows lists twelve places to intervene in a system, in order of effectiveness (Meadows 1999). The least effective, but easiest to implement, solutions are changing numbers, buffers, and materials—increasing the "N" of direct reports in data samples, providing more community funding, etc. Harder to implement but more impactive are solutions that influence feedback; for example, by strengthening the information flow between autism researchers and the Autistic community. The next level solutions are structural; they might change who or what is driving the information in the system, add new structures or dismantle old structures—they change the parts and relationships, thereby changing the system itself. Structural solutions are far more difficult to understand and implement but if the new structures are different from the old structures, the system's behavior will change. At the furthest and most impactful (but often hardest to implement) end of the lever, lies shifting mental models and paradigms.

Applying the concept of leverage to the knowledge-power dynamic that excludes autistic people from knowledge production in autism research, participatory or community-engaged research provides a feedback solution. It doesn't change existing structures but does change the nature of the information flow between autistic people and autism researchers. If researchers are working with the community to use nondiscriminatory language in their papers, then when science journalists report on the findings to the public, they may also use nondiscriminatory language; when policy makers seek legitimized sources of knowledge from the scientific community, they will read work that was co-produced by people with lived experience, etc. This is the type of leverage I described in my "Shifting the System" chapters (Raymaker 2013, 2020).

But that higher-order and more powerful leverage is already working its way through the world in the form of neurodiversity. Neurodiversity operates on all four areas of ethical tension. It replaces pathology models of disability, redefines normality, expands inclusion, and radically shifts research priorities. This type of paradigmatic leverage is what is now causing a shift from pathology-focused autism awareness month (a time when many of us hide because it's a month of relentless public discrimination and stigmatization (Praslova 2022) to civil-rights focused autism acceptance month (Moya 2022). It is also what is fueling new infrastructures to support inclusive autism research, like the Autism Intervention Research Network on Physical Health (AIR-P 2023) which requires neurodiversity training and framing to join. However, like inclusive research, it is critical to use the term neurodiversity correctly—it is not a synonym for "strengths-based" approaches, but a human rights concept with teeth.

## 11.6 Actualizing Inclusive and Empowered Futures

Futurists and futures thinkers say, "the future is plural." By this they mean there are many possible futures—some preferable, others abhorrent, some probable, others preposterous (but not impossible) (Bengston 2017). There exists both a world in which, a hundred years from now, the medical model retains dominance and autistic people are eradicated, and a world in which the neurodiversity paradigm becomes dominant, and people move fluidly between different neuro-sensory experiences as the new dominant cultural norm. In examining the future of inclusion and empowerment, I encourage imagination and dreaming big. What do the more preferable, more probable, and more authentically inclusive and empowered futures look like? Then—how might we find and use leverage to realize them?

Continuing to locate ways to expose, disrupt, and restructure the power-knowledge dynamic, use holistic thinking, and center autistic people fully in autism research will help. But I am particularly encouraged by the potential of new frameworks that suggest less looking toward how to include people with disabilities in existing, dominant systems and more looking toward how we, as disabled people, can center our own culture and create entirely novel solutions from within. In particular, the Sins Invalid Disability Justice Framework holds substantial promise, and provides a concrete pathway to action for those of us within disability communities to take hold of our inherent power, trust and own the legitimacy of our precious knowledges, and direct our own empowered futures (Berne 2018).

## References

AASPIRE (Academic Autism Spectrum Partnership in Research and Education). 2023. AASPIRE inclusive research toolkit. https://aaspire.org/inclusion-toolkit. Accessed 9 Apr 2024.

AIR-P (Autism Inclusive Research Network on Physical Health). 2023. https://airpnetwork.ucla.edu. Accessed 9 Apr 2024.

Anonymous. 2010 Autism research program: Peer review panel summary statement, US Department of Defense.

Aylward, Brandon S., Diana E. Gal-Szabo, and Sharief Taraman. 2021. Racial, ethnic, and sociodemographic disparities in diagnosis of children with autism spectrum disorder. *Journal of Developmental and Behavioral Pediatrics* 42 (8): 682–689.

Bengston, David N. 2017. *Ten principles for thinking about the future: A primer for environmental professionals*, U.S. Forest Service.

Berne, Patricia, Aurora Levins Morales, David Langstaff, and Sins Invalid. 2018. Ten principles of disability justice. *Women's Studies Quarterly* 46 (1–2): 227–230.

Bradley, Louise, Rebecca Shaw, Simon Baron-Cohen, and Sarah Cassidy. 2021. Autistic adults' experiences of camouflaging and its perceived impact on mental health. *Autism in Adulthood* 3 (4): 320–329.

Brown, Lydia X.Y. 2011. What is self-advocacy? *Autistic Hoya*. https://www.autistichoya.com/2011/07/what-is-self-advocacy.html. Accessed 9 Apr 2024.

Caldwell, Raymond. 2012. Leadership and learning: A critical reexamination of Senge's learning organization. *Systemic Practice and Action Research* 25 (1): 39–55.

Cassidy, Sarah A., Kezia Gould, Ellen Townsend, Mirabel Pelton, Ashley E. Robertson, and Jacqui Rodgers. 2020. Is camouflaging autistic traits associated with suicidal thoughts and behaviours? Expanding the interpersonal psychological theory of suicide in an undergraduate student sample. *Journal of Autism and Developmental Disorders* 50 (10): 3638–3648.

Checkland, Peter. 1999. *Systems Thinking, Systems Practice: Includes a 30-Year Retrospective*. Chichester: Wiley.

Cho, Sumi, Kimberly W. Crenshaw, and Leslie McCall. 2013. Toward a field of intersectionality studies: Theory, applications, and praxis. *Signs: Journal of Women in Culture and Society* 38 (4): 785–810.

Churchman, C. West. 1979. *The Systems Approach and Its Enemies.* New York: Basic Books.

Foucault, Michel. 1980. *Power/Knowledge: Selected Interviews and Other Writings, 1972–1977.* New York: Vintage.

IACC (Interagency Autism Coordinating Committee). 2017. Interagency Autism Coordinating Committee Strategic Plan for Autism Spectrum Disorder Research 2016–2017 Update. U.S. Department of Health and Human Services Interagency Autism Coordinating Committee website. http://iacc.hhs.gov/strategic-plan/2017/index.shtml. Accessed 9 Apr 2024.

Israel, Barbara A., Eugenia Eng, Amy J. Schulz, and Edith A. Parker. 2005. *Methods in community-based participatory research for health.* San Francisco: Wiley.

Jackson, Michael C. 1990. The origins and nature of critical systems thinking. *Systems Practice* 4 (2): 131–149.

Jivraj, Jamil, Lori-Ann Sacrey, Amanda Newton, David Nicholas, and Lonnie Zwaigenbaum. 2014. Assessing the influence of researcher–partner involvement on the process and outcomes of participatory research in autism spectrum disorder and neurodevelopmental disorders: A scoping review. *Autism* 18 (7): 782–793.

Lendaris, George G. 1986. On systemness and the problem solver: Tutorial content. *IEEE Transactions on Systems, Man, and Cybernetics* 16 (4): 603–610.

Lewin, Kurt. 1946. Action Research and Minority Problems. *Journal of Social Issues* 2 (4): 34–46.

Meadows, Donella. 1999. *Leverage points: Places to intervene in a system.* Hartland.

Meadows, Donella. 2002. Dancing with systems. *The Systems Thinker* 13 (2). https://donellameadows.org/archives/dancing-with-systems/. Accessed 9 Apr 2024.

Moya, Maria J. 2022. Autism Acceptance Month is underway. Here's why the name is important. USA Today, April 2. https://www.usatoday.com/story/news/health/2022/04/02/autism-acceptance-month-name-change/7243001001/. Accessed 9 Apr 2024.

Nicolaidis, Christina, Dora M. Raymaker, Elesia Ashkenazy, Katherine E. McDonald, W. Cody Boisclair, Amelia E.V. Baggs, Steven Kapp, Michael Weiner, and Sebastian Dern. 2015. "Respect the way I need to communicate with you": Healthcare experiences of adults on the autism spectrum. *Autism* 19 (7): 824–831.

Nicolaidis, Christina, Dora M. Raymaker, Steven K. Kapp, Amelia Baggs, Elesia Ashkenazy, Katherine McDonald, Michael Weiner, Joelle Maslak, Morrigan Hunter, and Andrea Joyce. 2019. The AASPIRE practice-based guidelines for the inclusion of autistic adults in research as co-researchers and study participants. *Autism* 23 (8): 2007–2019.

Nicolaidis, Christina, Dora M. Raymaker, Katherine E. McDonald, Sebastian Dern, Elesia Ashkenazy, W. Cody Boisclair, Scott M. Robertson, and Amelia Baggs. 2011. Collaboration strategies in non-traditional CBPR partnerships: Lessons from an academic-community partnership with autistic self-advocates. *Progress in Community Health Partnerships: Research, Education, and Action* 5 (2): 143–150.

Nicolaidis, Christina, Dora M. Raymaker, Katherine E. McDonald, Sebastian Dern, W. Cody Boisclair, Elesia Ashkenazy, and Amelia Baggs. 2013. Comparison of healthcare experiences in autistic and non-autistic adults: A cross-sectional online survey facilitated by an academic-community partnership. *Journal of General Internal Medicine* 28 (6): 761–769.

Nicolaidis, Christina, Dora M. Raymaker, Katherine E. McDonald, Emily M. Lund, Sandra Leotti, Steven K. Kapp, Marsha Katz, Leanne M. Beers, Clarissa Kripke, Joelle Maslak, Morrigan Hunter, and Kelly Y. Zhen. 2020. Creating accessible survey instruments for use with autistic adults and people with intellectual disability: Lessons learned and recommendations. *Autism Adulthood* 2 (1): 61–76.

Patient-Centered Outcomes Research Institute. 2013. Engagement resources. https://www.pcori.org/engagement/engagement-resources. Accessed 10 May 2023.

Praslova, Ludmila N. 2022. Your “Autism Awareness Day” might be excluding autistic people. *Harvard Business Review.* https://hbr.org/2022/04/your-autism-awareness-day-might-be-excluding-autistic-people. Accessed 9 Apr 2024.

Raymaker, Dora M. 2020. Shifting the system: AASPIRE and the loom of science and activism. In *Autistic Community and the Neurodiversity Movement: Stories from the Frontline*, ed. Steven Kapp, 133–144. Singapore: Palgrave Macmillan.

Raymaker, Dora M., and Katherine E. McDonald. 2013. Paradigm shifts in disability and health: Toward more ethical public health research. *American Journal of Public Health* 103 (12): 2165–2173.

Raymaker, Dora M., and Christina Nicolaidis. 2013. Participatory research with autistic communities: Shifting the system. In *Worlds of Autism: Across the Spectrum of Neurological Difference*, eds. Joyce Davidson and Michael Orsini. Minneapolis: University of Minnesota Press.

Raymaker, Dora M., Alan R. Teo, Nicole A. Steckler, Brandy Lentz, Mirah Scharer, Austin Delos Santos, Steven K. Kapp, Morrigan Hunter, Andrea Joyce, and Christina Nicolaidis. 2020. “Having all of your internal resources exhausted beyond measure and being left with no clean-up crew”: Defining autistic burnout. *Autism Adulthood* 2 (2): 132–143.

Raymaker, Dora M., Mira Scharer, Joelle Maslak, Laurie E. Powers, Katherine E. McDonald, Steven K. Kapp, I. Moura, Anna “Furra” Wallington, and Christina Nicolaidis. 2023. “[I] don’t wanna be just a cog in the machine:” Narratives of autism and skilled employment. *Autism* 27 (1): 65–75.

Richards, Aisha. 2023. Evolution of disability models. https://shadesofnoir.org.uk/content/evolution-of-disability-models/. Accessed 9 Apr 2024.

Rittel, Horst W.J., and Melvin M. Webber. 1973. Dilemmas in a general theory of planning. *Policy Sciences* 4 (2): 155–169.

Senge, Peter. 1990. *The fifth discipline: The art and practice of the learning organization.* New York: Doubleday/Currency.

Shefcyk, Allison. 2015. Count us in: Addressing gender disparities in autism research. *Autism* 19: 131–132.

Sinclair, Jim. 2005. History of ANI. http://www.autreat.com/History_of_ANI.html. Accessed 9 Apr 2024.

Van Hilvoorde, Ivo, and Laurens Landeweerd. 2010. Enhancing disabilities: Transhumanism under the veil of inclusion? *Disability Rehabilitation* 32 (26): 2222–2227.

Walker, Nick, and Dora M. Raymaker. 2021. Toward a neuroqueer future: An interview with Nick Walker. *Autism in Adulthood* 3 (1): 5–10.

# Chapter 12
# Social Validity and the Consideration of Neurodiversity in Relation to Ethics

**Joy F. Johnson and Jacob A. Sadavoy**

**Abstract** The opinions and values of patients and clients are critical data for healthcare practitioners to evaluate the quality of services provided. In behaviour analysis, social validity is one of, if not the most important element in evaluating treatment efficacy and effectiveness. Montrose Wolf's seminal paper offered that the work of a behaviour analyst should be validated on three criteria: (1) social significance of the goals, (2) social appropriateness of the procedures, and (3) the social importance of the effects (Wolf in J Appl Behav Anal 11:203–214, 1978). This chapter discusses the value of social validity and the effects of systemic ableism that limit a practitioner's ability to provide quality care that is both meaningful and individualized for the client.

**Keywords** Social validity · Neurodiversity · Ableism · Applied behaviour analysis · Autism

## 12.1 Introduction

Ableism refers to bias, prejudice, and discrimination against disabled individuals in which those with disabilities are deemed inferior to nondisabled people. This marginalization can take on many forms, such as the belief that nondisabled people feel that those with a disability need saving, or are incapable of contributing to society like an able-bodied individual. Others may feel that those with a disability should try to alleviate their "affliction" and would benefit from a cure for their blight. In other cases, the lived experience of those with disabilities are dismissed and replaced with assumed experiences by the able community. Take, for instance, the debate

J. F. Johnson
Spectrum Support, LLC, Timonium, USA
e-mail: spectrumsupport@joyfjohnson.com

J. A. Sadavoy (✉)
CB Consultants, Sag Harbour, USA
e-mail: jacob@committedbehavior.com



A. Bianchi and J. A. Vogt (eds.), *Intellectual Disabilities and Autism: Ethics and Practice*, The International Library of Bioethics 108,
https://doi.org/10.1007/978-3-031-61565-8_12

between person-first language (e.g., person with disability, person with deafness, person with autism) and identity-first language (e.g., disabled person, deaf person, Autistic person). Many in the disabled community have informed those that are not disabled that their preference is identity-first language (Bury et al. 2023; Kenny et al. 2015; Organization for Autism Research 2020; Taboas et al. 2023). Despite this call for change in terminology, many, especially in government institutions and within academia (Centers for Disease Control and Prevention 2022; Dunn and Andrews 2015), have ignored the resounding data from the disabled community due to familiarity with person-first language. There is also, perhaps, a sense of entitlement that those who are nondisabled or in academia maintain a sense of knowing what is best for the disabled community as experts, despite not having a disability.

Within the disabled community, there is a growing sense of disability as being natural and an essential component of human diversity (Barnes 2016) that should be celebrated as an enrichment to humanity (Dwyer 2022). Unfortunately, like many diversity efforts, those who are in positions of power (e.g., privileged identities) maintain power through reinforcing those who share certain similarities (e.g., race, socioeconomic status, gender identity, sexual orientation, religion, disability). Many in the disabled community are often undermined and learn that those without a disability are superior. This creates a system where the disabled are silenced, their opinions are inconsequential, and their experiences are invalidated. In this context, self-advocacy and ideation of goals is stunted, further perpetuating a reliance on nondisabled people. Despite the ethical imperative for all health practitioners to practise without discrimination, ableism is rampant and a barrier to providing effective care for the disabled community (Lindsay et al. 2022).

Historically, the allistic population has viewed autism as an identity that is unsafe, unhealthy, and unwanted for the Autistic individual. There have been atrocities committed in state-funded residential programs and schools (e.g., Willowbrook, Pennhurst, Sunland Hospital, Fernald School) where residents with cognitive delays were dehumanized and suffered abuse from allistic caregivers who thought they were providing care. In Ole Ivar Lovaas' seminal paper (1987, 8), he sought to demonstrate that Autistics are educable and celebrated their "recovery" from autism as they now presented as "indistinguishable from their normal friends". There are many examples of applied behaviour analysis (ABA) erasing Autistic traits as a primary indicator of social significance because (a) social success can be assessed via social acceptance and appearing indistinguishable allows for greater social opportunities, and (b) Autistic normative behaviours are aberrant according to a model of normalcy that does not consider neurodivergence. Autistic traits are symptoms and indicate disability within the medical context and as such, these traits are targeted for reduction.

Many Autistic self-advocates argue that trying to make Autistic people appear allistic or indistinguishable is harmful and may result in adverse effects to the well-being of the Autistic individual. Often the real disabling aspect of being Autistic resides in the lack of acceptance and support by the Allistic community. Health practitioners can mitigate ableist norms by obtaining social validity assessment data from the patients themselves and using that data to better serve their patients.

## 12.2 Systemic Ableism in Healthcare

The World Health Organization acknowledges quality health services as being effective (evidence-based), safe (harmless), and person-centred (individualized to patients' preferences, needs, and values) (Kieny et al. 2018). Improving health outcomes to maintain an optimal quality of life should be accessible to all; however, health care often looks different contingent on one's ability to participate actively in care decisions. Health care available for the Autistic and intellectually disabled populations are at times pseudoscientific, ineffective, harmful, and not person-centred. It is essential that disabled and Autistic patients are the ones guiding their treatment; however, in ableist healthcare services, the patient's input is often ignored based on the assumption that their "affliction" is abnormal or deviant.

As healthcare professionals, listening to our patients and discussing treatment options with them is of ethical significance. Affording this autonomy and right to our patients should not be contingent on ability. Ideally, our patients should feel empowered to select a treatment devoid of undue pressure, guilt, or coercion and access care that is person-centred. Unfortunately, in many healthcare services, a medical model of care is often employed. The medical model, an ableist approach, prioritizes resolving the health problem with the terminal goal to "fix" the patient to be a nondisabled member of society. Within this model the patient is silenced as it is the healthcare provider who is the expert and knows what is best and challenges the patients' experiences in favour of treatment that will reduce symptoms of the disability while inadvertently diminishing the patients' identity. The medical model does not accommodate mental or social problems.

The medical model for autism therapy reinforces the ableist notion that Autistic traits are inherently incompatible with normal behaviour (i.e., behaviour exhibited by Allistic individuals) and thus require intervention. Autistic traits are largely classified as deficits and maladaptive, rather than variations of human behaviour. Applied behaviour analysis is the science of human behaviour and could be quite beneficial to the general population in various contexts; however, it is primarily prescribed for the "treatment of autism" (i.e., 72.1 percent of certificants as of 2022) (Behavior Analyst Certification Board 2022; Dillenburger and Keenan 2009). This further leans toward the ableist notion that Autistic people need behaviour analytic services because they fail to meet neurotypical standards as determined by the Allistic community. What one deems as a strength, a weakness, appropriate, inappropriate, maladaptive, or dysfunctional is largely subjective due to the standard of measure aligning with general societal norms. Norms favour the privileged class and fall within a spectrum of ableness; it is the nondisabled that select the behaviour that is normed. Thus, allistic behaviour is considered "normal" which further entrenches systemic ableism at the expense of the Autistic population.

By contrast, the social model recognizes these same deficits not as a weakness or a challenge inherent with the disorder itself, but as a challenge imposed by the way society is organized (Bottema-Beutal et al. 2021; Graber and Graber 2023). For example, stereotypy can be socially stigmatizing, and one who adopts the medical model may choose to treat stereotypy for a variety of reasons: it impedes social relationships, disrupts work performance, and is annoying, loud, disruptive, etc. Someone who adopts the social model would assess the function of the stereotypy and how it serves the individual. Many Autistics share that stereotypy serves as a coping strategy (Trevisan et al. 2017) and targeting stereotypy for reduction could result in a harmful transference of topography. The punishment that exists for stereotypy in the allistic community is a social mechanism that perpetuates ableism. If society embraced Autistic identity and the behaviours associated with Autism (social model), stereotypy would only be targeted for reduction if it was deemed harmful or if the Autistic individual freely sought for alternative replacement behaviours.

ABA ethics dictate a humanistic, client-centred approach that is necessary for sound ethical practice (Behavior Analyst Certification Board 2020). If humanistic approaches were utilized, the focus would shift from attaining neurotypical behavioural standards to programs that support an ability to manoeuvre the world authentically as an Autistic individual. This is more of a service model in which practitioners take on a role of servants as opposed to a treatment model where the practitioners take the lead. Approaching interventions from a consumer-led support model can also aid in facilitating societal wide inclusion, as differences are not "treated", and normative behaviour is not favoured as superior.

## 12.3 Current Social Validity Assessment in Applied Behaviour Analysis

ABA as per Baer et al. (1968, 91) can be defined as "the process of systematically applying interventions based upon the principles of learning theory to improve socially significant behaviours to a meaningful degree, and to demonstrate that the interventions employed are responsible for the improvement in behaviour". Further, the meaningfulness of socially significant behaviours, procedures, and outcomes can only be assessed and evaluated by the consumers themselves (Schwartz 1991). Behaviour analysts are required to evaluate the social significance of behaviours by using assessment tools, observations, and interviewing consumers and/or other interested parties based on the Behavior Analyst Certification Board (BACB®) Code of Ethics[1] (Behavior Analyst Certification Board 2020). Additionally, sourcing the values and interests of the client is essential as one of three tenets of evidence-based practice (Slocumb et al. 2014).

However, oftentimes ABA treatment is designed to promote the client's strengths and interests, as defined by the care team, while targeting skill deficits and areas of need by addressing behaviours that are deemed inappropriate, non-contextual, dysfunctional, and/or maladaptive. An ableist, medical model approach employs behaviour scrutinization largely based on what is considered acceptable as per typical behaviours within the framework of general societal norms and values (Nicolson

[1] See all four core principles plus Code Elements 2.01, 2.09, 2.13, 2.14, and 3.01.

et al. 2020). Using assent-based procedures, promoting choice, and limiting coercion are current trends in ABA literature. These practices can be used to promote patient participation within a treatment program and replace prescriptive delivery of health services by practising more collaborative methods (Morris et al. 2021; Rajaraman et al. 2022; Staubitz et al. 2022). Gaining participants' feedback about their perception of the appropriateness, value, efficiency, and effectiveness of interventions is critical in providing informed, client-centred care (Hanley 2010; Potter et al. 2013; Slocum and Tiger 2011). Using these methods, the acceptability of treatment and outcomes are assessed and determined (Baer et al. 1968). This aligns with a person-centred therapeutic approach where the practitioner takes a collaborative, non-authoritative role, providing an opportunity for the consumer to take a greater responsibility in the process and treatment planning (Holburn 2002; Holburn et al. 2007).

Seminal behaviour analytic researchers Kazdin (1977) and Wolf (1978) proclaimed the necessity of social validity data in experimental research and behaviour change programs, to ensure that interventions yield meaningful and socially significant contributions to consumers' lives. Despite these seminal works being written more than four decades ago, social validity continues to be understudied and undervalued in behaviour analytic research (Carr et al. 1999; Ferguson et al. 2018). One hypothesized reason is the subjective nature of social validity measures, typically questionnaires or rating scales, which may be considered too subjective to provide valid and reliable measures, often resulting in their rejection by many behaviour analysts (Carr et al. 1999; Carroll and Peter 2014; Ferguson et al. 2018). These findings show that researchers and practitioners are not necessarily informed when it comes to ensuring the study of behaviour change programs is of value for the participant or client. Despite its value, social validity is largely absent from behaviour analytic research, and effective modalities for collecting social validity data from consumers remain arbitrary in the applied setting. However, without ongoing feedback from the individual being served, the treatment goals are in jeopardy of being biased (i.e., solely based on the professionals' experiences and opinions) and, as a result, socially insignificant.

Upon the introduction of social validity in the literature, Wolf (1978) cautioned the absence of consumer acceptance would result in consumer rejection. When social validity is collected by behaviour analysts, these data are often solicited from caregivers,[2] educators, and other interested parties and rarely, if ever, from the individual receiving behavioural health care. This data must be sought from the individual receiving services. Ignoring their experience within treatment does not allow their participation in the therapy and the practitioner is unable to assess the social validity, social significance, and value of their work. In some cases, the client may have a

---

[2] The term "caregiver" is used noting that preferences exist for other terms, e.g., "parent", "family members", "support networks", "guardians", "legally authorized representatives", "informal support networks". In all cases, learning and using the individualized preferences should be adopted by all practitioners.

cognitive impairment that limits their ability to conveniently provide service feedback; however, it is essential for these clients to inform treatment based on their assessment of the quality of service they are receiving. Given the limitations clinicians have for obtaining this information from some clients, alternative measures can be used. For example, Parsons and colleagues (2012) used indices of happiness by observing affect and duration of engagement without support as mechanisms to assess willingness to participate. A replication was completed with mostly children using the picture exchange communication system (PECS) to assess levels of happiness and unhappiness which can be used to inform programming (Ramey et al. 2023). These data, coupled with providing choice and preference assessments, can provide an access to gain treatment assent for those that may be unable to voice their opinion through more conventional modalities.

## 12.4 Current Social Validity Assessment in Health Care

Similarly, social validity or patient satisfaction also remain largely uncaptured in health services. Gill and White's (2009) meta-analysis found that patient satisfaction has been studied extensively with little to no agreement on best practices for capturing feedback. They acknowledge a central focus on patient satisfaction as it relates to perceptions of health care and its delivery (e.g., did the health care received exceed the patient's anticipated expectations or was the experience worse than expected). In response, federal health agencies designed a standardized survey, the Hospital Consumer Assessment of Healthcare Providers and Systems (CAHPS) to assess patients' perceptions of hospital care. This survey measures patients' perceptions of their care related to communication with doctors and nurses, hospital cleanliness, pain management, and attention to patient needs (Mazurenko et al. 2017). A review by Ferrand and colleagues (2016) found a glaring need for studies that move away from a confined assessment of patient satisfaction relative to health care, to a focus on understanding patients' perceptions of their individual experiences. This would entail research to investigate how the perceived quality of services received relates to patient satisfaction. Thus, identifying patients' expectations of deliverables and outcomes of healthcare services prior to commencing said services would be tremendously valuable; however, it is rarely done (Oben and Corliss 2021). Within behaviour analysis, this type of assessment should be common practice as intakes and service agreements require practitioners to discuss outcomes in collaboration with the client; as all plans and services should be guided by the one receiving services.

Two primary barriers associated with social validity are: (1) patients' opinions are not sought, and (2) patients consider their opinions inferior to the professional or the perceived "expert". Regardless of the health services provided, professionals should plan services with their patients. Feedback should be sought for all aspects of the services rendered on an ongoing basis and used to inform modifications resulting in improved patient experiences. The latter barrier can only be overcome if the power imbalance between the professional and the patient is addressed (Beirne and

Sadavoy 2021). The practitioner must be aware that prospective clients: may feel indebted (Fouka and Mantzorou 2011), may fear negative repercussions for refusing health services (LeBlanc et al. 2018), may believe there are insufficient alternatives, may have feelings of inadequacy, or may have feelings that subordinate their own participation at the expense of perceived lofty clinical expertise (Beirne and Sadavoy 2021). Thus, it is the responsibility of the healthcare provider to explain procedures and options colloquially, to ensure that the patient understands both the treatment being discussed, as well as their role in ultimately being the deciding factor as to whether to move forward or investigate alternatives. The patient's decision must be respected, as it is *their* care, whilst the healthcare professional should empathically support the patient and ensure they are heard and appreciated, even if their decision goes against the professional's better judgement (Beirne and Sadavoy 2021). For example, Derksen et al. (2013) found that perceived empathetic relationships by patients correlated with improvement of patient satisfaction and adherence to procedures, decreased anxiety and distress, as well as better diagnostic and clinical outcomes. Even when faced with challenging situations, patients reported their care as favourable if they felt heard and supported. In subsequent studies, patients often report better overt outcomes and greater satisfaction in their care when they are heard and respected (Chaitoff et al. 2017; Sinclair et al. 2016). Being proactive, honest, and explaining services and interventions is critical in making sure the patient is an engaged and informed participant in their own health care, but it also ensures that the patient is involved and the treatment is responsive to the environment (two criteria that lead to optimal outcomes) (Williams 2021).

## 12.5 Masking

Masking (a component of social camouflaging) is the act of consciously or unconsciously suppressing atypical behaviours, often related to neurodivergence, as well as displaying neurotypical expected behaviours to avoid discrimination or gain social acceptance (Bradley et al. 2021; Cook et al. 2022; Hull et al. 2017). Current research suggests that masking is harmful to the Autistic population with a detrimental impact on health and wellness, including experiences of exhaustion, challenges navigating stereotypes, and threats to identity and self-perception (Hull et al. 2017). In a society in which normative behaviours are favoured and selected as appropriate, good, and socially significant, masking should be assumed, especially as many Autistic traits are punished in an effort to shape human behaviour to normative ideals.

Furthermore, the assessment tools used to guide interventions are also based on neurotypical norms, resulting in a lack of discernment of common behaviours by those being assessed. Current assessments that are widely used often neglect marginalized identities (Henrich et al. 2010) and identify neurodivergent norms (differences) as deficits. For instance, non-harmful self-stimulatory behaviours are identified as maladaptive behaviour by most assessment tools. However, many Autistic people report self-stimulatory behaviour as beneficial for both emotional and sensory regulation (Kapp et al. 2019). Therefore, targeting a trait for reduction that serves as a coping strategy for emotional and physical self-regulation can be a source of harm. Another example is a common assessment target of providing eye contact when one's name is called; a target that may be deemed as a socially significant, appropriate social goal and prerequisite for making friends or performing well at a job interview. Again, through self-report of Autistic people's experiences, many share that eye contact causes overstimulation and as such an effective coping strategy that is developed intrinsically is to avoid eye contact (Trevisan et al. 2017). Moreover, in many cultures providing eye contact may be perceived as offensive, thus it is also culturally unresponsive to target eye contact blindly. Imposing eye contact at the expense of the listener's comfort reinforces a preference to present as allistic or mask (i.e., the Autistic individual is forced to comply with allistic normative behaviour at the expense of their own needs) despite the fact that across the autism population, avoiding eye contact would be considered normative (and in many cultures outside of North America). The treatment of differences has resulted in the systemic promotion of ableism and exclusionary practices in behavioural health programs, but this is wholly avoidable.

## 12.6 Social Validity and Caregiver Training

A central component to the assessment of social validity for children is caregiver concerns and requests. Although unintentional, caregiver requests are often steeped in ableist societal norms and expectations. When a young child is diagnosed, this may be the caregiver's initial introduction to neurodiversity and the Autistic community. Lack of exposure, unawareness, and societal systems that reinforce ableism may frequently result in allistic caregivers expecting their child to behave and appear in alignment with their own norms and those of their neurotypical-aged cohort. It is imperative to educate caregivers about ableism when assessing and addressing socially significant behaviours. Behaviour analysts must recognize when requests do not align with Autistic community values and be able to address these occurrences via caregiver training. For example, if a caregiver requests the reduction of non-harmful self-stimulatory behaviour, this request must be respectfully denied. The caregiver must be presented with relevant information evidencing potential harms that may ensue and why the behaviour is beneficial for their child. Additionally, caregiver training should include how to support the generalization of relevant skills and the generalization of neurodiversity-affirming practices within their own behavioural repertoire.

## 12.7 Considering Neurodiversity and the Future of Social Validity Assessment

What may be more beneficial to Autistic people are interventions that do not seek indistinguishability or conformity but rather support and affirm Autistic identity. This may be in the form of teaching the components that encompass: (1) self-determination (see ARC Self-Determination Scale) (Wehmeyer 1995); (2) self-advocacy, which supports the exploration and development of advocating for individualized accommodations; and (3) self-management, which aids in tolerating the challenges Autistic people have to endure navigating systemic ableism. When assessing effectiveness of a behaviour analytic program, it is critical that the targets are not ones that focus on behaviour reduction for Autistic traits unless those traits are unsafe, unhealthy, or unwanted as determined by the consumer. A practitioner cannot know what is best for the client unless their interventions and goals are informed by the client.

Behaviour analysts can assist society in being more inclusive. Rather than simply treating differences, practitioners can focus treatment on the identification of effective accommodations and target the three concepts which, according to Peterson and colleagues (2020), define self-determination—choice, self-control, and self-management. These behaviours can be measured and promoted within a behaviour analytic program. Choice needs to be available, otherwise the client is passively participating in their own life and following the lead of the allistic practitioner or caregiver which will promote prompt dependency and learned helplessness (Staubitz et al. 2022). For self-control, understanding consequences of choices made based on informed decisions may require accepting "larger delayed reinforcers" and overcoming impulsivity (Lerman et al. 2006, 229–232). Finally, self-management is the ability to organize one's own behaviour to access reinforcers and avoid punishers (Peterson et al. 2020). As a result, the Autistic learner will gain the skills to validate their behaviours and better understand themselves as they navigate an ableist world as an Autistic human. What is critical is *not* to treat the individual as inferior or even as Autistic; the key is to respect the autonomy of the individual. Social validity is one way in which a practitioner can truly understand the ongoing needs of those they serve.

The goal of behavioural health services for consumers within the Autistic community is not to appear physically as neurotypical (i.e., effectively mask) but to learn to manoeuvre the world authentically as an Autistic human. Instead of targeting behaviours based on what will pass as allistic, behaviour analysts should commit to teaching behaviours that will support self-advocacy and self-determination. The procedures of service should be person-centred and supportive and not dictated by an allistic-normed assessment. Social validity is our gateway into assessing the quality of treatment in the opinion of the person who matters most in the treatment—the client. Without their ongoing feedback, good intentions may cause harm. Other considerations include having Autistics on your organizations' advisory boards, paying for mentorship to become an effective ally, and recognizing your personal biases by listening to others of different cultural backgrounds. As an evidence-based practice,

ABA must source client values, be responsive to the client's needs, wants, and wishes, and continually assess the efficacy of program objectives. Doing any less is not at the heart of the science. ABA must be socially valid, which entails the behaviour analyst acknowledging and confronting societal ableism in its many forms. If they hope to develop behaviour analytic interventions that are ethically sound, it is essential to recognize that the Autism identity of the client makes them the amazing person they are.

## References

Baer Donald M., Montrose M. Wolf, and Todd R. Risley. 1968. Some current dimensions of applied behavior analysis. *Journal of Applied Behavior Analysis* 1 (1): 91–97.

Barnes, Elizabeth. 2016. *The Minority Body: A Theory of Disability*. New York: Oxford University Press.

Behavior Analyst Certification Board. 2020. *Ethics Code for Behavior Analysts*. https://bacb.com/wp-content/ethics-code-for-behavior-analysts.

Behavior Analyst Certification Board. 2022. *BACB certificant data*. https://www.bacb.com/BACB-certificant-data/. Accessed 3 Sept 2022.

Beirne, Ann, and Jacob A. Sadavoy. 2021. *Understanding Ethics in Applied Behavior Analysis: Practical Applications*. London: Routledge.

Bottema-Beutel, Kristen, Steven K. Kapp, Jessica N. Lester, Noah J. Sasson, and Brittany N. Hand. 2021. Avoiding Ableist language: Suggestions for autism researchers. *Autism in Adulthood* 3 (1): 18–29. https://doi.org/10.1089/aut.2020.0014.

Bradley, Louise, Rebecca Shaw, Simon Baron-Cohen, and Sarah Cassidy. 2021. Autistic adults' experiences of camouflaging and its perceived impact on mental health. *Autism in Adulthood* 3 (4): 320–329. https://doi.org/10.1089/aut.2020.0071.

Bury, Simon M., Rachel Jellett, Jennifer R. Spoor, and Darren Hedley. 2023. "It Defines Who I Am" or "It's Something I Have": What language do [autistic] Australian adults [on the autism spectrum] prefer? *Journal of Autism and Developmental Disorders* 53 (2): 677–687. https://doi.org/10.1007/s10803-020-04425-3.

Carr, James E., Jennifer L. Austin, Lisa N. Britton, Karen K. Kellum, and Jon S. Bailey. 1999. An assessment of social validity trends in applied behavior analysis. *Behavioral Interventions* 14 (4): 223–231. https://doi.org/10.1002/(sici)1099-078x(199910/12)14:4%3c223::aid-bin37%3e3.0.co;2-y.

Carroll, Regina A., and Claire C. Peter. 2014. Methods for assessing social validity of behavioral intervention plans for children with attention deficit hyperactivity disorder. *Acta De Investigación Psicológica* 4 (3): 1642–1656. https://doi.org/10.1016/s2007-4719(14)70971-8.

Centers for Disease Control and Prevention. 2022. *Communicating with and about people with disabilities*. https://www.cdc.gov/ncbddd/disabilityandhealth/materials/factsheets/fs-communicating-with-people.html. Accessed 12 Apr 2024.

Chaitoff, Alexander, Bob Sun, Amy Windover, Daniel Bokar, Joseph Featherall, Michael B. Rothberg, and Anita D. Misra-Hebert. 2017. Associations between physician empathy, physician characteristics, and standardized measures of patient experience. *Academic Medicine* 92 (10): 1464–1471. https://doi.org/10.1097/acm.0000000000001671.

Cook, Julia, Laura Crane, Laura Hull, Laura Bourne, and William Mandy. 2022. Camouflaging behaviours used by autistic adults during everyday social interactions. *Autism* 26 (2): 406–421. https://doi.org/10.1177/13623613211026754.

Derksen, Frans, Jozien Bensing, and Antoine Lagro-Janssen. 2013. Effectiveness of empathy in general practice: A systematic review. *British Journal of General Practice* 63 (606): e76–e84. https://doi.org/10.3399/bjgp13x660814.

Dillenburger, Karola, and Mickey Keenan. 2009. None of the A's in ABA stand for autism: Dispelling the myths. *Journal of Intellectual & Developmental Disability* 34 (2): 193–195. https://doi.org/10.1080/13668250902845244.

Dunn, Dana S., and Erin E. Andrews. 2015. Person-first and identity-first language: Developing psychologists' cultural competence using disability language. *American Psychologist* 70 (3): 255–264. https://doi.org/10.1037/a0038636.

Dwyer, Patrick. 2022. The neurodiversity approach(es): What are they and what do they mean for researchers? *Human Development* 66 (2): 73–92. https://doi.org/10.1159/000523723.

Ferguson, Julia L., Joseph H. Cihon, Justin B. Leaf, Sarah M. Van Meter, John McEachin, and Ronald Leaf. 2018. Assessment of social validity trends in the journal of applied behavior analysis. *European Journal of Behavior Analysis* 20 (1): 146–157. https://doi.org/10.1080/15021149.2018.1534771.

Ferrand, Yann B., Jennifer Siemens, Danny Weathers, Lawrence D. Fredendall, Yunsik Choi, Ronald G. Pirrallo, and Matthew Bitner. 2016. Patient satisfaction with healthcare services: A critical review. *Quality Management Journal* 23 (4): 6–22. https://doi.org/10.1080/10686967.2016.11918486.

Fouka, Georgia, and Marianna Mantzorou. 2011. What are the major ethical issues in conducting research? Is there a conflict between the research ethics and the nature of nursing? *Health Science Journal* 5 (1): 3–14.

Gill, Liz, and Lesley White. 2009. A critical review of patient satisfaction. *Leadership in Health Services* 22 (1): 8–19. https://doi.org/10.1108/17511870910927994.

Graber, Abraham, and Jessica Graber. 2023. Applied behavior analysis and the abolitionist neurodiversity critique: An ethical analysis. *Behavior Analysis in Practice* 16 (4): 921–937. https://doi.org/10.1007/s40617-023-00780-6.

Hanley, Gregory P. 2010. Toward effective and preferred programming: A case for the objective measurement of social validity with recipients of behavior-change programs. *Behavior Analysis in Practice* 3 (1): 13–21. https://doi.org/10.1007/bf03391754.

Henrich, Joseph, Steven J. Heine, and Ara Norenzayan. 2010. Beyond WEIRD: Towards a broad-based behavioral science. *Behavioral and Brain Sciences* 33 (2–3): 111–135. https://doi.org/10.1017/s0140525x1000072.

Holburn, Steve. 2002. How science can evaluate and enhance person-centered planning. *Research and Practice for Persons with Severe Disabilities* 27 (4): 250–260. https://doi.org/10.2511/rpsd.27.4.250.

Holburn, Steve, Anne Gordon, and Peter Vietze. 2007. *Person-Centered Planning Made Easy: The PICTURE Method*. Baltimore: Brookes Pub.

Hull, Laura, K.V. Petrides, Carrie Allison, Paula Smith, Simon Baron-Cohen, Meng-Chuan Lai, and William Mandy. 2017. "Putting on My Best Normal": Social camouflaging in adults with autism spectrum conditions. *Journal of Autism and Developmental Disorders* 47 (8): 2519–2534. https://doi.org/10.1007/s10803-017-3166-5.

Kapp, Steven K., Robyn Steward, Laura Crane, Daisy Elliott, Chris Elphick, Elizabeth Pellicano, and Ginny Russell. 2019. "People should be allowed to do what they like": Autistic adults' views and experiences of stimming. *Autism* 23 (7): 1782–1792. https://doi.org/10.1177/1362361319829628.

Kazdin, Alan E. 1977. Assessing the clinical or applied importance of behavior change through social validation. *Behavior Modification* 1 (4): 427–452. https://doi.org/10.1177/014544557714001.

Kenny, Lorcan, Caroline Hattersley, Bonnie Molins, Carole Buckley, Carol Povey, and Elizabeth Pellicano. 2015. Which terms should be used to describe autism? Perspectives from the UK autism community. *Autism* 20 (4): 442–462. https://doi.org/10.1177/1362361315588200.

Kieny, Marie-Paule, Timothy G. Evans, Stefano Scarpetta, Edward T. Kelley, Niek Klazinga, Ian Forde, Jeremy H. Veillard, Maurice Henri, Sheila Leatherman, Shamsuzzoha Syed, Su Mean Kim, Sepidek Bagheri Nejad, and Laim Donaldson. 2018. *Delivering quality health services: A global imperative for universal health coverage (English)*. Washington, D.C.: World Bank Group. http://documents.worldbank.org/curated/en/482771530290792652/Delivering-quality-health-services-a-global-imperative-for-universal-health-coverage.

LeBlanc, Linda A., Melissa R. Nosik, and Anna Petursdottir. 2018. Establishing consumer protections for research in human service agencies. *Behavior Analysis in Practice* 11 (4): 445–455. https://doi.org/10.1007/s40617-018-0206-3.

Lerman, Dorothea C., Laura R. Addison, and Tiffany Kodak. 2006. A preliminary analysis of self-control with aversive events: The effects of task magnitude and delay on the choices of children with autism. *Journal of Applied Behavior Analysis* 39 (2): 227–232. https://doi.org/10.1901/jaba.2006.90-05.

Lindsay, Sally, Kristina Fuentes, Sharmigaa Ragunathan, Luiza Lamaj, and Jaclyn Dyson. 2022. Ableism within health care professions: A systematic review of the experiences and impact of discrimination against health care providers with disabilities. *Disability and Rehabilitation* 45 (17): 2715–2731. https://doi.org/10.1080/09638288.2022.2107086.

Lovaas, O. Ivar. 1987. Behavioral treatment and normal educational and intellectual functioning in young autistic children. *Journal of Consulting and Clinical Psychology* 55 (1): 3–9. https://doi.org/10.1037/0022-006x.55.1.3.

Mazurenko, Olena, Taleah Collum, Alva Ferdinand, and Nir Menachemi. 2017. Predictors of hospital patient satisfaction as measured by HCAHPS. *Journal of Healthcare Management* 62 (4): 272–283. https://doi.org/10.1097/jhm-d-15-00050.

Morris, Cody, Jessica J. Detrick, and Stephanie M. Peterson. 2021. Participant assent in behavior analytic research: Considerations for participants with autism and developmental disabilities. *Journal of Applied Behavior Analysis* 54 (4): 1300–1316. https://doi.org/10.1002/jaba.859.

Nicolson, Amanda C., Junelyn F. Lazo-Pearson, and Jackie Shandy. 2020. ABA finding its heart during a pandemic: An exploration in social validity. *Behavior Analysis in Practice* 13 (4): 757–766. https://doi.org/10.1007/s40617-020-00517-9.

Oben, Patrick, and Mandy Corliss. 2021. The difference between HCAHPS and the patient experience and its implications for physicians, experience leaders, and health executives. *Journal of Patient Experience* 8: 1–2. https://doi.org/10.1177/23743735211034026.

Organization for Autism Research. 2020. *1,000 people surveyed, survey says…* https://researchautism.org/oaracle-newsletter/1000-people-surveyed-survey-says/. Accessed 12 Apr 2024.

Parsons, Marsha B., Dennis H. Reid, Erik Bentley, Amy Inman, and L. Perry Lattimore. 2012. Identifying indices of happiness and unhappiness among adults with autism: Potential targets for behavioral assessment and intervention. *Behavior Analysis in Practice* 5 (1): 15–25. https://doi.org/10.1007/bf03391814.

Peterson, Stephanie M., Elian Aljadeff-Abergel, Rebecca R. Eldridge, Nathan J. VanderWeele, and Nicholas S. Acker. 2020. Conceptualizing self-determination from a behavioral perspective: The role of choice, self-control, and self-management. *Journal of Behavioral Education* 30 (2): 299–318. https://doi.org/10.1007/s10864-020-09368-4.

Potter, Jacqueline N., Gregory P. Hanley, Matotopa Augustine, Casey J. Clay, and Meredith C. Phelps. 2013. Treating stereotypy in adolescents diagnosed with autism by refining the tactic of "using stereotypy as reinforcement." *Journal of Applied Behavior Analysis* 46 (2): 407–423. https://doi.org/10.1002/jaba.52.

Rajaraman, Adithyan, Gregory P. Hanley, Holly C. Gover, Johanna L. Staubitz, John E. Staubitz, Kathleen M. Simcoe, and Rachel Metras. 2022. Minimizing escalation by treating dangerous problem behavior within an enhanced choice model. *Behavior Analysis in Practice* 15 (1): 219–242.

Ramey, Devon, Olive Healy, and Emma McEnaney. 2023. Defining and measuring indices of happiness and unhappiness in children diagnosed with autism spectrum disorder. *Behavior Analysis in Practice* 16: 194–209.

Schwartz, Ilene S. 1991. The study of consumer behavior and social validity: An essential partnership for applied behavior analysis. *Journal of Applied Behavior Analysis* 24 (2): 241–244. https://doi.org/10.1901/jaba.1991.24-241.

Sinclair, Shane, Jill M. Norris, Shelagh J. McConnell, Harvey M. Chochinov, Thomas F. Hack, Neil A. Hagen, Susan McClement, and Shelley R. Bouchal. 2016. Compassion: A scoping review of the healthcare literature. *BMC Palliative Care* 15: 1–16. https://doi.org/10.1186/s12904-016-0080-0.

Slocum, Sarah K., and Jeffrey H. Tiger. 2011. An assessment of the efficiency of and child preference for forward and backward chaining. *Journal of Applied Behavior Analysis* 44 (4): 793–805. https://doi.org/10.1901/jaba.2011.44-793.

Slocum, Timothy A., Ronnie Detrich, Susan M. Wilczynski, Trina D. Spencer, Teri Lewis, and Katie Wolfe. 2014. The evidence-based practice of applied behavior analysis. *The Behavior Analyst* 37 (1): 41–56. https://doi.org/10.1007/s40614-014-0005-2.

Staubitz, Johanna L., John E. Staubitz, Marney S. Pollack, Rachel A. Haws, and Michelle Hopton. 2022. Effects of an enhanced choice model of skill-based treatment for students with emotional/behavioral disorders. *Journal of Applied Behavior Analysis* 55 (4): 1306–1341. https://doi.org/10.1002/jaba.952.

Taboas, Amanda, Karla Doepke, and Corinne Zimmerman. 2023. Preferences for identity-first versus person-first language in a US sample of autism stakeholders. *Autism* 27 (2): 565–570.

Trevisan, Dominic A., Nicole Roberts, Cathy Lin, and Elina Birmingham. 2017. How do adults and teens with self-declared autism spectrum disorder experience eye contact? A qualitative analysis of first-hand accounts. *PLoS ONE* 12 (11): e0188446. https://doi.org/10.1371/journal.pone.0188446.

Wehmeyer, Michael L. 1995. *ARC's Self Determination Scale*. PsycTESTS Dataset. https://doi.org/10.1037/t03340-000.

Williams, Cherelle M. 2021. Cultural responsiveness. In *A Scientific Framework for Compassion and Social Justice: Lessons in Applied Behavior Analysis*, eds. Jacob A. Sadavoy and Michelle L. Zube, 89–93. London: Routledge.

Wolf, Montrose M. 1978. Social validity: The case for subjective measurement or how applied behavior analysis is finding its heart. *Journal of Applied Behavior Analysis* 11 (2): 203–214. https://doi.org/10.1901/jaba.1978.11-203.

# Chapter 13
# Disability and Subsidiarity: Toward Social and Political Inclusion

**Parker Gamble and Lorraine Krall McCrary**

**Abstract** While the ability to access social and political rights is essential to justice for people with disabilities, so is the ability to participate in local associations and local government, which can serve as agency-generating groups, and can provide more widespread opportunities for people to participate in politics. The requirements of a just society, then, cannot only be met by the decisions of national government or by international declarations, but also require attention to local politics and associations. But local politics and local associations are sites of power, and, like any sites of power, also hold the potential for abuse. As a result, we turn to the principle of subsidiarity, which not only attends to local governance as a way of empowering individuals, but also attends to the need for oversight of local governance in order to provide support and assistance, as well as protection against abuse. Our work considers the insights of the principle of subsidiarity for theorizing a just society for people with intellectual disabilities that takes seriously the responsibilities of different levels of politics and society.

**Keywords** Disability · Disability rights · Social inclusion · Subsidiarity · Local community

## 13.1 Introduction

Through the Disability Rights Movement and the Independent Living Movement in the U.S., people with disabilities have been fighting for essential rights to education, transportation, employment, and care. Though the American with Disabilities Act (ADA) made large and crucial strides toward equality for people with disabilities, the

P. Gamble
Wabash College, Crawfordsville, USA
e-mail: prgamble23@wabash.edu

L. K. McCrary (✉)
Political Science, Wabash College, Crawfordsville, USA
e-mail: mccraryl@wabash.edu



A. Bianchi and J. A. Vogt (eds.), *Intellectual Disabilities and Autism: Ethics and Practice*, The International Library of Bioethics 108,
https://doi.org/10.1007/978-3-031-61565-8_13

lack of financial support for the ADA and uneven implementation of the act means that equal rights for people with disabilities have not been secured. ADAPT (American Disabled for Attendant Programs Today) engages in activism seeking supports for people with disabilities in community settings, rather than through institutions such as nursing homes. This chapter argues that in addition to the crucial work of activist organizations in fighting for access to social and political supports for people with disabilities, there is theoretical work to be done. The responsibilities of a just society toward people with disabilities are even more extensive than securing these essential rights. We argue that, in addition to the right to access education, transportation, employment, and care, a just society must also involve people with disabilities in the collaborative decision-making processes that are essential to self-governance—at the national, state, and local levels—as well as reflect on the importance of inclusion not only in societal spaces, but in the community life that fills those spaces. The mantra of the disability rights movement— "Nothing about us without us"—has implications not only for choice by disabled people about the providers of care, but for the involvement of people with disabilities in the governments and communities that make up our world.

Local communities and local politics are one location of care and decision making that this chapter takes up. Local communities and local politics provide the opportunity for substantial social and political engagement by a large number of people and also offer variation in experimental approaches. But local politics and local communities also have significant drawbacks: They often fail to fully incorporate disabled members and present a grave risk for mistreatment and abuses of power to occur. They can lack elite knowledge and consistent approaches to disability, as well. Rather than seeing the potential for abuse and mistreatment of marginalized people as a reason to move beyond the biases and prejudices that often accompany local communities, we see the potential of local communities as a site for supporting people with disabilities in the attainment of interdependence, as well as for practicing reciprocal practices of care. Yet the potential for abuse must be acknowledged and addressed. We turn to the principle of subsidiarity, as articulated first in Catholic social thought and as utilized within the European Union, to address this tension. Subsidiarity proposes that valuable local community should be supported by the larger levels of government, and those larger levels of government should also provide a check on community abuses and biases.

Political theorists have only recently begun to theorize questions of justice in a way that is inclusive of people with disabilities (Nussbaum 2006; Simplican 2015; Arneil and Hirschmann 2016). Some previous theorists have primarily viewed disabled people as welfare recipients—essentially a complex burden for society to grapple with, rather than fully participating and deserving members of society in their own right. Notably John Rawls (1993, 2005) excludes both those with physical and intellectual disabilities from the original position.[1] Iris Marion Young (1990, 2011), on

[1] Rawls develops the social contract approach, developing an idea of an imagined "original position," where people pass through the "veil of ignorance," forgetting or ignoring facts about themselves in

the other hand, argues that welfare distribution for marginalized people is insufficient and that justice requires "participation in public discussion and processes of democratic decision making" (1990, 91). We see participation in local community, embedded in a robust federal structure, as one essential aspect of public discussion and democratic decision making.

Other thinkers affirm the importance of local government within other levels of governance; however, the mechanisms of interaction between different levels of government, particularly insofar as it pertains to the need for a just and inclusive society for people with disabilities, has been undertheorized. MacIntyre (1999) argues that local communities are uniquely capable of nurturing virtue and the capacity for independent practical reasoning, elevating the local above other forms of governance. Kittay (2011) views care as crucial for a society that is attentive to people with disabilities, observing such care as best achievable with the development of interpersonal relationships. These arrangements, Kittay argues, should be supported by higher levels of government, though the mechanisms of interaction are largely implicit and unexplored. Nussbaum's (2006) capabilities approach also implies the need for national and international political environments, as well as local political developments, again opening the conversation to a more robust understanding of the interactions between these different political levels. We aim to explore and theorize local community in relationship to plural other forms of society and government. Whilst many argue either for the exclusive domain of the community, family, or local government in providing the best treatment for people with intellectual disabilities, and others advocate for wider state administration of benefits, we argue that a bridge can be made, enabling the observed benefits of local engagement and the potential to utilize the power of the state. Subsidiarity, a principle articulated in Catholic social thought and present at the heart of the European Union's organization, enables the most local level to wield the most dominant power, yet allows space for interventions and involvement by larger associations when necessary. First, we will provide an overview of the principle of subsidiarity, after which we will theorize its application to disability.

---

order to theorize about what justice demands in an unbiased way. However, people with disabilities are not, in his conception, allowed into the original position. He writes:

> We do not mean to say, of course, that no one ever suffers from illness and accident; such misfortunes are to be expected in the ordinary course of life, and provision for these contingencies must be made. But given our aim, I put aside for the time being these temporary disabilities and also permanent disabilities or mental disorders so severe as to prevent people from being cooperating members of society in the usual sense. Thus while we begin with an idea of the person implicit in the public political culture, we idealize and simplify this idea in various ways in order to focus first on the main question. (Rawls 1993, 20)

Rawls reintroduces people with disabilities later as the recipients of redistribution of wealth by the government.

## 13.2 Subsidiarity

The principle of subsidiarity supports the empowerment of local institutions, opposing the consumption thereof by larger organizations when possible. The principle of subsidiarity originates from Catholic social thought and was first outlined by Leo (1891) in the encyclical *Rerum Novarum*, which was written to defend the rights of the workers to associate and form unions, attempting to articulate a third way between unfettered capitalism and socialism. Leo (1891) writes that individuals and communities should have freedom to act, but:

> Whenever the general interest or any particular class suffers, or is threatened with harm, which can in no other way be met or prevented, the public authority must step in to deal with it. … The limits must be determined by the nature of the occasion which calls for the law's interference—the principle being that the law must not undertake more, nor proceed further, than is required for the remedy of the evil or the removal of the mischief.

Here he maintains that the role of the state is to step in when the general interest or the interests of a particular group of people is harmed by individuals and communities. However, he is quick to circumscribe the power of the state, holding it to the power to remove what is threatening or harming the general interest or the interests of a particular group. The principle of subsidiarity was expanded upon by Pius (1931) in *Quadragesimo Anno*. He writes:

> Just as it is gravely wrong to take from individuals what they can accomplish by their own initiative and industry and give it to the community, so also it is an injustice and at the same time a grave evil and disturbance of right order to assign to a greater and higher association what lesser and subordinate organizations can do. For every social activity ought of its very nature to furnish help to the members of the body social, and never destroy and absorb them.
>
> The supreme authority of the State ought, therefore, to let subordinate groups handle matters and concerns of lesser importance, which would otherwise dissipate its efforts greatly. Thereby the State will more freely, powerfully, and effectively do all those things that belong to it alone because it alone can do them: directing, watching, urging, restraining, as occasion requires and necessity demands.

Pius XI elaborates an idea of subsidiarity here—that functions should be fulfilled at the lowest level that can fulfill them, though this may well change over time as a result of changing circumstances (1931, §79). Higher levels of government should help those that are lower; higher governments should also direct, watch, urge, and restrain, depending on what is needed. Interpretations and conceptions of subsidiarity have varied over time, though most have consistently supported what Andreas Føllesdal (1998, 211) explains as "minimal intervention—for instance, directives rather than regulations—allowing local accommodation both to circumstances and to expectations." This principle has been further developed and implemented by the European Union, which utilizes the idea as an organizational principle in establishing a union of independent states.

The principle of subsidiarity has come to be a founding principle of the European Union, which utilizes the idea to govern relations among states and between states and the central government of the European Union. Introduced and defined in the 1992

Treaty on European Union and the 1997 Treaty of Amsterdam, subsidiarity serves as a guiding principle to encourage local control. In the context of the European Union, subsidiarity encourages decisions to be made as closely to the citizen as possible, preferring the action of nations to the action of the European Union (Pavy 2022; Subsidiarity, n.d.). The Treaty on European Union (1992) states:

> In areas which do not fall within its exclusive competence, the Community [European Union] shall take action, in accordance with the principle of subsidiarity, only if and in so far as the objectives of the proposed action cannot be sufficiently achieved by the Member States and can therefore, by reason of the scale or effects of the proposed action, be better achieved by the Community.

The principle of subsidiarity envisions a strong local unit supported and enabled by larger associations. Decisions should be made as locally to the individuals involved as possible, with intervention when a smaller actor is unable to fulfill the aims or when the aims are, due to the scale or effects, "better achieved by the [larger] Community" (Ibid.). Intervention from the larger actor may occur at the request of the national government—likely due to an acknowledged inability to effectively conduct a given action—or in response to local failures, such as a violation of rights or liberties of members of the community.

Nearly all people find themselves situated in a complex network of communities and associations. From their nation to their family, from social clubs to their city, people are, at any given time, members of and participants in a plethora of partnerships. This complex membership creates a network of networks, activated and called upon by an individual at various times based upon which is most appropriate to respond to a given need. Individuals continually navigate and draw on different resources from this web of interconnected associations. The modern state, likewise, must recognize that it is part of this network when it relates to its citizens; it relates to citizens who are themselves enmeshed within a web of relationships and associations. These relationships and associations include a variety of voluntary associations, as well as citizenship in different levels of government, from the local to the state and national. The principle of subsidiarity offers insights into the relationships among different types of government in this complicated arrangement, emphasizing the power of local communities to act, whilst recognizing the value of such a network of intertwined networks in achieving broader aims.

The principle of subsidiarity recognizes the unique capability of the levels most local to the individual in encouraging human flourishing. To this extent, the principle of subsidiarity states the most local level possible should be prioritized, with intervention occurring when such a level is incapable of effectively achieving an aim. Whether a goal can be achieved by an individual, a family, a form of local community, or needs the support of the broader collective—or even international cooperation—the effective practice of such is able to be achieved. The principle of subsidiarity does not itself set out in advance the best location for a particular goal to be achieved; rather, it recognizes that making this determination will involve the use of practical reasoning that considers the specific situation and context. Based on our interpretation of the principle of subsidiarity, the location at which a particular goal may be achieved may well change as the social and political context changes. The principle, then, encourages primarily the development of local communities, interpersonal relationships, and associations, fostering an environment of local control. The approach, however, importantly retains space for the intervention of larger-scaled actors, appearing when missteps occur within a community or goals are not met. The appropriate level from which to act may vary from aim to aim, and even from community to community, dependent upon the particularities thereof. Even within a single goal, various subgoals may be most optimally achieved through various sizes of association. The default of the principle of subsidiarity is, throughout, the most local level.

In emphasizing a strong role for the most local level, the principle of subsidiarity accurately reflects the capability of local communities to foster the greatest engagement and achievement of involved individuals. As MacIntyre (1999, 94) explains, the virtues of acknowledged dependence can best be achieved within "some form of local community within which the activities of families, workplaces, schools, clinics, clubs dedicated to debate and clubs dedicated to games and sports, and religious congregations may all find a place." MacIntyre argues that local community achieves the greatest outcomes for the development of individuals as political and ethical beings, allowing spaces for people to act. When operating within a local community, individuals are most encouraged to be involved and may most directly see the results of decisions made. The ability to engage in politics locally is of particular importance to individuals with intellectual and developmental disabilities. The local level can provide unique adaptability and accommodations, enabling those with disabilities to become more involved and engaged in the political process and in social life. This unique ability of the local enables disabled members of the community to become effective participants, engaging in, receiving care from, and providing care for other members of the community. By encouraging the development of interpersonal connections and widespread involvement in various associations, the local community encourages all to become actively engaged in the social and political structure of the whole.

The principle of subsidiarity, however, recognizes the potential for these local communities to fail, go astray, or otherwise betray the interests of any or all of their members. In such instances, higher levels of the state may be activated to respond. In observing the effects of local community, one must remain cognizant of the capacity for wrongdoing—a threat which can be responded to through the input of and oversight by broader social structures. If, for instance, a particular community fails to uphold basic human rights, an intervention may occur in which some of the local group's authority is stripped and provided instead to a larger group. This capability for particular kinds of intervention provides a buffer against the potential for local communities to act in ways that mistreat or abuse their members. Local communities that conduct themselves poorly may see their ability to act in the future restricted or rescinded. A local community which acts unjustly would ideally experience a swift pushback, receiving a thorough intervention and losing some autonomy until such wrongs are righted. Accordingly, structures that provide oversight and accountability gain great importance. These structures can, for instance, check in on the work of local associations, interview members about their experience, and ensure that there is a clear process available through which complaints may be made and violations may be reported. Such structures may take a variety of forms, including oversight by state and federal government, as well as the formation of new community organizations to provide oversight. Local community can at times present bad actors with an opportunity to mistreat their community; the principle of subsidiarity enables an intervention from larger authorities to dampen this impulse.

Not only can broader levels of association provide needed help against abuses and incapacity in more local levels of association, but they may also prove helpful in supportive ways. From securing funding to establishing common principles, the larger association can provide something of an economy of scale to benefit local communities in conducting their activities. Such networked benefits can enable the local community to effectively achieve their aims regardless of individual size or scope, given the support of larger communities. The broader level, consisting of associations of associations, may provide support in setting the goals, locating funding, and other aspects which are often efficient at scale, but allow implementation, usage, and other locally important functions to differ between communities as appropriate. In addition to allowing for agency to be generated, maximizing the flexibility and responsiveness of care, and ensuring oversight against abuses of care, subsidiarity also proliferates the sites of justice. While the principle of subsidiarity is often taken to point to the need for local communities to be supported, rather than usurped, we take the principle of subsidiarity to have further implications, as well. The responsibility for forming a just polity and society is present not only at the national level, but also at the state and local level, as well as at the level of community associations. By proliferating the sites at which we theorize justice, subsidiarity also opens up the need to theorize the relationships among these different levels of government. Turning to care with, by, or for people with disabilities gives insight into possible benefits of subsidiarity, as well as examples of how it might work in practice.

## 13.3 Subsidiarity and Disability

One example (though by no means the only one) of what subsidiarity might look like with regard to disability is present in the town of Geel, Belgium, where adults with mental illness, "boarders," have been welcomed into the homes of families in the town since the Middle Ages; this practice emerged organically out of a pilgrimage tradition (McCrary 2017). Families in the town offer care in foster family homes for adults with mental illness. They are supported by the OPZ (Openbaar Psychiatrisch Zorgcentrum), a hospital for people with psychiatric illness that administers the program. The adult foster care program is funded by Belgium's federal government and is supported by programs in the city of Geel. When boarders experience psychiatric crises, the OPZ provides hospital care until they are able to return to the family with which they live. The OPZ supports, rather than usurps, care in the family—in fact in Geel, care is largely demedicalized, though many patients take medication.

The multi-level care offered in Geel for adults with mental illness has led to a town that sees care as central to its identity. Furthermore, living in a foster family is only one choice with which people with mental illness who are not able to live alone are presented—there are opportunities to live in apartments with the support of nurses or to live in group homes, among other options. In Geel, the practice of subsidiarity enables community support, provides greater ability for care to be individualized and flexible, allows for reciprocal relationships of care to develop between boarders and foster families, and fosters accommodating communities—both within homes and within the town of Geel. Of course, abuse is a threat in Geel, as in any community, and is something for which the OPZ looks out.

Subsidiarity can establish an environment in which the local community is optimally positioned to develop widespread buy-in and support for people with disabilities; moreover, in reciprocal relationships of care, the contributions of people with disabilities can be welcomed and received. This can support the generation of agency and the widespread social and political participation of people with disabilities. Within a model of subsidiarity, individuals are most commonly contributing to their own communities and engaging primarily with those they know, collaborating with those who have the most interest in a well-functioning community. As such, individuals may also observe the direct results of their social and political contributions, and are likely to stay in the community, which allows them to continue to see the impact of their contributions over time. In operating with others in the community, including people with intellectual and developmental disabilities, citizens gain first-hand insight and an increased valuation of their political utility. Likewise, people with intellectual and developmental disabilities become integrated, engaged members of the community, even in some cases developing their own political capacity. This direct, collaborative level of engagement encourages the development of strong and nurturing interpersonal relationships. In Geel, foster families care about the identity of their town as a merciful city. They see the impact that their parents have had by bringing boarders into their homes and so do the same themselves. And boarders themselves are understood as having much to offer the foster families with whom they live. Because people with disabilities are recognized as having the ability to give care and not only to receive it, their ability to act and contribute to the family and the town is supported and developed.

Providing support for local initiatives that offer care for people with disabilities, rather than standardizing care at the federal level of government, allows for the development of different forms of care in different places that are responsive to the immediate context and which can change as that context changes. In Geel, for example, foster family care for adults with mental illness emerged out of an agrarian society and has been able to adjust and adapt to accommodate modern society in which men and women typically both work outside of the home. And Geel is only one example of community care for people with disabilities—in different places, different forms of community care have emerged in response to the needs and abilities of different people in different contexts. The principle of subsidiarity views large, centralized, one-size-fits-all solutions as problematic. Within a subsidiarity-based model, each community may provide a unique set of care services to those with intellectual and developmental disabilities on the basis of community cultural needs and desires. Within some communities, then, some families may receive support to care for loved ones, whilst others may arrange for the provision of professional caretakers. State and federal services can often be difficult to navigate and particularly complicated to change and individualize; local communities, however, can be flexible, developing unique and adaptable arrangements to incorporate, include, and involve every member, regardless of ability. The principle of subsidiarity recognizes each person, community, and context as unique and enables treating them as such. For any individual, ability represents a spectrum—subsidiarity enables a system in which a case-by-case analysis of needs can occur and care can be provided at the most optimal level. Subsidiarity, then, furthers the development of a world in which individuals and their surrounding community are free to express and pursue their desires, receiving care in a nature most suitable to their situations, with the capability of an intervention should rights or liberties be violated.

The local, while empowering people with disabilities to engage civically, is also a site of power where grave abuses may occur. Having such segmented and diverse points of power raises the number of points at which these events may occur. Yet, the local level retains an importance which, if stripped away, risks diminishing opportunities for the practice of agency for people with disabilities and all people. This reality creates a need for a thorough and strong mechanism to conduct oversight of the conduct of communities, preventing potential wrongs and responding quickly when they may occur. Such a mechanism will look different in different situations. Mechanisms for reporting concerns must be set up, and other safeguards against abuse should be implemented. If an accusation of abuse occurs, it should be thoroughly investigated with repercussions delivered as appropriate. Oversight mechanisms will work differently in different contexts: states and federal governments can provide oversight of local governments; the EU can provide oversight of the national governments that are within it; associations can also ensure that they are supported and overseen by different forms of governmental or non-governmental organizations. In Geel, for example, the federally funded and legislated OPZ provides oversight for families practicing foster family care.

If a community acts against the interests of some members, the principle of subsidiarity provides room for intervention. When such a violation occurs, the larger levels of government may impose some form of sanction, restricting the local exercise of power on such matters. Violating a member's rights, then, risks severe consequences for a community operating under the principle of subsidiarity. At Geel, the staff of the OPZ frequently visit with boarders and foster families in their home, attentive to any signs of conflict in the home, for which staff can offer support, as well as to any signs of abuse, which, when found, lead to the removal of the boarder from the home. Other members of the town also know how to reach the OPZ if they have any concerns to report. The removal of a boarder from a foster family home results in stigma for foster families from the society in Geel. Subsidiarity, then, enables the needs of each individual to be served primarily by a local, adaptable community with mechanisms to prevent abuses of power and to support those whom the local community may be unable to adequately support. If this model were implemented for citizens with intellectual and developmental disabilities, then they could receive support within their community whenever possible, whilst retaining the availability of outside care should their local community fail to meet their needs.

A board is one form of oversight mechanism for local associations. The members on such a board would come from the communities which they are overseeing, alongside board members from similar local associations in neighboring communities, to conduct active oversight of the local associations. The board would serve a critical role, identifying potential missteps and responding to wrongs, as well as ensuring that internal mechanisms exist to avoid abuse and safely report concerns about abuse. Comprising the group of community members further aligns with the principle of subsidiarity, empowering local communities to act on matters that they are capable of handling, thereby strengthening the implementation of subsidiarity and creating a buffer against the worst impulses of the human spirit. Particularly important to the composition of such a board, then, are representatives of the groups for whom the threat of abuse is most substantial. People with disabilities should be present in the process of oversight of local communities whenever possible. Leaders of marginalized communities may come to play an important role in addressing the wrongs committed against their communities. Responses, should wrongdoing be discovered, may take the form of proportional, agreed upon repercussions against involved individuals or a removal of some of the local association's autonomy for group-level conduct.

## 13.4 Toward Robust Social and Political Inclusion

The principle of subsidiarity provides a way to theorize the responsibilities of different groups for the social and political inclusion of individuals with disabilities—the responsibility for inclusion occurs not only at the state or national level, but also at the level of local government and through local associations. In fact, local government and local associations have the potential to involve a far higher percentage of citizens than state or federal government does and to involve them more deeply, thereby widening the opportunity for people to develop their agency and practice political participation. The flexibility and particularity of the local level holds promise as a realm for including people with disabilities in a way that supports their individuality and accommodates their particular needs; on the other hand, local associations and local government may be the loci of abuse, exclusion, and oppression, perhaps especially for people in positions of vulnerability. But advocating for subsidiarity differs from simply advocating for localism. Subsidiarity recognizes that not all local governments and local associations are good and asserts the need for higher levels of government to support them and to provide oversight against abuses. Subsidiarity implies the need for the construction of oversight bodies, which serve as a protection against the potential downfalls of local government and local associations, while keeping alive the possibility of greater social inclusion by supporting, rather than eliminating, local government and local associations. Ultimately, if this principle were implemented, then people with intellectual and developmental disabilities, who are typically not involved in local communities and political structures, could meaningfully participate in and contribute to the society in which they live.

## References


Arneil, Barbara, and Nancy Hirschmann. 2016. *Disability and Political Theory*. Cambridge: Cambridge University Press.

Føllesdal, Andreas. 1998. Survey article: Subsidiarity. *The Journal of Political Philosophy* 6 (2): 190–218.

Kittay, Eva Feder. 2011. The ethics of care, dependence, and disability. *Ratio Juris* 24 (1): 49–58.

Leo XIII. 1891. *Rerum Novarum*. https://www.vatican.va/content/leo-xiii/en/encyclicals/documents/hf_l-xiii_enc_15051891_rerum-novarum.html. Accessed 9 Apr 2024.

MacIntyre, Alasdair. 1999. *Dependent Rational Animals*. Chicago, IL: Open Court.

McCrary, Lorraine. 2017. Geel's family care tradition: Toward the social inclusion of persons with disabilities. *Disability and Human Rights Special Issue of the Journal of Literary and Cultural Disability Studies* 11 (3): 285–301.

Nussbaum, Martha C. 2006. *Frontiers of Justice: Disability, Nationality, Species Membership*. Cambridge, MA: Harvard University Press.https://doi.org/10.4159/9780674041578.

Pavy, Eeva. 2022. The principle of subsidiarity. European Parliament: Fact sheets on the European Union. https://www.europarl.europa.eu/factsheets/en/sheet/7/the-principle-of-subsidiarity.

Pius XI. 1931. *Quadragesimo Anno*. https://www.vatican.va/content/pius-xi/en/encyclicals/documents/hf_p-xi_enc_19310515_quadragesimo-anno.html. Accessed 9 Apr 2024.

Rawls, John. 2005. *Political Liberalism*. New York: Columbia University Press.

Rawls, John. 1993. *Political Liberalism*. New York: Columbia University Press.

Simplican, Stacy. 2015. *The Capacity Contract: Intellectual Disability and the Question of Citizenship*. Minnesota Press.

Subsidiarity. European Commission Glossary. https://ec.europa.eu/regional_policy/policy/what/glossary/subsidiarity_en. Accessed 9 Apr 2024.

*Treaty on European Union*. 1992. https://eur-lex.europa.eu/legal-content/EN/TXT/HTML/?uri=CELEX:11992M/TXT&from=EN. Accessed 9 Apr 2024.

Young, Iris M. 1990/2011. *Justice and the Politics of Difference*. Princeton: Princeton University Press.

# Part IV
# Inclusion and Empowerment: Practical Approaches

# Chapter 14
# How Does the WHO's Framework for Functioning, Disability and Health (ICF) Provide an Ethical Foundation for 21st Century Clinicians?

**Gabriel M. Ronen and Peter L. Rosenbaum**

**Abstract** The WHO's twenty-first century framework for health, the International Classification of Functioning, Disability and Health (the ICF), integrates many elements of people's lives to enable people to create a unique self-portrait. ICF language illustrates the evolution of concepts like "impairment" and "disability". In this essay we illustrate how the ICF's approach to health ranges well beyond traditional biomedical diagnosis as the entry point for services and offers opportunities to contextualize a person with any health "impairment" in terms of their functioning, personal values and preferences, and environments. We argue that *the use of the ICF makes it possible for health services to provide care that addresses Beauchamp and Childress's four ethical principles. Justice* is served by the universality of the ICF. Incorporating the voices and values of people respects individual *autonomy.* Codesigning approaches to care with providers facilitates achievement of *beneficence* (doing the right things in the context of people's own voices and values) and avoids *maleficence* (doing things judged not to be in the best interests of the person). We believe that this ICF-based approach to health and health care offers a unique ethical framework.

**Keywords** Biopsychosocial health · Classifications of health systems · Ethical principles · F-words · Human rights · Taxonomy

---

G. M. Ronen (✉) · P. L. Rosenbaum
Department of Pediatrics and CanChild Centre for Childhood Disability Research, McMaster University, Hamilton, ON, Canada
e-mail: roneng@mcmaster.ca

P. L. Rosenbaum
e-mail: rosenbau@mcmaster.ca



A. Bianchi and J. A. Vogt (eds.), *Intellectual Disabilities and Autism: Ethics and Practice*, The International Library of Bioethics 108,
https://doi.org/10.1007/978-3-031-61565-8_14

## 14.1 Background to the ICF and F-Words

In 2001 the World Health Organization (WHO) published a revision of its original 1980 effort to classify and discuss the consequences of diseases, injuries, or disorders. Twenty-one years after the linear ICIDH framework (WHO 1980), the *International Classification of Functioning, Disability and Health* (known as the *ICF*) (WHO 2001) offers an integrated approach to health that brings together in a single framework what are often presented as the disparate threads of the biomedical, psychological, and social dimensions and concepts about health (Fig. 14.1). The classification component of the ICF is an elaborate alphanumeric analogue of WHO's International Classification of Diseases (WHO 2019/2021); however, for this chapter it is the framework rather than the classification that we find most compelling, and on which the ideas in this chapter are focused.

> The ethical importance of the ICF concepts is that they offer universality, with the corollary of providing a lens of justice and equity for everybody.

There are many healthcare classifications, but few if any are universally applicable to all individuals and populations irrespective of their demographic characteristics or underlying biomedical pathology. Thus, a first argument about the ethical importance of the ICF concepts is that they offer universality, with the corollary of providing a lens of justice and equity for everybody. As a classification system, the ICF can enable governments, researchers, and the health care sector worldwide to exchange perspectives on individuals' health, and aggregate health data, using an *internationally consistent taxonomy* familiar to all. Because of its universal applicability, the ICF *can be equally applied to vulnerable individuals and populations* of all ages (the ethical principle of justice); it is also a valuable tool to assess an integrated picture of the *biopsychosocial functioning of people with disabilities (PWD)* and to follow their progress with appropriate management plans and interventions based on

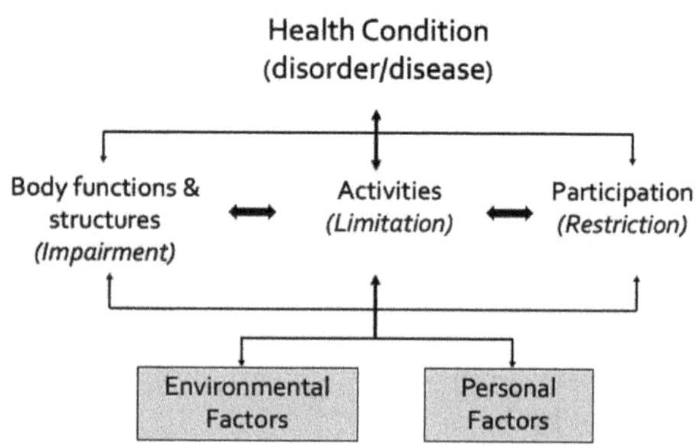


**Fig. 14.1** The ICF framework. Reprinted from WHO 2001, p. 18 © 2001

patient perspectives (the ethical principle of autonomy). In this chapter we focus on the ethical implications of adopting the ICF framework in the context of children and young people with long-term neurological impairments who experience disabilities. We also reflect on how the ICF contributes to the understanding and application of ethics in the world of disability.

### *14.1.1 The ICF Concepts Animated and Adapted into F-Words*

A popular adaptation and animation of the ICF concepts, created in the context of child and youth disability, their families, and close social networks, is the work of CanChild Centre for Childhood Disability Research (to which both authors belong). Broadly branded as *The F-words for Child Development*, these ideas (Fig. 14.2) are meant to bring the ICF concepts to life with whimsical ideas that enable people with impairments and their families to personalize their reality in the context of this integrated universal ICF framework for health (Rosenbaum and Gorter 2012). A useful additional component to the ICF framework is the introduction of the dimension of *future* to the *F words* concepts. This addition alerts people to the realization that children's lives expand beyond the present once we consider the ICF (and particularly the F-Words) as a functioning model of the family as a unit, and that *future concerns and planning* are, or should always be, present in parents' and professionals' minds. By bringing the dimension of time into focus, we encourage families and service providers to consider how advice about current issues may have longer-term impacts, and how aspects of the child's or family's future may impact current advice and interventions.

These ideas, and the countless tools to enable families and service providers to use them (see CanChild 2023), have captured people's attention around the world and made this presentation of the ICF widely accessible and understandable.

## 14.2 The ICF Through an Ethical Lens

There are myriad ethical challenges in the clinical arena, in health services research activities, and in the creation of health policies for PWD (Rosenbaum et al. 2016). Many are enshrined in international and national conventions (e.g., for the rights of persons with disabilities) (United Nations 2006) and those of children (United Nations 1989). When contemplating ethical approaches in the context of PWD one has to consider the right of PWD to have their voices heard and to be empowered to participate actively in their own healthcare decisions (Ronen et al. 2022; Ronen 2023). In addition, one needs to identify potential disparities in health autonomy and representation among PWD, and to evaluate potential risks and benefits of any

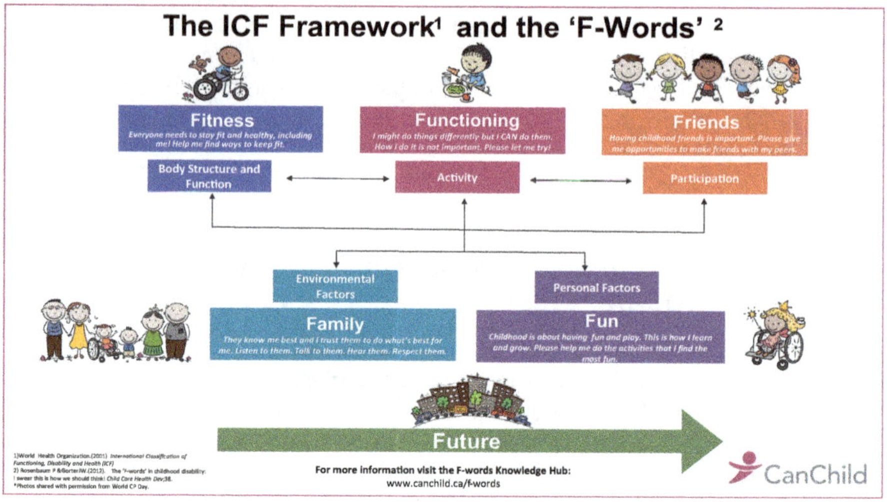


**Fig. 14.2** The F-words for Child Development, integrated with the WHO's ICF framework [Reproduced with permission of CanChild Centre for Childhood Disability Research, McMaster University (See www.canchild.ca/f-words)]

intervention strategy or technique. Consequently, one would seek to build rapport in a flexible and creative atmosphere, as is possible with the F-Words initiative. This would include developing and applying appropriate non-verbal communication techniques to enhance communication and to identify diverse person-focused opportunities for biopsychosocial interventions consistent with each individual's values, wishes, and capacities (Cruz Rivera et al. 2022), illustrating the ethical principle of justice.

Several key attributes of the ICF prompt us to encourage colleagues to consider ethical implications across many dimensions of our work in healthcare. To begin with, the ICF framework is an internationally crafted interdisciplinary universal approach to health and functioning that is practical, non-categorical, and etiology-neutral. The ICF framework makes it imperative to "rule in" personal and environmental factors to describe the *person* with a chronic impairment beyond their health issues. These highly individualized perspectives then contribute to an inclusive and useful understanding of the possible *impacts* of the health condition on the individual experiencing it, complementary to the implications of how the condition is viewed and managed biomedically. The expectation to consider the social determinants of health in each situation is incorporated into the ICF profile and is not assumed to exist and then forgotten. This way of thinking about the person-with-the-condition facilitates the identification and organization of issues, problems, strengths, and resources into a coherent "whole". In light of this novel framework for health one might then ask whether it is ethical to focus solely on the biomedical components of a person's health issues, as has been the tradition in health care?

Is it ethical to focus solely on the biomedical components of a person's health issues?

With our current expectation of seeking, hearing, and listening to the voices of people with lived experience of the condition, this ICF-based approach (with the F-words tools and language) empowers the patient and their family as partners in rational personal decision-making about interventions and services (autonomy). This is ultimately helpful to service providers and families, insofar as it guides providers toward issues important to the individual, in what can be considered an expression of "personalized medicine". Furthermore, the use of an ICF-informed approach to health-care encourages families to express their own ethical dilemmas. An example might be an individual's or family's decision to refuse interventions (e.g., mechanical ventilation for severe respiratory difficulties in people with progressive muscular dystrophies)—something they deem to be risky or otherwise not consistent with their values and wishes—even if such intervention is known to be "evidence-based" in other circumstances.

The use of an ICF-informed approach to health-care encourages families to express their own ethical dilemmas.

This integrated framework opens multiple points of entry for interventions that are important to the individual, moving far beyond the health care system's traditional focus on biomedical interventions. This way of thinking has increasingly been taken up by providers and policy makers in many areas of medicine, including interprofessional education and collaboration for both clinical care and health services research (Kraus de Camargo et al. 2018; Ronen et al. 2020; Kauff et al. 2023). Jennifer Johannesen, the mother of a child with multiple disabilities, wrote *No Ordinary Boy* (Johannesen 2011), a book about her journey with her son. After his death she learned about the ICF and concluded that when the ICF concepts are well used, members of the healthcare team collaborate, aligning their care toward shared goals. She also noted that families may discover that adherence to treatment regimens designed to achieve "normal development" of their impaired child may impede explorations for opportunities for fun participation, and that the ICF (and the F-Words) may support parents of disabled children to prioritize their goals and decisions. Johannesen recognized that for families the ICF can be helpful in multiple ways, namely: (i) as a *dialogue tool* (stimulating creative problem solving); (ii) as a *decision aid* (discussing the potential holistic impacts of interventions beyond the biomedical ones); (iii) as a *planning and communications support tool for families* (reflecting on their own situation); and (iv) to promote *compassionate intervention* (bringing a sense of possibilities to a family when the biomedical approaches may be futile) (Johannesen 2018).

## 14.3 Ethical Aspects of Classification Systems

As noted earlier, the ICF is a classification system of health functioning and not a priori an ethical tool. However, the ICF offers additional specific advantages that can assist in ethical deliberation.

In general, reliable and valid classification systems intend, and are developed, to apply a unifying language designed to improve the quality, clarity, and consistency of communication. This in turn facilitates comparison and transfer of knowledge and specific data among studies, decreasing the need to re-create parallel systems or misuse existing ones (Towns et al. 2017). This should lead to more accurate and consistent definitions of constructs, allowing development of hierarchical levels within the system, helping with integrating data, identifying new elements, and enhancing understanding of how components interrelate. These consistent approaches may help to identify additional stakeholders and allow measurements and meta-analyses. Building sound valid and reliable classifications into our daily healthcare systems would improve medical practice, clinical research, health policy, and allocation of funding, and indirectly assist ethical reflections and deliberations. As one example, validated classification systems (i) improve communication because they have consistent meaning, and (ii) get past the traditional unreliable and value-laden words like *mild*, *moderate*, and *severe* to describe professionals' judgments of the impact of impairments (Rosenbaum et al. 2014). Potentially more important would be the consequences of disregarding or misrepresenting critical classification systems, analogous to disregarding evidence-based medicine, such as ignoring the presence of an allergy when prescribing antibiotics, that may impede individuals or subpopulations from receiving the best available interventions and services. Such disregard or misrepresentation may even directly or indirectly lead to injustice and harm (violating *justice* and *non-maleficence,* two of the four key ethical principles articulated by Beauchamp and Childress (2019)).

## 14.4 The ICF Language Illustrates Its Utility as an Ethical Tool

The ethical significance of the ICF's universality is that it is inclusive: there is no legitimate reason why any minority group should be denied respect, services, or intervention. The language promoted in the ICF reflects the evolution of our thinking about functional differences. In current ICF terminology, "*impairments*" remains the term to refer to the biomedical manifestations of a health condition at the level of body systems; the external consequences of an impairment on a person's *activity* reflect a formal interest in *functioning* (as in *activities* of daily living). The extent that an impairment interferes with a person's ability to undertake their usual social roles (restricting their *participation*) speaks to the *impact* of the impairment on the person as a social being. The ICF thus presents a dynamic interconnected and ultimately transactional way of considering not only the several elements of the ICF but equally their likely impact on each other.

Traditional conceptualizations of people's limitations reflected a belief that there was a *within-the-person* problem to be "fixed"; by contrast, in the ICF formulation "disability" is a *social* construct that highlights barriers in the external world that prevent a person with an "impairment" from being fully engaged in life in ways important to them. Consider the many physical, attitudinal, political, financial, and other impediments to people who function differently from usual, and whose horizons may be constrained by factors outside themselves. However, provision of *environmental* interventions—be they physical (wheelchairs and ramps) or policy-related (legislating against discrimination in employment for people with impairments)—can empower people to be fully engaged in society with their individual capabilities *despite* irremediable impairments in body structure and function.

Utilizing the ICF may direct service providers to appropriate child- and family-centered intervention(s), such as improving the family's social or physical environment and providing guidance towards national legislation mandating accessibility to public places (Basu et al. 2023).

Impairment alone is not responsible for the personal and social disadvantages often associated with it.

Health professionals should abandon the assertion that privileges biomedical interventions, such as genetic engineering, as primary approaches to overcome neurological and developmental disability by reducing the level of impairments and prolonging survival. The "disability" is as much the product of the individual's physical, social, and attitudinal environments as the biomedical impairment (World Health Organization 2023). Indeed, in many instances, the bulk of the disvalue of living with an impairment—too often alluded to pejoratively over the last two centuries with terms such as as "handicap", "unfit", "moral delinquency", "mental defective" or "degenerate" that create isolation, poverty, lack of education or employment opportunities—can be overcome by employing the ICF concepts as a guide toward interventions to enhance functioning and engagement in life. The ICF identifies that there is no place for societal hierarchies or eugenics in any society since reality shows that a person's environment plays a major role in mediating many of the traits proponents of eugenics argue can be eradicated by sterilization or extermination (Ronen et al. 2009). One of the most notorious decisions of the US Supreme Court was its 1927 decision to uphold the involuntary sterilization of a woman deemed "feeble-minded" under *Buck v. Bell*, which paved the way for over 60,000 sterilizations of the "unfit" in the US. Sadly, *Buck v. Bell* has never been overturned (Cohen 2016). The eugenic movements have missed what the ICF can illustrate of anybody's ability to enrich their close society, whether it is a source of laughter, observed joy, or enriching the family experience.

The ICF's glossary, cardinal components and contextual factors may help clarify what factors are associated with providers' communication difficulties and point to skills and processes that professionals can acquire to become effective communicators from the onset of their career, much earlier than (possibly) acquiring these skills in the course of a lifelong experience. Using the holistic ICF framework to guide our thinking, history-taking, and formulation of the issues—if these have been formally identified rather than simply assumed to be part of a person's reality—would allow the person and their issues to be distinguished from the impairments that have been the traditional focus of medical care. Through the ICF framework, one may identify the nature of problems being experienced by the individual and the family and detect reversible causes of distress. By using the ICF terminology one can avoid any perceived stigmatizing language or misconceptions. Furthermore, using the ICF framework has the potential to foster realistic hope, helping to explain the potential benefit or futility of certain interventions or alternative treatments. This kind of conceptual illumination can be liberating for everyone—it impacts not only people with impairments but also their families, service providers, clinical programs, and even policy-makers (Soper et al. 2019).

## 14.5 The ICF Concepts as an Ethical Tool

Impairments are now recognized as traits integral to all human experience, to which everyone is likely to be exposed at some point in their lives. Disability results from the interaction between health conditions such as body or mental health impairment, and a range of environmental and personal factors. An estimated 1.3 billion people—or 16% of the global population—experience a significant disability today. This number is growing because of an increase in noncommunicable diseases and people living longer (WHO 2023). Thus, universal "justice in the allocation of resources requires equal consideration to all people with disabilities" worldwide (Bickenbach 2012, S165).

> Impairments are integral to all human experience. Disability results from the interaction between health conditions and a range of environmental and personal factors.

According to Bickenbach (2012, S164) "the ICF carries with it what might be called *instrumental ethical issues* associated with the application of the classification to human beings". To our way of thinking, core ethical elements within the ICF framework promote and foster several perspectives: (i) early detection and intervention, specifically where such services are not routinely available (Basu et al. 2023); (ii) a positive approach, emphasizing the child's and family's strengths rather than focusing on deficits and problems; (iii) autonomy of the individual who contributes to the identification of their issues in their own words, leading to (iv) respect and a dignified approach for the person with the impairment; (v) promoting a person- and family-centred approach to all aspects of service delivery; (vi) thus enabling service providers to immerse ourselves in the patient's narrative and environment to create (vii) an inclusive, integrated and holistic approach to care.

Relevant implications of these arguments include an organized, non-didactic, and non-prescriptive guide on how ethical questions that arise can be identified and justly addressed (Dan 2016). This then offers learners a framework to understand and apply the intersecting fields of good clinical practice, morality, and ethics and helps them develop practical ethical thinking and empathy, hopefully enabling detection and prevention of the potential to create harm. Ethically, any intervention that improves functioning as assessed *objectively* (from the "outside") needs also to be able to show improvement in *relevant person-reported outcome measures*, including quality of life, well-being, and satisfaction with the care received. This reflection is paramount when one contemplates introducing novel (and usually expensive) technologies such as artificial intelligence equipment into the health system of PWDs, in order to guard against the risk of a "technological imperative" that implies that "if something can be done, then it must be done" (Dan 2021). Here the ICF, in combination with person-reported quality of life, well-being, and satisfaction, should serve as a benchmark to decide about the usefulness of introducing any technology into the health system, and more specifically to guide the development of novel technologies for PWDs. This concept can and should be applied to give parents permission to decline available and approved novel therapies recommended by overly optimistic clinicians, such as the emergent gene therapies of infantile muscular atrophy (Hartsock 2023). Indeed, while the field of genetics has concentrated on how to achieve better gene therapies there has been less focus on whether, when, and why one should use such technologies (Kratzer et al. 2022).

Articulating fundamental questions becomes especially meaningful for clinicians working with populations complexly impaired by genetic conditions, where novel interventions are in sight or already exist. These fundamental questions include: (i) What do we mean by the person's best interest and by advocacy—and who should make these judgements? Similarly, (ii) Who should decide whether people are indeed "better" after an intervention? Next, (iii) by what yardsticks can we evaluate the person's changes in their health over time in relation to the interventions? These questions lead to additional considerations, namely: (iv) Do individuals with a health condition share the researchers' perspectives, or do they have additional or different concerns about their functioning and health? If so, (v) What outcomes should be assessed and with what frames of reference as our guide (Ronen et al. 2022; Ronen 2023)? These questions are anchored within the ICF framework and correspond to the basic ethical principles of autonomy, justice, beneficence, and non-maleficence (Beauchamp and Childress 2019). However, these principles have not always been applied to gene therapy development, research, and trial design (Kratzer et al. 2022). For example, to adhere to the principle of *justice*, gene therapies should be cost-effective and equally accessible to all patients; they should be designed to limit potential negative side-effects under the principle of *non-maleficence*; they should be expected to improve the patient's functioning as realized in the ICF concepts and quality of life under the principle of *beneficence*; and they should be offered in a manner that includes disclosure of all conflicts of interest and potential side-effects so that the patient may make an informed decision under the principle of *respect for autonomy* (Kratzer et al. 2022).

It is important to remind readers of the complexity of intersectionality when considering both the extremes of age and the presence of impairments in health and development that affect people's functioning. For instance, while older people are at greater risk from storms, heat and air pollution (Brunkard et al. 2008; Basu and Ostro 2008), individuals with disabilities, especially those with reduced adaptive capacity, are among the most vulnerable people in the world—the most likely to suffer, and the least likely to survive the consequences of environmental and climate change events (Bell et al. 2018; Kreslake et al. 2016). Systems designed to assist people in times of emergency are often inaccessible to disabled people, as critically illustrated in the Pulitzer Prize-winning book *Five Days at Memorial* by Sheri Fink about the catastrophe during hurricane Katrina (Fink 2013). Furthermore, in many parts of the world, emergency management does not even recognize disability to be of priority concern. The most accurate prevalence data on the impact on health come from Japan's coastal area following the 2011 tsunami, where despite a one-hour interval between the earthquake and the warning of the tsunami, the death rate was at least twice as high for persons registered as disabled compared to the rest of the population (Japanese Society for Rehabilitation of Persons with Disabilities 2015). Adverse effects of malaria, diarrhea, undernutrition, environmental events, and air pollution are presently concentrated among children, for reasons of physiological susceptibility (Perera and Nadeau 2022). Children are more vulnerable to heat-related illnesses, owing to their small body mass to surface area ratio, although evidence of excess heat-related mortality in this age group is mixed (Basu and Ostro 2008; Sari Kovats and Hajat 2008). After a natural disaster children experience a range of aggravations such as fear of death or loss of a loved one, the loss of a home and community, displacement to a strange neighborhood or school, and even separation from their family (Powell and Holleran-Steiker 2017).

The full impacts of climate change on biological and human systems are probably being seriously underestimated (Thornton et al. 2014). Healthcare professionals and advocacy groups should recognize the perils resulting from climate change and our responsibility to become involved to ensure the safety and preparedness of disabled people and their families. Our primary responsibilities toward disabled people and their families are to recognize the existing and exacerbating problems arising from climate change, to avoid creating new problems for vulnerable populations, and to mitigate existing problems (Levy and Patz 2015). As numerous environmental factors are playing a role in the onset and severity of human diseases, understanding the modulatory effect of climate change is a priority.

This reality is highlighted because an emerging contemporary application of the ICF as an instrumental ethical tool for disability-related health issues is its potential role in environmental and climate change events. In their systematic review to identify factors associated with people's vulnerability to climate change events, Gaskin et al. (2017) used the ICF framework to both tabulate the factors of susceptibility to climate change and put forward their recommendations for survival plans for vulnerable populations. Factors involving impairments of body functions and structures included cognitive and hearing impairments, thermoregulation difficulties, and the relapse, exacerbation, or progression of impairments. Factors relating to activity limitations and participation restrictions included limited preparedness, difficulties with evacuation, and difficulties reassembling individual accommodations and repairing or replacing adaptive equipment, some of which (like ventilators) may literally be life-sustaining. Common environmental factors included limited practical support from government agencies and disability organizations. Personal factors that increase vulnerability included being female, uncoupled or living alone, nonwhite ethnicity, and having a low income. The adaptive capacity of people with disability seems to be promoted through the availability of information (education) and practical support from many sources of disability and mainstream organizations, family and friends, television, teachers, and the accessibility of shelters. Unfortunately, there is a lack of studies specific to people with intellectual disability.

By utilizing the ICF framework, Gaskin et al. (2017) recommended survival plans based on four event phases: (i) preparedness, (ii) evacuation, (iii) during the event, and (iv) recovery (Table 14.1). This example illustrates the usefulness and applicability of the ICF framework and concepts in climate and environmental justice.

As with any other classification, one needs to guard against abuse in the application of the ICF—where dignity can be ignored and confidentiality offended. Patients and families should therefore be informed and provide consent when the intention is to collect and disseminate data. Researchers and clinicians should also guard against misusing the ICF as a justification for what Kratzer et al. (2022) cautioned as a serious ethical hurdle to heritable human genome editing, thus opening a "slippery slope association with eugenics". The authors warned against developing or pursuing unnecessary "desirable" enhancement technics of natural human traits such as cognition, height, physical ability, or attractiveness, and in so doing risking the creation of a new class system of genetic haves and have-nots.

## 14.6 Conclusion

The paradigm shift brought about by the underlying *intrinsic ethical issues* and *protective practices* inherent in the ICF is no longer in dispute. Furthermore, *the ICF should provide an impetus for developing further disability-friendly, ethically-informed, and protective policies* (Bickenbach 2012). In our opinion the ICF should be taught to all health professionals-in-training, and used routinely in all clinical settings, and in an inclusive and equitable fashion for all underserved vulnerable populations. These recommendations parallel those for the use of patient-reported outcomes (Calvert et al. 2022).

**Table 14.1** Adapted from Gaskin et al. (2017) Mapping the factors associated with adaptive capacity onto the ICF framework. © American Meteorological Society. Used with permission (The items in italic were added to the original table)

| Event phase | Body structure and function | Activity and participation | Environmental factors | Personal factors |
|---|---|---|---|---|
| Preparedness | *Impaired structures and/or functions* | Emergency planning, Keeping an emergency pack Seeking information from family, caregivers, and friends *Keep means to assist communication* | Emergency planning, support from disability organizations, Information from television and teachers *Assess health and functional status of individuals with impairments to determine their needs of support* *Developing warning systems, communicating helpful practices (i.e., cooling practices)* (Kreslake et al. 2016) *Disaster risk management* (Ebi et al. 2021) Living close to key services *Keep contact list at a central community site* *Assistive Communication means availability of essential prescription medications, household medicines and list of prescribed medications at hand* | *Health literacy about own condition and needs* Formal emergency education, Faith, *Building resilience* (Ebi et al. 2021) |
| Evacuation | | *Keep open line of communication with rescue services* | Support to evacuate from family and friends | |
| During the event | | *Communication* | Accessible shelters | *Potential traumatic events* |
| Recovery from the event | *Potential post-traumatic impaired mental functions* Potential additional injuries | | *Reassess health, functional status, and need for medications. Any technical equipment (e.g., ventilator, mobility device, communication aids…)* Practical support from: (i) mainstream organizations (ii) disability organizations (iii) family and friends | |

## References


Basu, Anna, Kathleen Friel, Bolajoko Olusania, and Mijna Hadders-Algra. 2023. Viewing disability: Seeing the other side. *Dev Med Child Neurol* 65: 443–445.

Basu, Rupa, and Bart D. Ostro. 2008. A multicounty analysis identifying the populations vulnerable to mortality associated with high ambient temperature in California. *Am J Epidemiol* 168: 632–637.

Beauchamp, Tom L., and James F. Childress. 2019. *Principles of Biomedical Ethics*, 8th ed. New York: Oxford University Press.

Bell, Jesse E., Claudia L. Brown, Kathryn Conlon, Stephanie Herring, Kenneth E. Kunkel, Jay Lawrimore, George Luber, Carl Schreck, Adam Smith, and Christopher Uejio. 2018. Changes in extreme events and the potential impacts on human health. *J Air Waste Manag Assoc* 68: 265–287.

Bickenbach, Jerome. 2012. Ethics, disability and the international classification of functioning, disability and health. *Am J Phys Med Rehabil* 91 (13 Suppl 1): S163–167.

Brunkar, Joan, Gonza Namulanda, and Raoult Ratard. 2008. Hurricane Katrina deaths, Louisiana, 2005. *Disaster Med Public Health Preparedness* 2 (4): 215–223. https://doi.org/10.1097/DMP.0b013e31818aaf55.

Calvert, Melanie J., Samantha Cruz Rivera, Ameeta Retzer, et al. 2022. Patient reported outcome assessment must be inclusive and equitable. *Nat Med* 28: 1120–1124.

CanChild Centre for Disability Research at McMaster University. 2023. *F-words in Childhood Disability*. www.canchild.ca/f-words. Accessed 10 Apr 2024.

Cohen, Adam. 2016. *Imbeciles: The Supreme Court, American Eugenics, and the Sterilization of Carrie Buck*. New York: Penguin Press.

Dan, Bernard. 2016. Epilogue looking back to the future. In *Ethics in Childhood Neurodisability: Cases, Principles and Clinical Practice*, eds. Peter L. Rosenbaum, Gabriel M. Ronen, Bernard Dan, Eric Racine, and Jennifer Johannesen, 353–358. London: Mac Keith Press.

Dan, Bernard. 2021. New ethical issues in cerebral palsy. *Front Neurol* 12: 650653.

de Camargo, Olaf Kraus, Liane Simon, Gabriel M. Ronen, and Peter L. Rosenbaum. 2018. *ICF: A hands-on approach for clinicians and families*. London: Mac Keith Press. 978-1-911612-04-9.

Ebi, Kristie L., Jennifer Vanos, Jane W. Baldwin, et al. 2021. Extreme weather and climate change: Population health and health system implications. *Annu Rev Public Health* 42: 293–315.

Fink, Sheri. 2013. *Five days at Memorial: Life and Death in a Storm-Ravaged Hospital*. New York: Crown Publishers.

Gaskin, Cadeyrn J., Davina Taylor, Susan Kinnear, et al. 2017. Factors associated with the climate change vulnerability and the adaptive capacity of people with disability: A systematic review. *Weather Climate Soc* 19: 801–814.

Hartsock, Jane A. 2023. The white coat cape: An ethical analysis of emerging therapies to treat spinal muscular atrophy. *Sem Ped Neurol* 45: 101036.

Japanese Society for Rehabilitation of Persons with Disabilities. 2015. Death rate of the citizen, death number and rate of people with disabilities by the disaster in MIYAGI prefecture. https://www.dinf.ne.jp/doc/english/resource/JDF/appendix1.html. Accessed 10 Apr 2024.

Johannesen, Jennifer. 2011. *No ordinary boy*. Toronto: Low to the Ground Publishing.

Johannesen, Jennifer. 2018. The ICF from the parent perspective. In *ICF: A hands-on approach for clinicians and families*, eds. Olaf Kraus de Camargo, Liane Simon, Gabriel M. Ronen, and Peter L. Rosenbaum, pp 123–134. London: Mac Keith Press, 978-1-911612-04-9.

Kauff, Mathias, Thorsten Bührmann, Friederike Gölz, et al. 2023. Teaching interprofessional collaboration among future healthcare professionals. *Front Psychol* 14: 1185730.

Kovats, R. Sari, and Shakoor Hajat. 2008. Heat stress and public health: A critical review. *Annu Rev Public Health* 29 (1): 41–45.

Kratzer, Kateryna, Landon J. Getz, Thibaut Peterlini, Jean-Yves Masson, and Graham Dellaire. 2022. Addressing the dark matter of gene therapy: Technical and ethical barriers to clinical application. *Hum Genet* 141 (6): 1175–1193.

Kreslake, Jennifer M., Katherine M. Price, and Mona Sarfary. 2016. Developing effective communication materials on the health effects of climate change for vulnerable groups: A mixed methods study. *BMC Public Health* 16: 946.

Levy, Barry S., and Jonathan A. Patz. 2015. Climate change, human rights, and social justice. *Ann Glob Health* 81: 310–322.

Perera, Frederica, and Kari Nadeau. 2022. Climate change, fossil-fuel pollution, and children's health. *NEJM* 386: 2303–2314.

Powell, Tara, and Lori K. Holleran-Steiker. 2017. Supporting children after a disaster: A case study of a psychosocial school-based intervention. *Clin Soc Work J* 45: 176–188.

Rivera, Samantha Cruz, Olalekan L. Aiyegbusi, Jonathan Ives, et al. 2022. Ethical considerations for the inclusion of patient-reported outcomes in clinical research: The PRO ethics guidelines. *JAMA* 327 (19): 1910–1919.

Ronen, Gabriel M. 2023. Capturing meaningful outcomes in pediatric neurology: Further reflections on principles, challenges, and opportunities. *Pediatr Neurol* 135: 38–43.

Ronen, Gabriel M., Olaf Kraus de Camargo, and Peter L. Rosenbaum. 2020. How can we create Osler's "Great Physician"? Fundamentals for physicians' competency in the twenty-first century. *Med Sci Educ* 30:1279–1284. https://doi.org/10.1007/s40670-020-01003-1.

Ronen, Gabriel M., Brandon Meaney, Bernard Dan, et al. 2009. From eugenic euthanasia to habilitation of disabled children: Andreas Rett's contribution. *J Child Neurol* 24: 115–127.

Ronen, Gabriel M., Peter L. Rosenbaum, and David L. Streiner. 2022. Patient perspectives in pediatric neurology: A critical shift in the paradigm of outcome measurement. *Dev Med Child Neurol* 64:149–155.

Rosenbaum, Peter L., and Jan W. Gorter. 2012. The 'F-Words' in childhood disability: I swear this is how we should think! *Child Care Health Dev* 38 (4): 457–463.

Rosenbaum, Peter L., Ann-Christin Eliasson, Mary Jo Cooley Hidecker, and Robert J. Palisano. 2014. Classification in childhood disability: Focusing on function in the 21st century. *J Child Neurol* 29 (8):1036–1045. https://doi.org/10.1177/0883073814533008.

Rosenbaum, Peter L., Gabriel M. Ronen, Bernard Dan, Eric Racine, and Jennifer Johannesen, eds. 2016. *Ethics in Child Health: Principles and Cases in Neurodisability.* London: Mac Keith Press.

Soper, Alice K., Andrea Cross, Pete L. Rosenbaum, et al. 2019. Descriptive title: Exploring the international uptake of the 'F-words in Childhood Disability': A citation analysis. *Child Care Health Dev* 45: 473–490.

Thornton, Philip K., Polly J. Ericksen, Mario Herrero, and Andrew J. Challinor. 2014. Climate variability and vulnerability to climate change: A review. *Glob Change Biol* 20: 3313–3328.

Towns, Megan, Peter L. Rosenbaum, Robert Palisano, et al. 2017. Should the gross motor function classification system be used in children who do not have cerebral palsy? *Dev Med Child Neurol* 60: 147–154.

UN (The United Nations). 2006. (CRPD) *Convention on the rights of persons with disabilities.* Treaty Series, 2515, 3, GA Res 61/106, 76th Plen. Mtg., UN Doc A/Res/61/106. https://social.desa.un.org/issues/disability/crpd/convention-on-the-rights-of-persons-with-disabilities-crpd. Accessed 10 Apr 2024.

UN (The United Nations). 1989. *Convention on the Rights of the Child.* Treaty Series 1577:3. New York: U.N.

World Health Organization. 1980. *International Classification of Impairments, Disabilities and Handicaps.* Geneva: WHO Press.

World Health Organization. 2001. *International Classification of Functioning, Disability and Health.* Geneva: WHO Press.

World Health Organization. 2019/2021. *International Classification of Diseases, Eleventh Revision (ICD-11)*. Licensed under creative commons attribution-noderivatives 3.0 IGO licence (CC BY-ND 3.0 IGO). https://icd.who.int/browse11. Accessed 10 Apr 2024.

World Health Organization. 2023. *Health Topics: Disability*. https://www.who.int/health-topics/disability#tab=tab_1. Accessed 10 Apr 2024.

# Chapter 15
# The Potential Contribution of Ethical Frameworks in Guiding the Development of Digitally Inclusive Support Practices for Adults with Learning Disabilities

**Jane Seale**

**Abstract** The focus of this chapter is adults with learning disabilities and the support practices that enable them to access and use technologies. These practices have been influenced by two conflicting risk discourses: the risk of digital exclusion and inequalities discourse and the safeguarding and risk aversion discourse. Attempts to mediate this conflict by proposing a third discourse focused on positive risk-taking have had limited success. The potential of ethical frameworks to offer a way forward is examined. Firstly, the extent to which ethical frameworks have been applied to adults with learning disabilities and their use of technologies is explored. Then the extent to which ethical frameworks contribute to each of the three risk discourses is discussed. The conclusion of this examination suggests that the ethics literature provides some support for a possibility-focused approach to supporting people with learning disabilities to use technology and that ethicists should collaborate with people with learning disabilities and digital inclusion researchers to develop detailed case examples, with accompanying guidance, of what such digitally inclusive support practice might look like in practice.

**Keywords** Learning disabilities · Technology · Digital exclusion · Risk · Ethics

## 15.1 Introduction

The focus of this chapter is the support practices that might enable people with learning disabilities to access and use technologies. I am using *person-first* language as this is the most commonly used term by self-advocacy groups in the UK, where I have worked both as a support worker and a researcher. The argument for seeing the person before the diagnostic label, is in part, a reaction to the dehumanizing labels

---

J. Seale (✉)
The Open University, Milton Keynes, UK
e-mail: jane.seale@open.ac.uk

 
A. Bianchi and J. A. Vogt (eds.), *Intellectual Disabilities and Autism: Ethics and Practice*, The International Library of Bioethics 108,
https://doi.org/10.1007/978-3-031-61565-8_15

of "mentally subnormal" or "mentally retarded" that dominated the field up until the 1980s. Some of the authors that I cite in this chapter use alternative terms for learning disabilities such as intellectual disabilities (ID) or intellectual and developmental disabilities (IDD).

In the context of this chapter, I will be focusing on adults rather than children where *adulthood* is understood as beginning when a person leaves compulsory education, recognizing that some people with learning disabilities leave compulsory education between the ages of eighteen and twenty-five. A major reason for focusing on adulthood rather than childhood is that evidence suggests that when people with learning disabilities leave compulsory education, much of the funding and support they receive to access and use technologies disappears (Seale 2019).

For the purposes of this chapter *technology* will be used as an umbrella term that includes Information and Communication Technologies (ICT); Information Technologies (IT) and Assistive Technology (AT). Examples include mobile phones; tablets; the Internet; social media; electronic communication aids; telecare equipment; environmental control systems and games consoles.

Most people with learning disabilities will need some kind of support from the people around them to access and use technologies. For the purposes of this chapter, I will label these people *support workers*. A support worker could be a parent, carer, teacher, advocate, or a health or social care professional. Evidence indicates that these support workers exert a huge amount of influence over whether and how people with learning disabilities use technologies (Ramsten and Blomberg 2019; Seale 2020; Rasouli et al. 2021). Supporting adults with learning disabilities to use technologies can be challenging because it requires skillful decision making that involves identifying the needs of the person, evaluating options, anticipating outcomes, and navigating a degree of risk.

Over the last twenty years, literature in the field of technology use by people with learning disabilities has been dominated by two conflicting risk discourses. Firstly, a digital exclusion and inequalities discourse that warns of potential harm if people with learning disabilities are not supported to access and use technologies. Secondly, a safeguarding discourse that warns of the potential harm if people with learning disabilities do use technologies, particularly those associated with the Internet. A third, positive risk-taking discourse has emerged in an attempt to encourage practitioners to balance the benefits and risks of technology use.

### *15.1.1 The Risk of Digital Exclusion and Inequalities*

When practitioners and researchers talk about the potential of technology and what it can offer people with learning disabilities, a wide variety of benefits are highlighted. Proposed educational benefits include learning independent living skills such as wayfinding, shopping, cooking, and managing money. Social benefits include being able to communicate and interact with others. Vocational benefits include the opportunity to learn skills that might lead to employment (e.g., catering and hospitality).

Health benefits include tracking and monitoring fitness and health (Seale 2022a). However, despite the potential for technology to facilitate positive experiences and outcomes for people with learning disabilities, we know that many of them are not able to benefit from this potential; they are digitally excluded (Normand et al. 2016; Seale and Chadwick 2017). There is evidence that people with learning disabilities are lagging behind the general population, in that they have less access to devices, are using the Internet in less varied ways, and therefore feel least included in the digital society (Ågren et al. 2020a; Johansson et al. 2021).

Addressing the digital exclusion of people with learning disabilities requires support workers to focus on four indicators of digital inclusion (Seale 2009, 2020; Chadwick et al. 2013; Cumming et al. 2014; Ågren et al. 2020b):

(1) Access
   (a) to technology
   (b) to support
(2) Use
   (a) knowing how to use technologies
   (b) meaningful use of technologies
(3) Empowerment
   (a) being able to exert control and choice over use
   (b) choosing to use or not to use technology
(4) Participation
   (a) being able to participate in society through the use of technologies
   (b) from civic participation through to participation in education and employment.

The Covid-19 pandemic highlighted numerous ways that people with learning disabilities experienced lack of *access* to technologies. For example, results from a survey of UK support workers revealed several access-related barriers such as not owning technology; not having access to the Internet; not being able to afford to use technology (e.g., mobile data charges); poor Internet connectivity; and non-functioning technology. The results also revealed that even if the person with learning disabilities had access to technology, there were a variety of barriers to them being able to *use* it including not knowing how to use the technology and lacking confidence to use the technology (Seale 2020). Cumming et al. (2014) shared two case studies in which two women with learning disabilities had been given iPads to use in a research project that they were participating in as co-researchers. One of the women shared how she was having difficulties with staff in her group home "borrowing" the device without permission. The other shared how her device had been taken away from her because her family thought that some of the photographs stored in the iPad were indecent. Cumming et al. (2014) suggest that these are examples of people with learning disabilities having choices forced on them. In addition to being *disempowered,* I would argue that the two women were being denied opportunities for *participation,* since without the iPad, their opportunities to participate in the research were significantly limited.

### *15.1.2 Safeguarding and Risk Aversion*

The new millennium has seen an emergence of a *risk culture* where perceptions of the vulnerability of people with learning disabilities have dominated the way those who support them think about their practice (Seale et al. 2013). This has resulted in a focus on safeguarding, particularly in relation to the perceived risks of using the Internet such as being the victim of online abuse or scams (Chadwick 2019; Seale 2014). Original iterations of safeguarding practices reflected a desire to balance the benefits of technology use, such as digital inclusion, with the risks. However, this very quickly mutated into practice that focused largely on risks rather than benefits.

For support workers, there is evidence that this risk aversiveness is heavily influenced by their perceptions of risk (e.g., Simplican et al. 2018; Ramsten and Blomberg 2019). There is also evidence that there are significant differences in the risk perceptions of people with learning disabilities and their support workers (e.g., Chiner et al. 2017) and that sometimes, in order to avoid being barred from using technology, young people with learning disabilities hide their technology use from their support workers (Löfgren-Mårtenson 2008). For service providers funded by local authorities, this risk aversiveness is motivated in part by a desire to avoid sanctions or recriminations if people with learning disabilities were harmed as a result of their online activities.

### *15.1.3 Positive Risk-Taking and Possibility-Focused Support*

In an attempt to counter the dominance of risk-averse support practices that can result in excluding people with learning disabilities from benefitting from using technologies I proposed that support practice should be underpinned by a positive risk-taking framework (Seale 2014) comprising of four key components:

(1) Resilience
   (a) believing that people with learning difficulties have the potential to be resilient or to live resilient lives
   (b) believing that support workers have the potential professional skills and experience to be resilient and cope with supporting actions that may have unpredictable outcomes

(2) Risk management

   (a) putting in place strategies that attempt to mitigate the risks of technology use, in the hope that there will be positive outcomes
   (b) risk is understood broadly as the possibility of digital exclusion, the possibility of failing to achieve a goal, or the possibility of a negative experience (e.g., online abuse, or technical difficulties)

(3) Shared decision making

   (a) a process that takes place between people with learning disabilities and their support workers in which the possible positive and negative outcomes of technology use are considered

(4) Creativity

   (a) identifying possibilities for positive outcomes as a result of technology use
   (b) refusing to give up when circumstances seem impossible
   (c) using imagination to solve the problem of how the risks related to using a particular technology can be managed in order to maximize the possibilities for a positive outcome.

I argue that digitally inclusive support practice needs to incorporate all four components in order to maximize possibilities for positive outcomes.

Since proposing the positive risk-taking framework, I have sought to find case examples that provide evidence for the importance of each of its four components. For example, in the 'Telling Tales of Technology' study, I captured the life histories of ten adults with learning disabilities, particularly in relation to their use of technologies from childhood through to adulthood (Seale 2022a). These life histories revealed that many of the participants were competent and confident technology users and that with support they had developed a resilient and successful approach to managing the risks of technology use.

Although my positive risk-taking framework has sparked interest in the field, I have noticed that the majority of interest has been focused on the risk management component. There has been very little attention paid to resilience, creativity, or shared decision making. In order to encourage a wider focus on all components of the framework I have renamed it the *possibility-focused* framework (Seale 2022b). In addition, I am keen to identify other discourses and areas of research that might be able to: (1) marry the tensions between the digital exclusion and inequalities discourse and the safeguarding and risk aversion discourse, and (2) illuminate characteristics of digitally inclusive support practices. In the following sections I will explore the potential of ethical frameworks to address these issues. In doing so, I will address two key questions:

(1) Have ethical frameworks been applied to adults with learning disabilities and their use of technologies?
(2) To what extent has the application of ethical frameworks to adults with learning disabilities and their use of technologies addressed:

   (a) the risk of digital exclusion and digital inequalities?

(b) risk aversiveness?
(c) the potential of positive risk-taking?

## 15.2 The Application of Ethical Frameworks to Adults with Learning Disabilities and Their Use of Technologies

There has been very little research or debate on the ethics of technology use by adults with learning disabilities. Of the small literature that exists, some report empirical research exploring ethics at either a service or individual level (e.g., Barnard and Beyer 2009; Chalgoumi et al. 2019). Some are discussion papers, debating the applicability of health ethics to the field (e.g., Perry et al. 2009; O'Brolcháin 2018). One of the earliest examples of empirical work in this field is the TATE project which ran between 2004 and 2007 and was led by a UK service provider for people with learning disabilities, with partners in Hungary, Spain, and Latvia. One strand of the project sought to work with assistive technology manufacturers to develop and install new assistive technologies into the homes of people with learning disabilities. In reflecting on the experiences of the TATE project, Barnard and Beyer (2009, 56) report that they were surprised at the lack of ethical frameworks for the provision of personalized technology by local authorities; noting that there was no "national or European standard, with an agreed ethical code to guide providers, commissioners and manufacturers in delivering solutions that balance safety, risk, independence and quality of life." In the absence of an ethical framework, the TATE project sought to develop an approach whereby AT would be provided if it fitted into the accepted assessment and support plan of the individual.

In order to address the lack of an ethical framework that is specifically designed to address the provision of technologies for people with learning disabilities, Perry et al. (2009) proposed that Beauchamp and Childress's (2001) health ethics framework be adapted. They also argued that because it has already been used to consider the ethical issues associated with the use of AT and telecare with people with dementia, it could be equally useful for consideration of the ethical issues for people with learning disabilities. Beauchamp and Childress (2001) described four major ethical principles: respect for autonomy, beneficence, non-maleficence, and justice. Autonomy relates to self and the right of the individual to make choices in alignment with their beliefs, values, and preferences. Beneficence is the principle of working for the benefit of the individual. Non-maleficence is the principle of not causing harm as a result of an intervention (or lack of). Justice relates to coming to fair and just decisions when balancing conflicting needs or principles. The question of distributive justice can arise when resources are limited within a health care setting and healthcare professionals find themselves trying to make challenging decisions about what constitutes fair access to resources.

In order to promote autonomy, Perry et al. (2009) suggest that AT and telecare could be used to facilitate the teaching of skills which might promote increased independence. In addition, smart home technology may increase independence if it results in reduced dependence on support staff for assistance. Since many people with learning disabilities have difficulties retaining information, O'Brolcháin (2018) suggests that technologies that can supply knowledge in forms that are more easily understood (e.g., maps, guides, reminders) will enhance autonomy for people with learning disabilities. However, if the information is overly simplified, there is a risk that knowledge will be misrepresented or distorted to such an extent that they will not be able to make genuinely informed decisions. With regards to beneficence, Perry et al. (2009) propose that AT and telecare may be beneficial to an individual if it promotes access to information, entertainment, social contact, skills development, and decreases dependence on support staff. With regards to non-maleficence, Perry et al. (2009) acknowledge that some AT and telecare may prevent harm (e.g., carbon monoxide detectors) and some may cause harm (e.g., faulty or unreliable technology). In illuminating the principle of justice, Perry et al. (2009) suggest that the costs of installing AT in a residential service could be justified if it frees staff time to support those whose needs are greatest. However, in a service where there are constrained resources, it may not be fair to meet all of the AT needs of one person if it means there will not be enough resources to meet the AT needs of other service users.

The Wales Personalised Technology Community of Practice (2019) led by the self-advocacy group 'Learning Disabilities Wales' drew on the work of Perry et al. (2009) to produce an ethical framework for what they called *personalised technologies*. With the exception of referencing *justice*, the document does not replicate the language of Beauchamp and Childress (2001). Instead of autonomy, the framework refers to independence, choice, control, and consent. Instead of referring to beneficence or non-maleficence, the framework calls for support workers to ask the questions 'who is benefitting?' and 'does it do any harm?'. Four examples of harm are offered: loss of privacy, stigmatization, risk, and loss of social contact. However, the framework offers no case examples to illustrate how ethical decisions would be made and then implemented. In addition, there is no evidence that the framework is being adopted by support workers.

### *15.2.1 To What Extent Do Ethical Frameworks Address the Digital Exclusion and Inequalities Experienced by Adults with Learning Disabilities?*

The literature on the application of ethics to the use of technologies by adults with learning disabilities does not reference any studies on the digital exclusion or inequalities they experience and therefore does not explicitly suggest that this might be a driver for ethical practice. However, it does offer some limited insights into addressing digital inclusion and digital inequalities.

With regards to digital exclusion, there is a subset of the ethics literature that may illuminate issues relating to the empowerment of people with learning disabilities through their use of technology. This literature discusses the challenge to autonomy presented by smart homes and telecare. A smart home is equipped with electronic devices that can be controlled remotely by smartphone or computer and can be used to control lighting, temperature, security, devices such as cookers, and entertainment systems such as TV and radio. Sometimes, smart home technology is combined with telecare. For example, voice assistants (e.g., Amazon Alexa) that can be programmed to remind a person to take their medication. It is commonly claimed that smart home and telecare technologies are not intended to substitute for human care. However, Perry et al. (2009) acknowledge that services may be tempted to use reductions in staff costs as a reason for implementing these technologies. If the introduction of these technologies is used to reduce the amount of staff contact with the individual, the gains to be made by supporting the person's capability to live independently could conceivably come at a cost in terms of lessened human contact, thereby contributing to their further social isolation. Often, the support workers who regularly visit individuals in semi-independent living situations are some of their closest social contacts. Thus, the principles of respect for autonomy (represented by use of assistive technologies to enhance the person's independence) and beneficence (represented by the importance of supporting and/or maintaining social connections and inclusion for the person) come into conflict should decisions be driven solely by cost-cutting concerns.

The use of monitoring and surveillance technologies in smart homes and telecare has prompted O'Brolcháin and Gordijn (2019) to discuss the challenges to privacy and the risks to a person's autonomy. They argue that people with learning disabilities may not be competent enough to give informed consent to loss of privacy. In addition to the employment of advocates or surrogates who can make decisions in the best interest of the person, O'Brolcháin and Gordijn (2019) propose that smart home designers build dynamic privacy protecting measures into smart homes. One element of dynamic privacy protection is that monitoring and recording systems would have adjustable privacy settings that are sensitive to different spatial, social and activity contexts. A second element is to determine who has access to the data and what level of access they have. They also argue that people with learning disabilities should be involved in making the decisions regarding the privacy protections they want programmed into the smart home systems. Finally, O'Brolcháin and Gordijn (2019) argue that for a dynamic privacy protection system to function ethically, it will be necessary for rolling consent to be acquired, for example, through accessible electronic consent forms that are built into tablets or computer screens and displayed in places such as hallways.

Although O'Brolcháin and Gordijn (2019) have raised some interesting issues, just a few empirical studies have explored whether these issues arise in reality, how people with learning disabilities and their support workers react to them, or what solutions have been trialed and evaluated (Woensdregt et al. 2020; Rasouli et al. 2021). In a study based in the Netherlands, Woensdregt et al. (2020) interviewed twenty stakeholders in order to explore experiences, priorities, and concerns relating to the use of sensors. Participants identified putting the interest of the clients first as an important ethical concern for using all kinds of sensor applications and the client's privacy was the most mentioned issue in this regard. Furthermore, participants suggested that sensor applications should only be used when there is informed consent from the user.

The literature on the ethics of technology use by adults with learning disabilities does not explicitly point to the research on digital inequalities when discussing the factors that might influence ethical decision making. However, some ethicists have considered whether and how people with learning disabilities should be treated differently to others. Wasserman (2019) discusses whether protective measures should be put in place to prevent people with learning disabilities going online and running the risk of loss of privacy. Wasserman (2019) suggests that norms of privacy are different online, compared to real life. What might have been considered inappropriate disclosure of information in the previous century, may be accepted, or even expected in current online situations. Wasserman (2019) therefore proposes that people with learning disabilities should be supported to make their own informed decisions about how much risk they are willing to take in revealing personal information about themselves. They argue that it would be "disrespectful to impose a more restrictive standard of personal disclosure on people with ID [intellectual disabilities] than on others" (Ibid., 225). Focusing on the lack of current market interest in providing useful or necessary technologies to people with learning disabilities, O'Brolcháin and Gordijn (2018) argue that they run the risk of being on the wrong side of the digital divide, which may further marginalize them. They position this as an issue of distributive justice.

### 15.2.2 *To What Extent Has the Application of Ethical Frameworks to Adults with Learning Disabilities and Their Use of Technologies Addressed Risk Aversiveness?*

The issue of adopting an excessively risk-averse approach to supporting technology use by people with learning disabilities prompts us to consider the extent to which the ethical principles of autonomy and non-maleficence (which in this context can be understood as avoidance of risk) come into conflict. The ethics literature does not explicitly address the pros and cons of adopting an overly risk-averse approach to supporting technology use by people with learning disabilities. However, one debate in the field does challenge the perception of people with learning disabilities as being vulnerable and not being able to manage risk. This perception underpins much of the risk-averse safeguarding discourse. An influential study by Chalghoumi et al. (2019) sparked an important debate about the potential of people with learning disabilities to manage their own risk when online (Wasserman 2019; O'Brolcháin and Gordijn 2018). Chalghoumi et al. (2019) conducted three semi-structured focus groups with six people with learning disabilities about their use of IT and their perceptions regarding the benefits of using IT. All of the participants perceived that using

IT had benefits, particularly with regards to increasing their autonomy. However, although they understood the concept of privacy and the need to be concerned about protecting privacy in real life settings, they did not translate this awareness to online settings. Despite these findings, Chalgoumi et al. (2019) did not call for a raft of protective measures that might run the risk of preventing people with learning disabilities from going online at all. Instead, they suggested that as they were aware of privacy issues and adopted adequate protective measures in real-life, with the right education and tools people with learning disabilities had the potential to transfer their privacy protecting behaviors to their online activities.

### 15.2.3 To What Extent Does the Application of Ethical Frameworks Support a Possibility-Focused Approach to Supporting Adults with Learning Disabilities and Their Use of Technologies?

The value of my original positive risk-taking framework in helping to promote ethical practice is highlighted by Chalghoumi et al. (2019, 213) who suggested that it offers "operational strategies and steps that may help take our reflection to a practical level." In addition, across the ethics literature, it is possible to distill out four potential components of ethical practice that resonate with the components of my possibility-focused framework.

With regards to resilience, the possibility-focused framework proposes that the decision regarding whether or not to support people with learning disabilities to use technologies is influenced by the extent to which support workers believe that people with learning difficulties have the potential to be resilient or to live resilient lives. This is similar to the arguments made by Chalgoumi et al. (2019) regarding the potential of people with learning disabilities to learn how to protect their privacy online with the right support.

Risk management involves putting in place strategies that attempt to mitigate the risks of technology use, in the hope that there will be positive outcomes. I would argue that this aligns well with the practice of *balancing the potential risks and benefits* proposed by Chalgoumi et al. (2019). Despite a call from Chalghoumi et al. (2019), for further research into how to support people with learning disabilities to understand and balance the benefits and risks of being online, to date empirical studies have only focused on how support workers understand and balance the benefits and risks of technology use (Simplican et al. 2018; Ramsten and Blomberg 2019).

Shared decision making between people with learning disabilities and their support workers involves considering the possible positive and negative outcomes of technology use. This reflects the argument made by O'Brolcháin and Gordijn (2019) that people with learning disabilities should be involved in making the decisions regarding the privacy protections they want programmed into the smart home systems as well as a call by Chalgoumi et al. (2019) for collaboration between all stakeholders. Chalghoumi et al. (2019) argued that making technology use more ethical for people with learning disabilities is not the sole responsibility of one stakeholder (e.g., technology developers or policy makers). Instead, they argued it is the shared responsibility of all stakeholders and they called for a collaboration of caregivers, professionals, and people with learning disabilities.

Finally, writing in the context of the pandemic and what might be considered *just* practice, Nelson and Francis (2020, 329) suggested that if the quality of life for people with learning disabilities is threatened due to lack of access to technologies, their needs may need to be addressed by "increased and imaginative use of technologies." The call for imaginative uses of technology reflects my argument for the need for creativity in support practices, where imagination is used to solve the "problem" of how the risks related to using a particular technology can be managed in order to maximize the possibilities for a positive outcome.

## 15.3 Conclusion

People with learning disabilities have less access to technologies and are using them less than people without disabilities. This impacts their opportunities to participate in society in ways that are meaningful to them. Efforts by support workers to address this risk of digital exclusion have been counteracted by efforts to eliminate the perceived risks of using technologies. In order to overcome this stalemate, we need creative, resilient support practices that can embrace risk and work in partnership with people with learning disabilities to identify possibilities for digital inclusion. Currently, this kind of practice is uncommon because researchers and practitioners are overly fixated on risk management to the detriment of other potential components of a digitally inclusive support practice.

The ethics literature provides some support for a possibility-focused approach to supporting people with learning disabilities to use technology through its endorsement, albeit limited, of shared decision making, balancing the benefits and risks of technology use, imagination, and recognizing the potential of people with learning disabilities to manage their own risk, with the right support.

It is important to explore what a possibility-focused approach to support might look like and offer insights into the decision-making process that might underpin this practice. In order to do this, it would be helpful if researchers in the field of technology and learning disabilities, ethicists, and people with learning disabilities worked together to develop detailed case examples with accompanying guidance.

## References

Ågren, Kristen Alfredsson, Annette Kjellberg, and Helena Hemmingsson. 2020a. Access to and use of the Internet among adolescents and young adults with intellectual disabilities in everyday settings. *Journal of Intellectual & Developmental* 45: 89–98. https://doi.org/10.3109/13668250.2018.1518898.

Ågren, Kristen Alfredsson, Annette Kjellberg, and Helena Hemmingsson. 2020b. Digital participation? Internet use among adolescents with and without intellectual disabilities: A comparative study. *New Media & Society* 22: 2128–2145. https://doi.org/10.1177/1461444819888398.

Barnard, Steve, and Steven Beyer. 2009. Barriers to using personalised technology with people with learning disabilities. *Journal of Assistive Technologies* 3: 50–57. https://doi.org/10.1108/17549450200900025.

Beauchamp, Tom L., and James F. Childress. 2001. *Principles of Biomedical Ethics*, 5th ed. Oxford: Oxford University Press.

Chadwick, Darren. 2019. Online risk for people with intellectual disabilities. *Tizard Learning Disability Review* 24: 180–187. https://doi.org/10.1108/TLDR-03-2019-0008.

Chadwick, Darren, Caroline Wesson, and Chris Fullwood. 2013. Internet access by people with learning disabilities: Inequalities and opportunities. *Future Internet* 5: 376–397. https://doi.org/10.3390/fi5030376.

Chalghoumi, Hajer, Virginie Cobigo, Crislee Dignard, Amélie Gauthier-Beaupré, Jeffrey W. Jutai, Yves Lachapelle, Johanna Lake, Rawad Mcheimech, and Maxine Perrin. 2019. Information privacy for technology users with intellectual and developmental disabilities: Why does it matter? *Ethics & Behavior* 29: 201–217. https://doi.org/10.1080/10508422.2017.1393340.

Chiner, Esther, Marcos Gómez-Puerta, and María Cardona-Moltó. 2017. Internet use, risks and online behaviour: The view of internet users with intellectual disabilities and their caregivers. *British Journal of Learning Disabilities* 45: 190–197. https://doi.org/10.1111/bld.12192.

Cumming, Therese, Iva Strnadová, Marie Knox, and Trevor Parmenter. 2014. Mobile technology in inclusive research: Tools of empowerment. *Disability & Society* 29: 999–1012. https://doi.org/10.1080/09687599.2014.886556.

Johansson, Stefan, Jan Gulliksen, and Catharina Gustavsson. 2021. Disability digital divide: The use of the internet, smartphones, computers and tablets among people with disabilities in Sweden. *Universal Access in the Information Society* 20: 105–120. https://doi.org/10.1007/s10209-020-00714-x.

Löfgren-Mårtenson, Lotta. 2008. Love in cyberspace: Swedish young people with intellectual disabilities and the Internet. *Scandinavian Journal of Disability Research* 10: 125–138.

Nelson, Ryan, and Leslie Francis. 2020. Justice and intellectual disability in a pandemic. *Kennedy Institute of Ethics Journal* 30: 319–338. https://doi.org/10.1353/ken.2020.0017.

Normand, Claude, Dany Lussier-Desrochers, Stéphanie-M. Fecteau, Valérie Godin-Tremplay, Marie-Ève Dupont, Jeannie Roux, Martin Caouette, and Alejandro Romero. 2016. A conceptual model of factors leading to the digital exclusion of people with neurodevelopmental disorders. *Annual Review of Cybertherapy and Telemedicine* 14: 23–29. https://www.arctt.info/volume-14-summer-2016.

O'Brolcháin, Fiachria. 2018. Autonomy benefits and risks of assistive technologies for persons with intellectual and developmental disabilities. *Frontiers in Public Health* 6: 296. https://doi.org/10.3389/fpubh.2018.00296.

O'Brolcháin, Fiachra, and Bert Gordijn. 2018. Persons with intellectual and developmental disabilities and information technologies. Some ethical observations—a comment on Chalgoumi et al. *Ethics Behavior* 29: 218–222. https://doi.org/10.1080/10508422.2018.1471998.

O'Brolcháin, Fiachra, and Bert Gordijn. 2019. Privacy challenges in smart homes for people with dementia and people with intellectual disabilities. *Ethics and Information Technology* 21: 253–265. https://doi.org/10.1007/s10676-019-09507-0.

Perry, Jonathan, Stephen Beyer, and Søren Holm. 2009. Assistive technology, telecare and people with intellectual disabilities: Ethical considerations. *Journal of Medical Ethics* 35: 81–86. https://doi.org/10.1136/jme.2008.024588.

Ramsten, Camilla, and Helena Blomberg. 2019. Staff as advocates, moral guardians and enablers—Using ICT for independence and participation in disability services. *Scandinavian Journal of Disability Research* 21: 271–281. https://doi.org/10.16993/sjdr.608.

Rasouli, Omid, Lisbeth Kvam, Vigdis Schnell Husby, Monica Røstad, and Aud Elisabeth Witsø. 2021. Understanding the possibilities and limitations of assistive technology in health and welfare services for people with intellectual disabilities, staff perspectives. *Disability and Rehabilitation: Assistive Technology.* https://doi.org/10.1080/17483107.2021.1963856.

Seale, Jane. 2009. *Digital Inclusion. Research briefing by the Technology Enhanced Learning Phase of the Teaching and Learning Research Programme.* University of Southampton. https://tinyurl.com/4kdjwz6r. Accessed 9 Apr 2024.

Seale, Jane. 2014. The role of supporters in facilitating use of technologies by adolescents and adults with learning disabilities: A place for positive risk taking? *European Journal of Special Needs Education* 29: 220–236. https://doi.org/10.1080/08856257.2014.906980.

Seale, Jane. 2019. Wilderness and resistance: Illuminating the digital inequalities experienced by adults with learning disabilities between 1970 and 1999. *Disability & Society* 34: 1481–1503. https://doi.org/10.1080/09687599.2019.1576504.

Seale, Jane. 2020. Keeping connected and staying well: The role of technology in supporting people with learning disabilities during the coronavirus pandemic. Milton Keynes: The Open University. http://oro.open.ac.uk/75127/.

Seale, Jane. 2022a. It's not all doom and gloom: What the pandemic has taught us about digitally inclusive practices that support people with learning disabilities to access and use technologies. *British Journal of Learning Disabilities* 51 (2): 218–228. https://doi.org/10.1111/bld.12497.

Seale, Jane. 2022b. *Technology Use by Adults with Learning Disabilities Past, Present and Future Design and Support Practices.* New York: Routledge.

Seale, Jane, and Darren Chadwick. 2017. How does risk mediate the ability of adolescents and adults with intellectual and developmental disabilities to live a normal life by using the Internet? *Cyberpsychology: Journal of Psychosocial Research on Cyberspace* 11 (1): Article 2. https://doi.org/10.5817/CP2017-1-2.

Seale, Jane, Melanie Nind, and Ben Simmons. 2013. Transforming positive risk-taking practices: The possibilities of creativity and resilience in learning disability contexts. *Scandinavian Journal of Disability Research* 15: 233–248. https://doi.org/10.1080/15017419.2012.703967.

Simplican, Stacy, Carolyn Shivers, June Chen, and Geraldine Leader. 2018. "With a Touch of a Button": Staff perceptions on integrating technology in an Irish service provider for people with intellectual disabilities. *Journal of Applied Research in Intellectual Disabilities* 31: 130–139. https://doi.org/10.1111/jar.12350.

Wales Personalised Technology Community of Practice. 2019. *Personalised Technology: An ethical framework.* https://www.ldw.org.uk/wp-content/uploads/2019/10/All-Wales-Personalised-Technology-CoP-Ethical-Framework.pdf. Accessed 9 Apr 2024.

Wasserman, David. 2019. A case for greater risk tolerance in internet use by adults with intellectual disabilities: A comment on Chalghoumi et al. *Ethics & Behavior* 29: 223–226. https://doi.org/10.1080/10508422.2018.1522256.

Woensdregt, Gerdien, Graziano D'Addabbo, Hans Schulten, Claudia van Alfen, and Paula Sterkenburg. 2020. Sensors in the care of persons with visual- or visual-and-intellectual disabilities: Use, needs, practical issues, and ethical concerns. *Technology & Disability* 32: 15–24. https://doi.org/10.3233/TAD-180212.

# Chapter 16
# Promoting Advocacy and Empowerment for Individuals with Intellectual Disabilities and Autism

**Meghan M. Burke, Chak Li, Amanda Johnston, and Adriana Kaori Terol**

**Abstract** Throughout time, advocacy has been needed to ensure individuals with disabilities and their families demonstrate positive outcomes. Specifically, advocacy has been conducted by parents of individuals with intellectual disabilities, including those with autism, as well as self-advocates. With the COVID-19 pandemic, parent advocacy and self-advocacy became even more critical. However, there are concerns about how to balance parent advocacy and self-advocacy, especially in cases where the goals of such advocacy efforts are different. We suggest using person-family interdependent planning to unite parent advocacy and self-advocacy to ensure positive outcomes for individuals with intellectual disability, including individuals with autism, and their families.

**Keywords** Advocacy · Self-advocate · Planning · Services · Family

---

M. M. Burke (✉) · C. Li · A. Johnston
Vanderbilt University, Nashville, TN, USA
e-mail: meghan.burke@vanderbilt.edu

C. Li
e-mail: chak.li@vanderbilt.edu

A. Johnston
e-mail: amanda.johnston@vanderbilt.edu

A. K. Terol
University of Illinois at Urbana-Champaign, Champaign, IL, USA
e-mail: aterol2@illinois.edu

 
A. Bianchi and J. A. Vogt (eds.), *Intellectual Disabilities and Autism: Ethics and Practice*, The International Library of Bioethics 108,
https://doi.org/10.1007/978-3-031-61565-8_16

## 16.1 Introduction

Advocacy is often needed for individuals with intellectual and developmental disabilities (IDD), including those with autism[1], across the lifespan. Indeed, advocacy is a powerful tool that has been used to create positive change for individuals with disabilities and their families around the world (World Health Organization 2009). Historically, there has been a need for advocacy with respect to deinstitutionalizing and transitioning individuals with disabilities into community living, developing disability policy, encouraging inclusion, and addressing health disparities for individuals with disabilities (Cohen 2013; Trainor 2010).

Parent advocacy has been particularly influential to forge meaningful change for individuals with disabilities. Parents tend to have unique perspectives on the needs of their offspring and, as a result, are often the most effective advocates for their children (Wright and Taylor 2014). Parents may advocate for their own children, for other families, and for systemic changes for all individuals with disabilities.[2] Thus, parent advocacy has the ability to impact individual as well as collective change.

At the same time, it is important to balance parent advocacy with self-advocacy by individuals with disabilities. Self-advocacy refers to the person with the disability advocating on their own behalf. Coined during the beginning of the self-advocacy movement, a common mantra among individuals with disabilities is "Nothing about us, without us" (Charlton 2000). The research underscores the importance of self-advocacy, noting that self-determination (i.e., a person making their own choices) is a best practice among individuals with IDD, including those with autism (Wehmeyer and Shogren 2016). When youth with IDD are supported to make their own choices, they demonstrate improved school (Shogren et al. 2012) and post-school outcomes (Shogren and Ward 2018).

Policy, including laws and educational guidelines, support both parent and self-advocacy. Consider the Individuals with Disabilities Education Act[3] (IDEA) of 2004—the federal special education law in the United States. According to IDEA, parents are equal members of educational decision-making teams. Thus, parent advocacy is emphasized in the IDEA legislation. However, IDEA also recognizes the importance of self-advocacy. For instance, it is required for youth with disabilities to be invited to attend their education planning meetings. Furthermore, when a youth turns eighteen years of age—unless there is a legal guardian—the youth becomes responsible for their own educational decisions. This shift in decision-making authority emphasizes the significance of both parent advocacy and self-advocacy for individuals with disabilities.

---

[1] For this chapter, we have chosen to use person-first language for consistency.

[2] Li., Chak, Meghan M. Burke, Julie Lounds Taylor, Leann Smith DaWalt, and Zachary Rossetti. Under review. *Establishing an advocacy activities scale for parents of individuals with intellectual and developmental disabilities*.https://www.aaidd.org/docs/default-source/prepressarticles/establishing-an-advocacy-activities-scale-for-parents-of-individuals-with-intellectual-and-developmental-disabilities.pdf?sfvrsn=6d0e0221_0

[3] US Congress. 2004. "Individuals with Disabilities Education Act" (IDEA), 20 U.S.C. § 1400.

Despite disability policies, it may be difficult to balance parent advocacy and self-advocacy, especially in situations where parent advocacy overshadows or undermines self-advocacy. For example, medical professionals may be more inclined to listen to a parent's advocacy requests instead of the youth's advocacy because of discrimination and preconceived notions about the inability of the youth to make decisions on their own behalf (Churchwell and O'Malley 2023). Also, balancing parent and youth advocacy may be challenging if the goal of a parent's advocacy differs from the goal of the self-advocate. Even if the parent's intent is to act in the best interest of the youth (i.e., from a position of beneficence), the parent may interfere with the youth's ability to exercise their autonomy. This may be especially true in relation to families of individuals with IDD, including those with autism. For example, a youth may want to live independently and only accept minimal community supports, whereas their parents may want the individual to reside in the family home and receive multiple supports. Given the importance of both parent advocacy and self-advocacy, the question then becomes, how do we promote the autonomy of the individual with IDD when their desires and goals oppose those of their family members; and how do we address opposing advocacy among individuals with IDD and their family members.

In this chapter, we argue that a balance between parent advocacy and self-advocacy must be attained to ensure that people with IDD experience positive outcomes and that their autonomy is respected. To this end, we begin the chapter by reviewing parent advocacy and self-advocacy across historical and international contexts. Then, we discuss the current need for parent advocacy and self-advocacy, especially as the world adapts to life after the COVID-19 pandemic. We follow with a review of the limited research that explores parent and self-advocacy within the same families and conclude with ways to move forward that empower both self-advocates and their families to conduct advocacy.

## 16.2 A Brief History of Disability Advocacy

### *16.2.1 Parent Advocacy*

At the beginning of the twentieth century, most advocacy efforts were individualized (Walmsley et al. 2017). Families were often isolated due to stigma, and there were challenges to meet other families of individuals with disabilities. Thus, parent advocacy efforts were largely focused on advocating for a family member with a disability. In the second half of the twentieth century, parent advocacy shifted to collective advocacy. Unlike individualized advocacy, collective advocacy is characterized by unified support by a group of individuals to forge systemic change for many people with disabilities—not just one individual (Burke et al. 2018; Walmsley et al. 2017). Historically, in the United States, collective advocacy efforts date back to the years following World War II with the formation of various advocacy organizations (Scotch 2009). One of the earliest examples of collective advocacy is the creation of the National Association for Retarded Children (now called The Arc) in 1950. Perhaps the leading national and international agency supporting the advocacy of individuals with IDD and their families, the creation of The Arc was followed by the creation of other advocacy organizations (Pfeiffer 1993).

There have been several impacts of parent advocacy. For example, parent advocacy efforts have led to the creation of special education laws in several countries around the globe. Parent advocacy prompted the initiation of the Education (Handicapped Children) Act in England and Wales[4] (1970), the Education (Mentally Handicapped Children) Act in Scotland[5] (1974), and the Education for All Handicapped Children Act (1975),[6] now called IDEA, in the United States (Trainor 2010; Walmsley et al. 2017). Parent advocacy has also been crucial in addressing disparities and promoting inclusion. For example, parent advocacy has been influential in many Latin American countries. According to Cohen (2013), parent advocacy groups have been instrumental in legislating inclusive practices in Latin American countries such as Mexico, Peru, and Paraguay. Internationally, parent advocacy has prompted systemic changes leading to tremendous improvements (e.g., deinstitutionalization) for individuals with disabilities and their families.

### *16.2.2 Self-Advocacy*

The self-advocacy movement amongst people with intellectual disabilities began in Sweden in the 1960s and focused on their rights as citizens in addition to equality, respect, and power (Baggs 2019; Smith and Mueller 2022; Walmsley and Central England People First History Project Team 2014). During the 1970s and 1980s, the disability rights movement gained momentum as the role of advocacy expanded beyond efforts by parents of individuals with disabilities and began to include individuals with disabilities as self-advocates. In the United States, self-advocates fought for disability rights via legislation such as Section 504 of the Rehabilitation Act of 1973 and the Americans with Disabilities Act in 1990. Through their collective power advocating for disability rights, the impact of self-advocates became widespread.

---

[4] UK Public General Acts. "Education (Handicapped Children) Act", Chap. 52. July 23, 1970. https://www.legislation.gov.uk/ukpga/1970/52/enacted#:~:text=1970%20CHAPTER%2052,and%20for%20purposes%20connected%20therewith. Accessed 9 April 2024.

[5] UK Public General Acts. "Education (Handicapped Children) (Scotland) Act", Chap. 27. July 17, 1974. https://www.legislation.gov.uk/ukpga/1974/27#:~:text=An%20Act%20to%20make%20provision,meaning%20of%20the%20Mental%20Health%20_ Accessed 9 April 2024.

[6] US Congress. "H.R.7217—94th Congress (1975–1976): Education for All Handicapped Children Act". PL No. 94–142. July 29, 1975. https://www.congress.gov/bill/94th-congress/house-bill/7217. Accessed 9 April 2024.

Self-advocacy efforts by individuals with disabilities occur in many countries around the world. In China, self-advocacy is a relatively new phenomenon as individuals with disabilities had limited opportunities historically. However, as recently as 2017, the self-advocacy of individuals with disabilities has led to legislative advances promoting disability rights (Zhang 2017). Similarly, self-advocacy in Iceland has been the result of significant societal shifts. Due to the efforts of individuals with IDD, the public now views individuals with disabilities as "active and able" (Björnsdóttir and Jóhannesson 2009, 443). In other Western countries (e.g., the Netherlands, Australia), individuals with disabilities are beginning to note small shifts in systems (e.g., healthcare) due to self-advocacy (Henderson and Bigby 2016; Waltz et al. 2015). Across the lows and highs of advances in human rights, self-advocacy has consistently improved the lives of individuals with disabilities across the globe.

## 16.3 Advocacy During the COVID-19 Pandemic

### *16.3.1 Parent Advocacy During the Pandemic*

Internationally, the need for parent advocacy and self-advocacy on behalf of people with disabilities became heightened following the outbreak of COVID-19. During the COVID-19 pandemic, disparities were amplified among individuals with disabilities, including limited access to healthcare services (Okonkwo et al. 2021), inequity in receiving remote instruction (Harris et al. 2021), and limited receipt of community and support services (Caldwell et al. 2022). As a result of increasing inequities, there was a greater need for advocacy. The fact that we typically advocate for things we need to enhance our capabilities (e.g., access to supports, resources, services, reasonable accommodations, etc.) underscores the close association between advocacy work and justice. According to parents of individuals with disabilities, they had to advocate more often for their offspring with disabilities during the pandemic in order to ensure they received just treatment. Specifically, in a study of Canadian parents of individuals with disabilities, parents reported an increase in their advocacy efforts during the COVID-19 pandemic due to professionals' limited understanding of the challenges of students with disabilities and their families (Francis et al. 2021). Notably, advocacy was especially needed, and more often conducted, among families who were marginalized (e.g., families of color and Indigenous families).

As the world returns to pre-COVID-19 policies and procedures, there continue to be reasons for heightened parent advocacy. Consider several examples in the United States. During the COVID-19 pandemic, the federal government paused all redeterminations of the Medicaid Home and Community-Based Services waivers which provide access to home and community-based supports and services to millions of individuals with IDD, including those with autism (Churchwell and O'Malley 2023). As the COVID-19 pandemic is no longer considered a public health emergency, the government is calling for redetermination of such services to begin again (i.e., redetermining whether individuals with disabilities are still eligible to receive services). As such, policymakers and advocates worry that millions of individuals will lose their Medicaid coverage (Churchwell and O'Malley 2023). Thus, parent advocacy may become increasingly crucial to ensure that individuals with IDD, including those with autism, retain access to essential services and support.

### 16.3.2 Self-Advocacy During the Pandemic

Similar to parent advocacy, self-advocacy became critical during the COVID-19 pandemic. Like parents, self-advocates also needed to address the extant disparities in access to services and supports that were widened during the pandemic. However, in the case of individuals with IDD, self-advocacy is also about respect for their own autonomy (i.e., the right to make choices for themselves and to pursue their interests unfettered). Actions that impede their ability to express their values, goals, and interests violate their autonomy. Consider Alice Wong, a self-advocate living in the United States who relies on a ventilator (Wong 2020). During the COVID-19 pandemic, many states and health systems rationed ventilators for their COVID-19 patients. Such discrimination forced Alice and other self-advocates to advocate for their right to access healthcare (a matter of justice) and for others to reconceptualize their definition of "quality of life".

Yet, even as the COVID-19 pandemic draws to a close, there is a continued need for self-advocacy in order to ensure the needs of people with IDD are met. For instance, during the pandemic, telehealth became relatively widespread. This method of receiving healthcare and various types of therapeutic interventions can be especially helpful for individuals with disabilities because it reduces exposure to risk, which is unavoidable with travel (Goldberg et al. 2021). From a disability perspective, it would make sense for telehealth to remain a viable option, but increasing calls for a return to "normalcy" may influence telehealth opportunities to reduce. If widespread telehealth opportunities *do* continue to exist, then advocacy will still be needed to ensure that individuals have the necessary equipment (e.g., phones) and resources (e.g., data plans) to participate. Individuals with IDD are often significantly less likely to have access to the necessary technology and resources for telehealth visits compared to those without disabilities (Steel 2023). Thus, self-advocacy will continue to be critical in the post-COVID-19 world.

## 16.4 The Relation Between Parent Advocacy and Self-Advocacy

While both parent advocacy and self-advocacy have been studied as separate constructs, there is growing recognition of the importance of considering these two forms of advocacy together (Burke et al. 2019a, 2019b). Combined advocacy efforts—from the parent and the person with IDD—can promote greater collaboration and communication between individuals with IDD, their families, and service providers, leading to more effective care (Friedman and Rizzolo 2018). Unfortunately, few studies to date have examined parent advocacy and self-advocacy within the same sample. Of the limited extant research, when parent and self-advocacy co-occur, there are improved outcomes for both individuals with disabilities and their families. For example, when individuals with disabilities and their parents engage together in advocacy on behalf of the person with the disability, both experience increased knowledge, empowerment, and feelings of control over their lives (Wehmeyer 1998). By advocating for their needs and rights, individuals with IDD and their families can become more informed about available services and support, and better equipped to access them.

In our own work, we have found that parent advocacy and self-advocacy lead to improved outcomes. In one study involving individuals with IDD and their families (Burke et al. 2019a, 2019b), both parent and self-advocacy was critical to obtaining services and support which allows greater opportunities for self-determination and community participation for the individual with IDD. In another project, we examined the impact of an advocacy intervention program for parents of transition-aged youths with autism (Taylor et al. 2022). Results show that parent advocacy was strongly associated with increased service access (Lee et al. 2022). These findings suggest that when parent advocacy does not overshadow self-advocacy, it may even further promote the advocacy efforts of individuals with autism and, by extension, those with IDD. Ultimately, based on the limited research, it seems that the combination of parent advocacy and self-advocacy may enhance the quality of life of people with IDD and their families. However, even well-meaning people acting with the best of intentions (including friends, family members, and others who want to see "what's best" for the person), may pursue a certain end which, if it does not align with the values, goals, and interests of the person with disability, can unintentionally thwart their independence. In these cases, people with IDD may primarily benefit from, and prefer, expressing their own autonomous preferences, needs, and desires in at least some circumstances. As such, it may be most appropriate for parents to advocate *for* these kinds of opportunities to exist for people with IDD instead of advocating *on* the individuals' behalf. One way to achieve this delicate balancing act between parent and self-advocacy is highlighted in the section below.

## 16.5 Moving Forward to Balance Parent Advocacy and Self-Advocacy

It is important to appropriately balance parent advocacy and self-advocacy. According to Kim and Turnbull's (2004) work regarding the transition of individuals with severe intellectual disabilities from adolescence to adulthood, planning should *consider* the choices and preferences of the youth to the greatest extent possible. Specifically, Kim and Turnbull assert the importance of the choices and preferences of the individual with a disability while qualifying it with the term "to the greatest extent possible" (Kim and Turnbull 2004, 56). To this end, we argue that using an approach of person-family interdependent planning is apt. Person-centered planning prioritizes the needs, desires, and strengths of the person with a disability, whereas family-centered planning focuses on the values and desires of the family as a unit. In contrast to both person-centered planning and family-centered planning, person-family interdependent planning recognizes the needs, desires, and strengths of the support network (often, the family) as well as those of the individual with a disability. Person-family interdependent planning recognizes that for transitions to be successful, decisions must consider the quality of life of the family and the quality of life of the individual with IDD and/or autism (Blacher 2001).

As identified by Kim and Turnbull (2004), there are several principles of person-family interdependent planning. The principles include: the transition of an individual with IDD is influenced by the family system; individuals with IDD and their families have choices to make in their lives; no person is fully competent in all of life's domains; future planning should consider the needs of the individual with IDD and their family members; and policies and programs should support the individual with IDD as well as their family. These principles provide some direction for balancing parent advocacy with self-advocacy, as well as underscore the interdependence of parents and their offspring with disabilities.

The person-family interdependent planning approach may be especially useful among certain families and at specific points in time. For example, families from racial minority backgrounds wherein collectivism (versus individualism) is more valued (Achola and Greene 2016) may better align with a person-family interdependent planning approach. A similar perspective is evidenced in the medical profession's concept of relational autonomy wherein the patient relies on family and sometimes community members such as religious leaders to help them reach critical decisions. Therefore, even if a person cannot make a decision entirely independently, they can still have a meaningful part in the decision-making process. Person-family interdependent planning may also help when considering other family members. Research has shown that siblings of individuals with IDD often make life decisions with their brothers or sisters with IDD in mind (e.g., marital and reproductive decisions) (Hodapp et al. 2010). Person-family interdependent planning may therefore be used to help understand decision making among individuals with IDD and their siblings (Burke et al. 2021). Further, since future planning is also a principle of person-family interdependent planning, such planning may be especially helpful in aging families (Burke and Lakin 2019).

Person-family interdependent planning may also be beneficial when considering how to engage in effective parent advocacy and self-advocacy. This type of planning may serve to unite the perspectives of the individual with IDD and the family by suggesting that they should aim to have compatible advocacy goals. Further, it should be noted that combined advocacy efforts are interdependent, wherein the advocacy of the parent may impact self-advocacy and vice versa. This principle is supported by extant research (Burke et al. 2019a, b) as parent advocacy can foster self-advocacy.

Moving forward, we recommend a person-family interdependent planning approach to be used in order to advocate for people with IDD in an ethically defensible manner. Future research should consider collecting information from the perspectives of individuals with disabilities and their parents from the same respective families. By having parent–offspring dyads, we can have a more holistic understanding of parent and self-advocacy, including how advocacy efforts work with or against one another. The principles of person-family interdependent planning can also be used to guide research. For example, research should be conducted to examine advocacy activities conducted during future planning. By doing so, future planning interventions can be developed to include advocacy efforts that support the goals of the person with IDD and their family.

In terms of practitioners, person-family interdependent planning can also be used to encourage parent advocacy and self-advocacy. Many agencies (e.g., Parent Training and Information Centers) educate individuals with disabilities and their families about how to advocate effectively (Burke 2015). These agencies may also want to consider involving individuals with disabilities and their parents in the same advocacy training (versus separate training for the individuals with disabilities and their families, respectively) if this is of interest to both parties. In doing so, the training may also serve as a forum wherein individuals with disabilities and their families can learn from one another, families can be informed of the ways that they may be able to empower and enable (or disempower and disable) self-advocacy, and ultimately unite to advocate in a compatible manner and in ways that best support the individual's autonomy.

## References

Achola, Edwin O., and Gary Greene. 2016. Person-family centered transition planning: Improving post-school outcomes to culturally diverse youth and families. *Journal of Vocational Rehabilitation* 45 (2): 173–183.

Baggs, Mel. 2019. *The meaning of self-advocacy. Thinking person's guide to Autism*, February 11. http://www.thinkingautismguide.com/2019/02/the-meaning-of-self-advocacy.html. Accessed 9 April 2024.

Björnsdóttir, Kristín, and IngólfurÁsgeir Jóhannesson. 2009. People with intellectual disabilities in Iceland: A Bourdieuean interpretation of self-advocacy. *Intellectual and Developmental Disabilities* 47 (6): 436–446. https://doi.org/10.1352/1934-9556-47.6.436.

Blacher, Jan. 2001. Transition to adulthood: Mental retardation, families, and culture. *American Journal of Mental Retardation* 106: 173–188.

Burke, Meghan M. 2015. A literature review of parent training and information centers in the United States: Developing collaborative family-school partnerships. *Center for Advancement in Inclusive and Special Education (CAISE) Review* 3: 2–13.

Burke, Meghan M., Linda Sandman, Beatrize Perez, and Meghann O'Leary. 2018. The phenomenon of legislative advocacy among parents of children with disabilities. *Journal of Research in Special Education Needs* 18 (1): 50–58. https://doi.org/10.1111/1471-3802.12392.

Burke, Meghan M., and K. Charlie Lakin. 2019. The need for research about services and supports for aging individuals with severe disabilities and their families. *Research and Practice for Persons with Severe Disabilities* 44: 207–210.

Burke, Meghan M., Chung Eun Lee, Sarah A. Hall, and Zach Rossetti. 2019a. Understanding decision making among individuals with intellectual and developmental disabilities (IDD) and their siblings. *Intellectual and Developmental Disabilities* 57 (1): 26–41.

Burke, Meghan M., Kristina Rios, and Chung Eun Lee. 2019b. Exploring the special education advocacy process according to families and advocates. *The Journal of Special Education* 53 (3): 131–141.

Burke, Meghan, Chung Eun Lee, Zach Rossetti, and Sarah Hall. 2021. Parent and sibling roles in decision making with individuals with intellectual and developmental disabilities. In *Decision making by individuals with intellectual and developmental disabilities: Integrating Research into Practice*, ed. Ishita Khemka and Linda Hickson, 95–113. Cham: Springer Nature Switzerland AG.

Caldwell, Joseph, Miriam Heyman, Michael Atkins, and Sandy Ho. 2022. Experiences of individuals self-directing Medicaid Home and community-based services during COVID-19. *Disability and Health Journal* 15: 110–113.

Charlton, James I. 2000. *Nothing about us without us: Disability oppression and empowerment*. Oakland CA: University of California Press.

Churchwell, Keith, and Erin O'Malley. 2023. Effects on patients of continuous eligibility for Medicaid coverage: Lessons learned and lessons to be learned from the pandemic. *Circulation* 147: 650–652.

Cohen, Shana R. 2013. Advocacy for the "Abandonados": Harnessing cultural beliefs for Latino families and their children with intellectual disabilities. *Journal of Policy and Practice in Intellectual Disabilities* 10 (1): 71–78.

Friedman, Carli, and Mary C. Rizzolo. 2018. Friendship, quality of life, and people with intellectual and developmental disabilities. *Journal of Developmental and Physical Disabilities* 30 (1): 39–54.

Francis, Grace L., Heather M. Aldersey, Sharon Gabison, and Colleen M. Davison. 2021. Family-school collaboration for students with disabilities in Ontario, Canada. *Network for International Policies and Cooperation in Education and Training* 48: 1–2.

Goldberg, Elizabeth M., Frances N. Jiménez, Kevin Chen, Natalie M. Davoodi, Melinda Li, Daniel H. Strauss, Maria Zou, Kate Guthrie, and Roland C. Merchant. 2021. Telehealth was beneficial during COVID-19 for older Americans: A qualitative study with physicians. *Journal of the American Geriatrics Society* 69: 3034–3043.

Harris, Bryn, Maryellen Brunson McClain, Sonja O'Leary, and Jeffrey D. Shahidullah. 2021. Implications of COVID-19 on school services for children with disabilities: Opportunities for interagency collaboration. *Journal of Developmental and Behavioral Pediatrics* 42: 236–239.

Henderson, David, and Christine Bigby. 2016. 'We were more radical back then': Victoria's first self-advocacy organisation for people with intellectual disability. *Health and History* 18 (1): 42. https://doi.org/10.5401/healthhist.18.1.0042.

Hodapp, Robert M., Richard C. Urbano, and Meghan M. Burke. 2010. Adult female and male siblings of persons with disabilities: Findings from a national survey. *Intellectual and Developmental Disabilities* 48: 52–62.

Kim, Kyeong-Hwa., and Ann P. Turnbull. 2004. Transition to adulthood for students with severe intellectual disabilities: Shifting toward person-family interdependent planning. *Research & Practice for Persons with Severe Disabilities* 29 (1): 53–57.

Lee, Chung Eun, Meghan M. Burke, Leann Smith DaWalt, Chak Li, and Julie Lounds Taylor. 2022. The role of parental advocacy in addressing service disparities for transition-aged youth on the autism spectrum. *Autism* 26 (4): 1001–1006.

Okonkwo, Nneoma E., Ugochi T. Aguwa, Minyoung Jang, Iman A. Barré, Kathleen R. Page, Patrick S. Sullivan, Chris Beyrer, and Stefan Baral. 2021. COVID-19 and the US response: Accelerating health inequities. *BMJ Evidence-Based Medicine* 26: 176–179.

Pfeiffer, David. 1993. Overview of the disability movement: History, legislative record, and political implication. *Policy Studies Journal* 21 (4): 724–734.

Scotch, Richard K. 2009. "Nothing about us without us": Disability rights in America. *OAH Magazine of History* 23 (3): 17–22. https://doi.org/10.1093/maghis/23.3.17

Shogren, Karrie A., Susan B. Palmer, Michael L. Wehmeyer, Kendra Williams-Diehm, and Todd D. Little. 2012. Effect of intervention with the self-determined learning model of instruction on access and goal attainment. *Remedial and Special Education* 33: 320–330. https://doi.org/10.1177/0741932511410072.

Shogren, Karrie A., and Michael J. Ward. 2018. Promoting and enhancing self-determination to improve the post-school outcomes of people with disabilities. *Journal of Vocational Rehabilitation* 48 (2): 187–196.

Smith, Ivanova, and Carlyn O. Mueller. 2022. The importance of disability identity, self-advocacy, and disability activism. *Inclusive Practices* 1 (2): 47–54. https://doi.org/10.1177/27324745211057155.

Steel, Emily J. 2023. Assistive technology is a resource for building capabilities, but is it just addressing the symptoms of inequality? *Societies* 13: 7–22.

Taylor, Julie Lounds, Florencia Pezzimenti, Meghan M. Burke, Leann Smith DaWalt, Chung eun Lee, and Carol Rabideau. 2022. Development, feasibility, and acceptability of a nationally relevant parent training to improve service access during the transition to adulthood for youth with ASD. *Journal of Autism and Developmental Disorders* 52: 2388–2399.

Trainor, Audrey A. 2010. Diverse approaches to parent advocacy during special education home-school interactions. *Remedial and Special Education* 31 (1): 34–47.

Walmsley, Jan, Liz Tilley, Sue Dumbleton, and Janet Bardsley. 2017. The changing face of parent advocacy: A long view. *Disability & Society* 32 (9): 1366–1386. https://doi.org/10.1080/09687599.2017.1322496.

Walmsley, Jan, and Central England People First History Project Team. 2014. Telling the history of self-advocacy: A challenge for inclusive research. *Journal of Applied Research in Intellectual Disabilities* 27 (1): 34–43. https://doi.org/10.1111/jar.12086.

Waltz, Mitzi, Karin van den Bosch, Hannah Ebben, Lineke Van Hal, and Alice Schippers. 2015. Autism self-advocacy in the Netherlands: Past, present and future. *Disability & Society* 30 (8): 1174–1191. https://doi.org/10.1080/09687599.2015.1090954.

Wehmeyer, Michael L. 1998. Self-determination and individuals with significant disabilities: Examining meanings and misinterpretations. *Journal of the Association for Persons with Severe Handicaps* 23 (1): 5–16.

Wehmeyer, Michael L., and Karrie A. Shogren. 2016. Self-determination and choice. In *Handbook of evidence-based practices in intellectual and developmental disabilities*, ed. Nirbhay N. Singh, 561–584. Cham: Springer Nature Switzerland AG. https://doi.org/10.1007/978-3-319-26583-4_21.

Wong, Alice. 2020. *Disability visibility: First-person stories from the twenty-first century*. New York: Vintage Books.

World Health Organization. 2009. *Improving health systems and services for mental health*. https://www.who.int/publications-detail-redirect/9789241598774. Accessed 9 April 2024.

Wright, Amy Conley, and Sarah Taylor. 2014. Advocacy by parents of young children with special needs: Activities, processes, and perceived effectiveness. *Journal of Social Service Research* 40: 591–605. https://doi.org/10.1080/01488376.2014.896850.

Zhang, Chao. 2017. 'Nothing about us without us': the emerging disability movement and advocacy in China. *Disability & Society* 32 (7): 1096–1101. https://doi.org/10.1080/09687599.2017.1321229.

# Chapter 17
# A Justice Perspective on Ecologically-Based Employment Pathways for Individuals with Disabilities

**Laura R. Bowman, Carolyn McDougall, and Dolly Menna-Dack**

**Abstract** Canada's workforce has historically marginalized individuals with autism and intellectual disabilities for reasons including lack of specialized training, accommodations, education for employers, and general stigma. Evidence supports that early engagement with employment is correlated with lifelong workforce attachment. Early employment interventions for this population traditionally focus on training and changing the individual with disabilities to fit into the employment environment, yet avoid consideration of how employment settings, policies, or contextual factors might adapt to the needs of these employees. In this chapter, we address questions surrounding how we might connect individuals with disabilities to a workforce that is not built with their abilities in mind. In addressing this topic, we introduce multi-level interventions that support meaningful inclusion and integration of young workers with disabilities into the workforce, with our own program provided as a guiding example. The chapter concludes with reflections on how to action practices that will facilitate a more just and inclusive employment context for individuals with autism and intellectual disabilities.

**Keywords** Employment participation · Workforce · Inclusion · Youth · Collaboration · Ecological systems

L. R. Bowman (✉) · C. McDougall · D. Menna-Dack
Holland Bloorview Kids Rehabilitation Hospital, Toronto, Canada
e-mail: lbowman@hollandbloorview.ca

C. McDougall
e-mail: cmcdougall@hollandbloorview.ca

D. Menna-Dack
e-mail: dmennadack@hollandbloorview.ca



A. Bianchi and J. A. Vogt (eds.), *Intellectual Disabilities and Autism: Ethics and Practice*, The International Library of Bioethics 108,
https://doi.org/10.1007/978-3-031-61565-8_17

## 17.1 Introduction

Employment participation for people with disabilities (PWDs) is as complex an issue as any other discussed in this book. The mix of ethical, moral, policy-based, and practical issues that go into creating environments for meaningful and equitable inclusion are multifaceted and often uncoordinated. In this chapter, we walk through justice-based arguments for employment inclusion, evidence supporting the benefits of employment participation for PWDs and society, and practical considerations required to bring employment participation to life. We conclude with an example of our own work in promoting employment inclusion through programming, collaboration, and advocacy.

### *17.1.1 Author and Chapter Positionality*

Given the topic of this book, we would like to locate ourselves as authors. We have crafted this chapter from a critical theoretical lens, based in the lived, policy-based, and economic realities of employment for individuals with intellectual disability, autism, and other forms of disability in our practice context. We make assumptions of the inherent value and worthiness of individuals with autism and intellectual disabilities, arguing that they have the right to participate in employment if and how they choose. In this chapter, we use person-first language to align with our organization's policies that aim to consider disability as one of many characteristics.

To further situate this chapter, most examples are drawn from and considered within the Canadian context. Our perspective—and the ways in which we consider employment participation and opportunities for people with autism and intellectual disabilities in general—also originate from a Canadian perspective. Two of the authors are registered occupational therapists, and approach our understanding of human engagement from an occupational participation-based perspective (ACOTRO, ACOTUP, and CAOT 2021). One author is a clinical bioethicist with a pediatric focus and substantial practical experience as a job coach. Our practice is located within a pediatric rehabilitation hospital, and we work with partners from the education, vocational, private, and policy sectors. As practitioners primarily in the pediatric system, we promote the importance of the individual as autonomous while simultaneously recognizing the role of families and support systems as critical during this time of growth and development. It is, in fact, best practice to involve the natural supports in an individual's life when training for and pursuing employment (Liu et al. 2014). Our team additionally operates from a solution-focused perspective (Seko et al. 2020), a habilitation and rehabilitation perspective, a perspective promoting autonomy and agency (as defined by the individual and family), and a stance of advocacy. Our interacting perspectives impact how we view and act toward our goals for employment participation for PWDs.

## 17.2 Employment Participation for People with Disabilities

Employment participation is commonly considered synonymous with paid employment, however the concept is broader, including various roles and skill-building opportunities that contribute to broader engagement with productive or citizenship-based initiatives. In the developmental stage of youth (ages fifteen through twenty-four) (United Nations, n.d.), these might include household duties, volunteering, school-based experiential learning, internships, and paid employment (Bowman et al. 2022). Employment participation holds many benefits for individuals, families, and society. For individuals, employment participation relates to reduced use of public health services, higher perceived quality of life (Hall et al. 2013), and greater financial stability (Brucker et al. 2015).

We also know that *lifetime* workforce attachment is improved with earlier intervention, exponentially increasing the impact of the previously mentioned benefits. Evidence indicates that early intervention to support employment participation for PWDs relates to improvement in the amount of time (number of years) that individuals access and participate in the workforce (Krahn and Chow 2016; Wehman et al. 2020), and their attachment to the workforce (Caliendo and Schmidl 2016). The benefits to the individual, as outlined above, include quality of life, funds, workplace benefits (e.g., extended healthcare, dental, pension), and enhanced senses of self and belonging. In addition to clear benefits, employment and the resulting income are themselves social determinants of health, with their own rippling effects on other areas of health and well-being (Froehlich-Grobe et al. 2021).

For employers and companies, benefits of employing PWDs include higher employee retention, increased profits due to inclusion and visibility thereof, increased consumer diversity, and creativity and innovation (Lindsay et al. 2018; ODENa, n.d.). Hiring PWDs also addresses current and anticipated labour market shortages by accessing domestic talent that is often untapped and un(der)recognized (ODENa, n.d.). On a broader scale, benefits of increased workforce attachment for PWDs to the taxpaying public include decreased funds spent on social programming and social services (e.g., disability support funding), and increased revenue from income taxes paid on salary (Beyer 2017; Bowman et al. 2024; Jacob et al. 2015; Tompa et al. 2021). While not easily measured, theorized benefits include decreases in mental and physical health funding spent on caregiver well-being and potential income taxes paid by parents and caregivers who can return to work (Tompa et al. 2021).

### *17.2.1 Current Context*

Despite the strong rationale for creating an inclusive and equitable workforce, PWDs consistently record among the lowest levels of employment in Canada. The level is even lower for those with intellectual disability and autism (Berrigan et al. 2023). This appears to be due to a number of practical, political, and cultural factors that are historically entrenched in our workforce. For PWDs, structured, government-funded workforce support is limited until people leave the formal education system. The rationale is that youth in the education system are served by that system, which has the primary goal of education and graduation. Some programs exist, such as cooperative education placements, but there is little structured support to enter the workforce during secondary or post-secondary education. Once youth leave secondary or post-secondary education, they are expected to move from a primary occupation as a student (often with no solid job skills) to that of a worker. The transition is further complicated for those with disabilities, who may need additional assistance with job skills training, accommodations, modifications, social skills training, and other areas not addressed in schools. Instead of an interwoven service model in which education and employment systems work collaboratively to prepare youth with disabilities for the workforce (i.e., best practice) (Rutkowski et al. 2006), youth and their families are left scrambling for resources at a time that is *already* rife with significant changes.

We note that even when services *are* available, they most frequently intervene at the level of the PWD or community, rather than the system. Increasing efforts to raise employer awareness of the business benefits of inclusion, along with outlining *how* to include employees with disabilities is essential in moving toward an equitable workforce. It is insufficient to expect PWDs to bend themselves into shapes that are palatable for the workforce; we must change the workforce itself.

## 17.3 A Justice Issue

The ability to participate meaningfully and equitably in the workforce is a right for all persons, including PWDs. The evidence outlined above aligns the right to employment (as related to issues of health equity) with the building of social and economic capital.

**Intersecting Identities**

We take a moment and ask readers to recognize that, like anyone else, PWDs hold intersecting identities. While we have not given space in this short chapter to the impacts of other identities and potentially marginalizing factors such as race, gender, ethnicity, religion, and sexual orientation, they unquestionably play a role in the employment participation processes of PWDs. In this section, however, we focus on exploring the justice elements specifically related to disability that should be considered *alongside* the abovementioned intersecting identities.

## 17.3.1 What Has Been Done

### 17.3.1.1 Policy Level

The right to work and access meaningful employment with reasonable accommodations has been recognized through policies and conventions such as Article twenty-seven of the United Nations Convention on the Rights of Persons with Disabilities (CRPD) (UN 2006), and national and state or provincial acts such as the Employment Equity Act of Canada (1995). Statutes such as these provide insights into workplace elements that will support the inclusion of PWDs when seeking to attract, hire, and retain them. They provide metrics and outcomes for which to strive, primarily guiding employers on including PWDs in their organizations. The CRPD also outlines enablers of employment engagement for PWDs. For example, articles twenty-six (*Habilitation and Rehabilitation*) and twenty-seven (*Work and Employment*) indicate the need for access to vocational guidance, placement, and training, that should "begin at the earliest possible stage, and [be] based on the multidisciplinary assessment of individual needs and strengths" (UN 2006, 19).

Statutes and acts represent a starting point in driving awareness of the issue, outcomes for which to strive, and a commitment to change the employment culture to consider and include PWDs. And that is what they are, a starting point. If PWDs are inadequately prepared to participate in the workforce, and employers and organizations are mandated to accommodate but not equipped with the knowledge of what that might look like or mean for their bottom line, the connections become brittle, short-lived, and devoid of meaningful collaboration. Policies offer a minimum starting point to drive change and inclusion, but cannot be the only benchmark.

### 17.3.1.2 Provincial and Community Level

In addition to strides in policy, there exist many platforms and social programs that promote inclusion for PWD into the workforce. Programs vary by country and region, and can include interventions in the school, the community, and organizations. School-based programming may include special education curricula and cooperative educational opportunities with the goal of preparing transition-aged youth for life beyond high school. Community programming may include education sessions, employment supports, and other vocational training. Community-based programming may also be provided via best practice education and support from organizations such as the Canadian Association for Supported Employment (CASE, n.d.), the Ontario Disability Employment Network (ODENb, n.d.), and Ready, Willing & Able (RWA 2015).

#### 17.3.1.3 Individual, Service Provider, and Organizational Level

As was previously described, best evidence is being mobilized and integrated into policy and practice. Programs are being made available to PWDs, with the onus primarily on the individual to identify such opportunities. In particular, *early* intervention programs outside of educational settings are inconsistently funded. We are therefore left with programs that either rely on highly dedicated facilitators or are short-lived. Access to such programs varies by geographic location, requires awareness of the program's existence, and often depends on the ability to self-fund participation, all of which weigh in favour of privileged positions. These realities combine to perpetuate privilege-based gaps within the disabled community as to who can and will access meaningful employment.

#### 17.3.1.4 Why is Employment Still an Issue

With so much evidence on employment participation best practices for PWDs and a general policy-based ethos favouring equity and inclusion in the workforce, why does the employment climate for this group remain so bleak?

We posit that despite knowledge and evidence, there remains a need for concerted ecologically-based initiatives that support individuals, employers, communities, and policy-makers in changing the training and employment climates. We propose the implementation of integrated models that consider, bridge, and concert efforts to maximize the availability and efficacy of employment outcomes for this population.

## 17.4 What Can Be Done

To facilitate employment participation for PWDs, we propose an approach that spreads responsibility across contributors at multiple ecological levels. Integration would be driven by knowledge mobilization and implementation that promotes ongoing change and re-evaluation based on emerging data; for example, Menear et al.'s (2019) value-creating learning health systems and Damschroder et al.'s (2022) Consolidated Framework for Implementation Research. Figure 17.1 outlines how such an integrated model *might* look, with the various levels of integration listed from macro (top) to micro (bottom), and sample opportunities for integration provided as the items with arrows and letters on the left. This is meant to serve as an illustrative example. We deliberately position PWDs at the foundation of the model, with all collaborators stemming from the individual. We acknowledge that readers' own approaches or positions may vary based on their own roles and goals (e.g., readers approaching from a policy lens might invert the model). In the upcoming sub-sections, we demonstrate how such a system might work using the Model of Embedded Employment Supports (MEES; Fig. 17.1), followed by a practical example.

**Fig. 17.1** Ecologically-based, collaborative "*Model of Embedded Employment Supports*" *(MEES)* for PWDs

### Honour and Prepare the Individual

In our conception of an integrated model, we position the PWD at the foundation of the collaboration, since they are the focus of the intervention and collaboration. At this level, interventions and services help the individual identify their strengths and employment goals, find local resources, and access evidence-based interventions (for examples, see Sect. 17.2 on *Employment participation for PWDs*). Appropriate interventions will work *with* the person's strengths to set them up for success within their context (as depicted in Fig. 17.1, the MEES).

### Integrations

For this level, we have not depicted integrations (see layers and arrows on the left-hand side). This is because we see the individual's abilities, choices, and journey as driving all other levels' involvement. They will, of course, be impacted by their direct supports' attitudes of employment, availability of community supports, and access to employers, funding, and policy. Refer to upcoming sub-sections for deeper exploration.

### Prepare and Educate Direct Supports

The people in the individual's immediate environment will act as direct supports. In our model, we consider direct supports in two main groups: (1) family, friends, and personal supports, and (2) professionals in support roles. The attitudes of members of both groups toward employment will impact the support that is provided to the person who wants employment participation opportunities.

Personal support is a major indicator for employment attachment for PWDs (Liu et al. 2014). Their belief and investment in the employment potential for PWDs will impact the support that they provide for employment participation preparation, thereby either promoting or limiting potential learning and practice opportunities. Involving personal supporters in the employment participation journey is considered best practice in supporting transitions to adult life and productive adult roles for PWDs (Kohler et al. 2017; Larson and Bolton 2019). Integrations for this group include education on realistic possibilities of employment and direct involvement in employment planning.

For service providers, such as educators, vocational rehabilitation specialists, occupational therapists, and job coaches, education and practice are essential to enable meaningful integration of PWDs in the workforce. Best practices in this group include development of person-directed plans, promotion of interagency collaboration, creation and implementation of evidence-based program practices, and identification and evaluation of outcome measures to contribute to improvement of both program and individual outcomes.

**Integrations**

As direct supports are immediately connected to the individual in this conception of employment participation preparation, they are assumed to impact all levels of collaborators. As depicted in Fig. 17.1, integration opportunity A demonstrates the unique potential impact the direct support has on the person's knowledge that they *can* participate in employment, and direct supports' abilities to identify and integrate best evidence to support the individuals' journey to employment participation.

**The Connective Role of Community-Level Supports**

Community-level supports (e.g., advocacy groups that provide training to individuals, employers, and direct supports), act as hubs that connect PWDs and direct supports with funding and business connections, and connect the level of actual work placement with advocacy and policy. Their role is essential, but also amorphous in that it is shaped by the very things upon which it acts (existing advocacy, funding, programs, community connections). Through educating, organizing, and connecting, they help integrate the multiple moving pieces to best serve PWD. We note that the volume of responsibility of community-level agencies reflects the current employment context's general reliance on them. Our hope is that responsibilities will be better distributed among the levels of the MEES over time.

**Integrations**

Community-level supports constitute the only group in Fig. 17.1 depicted as having bi-directional impact. While all groups influence one another, most follow the same centrally-directed flow. This group must advocate outwardly in order to facilitate the other groups' inward flow. Starting with opportunity B, community-level supports provide services, oversight, education, and funding facilitation, thereby impacting the individual and direct supports. They provide education and connection to employers and organizations looking to employ PWDs or to review policies and practices, indicated via opportunity C. Opportunity D represents their advocacy work to build relevant and realistic systems that consider current employment needs and experiences of PWDs.

### Prepare and Educate Employers

The implication of having an employer/organization in the diagram is that they will employ PWDs as part of their workforce. Yet, if hiring practices are to be profitable and sustainable, we must build strong, intentional practices that consider both the needs of employers and prospective employees. In addition to hiring diverse talent (including PWDs), employers may benefit from interventions regarding human resource practices (i.e., accommodation, modification, recruitment, hiring, training, and retention). With education and assistance, employers have reported benefits of hiring PWDs as part of their organizations (Lindsay et al. 2018; ODENa, n.d.).

### Integrations

In the diagram, employers are depicted primarily as recipients of the opportunities presented. We acknowledge this as an oversimplification, as employers can and do influence industry partners and act as champions for inclusion after developing successful practices. For the purposes of our diagram, we assume that many employers are in the early stages of inclusive hiring and will increase their influence if and when they solidify their practices.

As was highlighted above, community-level partners can support employers looking to hire and train PWDs (opportunity C). This may include connecting employers with community agencies who support work placements for PWDs, engaging job coaches or skill trainers, and providing reviews and adaptations on organizational practices. This, in turn, may include identifying workplace practices, policies, or structures associated with recruitment and hiring processes, training, accessibility, or performance management, that are discriminatory or marginalizing for PWDs and others. Employers may then work with community organizations, direct supports, and individuals to support the process of employing PWDs (opportunity E). As will be elaborated in upcoming sections, employers will benefit from modifying policies and funding to support the inclusion of PWDs into their workforce, be they physical, procedural, management-based, or other.

### Fund Early Intervention

At a broader level, we acknowledge that change will unlikely occur or be sustained without ongoing, reliable, and equitable funding for early intervention and workplace integration. Currently, many students with disabilities are ineligible for funding related to direct workplace supports. Changing this would align with best evidence in supporting a "start early" approach and lead to greater lifetime gains. We advocate for sustained *public* funding to support equitable access to programming for individuals who may not currently pursue employment (by circumstance or by choice) due to funding and access barriers (Bowman et al. 2024).

### Integrations

As depicted in Fig. 17.1, funding will primarily support three main endeavors. Funds for employers to undertake education, policy reviews, and support hiring, will spur long-term change (opportunity F). Opportunity G depicts funding specifically supporting individual participation in the workforce, or what is traditionally associated with this type of public-level funding. Funding may also support community agencies or direct supports in providing programs, in job coaching that helps individuals integrate into the workforce, or other incentives that support workforce connection.

In regard to the point of equity discussed above, funding may need to be distributed differently in different areas in order for an equitable outcome to be achieved. It seems plausible that within a single territory (e.g., province, region), populations may be dense and need concentrated services, or may be rural and require focus on transport and access to service. Similarly, the business profile of a region may require programs and skills that focus on a specific industry. Additionally, cultural differences may necessitate services and supports offered with different approaches and foci. There are many considerations that support the case for flexible funding to meet local needs.

**Build Supportive Policy**

Policies that support employment participation for PWDs provide information on what employers, communities, and direct supports should be doing, and how they will be held accountable. Such policies should be drafted and updated with key community stakeholders (e.g., PWDs, direct supports, employers, funding bodies). Areas of focus for existing policies include the Accessible Canada Act (2019) and the Accessibility for Ontarians with Disabilities Act (2005), which describe expectations for relevant employers.

**Integrations**

In the flow depicted through Fig. 17.1, in which the individual is at the centre, policies are viewed as impacting the actors around the individual, putting practices in place to support the individual's move toward employment. Opportunities H and I represent policies that outline how public funding should be distributed and managed, and how employers are expected to engage, hire, train, and retain PWDs. Opportunity I depicts policies that guide how community and direct supports should work with PWDs, employers, and the connection between the two.

**Build and Integrate Evidence**

The identification, development, and mobilization of reliable evidence allows for the creation and sustainment of best practices that keep all contributors accountable and promote meaningful and relevant initiatives for all groups involved. Knowledge mobilization can follow a value-creating learning (health) system approach (Menear et al. 2019), whereby evidence is used to create programs that represent efficient and effective practices as supported by research and practice. The programs and services are then evaluated to determine whether and how the programs lead to outcomes that are worthwhile to those involved at all levels of Fig. 17.1. Additionally, the data that emerges *over time* should be evaluated in order to understand when and how fluctuations occur, and whether practices should change.

Each of the items discussed in Fig. 17.1, as well as research and evaluation, must integrate within and across all levels to create more equitable and inclusive work systems. So how might we integrate such best practices?

## 17.5 A Case Example: Holland Bloorview Kids Rehabilitation Hospital's Employment Participation Pathways Model

In order to demonstrate how an ecologically-based collaboration for employment participation for PWDs can be put into action, we present a model that is currently implemented in Ontario, Canada. The model represents a structure that allows the principles outlined in Fig. 17.1 to be implemented, with practical, actionable steps. Through this example, we intend to demonstrate that the levels presented in Fig. 17.1 do not necessarily need to be intervened upon independently, but each must be considered when creating a resource, program, policy, or initiative.

Since 2007, Holland Bloorview Kids Rehabilitation Hospital has offered programs that provide early employment participation opportunities for youth with physical, neurological, developmental, and intellectual disabilities. In 2020, the growing profile of programs were consolidated into an Employment Participation Pathway (EPP) (Bowman et al. 2022). The EPP can be viewed in Fig. 17.2. The EPP, like Fig. 17.1, was built around the individual and from the perspective of rehabilitation professionals that focus on funding, facilitating, and implementing interventions for youth and families. The model includes three levels: (1) an individual-focused pathway of interventions (top row); (2) initiatives directed toward community and direct supports (middle row); and (3) initiatives focused on building more inclusive supports, communities, and systems (bottom row). We will briefly describe each level, and then align the pathway with our MEES model.

The individual-focused pathway (top row) is at the core of the EPP model. A number of specific interventions (programs or services), aligning with best practices, support the individual through progressive skill-building and experiential opportunities for engagement in employment. These start with facilitated and supported employment experiences such as volunteering or co-operative education and also incorporate self-discovery and skill-building educational components. Individuals and their direct supports then attend employment action coaching, through which they create a specific plan for achieving their short- and long-term employment goals. This process includes one-to-one coaching with a service provider and group mentorship with families who have lived experience with the employment participation process. Equipped with their plan, individuals then participate in intensive supported work experiences, including paid part-time or seasonal jobs with the support of a job coach/

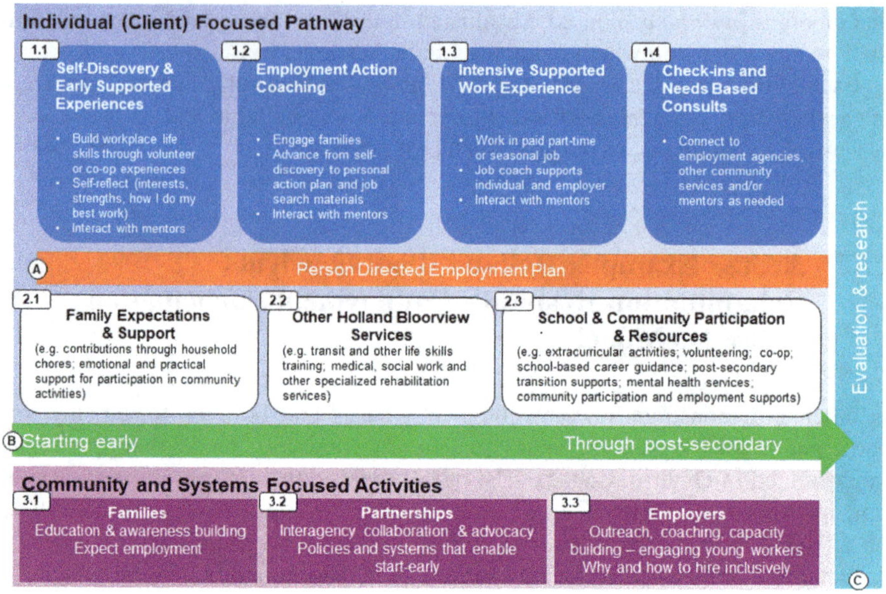


**Fig. 17.2** Holland Bloorview's Employment Participation Pathway model available from Bowman et al. 2022 (Taylor & Francis) via www.tandfonline.com

service provider. These opportunities for engagement in the workforce provide real-world experience and confidence for youth, as well as opportunities for employers to engage with workers with disabilities and job coaching professionals. Finally, as youth age out of EPP services, we provide check-ins and needs-based consults to connect with employment agencies and other community services. We also provide mentorship as they take steps toward employment. The top level of the EPP aligns with the individual, direct supports (opportunity A), community (opportunity B), and employer (opportunity E) levels of the MEES (Fig. 17.1) and is impacted by the policies (opportunity I) and funding available to facilitate such programming (opportunity G).

The middle level of the EPP includes initiatives geared toward community and direct supports within the youth's environment, including family, peers, service providers, educators, community service organizations, programs, schools, and the community at large. This level is often *not* the focus of direct interventions. We argue that in order to change the context of employment participation, including opportunities, understanding of possibilities, and the cultural perceptions and practices, there must be active intervention at this community level. In the EPP model, this is accomplished via two main methods: education and interagency collaboration. Education is essential and directed toward both the youth's personal and professional supports. For personal supports, education is provided on realistic expectations for youths' employment participation, which often has greater potential than they anticipate. Expertise and tailored intervention is then provided by our team regarding *how* to build employment skills and hold the youth to expectations in the home (e.g., chores, discussion of expectations) and the community (e.g., supporting youth to volunteer and participate in work). Engagement with personal supports reflects opportunity A in Fig. 17.1, whereby the family is expected to provide support and hold high expectations for their youth, and also to *receive* training by other direct supports. Supports for professionals at this level may include connection to other organizational, community, and school resources, such as programs, medical oversight, life skills training, mental health services, community participation, and employment supports. In addition to opportunity A, opportunity B occurs through community connections to programs, services, and resources to build skills related to the individual's planned employment trajectory.

The final level of the model is focused on building more inclusive supports, communities, and systems. Initiatives include partnerships and connections between agencies and organizations at multiple levels and across systems in order to direct advocacy and inclusive policies. It also includes high-level information creation and sharing for families who need to expect employment for their youth with disability, and for employers who may need to understand how and why to hire inclusively. At this level, opportunities B, C, and D are the focus, through which advocacy and education are undertaken by community actors. Such advocacy, however, is highly impacted by opportunities F, G, H, I and J, being the policies and funding that impact how services are made available, to whom, when, and with what expectations. Engagement at this level is crucial, but not always clear-cut, and therefore often overlooked.

Like the proposed MEES in Fig. 17.1, the EPP model has knowledge mobilization woven through all levels. The evaluation perspective stands similarly in both models, whereby data is used to measure and monitor changes that occur at all levels to swiftly identify and implement changes to maximize outcomes according to changing population and contextual needs.

## 17.6 Summary and Reflections

Evidence, practice outcomes, and research all indicate that individuals with autism and intellectual disabilities can and should be supported in their right to participate in employment. Yet, their participation rates are ranked among the lowest in Canada, with similar international trends. In order to change the practices and narrative of employment participation, we need to broaden the efforts guiding current activities and opt for interventions that integrate opportunities at multiple ecological levels, as highlighted through the case example and MEES. Such approaches will target the distinct areas of consideration (e.g., the individual, direct supports, employers, policies, funding) and encourage communication and flexibility so that when there are changes or needs at one level, the others can agilely consider and adapt.

Employment participation for PWDs is a complex issue that requires attention, coordination, and collaboration between many multifaceted groups at multiple ecological levels. There exist opportunities and avenues for improvement and engagement across sectors and settings; we look to the future where they can be integrated, and more equitable employment environments can be realized.

**Acknowledgements** Funding for this chapter's authors was provided by donors to Holland Bloorview Kids Rehabilitation Hospital Foundation including the Azrieli Foundation and RBC Future Launch initiative.

## References

*Accessibility for Ontarians with Disabilities Act*. S.O. 2005, c.11. https://www.ontario.ca/laws/statute/05a11. Accessed 9 April 2024.

*Accessible Canada Act*. 2019. c.10. https://laws-lois.justice.gc.ca/eng/acts/a-0.6/. Accessed 9 April 2024.

ACOTRO, ACOTUP, and CAOT. 2021. *Competencies for Occupational Therapists in Canada/ Référentiel de compétences pour les ergothérapeutes au Canada*. (Canada). https://acotro-acore.org/sites/default/files/uploads/ot_competency_document_en_web.pdf. Accessed 9 April 2024.

Berrigan, Patrick, Craig WM. Scott, and Jennifer D. Zwicker. 2023. Employment, education, and income for Canadians with developmental disability: Analysis from the 2017 Canadian survey on disability. *Journal of Autism and Developmental Disorders* 53 (2): 580–592. https://doi.org/10.1007/s10803-020-04603-3.

Beyer, Stephen. 2017. *Economic impact of inclusion of disabled persons in the labour market*. Éditions universitaires européennes.

Bowman, Laura R., Carolyn McDougall, Danielle D'Alessandro, Jeannette Campbell, and C.J. Curran. 2022. The creation and implementation of an employment participation pathway for youth with disabilities. *Disability and Rehabilitation* 45 (24): 4156–4164. https://doi.org/10.1080/09638288.2022.2140846.

Bowman, Laura R, Carolyn McDougall, Rene Doucet, Brendon Pooran, Yin Xu and Jeannette Campbell. 2024. Funding employment inclusion for Ontario youth with disabilities: A theoretical cost-benefit model. *Frontiers in Sociology* 9: 1281088. https://doi.org/10.3389/fsoc.2024.1281088.

Brucker, Debra L., Sophie Mitra, Navena Chaitoo, and Joseph Mauro. 2015. More likely to be poor whatever the measure: Working-age persons with disabilities in the United States. *Social Science Quarterly* 96 (1): 273–296. https://doi.org/10.1007/s10803-019-04244-1.

Caliendo, Marco, and Ricarda Schmidl. 2016. Youth unemployment and active labor market policies in Europe. *IZA Journal of Labor Policy* 5 (1): 1–30. https://doi.org/10.1186/s40173-016-0057-x.

CASE (Canadian Association for Supported Employment). n.d. About Us. https://supportedemployment.ca/about-us/. Accessed 9 April 2024.

Damschroder, Laura J., Caitlin M. Reardon, Marilla A. Opra Widerquist, and Julie Lowery. 2022. The updated consolidated framework for implementation research based on user feedback. *Implementation Science* 17 (1): 1–16. https://doi.org/10.1186/s13012-022-01245-0.

*Employment Equity Act*. 1995. S.C 1995, c.44. https://laws-lois.justice.gc.ca/eng/acts/e-5.401/.

Froehlich-Grobe, Katherine, Megan Douglas, Christa Ochoa, and Andrea Betts. 2021. Social determinants of health and disability. In *Public health perspectives on disability: Science, social justice, ethics, and beyond*, ed. Donald J. Lollar, Willi Horner-Johnson, and Katherine Froehlich-Grobe, 53–89. New York: Springer. https://doi.org/10.1007/978-1-0716-0888-3_3.

Hall, Jean P., Noelle K. Kurth, and Suzanne L. Hunt. 2013. Employment as a health determinant for working-age, dually-eligible people with disabilities. *Disability and Health Journal* 6 (2): 100–106. https://doi.org/10.1016/j.dhjo.2012.11.001.

Jacob, Andrew, Melissa Scott, Marita Falkmer, and Torbjörn Falkmer. 2015. The costs and benefits of employing an adult with autism spectrum disorder: A systematic review. *PLoS ONE* 10 (10): e0139896. https://doi.org/10.1371/journal.pone.0139896.

Kohler, Paula D., June Gothberg, and Jennifer Coyle. 2017. Using the taxonomy for transition programming 2.0 to guide transition education. In *Transitioning Children with Disabilities*, ed. Antonio L. Ellis, 169–182. Leiden, The Netherlands: Brill Sense.

Krahn, Harvey, and Angela Chow. 2016. Youth unemployment and career scarring: Social-psychological mediating effects? *Canadian Journal of Sociology* 41 (2): 117–138.

Larson, Mindy, and Alicia Bolton. 2019. Guideposts for success 2.0: A framework for successful youth transition to adulthood. *National Collaborative on Workforce and Disability for Youth*.

Lindsay, Sally, Elaine Cagliostro, Mikhaela Albarico, Neda Mortaji, and Leora Karon. 2018. A systematic review of the benefits of hiring people with disabilities. *Journal of Occupational Rehabilitation* 28 (4): 634–655. https://doi.org/10.1007/s10926-018-9756-z.

Liu, Songqi, Jason L. Huang, and Mo Wang. 2014. Effectiveness of job search interventions: A meta-analytic review. *Psychological Bulletin* 140 (4): 1009–1041. https://doi.org/10.1037/a0035923.

Menear, Matthew, Marc-André Blanchette, Olivier Demers-Payette, and Denis Roy. 2019. A framework for value-creating learning health systems. *Health Research Policy and Systems* 17: 79. https://doi.org/10.1186/s12961-019-0477-3.

ODENa (Ontario Disability Employment Network). n.d. Business Benefits. https://www.odenetwork.com/businesses/business-benefits/. Accessed 9 April 2024.

ODENb (Ontario Disability Employment Network). n.d. About ODEN. https://www.odenetwork.com/about/about-oden. Accessed 9 April 2024.

RWA (Ready, Willing & Able). 2015. Who we are. https://readywillingable.ca/about/. Accessed 9 April 2024.

Rutkowski, Susie, Maryellen Daston, Debra Van Kuiken, and Erin Riehle. 2006. Project SEARCH: A demand-side model of high school transition. *Journal of Vocational Rehabilitation* 25 (2): 85–96.

Seko, Yukari, Gillian King, Sarah Keenan, Joanne Maxwell, Anna Oh, and CJ Curran. 2020. Impact of solution-focused coaching training on pediatric rehabilitation specialists: A longitudinal evaluation study. *Journal of Interprofessional Care* 34 (4): 481–492. https://doi.org/10.1080/13561820.2019.1685477.

Tompa, Emile, Amirabbas Mofidi, Arif Jetha, Pamela Lahey, and Alexis Buettgen. 2021. Development and implementation of a framework for estimating the economic benefits of an accessible and inclusive society. *Equality, Diversity and Inclusion: An International Journal* 41 (3): 318–339. https://doi.org/10.1108/EDI-07-2020-0186.

United Nations. n.d. Global issues: Youth. Accessed 5 June 2023. https://www.un.org/en/global-issues/youth#:~:text=Who%20Are%20the%20Youth%3F,of%2015%20and%2024%20years. Accessed 9 April 2024.

United Nations. 2006. Convention on the rights of persons with disabilities. Edited by General Assembly. Geneva. https://www.un.org/disabilities/documents/convention/convoptprot-e.pdf. Accessed 9 April 2024.

Wehman, Paul, Carol Schall, Jennifer McDonough, Adam Sima, Alissa Brooke, Whitney Ham, Holly Whittenburg, Valerie Brooke, Lauren Avellone, and Erin Riehle. 2020. Competitive employment for transition-aged youth with significant impact from autism: A multi-site randomized clinical trial. *Journal of Autism and Developmental Disorders* 50 (6): 1882–1897. https://doi.org/10.1007/s10803-019-03940-2.

# Chapter 18
# Self-Advocacy and Community Advocacy Amongst Youths with Disabilities

**Dolly Menna-Dack and Laura R. Bowman**

**Abstract** The ability to advocate for oneself and one's community requires skills, practice, and opportunities. For youth with disabilities, self-advocacy is a skill they will need throughout their lives to enhance access to necessary services and supports. Community-level advocacy is also essential to promote structural and policy changes. In this chapter, we explore the function and importance of self-advocacy and community advocacy, as well as the need for strategies that engage youth with disabilities in building advocacy skills. Included is a practical example of our Youth Engagement Strategy that provides opportunities for youth to build and practice advocacy skills. We conclude the chapter by underpinning global benefits of self- and community advocacy for youth with disabilities.

**Keywords** Self-advocacy · Community advocacy · Youth with disabilities · Building advocacy skills · Health system accountability

## 18.1 Introduction

Self- and community advocacy are essential practices that facilitate individuals with disabilities to navigate their lives meaningfully. Effective advocacy requires various skills, such as self-awareness, communication, and collaboration. These, like any skill, require training and opportunities for practice. For individuals with intellectual disabilities and/or autism, advocacy-related skills need to be taught in relevant and developmentally appropriate ways. Introducing these skills at a young age can be beneficial in preparing youth to navigate the world and advocate for their needs within communities, systems, and societies that were not built with them in mind (King et al. 2023).

D. Menna-Dack (✉) · L. R. Bowman
Holland Bloorview Kids Rehabilitation Hospital, Toronto, Canada
e-mail: dmennadack@hollandbloorview.ca

L. R. Bowman
e-mail: lbowman@hollandbloorview.ca



A. Bianchi and J. A. Vogt (eds.), *Intellectual Disabilities and Autism: Ethics and Practice*, The International Library of Bioethics 108,
https://doi.org/10.1007/978-3-031-61565-8_18

Beyond the lifelong need for self-advocacy, the community that supports individuals with intellectual disabilities and/or autism must advocate for policies, practices, and systems that consider the needs of this group and our whole population. The disability community has historically engaged successfully in advocacy, with a particular swell of activity in the last half century (Galer 2015). Despite—or perhaps because of—the momentum gained by community-level disability advocacy, ongoing advocacy is needed to create increasingly equitable, inclusive environments, and maintain gains that support the citizenship of people with disabilities.

The evolving needs of this population requires perpetual advocacy. Identity, needs, and preferences develop over time, as do individuals' connection with one's identity. As individuals build skills such as self-reflection, and develop a critical perspective, they can begin to better identify their needs and impetus for advocacy. Through growth, maturity, experience, and the same reflective practices that spark acts of advocacy, individuals' understandings of their needs may evolve. They may experience shifts in identities that will require new approaches and pivots in their advocacy action. The only seemingly stable factor may be the ongoing need to reflect and advocate. It is therefore imperative to create practices and supports for training and learning self-advocacy for young persons with disabilities and the communities that support them. This chapter explores the importance of advocacy, what is involved, and why it is essential to prepare youth with disabilities to advocate for themselves and their futures.

## 18.2 Positioning Language and Context

The authors of this chapter work at Holland Bloorview Kids Rehabilitation Hospital (henceforth Holland Bloorview); Canada's largest teaching paediatric rehabilitation hospital that supports children and youth with physical, developmental, neurological, and intellectual disabilities. This hospital uses person-first language (e.g., youth with disabilities, individual with autism), and we will use person-first language throughout this chapter. We respect that some individuals prefer disability-first language and some may choose not to identify with disability at all, as per their culture or understanding of disability in their own worldview. How an individual chooses to self-identify is deeply personal and we wish to let the reader know that our use of person-first language in this chapter is based in our context and is intended with respect and openness.

The paediatric hospital wherein we are employed provides medical, rehabilitative, and life skills services and support. Among the life skills services offered are youth engagement programs, which include essential skills for youth to navigate their medical, physical, social, employment, and cultural realities. Our organization promotes inclusive, diverse, and collaborative environments for program creation and evaluation.

When we speak about our work with youth in this chapter, we are referring to individuals between the ages of fourteen and twenty-nine. The term is inconsistently defined and sometimes used interchangeably with adolescence and/or young adulthood. Our use of the broad age range is based on two considerations. First, from a developmental perspective, the trajectories for youth with disabilities to move into a self-defined adult life are impacted by reduced societal accessibility, fewer developmentally appropriate social opportunities, and increased general stigma as compared to their same-aged peers (Argan et al. 2017; Pandey and Agarwal 2013). With reduced access to opportunities to engage in the world, individuals with disabilities tend to take longer to participate in the adult world in ways considered "successful" by society (Pearson et al. 2021). The second consideration for a broad age range is based on our community partners' varied definitions of the term "youth". Additionally, we note that the age of twenty-nine is the maximum age for youth in the city of Toronto's adapted recreation programming (City of Toronto, n.d.), which is the city wherein we work. Finally, programs within our hospital's Youth Engagement Strategy range up to age twenty-nine, which also aligns with the United Nations' (2017) definition of "youth volunteers".

Throughout our work, we provide programming to prepare youth with disabilities to navigate the Canadian systems and culture through building life skills (e.g., social interaction, volunteering, community navigation, employment). By facilitating these experiences and engaging youth in building relevant life skills, we support the growth of self-awareness, critical social reflection, and ability to navigate legal, medical, and cultural policies and practices that govern access to services and supports. With increased understanding of their environments *and* contextually situated needs, we support their growth and ability to identify how to advocate for change. This work is done with consideration of the lived experiences of our clients and families, and also the local data narrative. According to Statistics Canada (2018), more than thirteen percent of Canadian youth aged fifteen to twenty-four identify as living with one or more disabilities. As practitioners who strive to work with the whole person, we help our clients consider their holistic identity when supporting their growth as advocates. There are several reasons that building advocacy skills may benefit youth with disabilities. For example, Canadian youth with disabilities are five to ten percent more likely to live in poverty than their non-disabled peers (Morris et al. 2018). Poverty represents just one compounding social factor that impacts overall well-being. In order to respond to this social factor (and related consequences) in a meaningful and developmentally appropriate way, the acts, skills, and practices of advocacy are essential.

## 18.3 Defining and Situating Advocacy for Youth with Disabilities

Advocacy is defined as "the act or process of supporting a cause or proposal" (Merriam Webster n.d.). By this definition, advocacy must be action-oriented in support of a specific outcome. The outcome is contextual and requires ongoing practice so the individual can recognize the need for advocacy, act, and subsequently reflect on whether the outcome met their needs and/or created change by taking action for themselves or others. Such outcomes might be meeting an individual or community need or, changing practices and policies. Advocacy can be done by and for individuals or communities. The goal of building a practice of advocacy is to foster the development of agency in youth with disabilities.

Advocacy can be a skill, an act or behaviour, or a practice. The *act* of advocating is most often associated with the term and might include behaviours such as directly asking for what one needs or lobbying for policy changes. Advocacy *skills* support individuals and communities in performing acts of advocacy, and these skills require training and practice. Relevant skills include self-reflection, critical awareness of personal, community, and societal positioning, and communication. Advocacy as a *practice* includes the ongoing and evolving thoughtful, intentional, and skilled undertaking of work to benefit oneself or one's community.

### *18.3.1 Self-Advocacy*

Self-advocacy occurs when an individual communicates their needs, concerns, or interests with the goal of having those needs, concerns, or interests met (Daly-Cano et al. 2015). In order for a person to advocate for oneself, critical self-reflective skills are required. This might include being able to identify one's own needs (e.g., help with attention and reduced distractions during testing) or determining accommodation supports that will maximize the chances of success (e.g., extended time for examinations). While seemingly simple and self-evident, youth with intellectual disabilities and/or autism can benefit from opportunities to name, learn, and practice the skills underpinning the act of advocating for oneself (Holland Bloorview 2018). Advocacy can be required for full participation in one's life, especially in the areas with which one interacts on a regular basis.

**What Might Self-Advocacy Look like?**

Self-advocacy occurs when an individual negotiates or makes known their needs to another person or party. For example, suppose Esme is a youth with intellectual disability. In order to build her social skills, gain work experience, and earn volunteer hours required for high school credits, Esme decides, with some support, to volunteer at a library. As a volunteer, Esme is assigned to work in the children's area with duties that include returning books to the shelves, tidying up toys, and cleaning up general messes. From her experiences in school, Esme knows that she needs to keep a written list for multi-step tasks and have a routine. She also knows that she needs regular breaks to calm her body and organize her thoughts. Following some critical self-reflection, Esme approaches a teacher at school to help her plan what she will need to do her best work at the library. Once the plan is developed, Esme takes it to her library supervisor. The supervisor helps Esme create a list of tasks, and agrees with Esme's proposed break schedule. The supervisor was glad to make accommodations and grateful that Esme advocated for what she needed. Additionally, they decided to meet after Esme's second shift so they could discuss other adaptations that may help Esme find success. After seeing the benefit of the adaptations, the supervisor began holding similar conversations with all new volunteers and staff.

### *18.3.2 Community Advocacy*

Community advocacy involves communicating needs to create change on behalf of a defined group (e.g., youth with disabilities, the disability community). In order to create change for a community, the group of individuals experiencing the same or similar challenge, alongside their supporters, may advocate for change that will positively impact all members. The goal of community advocacy typically includes changed policies and practices and/or a changed culture through increased public knowledge of the community's experience. Both self- and community advocacy are important in the lives of youth with disabilities so that they, and everyone who identifies as members of the relevant community, may have a better experience than that which is currently available. Both are valuable and necessary in different situations.

**What Might Community Advocacy Look like?**

Planned Lifetime Advocacy Network (PLAN) is an example of a national, non-profit advocacy-based group that aims to support positive outcomes for people with disabilities and their families. Among their many endeavours, they promote financial security as a priority for individuals and families with disabilities. Representatives of PLAN state, "We want to be sure we save enough money to handle emergencies and unforeseen events, but we also want to have enough to live comfortably. For our sons and daughters with disabilities, many live on low-incomes with limits to gifts, income and assets... Simply existing isn't good enough—we want our loved ones to live a good life" (PLAN 2023). In 2008, community advocacy work led by PLAN resulted in the Government of Canada introducing the Registered Disability Savings Plan (RDSP), a national tax-deferred savings vehicle for individuals with disabilities to promote financial well-being. PLAN's community advocacy has impacted the financial, socioeconomic, and emotional well-being of Canadian individuals and families living with disability.

For the community of persons with disabilities and their supporters, advocacy has led to important policy, practice, and cultural changes. A meaningful, inclusive existence for individuals with disabilities is a relatively new societal consideration. Consideration of individuals with disabilities' full personhood, with them as "equal" members of Canadian society only gained widespread attention in the latter half of the twentieth century. It was not until 1980 that Bill 82 was introduced, which amended the province of Ontario's Education Act to recognize the rights of students with disabilities in relation to publicly funded education (Burge et al. 2008). It was not until much later, in 2009, that we saw the last Ontarian disability institutions closed (Government of Ontario 2009). Through increased community advocacy for individuals with disabilities, we have seen increased momentum in policy, research, education, and inclusion that highlights the importance and general benefits of more pervasive social equity and diversity.

While community advocacy has benefitted people with disabilities, there remains a long journey ahead toward equity and inclusion. Greater societal inclusion will perpetually unearth inequitable practices and environments, which will in turn require additional self- and community advocacy to support ongoing inclusion. A strong advocacy practice is therefore essential for individuals with disabilities and supporters. We propose that the skills required for an advocacy practice be taught early and in developmentally appropriate ways so that youth can integrate them as they enter adulthood.

### *18.3.3 Advocacy for Youth with Disabilities*

Insofar as people with intellectual disability and autism should be taught advocacy skills at a young age, it is important to ask who may be responsible for teaching them. We propose that the institutions responsible for their well-being, development, and health (e.g., education systems, community programs, paediatric programs) have a moral imperative to teach advocacy skills as part of their work. Although there exists an increased need to advocate for themselves as a group for which society was not designed, youth with disabilities tend to have limited access to normative settings for social and life skill development (e.g., school, sports, clubs) (Argan et al. 2017; Askari et al. 2015; King et al. 2023). A great deal of the participation literature related to social interaction and skill building for this group is based on clinical programming, indicating that even their extracurricular interactions are largely based in disability-related settings (Askari et al. 2015). This means that youth with disabilities, who may already encounter physical, communication, and cognitively-based barriers to social interaction have less access to naturalistic environments in which to practice and grow advocacy-related skills.

Readers may pause here and reflect on the fact that youth with disabilities should not have to self-advocate at all, that they should be treated equitably by the systems and societies within which they interact. The authors agree and only make our current argument in light of the present reality, which requires advocacy practices (and associated skills) for successful interaction with systems and societies. It is for these reasons that we propose there exists an ethical obligation for paediatric rehabilitation institutions (along with other institutions, as listed above) to facilitate advocacy training and practice opportunities for youth with disabilities. As the settings in which these youth spend extra-curricular time and learn about navigating the world with a disability, it is incumbent upon these organizations to enable youth to succeed both within and outside of them. This becomes even more pertinent when considering that youth with disabilities are expected to navigate systems that require self-advocacy, and we must therefore *train* the necessary skills so that all individuals can advocate effectively.

In the following section, we present an organizational strategy for integrating self- and community advocacy into programs and operations. But first, we end this section with a disclaimer. Although we argue that it is essential for youth with disabilities to build a practice of self-advocacy into their lives (as supported and facilitated by relevant institutions), we do not mean to imply that it should be the *responsibility* of youth with disabilities to ensure that societies are inclusive and knowledgeable about equity and disability. Changing the way our systems and beliefs are structured is the role and responsibility of collective community advocacy.

## 18.4 Supporting Self-Advocacy for Youth with Disabilities: A Practical Example

As an organization that supports the development and well-being of youth with disabilities, Holland Bloorview has several initiatives that promote the building of essential life skills for clients. Recognizing the importance of advocacy, we acknowledge the need to create an environment that supports both community-level advocacy and self-advocacy to realize our hospital's vision of "the most meaningful and healthy futures for all children, youth and families" (Holland Bloorview a, n.d.). To achieve this aim, we have embedded a sustainable strategy for training individual advocacy skills in our youth. Our strategy allows their practice and growth of advocacy skills to result in effective community-level advocacy, in addition to individual advocacy. The result is an organizational commitment that creates opportunities for youth with

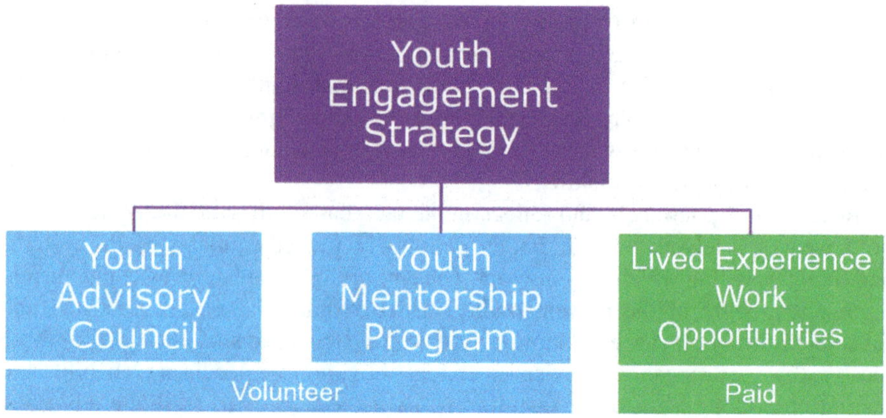


**Fig. 18.1** Holland Bloorview's Youth Engagement Strategy (Holland Bloorview 2012)

disabilities to build skills and practice communicating disability-related needs in a supported environment so that they are prepared to enter the adult world.

Holland Bloorview's Youth Engagement Strategy (YES) (Holland Bloorview 2012) is a model that builds essential life skills for youth with disabilities as they learn to navigate their worlds. The goal of the YES is to enable current and former youth-aged clients to turn their lived experience of disability into change for future generations. The YES (Fig. 18.1) has three arms, two of which are volunteer based, and one that provides paid work opportunities within the hospital. These initiatives provide multiple, progressive opportunities for youth to build and practice advocacy skills so they can promote individual and community-level change in their current context and the future.

The first arm of the YES, the Youth Advisory Council (YAC), was established in 1996 as a forum through which youth with lived experience of disabilities engage with one another, build leadership skills, and advocate for youth needs within the hospital and community. Approximately twenty youth participate annually as members. The YAC acts as a sounding board for the hospital and the community of individuals with disabilities on matters including program development, services, community design, research design and dissemination. The Youth Mentorship program is a cross-age, cross-disability peer mentorship program that provides intensive training for youth with disabilities to offer their lived experience and insight. Since 2012, over one hundred Youth Mentors have been trained, representing a cross-section of disabilities, ages, and genders. Youth Mentors provide connection and support for younger individuals with disabilities and their families. Youth Mentors also support educational opportunities for clients, families, healthcare providers, and healthcare students based on their lived experiences of disability. Additionally, these youth provide mentorship by collaborating with researchers, clinicians, and organizational-level project teams as they work to make meaningful changes for youth and families with disabilities. The final stream in the YES is Lived Experience Work Opportunities in the hospital.

The roles range from information navigation, to a unique clinical role known as the Youth Facilitator. Youth Facilitators are paid, embedded members of our hospital's rehabilitation teams who receive further training on mentorship and clinical interactions (Seko et al. 2021). Our hospital's commitment to value youth's lived experience has allowed us to embed advocacy and client perspectives into the healthcare process. This model has led to the immediate outcome of clients' needs and preferences being organically infused into their interdisciplinary care.

Through the YES, our hospital has embedded a process that supports the ongoing training and development of advocacy skills in youth. Youth who participate in each arm receive training and developmentally appropriate opportunities to grow their critical thinking, self-reflection, and communication skills. Arms of the YES are also embedded into broader organizational and community processes (e.g., Research Ethics Board consultation, program development) to ensure the ongoing inclusion of lived experience advocacy in our organizational and community endeavours.

## 18.5 Ethical Principles that Ground Self- and Community Advocacy Practices

The moral imperative to promote and build advocacy skills in our youth with intellectual disabilities and/or autism is grounded in ethical principles of trust, emerging autonomy, agency, and accountability. When environments that support advocacy exist, people with disabilities of all ages are better equipped to recognize and advocate for their needs so that their meaningful participation can be supported. Insofar as we want our communities and society to learn about and support persons with disabilities, advocacy is essential. By operationalizing the ethical principles below, such advocacy can be realized.

### *18.5.1 Trust*

Trust is an integral facilitator that enables a relationship to exist between the advocate (self-advocate or community) and the environment in which they are advocating. A trusting relationship can contribute to open communication, mutual respect, and collaborative solution finding, all of which facilitate successful advocacy. A sense of trust can signal a safe space for youth with disabilities to learn and practice advocacy skills. The trust to which we are referring is that the message from the advocate will be heard, acknowledged, and considered by those with the scope of authority to action change.

Trust is particularly important when considering youths' critical developmental stage and the need for safe environments in which they can explore and practice how they identify as an individual with a disability (and beyond) and what their personal advocacy might look like. Potential considerations for fostering trust between an organization and youth self-advocates include ensuring opportunities to select and/or take on advocacy projects that have value to the youth. Self or group selection of advocacy work affirms the youth's lived experience that an issue requires change, and an organization engages in a reciprocal trusting relationship when they acknowledge the need for change. When a youth takes on developmentally appropriate advocacy, they are also necessarily taking on responsibilities (e.g., communicating their needs), developing an understanding of their rights (e.g., the right to a disability accommodation that would enable participation), and gaining a sense of control.

### 18.5.2 Emerging Autonomy

Respect for autonomy is a key factor in fostering a trusting environment for youth to build and practice advocacy skills. According to Beauchamp and Childress (2013), the autonomous individual acts freely in accordance with a self-chosen plan. Youth with intellectual disabilities and/or autism are frequently stigmatized, resulting in reduced opportunities for meaningful choice making (Ditchman et al. 2013). Fostering a practice of advocacy creates critical opportunities for youth with disabilities to engage in exploration of interest and values, choice making, taking responsibility, and cooperative action. Organizations and systems tasked with supporting the development of youth with disabilities can leverage opportunities for advocacy to build necessary and transferable autonomy-enhancing skills (Holland Bloorview b, n.d.).

The concept of *emerging autonomy* is commonly used in clinical and educational practices to describe youths' continual development of skills that move them toward being increasingly autonomous (Complex Care at Home for Children n.d.; Holland Bloorview c, n.d.). The concept responds to the fact that some youth may, one day, be making choices for themselves independently. Although complete independence is not necessarily the ultimate goal of emerging autonomy, nor the appropriate or desired outcome, recognizing emerging autonomy is critical in combating the abovementioned stigma regarding the capabilities of youth with disabilities. The concept of emerging autonomy aligns with the United Nations' (1989) Convention on the Rights of the Child, Article 29 (d) that supports the notion of preparing children for participation in society. By placing importance on building skills that support autonomous choice making, we can shift the perception of independence away from an all-or-nothing approach and toward a more fulsome understanding of developmentally appropriate autonomous choice making and self-determination. Supporting emerging autonomy is important for service providers, organizations, and institutions to undertake via creating opportunities for youth with disabilities to build self-advocacy skills.

### 18.5.3 Agency

In addition to autonomy, self- and community advocacy contribute to the *agency* of youth with disabilities. Here we consider Montreuil and Carnevale's (2016, 510) definition of children's agency as meaning "…capacity to act deliberately, speak for oneself, (…) shaping their lives and the lives of others." Respecting and fostering agency can be accomplished by engaging in a reciprocal relationship with the youth advocate. Chapman and Tilley (2013) highlight the multi-faceted nature of the relationship between self-advocates and those that support them—the tension between ensuring that individuals with disabilities have the space to make choices so they can exercise agency, and their need for support. By developing a practice of self-advocacy, and through acts of advocacy, youth exercise agency.

### 18.5.4 Accountability

The act of advocacy necessarily assumes there is a person or institution to which one can advocate. The reason that every advocacy effort is directed toward a particular party is because they are perceived as having some influence, ability, or authority that may respond to the relevant cause. For example, if a youth with intellectual disability and/or their family feels as though a wait time for services is unjustifiably long, then they may express their desire for change by advocating to a particular organization or leader. In some cases, government figures or organizations may be the recipients of advocacy initiatives.

Some may not consider accountability as a principle that has an obvious connection to advocacy. Part of this reason is presumably because the act of the advocate is typically the focus, as opposed to the reciprocal relationship between advocates and the systems and environments in which they are situated. However, when a person or community undertakes acts of advocacy, there is inherent trust on their behalf that someone, at some point, will hear and respond. We therefore position accountability as a requirement for those who sit in positions of influence or authority with respect to the spheres in which youth with disabilities live. Organizations must hold themselves accountable to those they serve by creating opportunities for individual and community advocacy, as well as having mechanisms in-place to ensure receptivity and understanding for the needs of their communities. Such actions will foster trusting environments, emerging autonomy, and agency of the individuals with whom they work to support successful opportunities for advocacy-based dialogue.

## 18.6 Who Benefits from Disability Advocacy?

Disability advocacy, especially at the community level, has radically changed the ways that individuals with disabilities are included in policy, healthcare, culture, built environments, and society (Galer 2015). Yet, we propose that disability advocacy has gone beyond the population of individuals with disabilities to benefit society at large. The benefits can be seen at multiple levels, including within organizations, communities, and globally. Consequently, we propose that disability advocacy should be considered a universal necessity.

At the organizational level, for example, disability advocacy has led to changes that benefit the healthcare industry's clients (Carman et al. 2013). By creating spaces that honour the lived experience of individuals with disabilities, the healthcare industry can identify areas of care that need attention, reduce practices that are harmful or simply not meaningful to clients, and bolster preferred and effective programs. The uptick in collaboration, codesign, and universal design in healthcare all reflect the importance and impact of user-centred approaches to improve acceptable outcomes.

At a societal level, advocacy can—and does—help communities and governments identify programs and services that are most needed so that resources (e.g., funds, personnel, space) can be invested in meaningful ways. This can be seen in the creation of direct programs and services for individuals with disabilities, but can take broader forms as well. The example provided above regarding community advocacy for tax-deferred savings for people with disabilities (the RDSP) (Sect. 18.3.2) illuminates this point. The immediate outcome of the RDSP benefits the individual with the disability, though the implications are widespread. When an individual with disabilities has the means to achieve greater financial security, there are also positive financial and mental health outcomes for their caregivers and families (Bonis 2016). Furthermore, there is less reliance on government subsidies, programs, or incentives, which benefits the tax-paying public by redirecting funds elsewhere (Bowman et al. 2024; Tompa et al. 2021).

Disability advocacy has also played a large role in general discussions related to equity and individual rights. In 2005, the Ontario government introduced the Accessibility for Ontarians with Disabilities Act (Government of Ontario 2005) with the purpose of creating a barrier-free Ontario by 2025. This barrier-free Ontario relates to universally designed environments that promote accessibility. This work has led to standards for accessibility and global trends in design that benefit aging individuals, those responsible for young children, and those with temporary situational needs.

When we, as a society, enable people with disabilities to advocate *and* consciously decide to listen to their voices, we create space for creative and novel solutions. Knowing that society at large can benefit, we invite all readers into the advocacy process. One's role in the advocacy process will evolve and shift across time and situation. It may include calling for action or attention to a need, or being the one who is receptive to hearing what is needed. With that in mind, we offer one last note to proceed in a manner that mitigates tokenism. When advocacy becomes a trend, inclusion can become an expectation (organizational, cultural, professional).

The intention of advocacy and inclusion is noble, but if it is actioned in a way that is tokenistic, we risk losing the trust of advocates and the benefits that would otherwise arise.

## 18.7 Conclusion

Individuals with intellectual disabilities and/or autism can develop their advocacy through skill-building and opportunities to practice. Community advocacy practices can be supported by organizations and systems that consciously promote the sharing of and listening to needs. Advocacy is not an exercise in sharing in and of itself. Rather, it is a thoughtful and practiced approach to supporting equity, accessibility, and diversity through trust, autonomy, and agency.

## References

Argan, Martin, Andrew Wojcik, Irina Cain, Colleen Thoma, Edwin Achola, Kira M. Austin, Cynthia Allen Nixon, and Ronald B. Tamura. 2017. Participation of students with intellectual and developmental disabilities in extracurricular activities: Does inclusion end at 3:00? *Education and Training in Autism and Developmental Disabilities* 52 (1): 3–12.

Askari, Sorayya, Dana Anaby, Melanie Bergthorson, Annette Majnemer, Mayada Elsabbagh, and Lonnie Zwaigenbaum. 2015. Participation of children and youth with autism spectrum disorder: A scoping review. *Review Journal of Autism and Developmental Disorders* 2 (1): 103–114. https://doi.org/10.1007/s40489-014-0040-7.

Beauchamp, Tom L., and James F. Childress. 2013. *Principles of biomedical ethics*, 7th ed. London: Oxford University Press.

Bonis, Susan. 2016. Stress and parents of children with autism: A review of literature. *Issues in Mental Health Nursing* 37 (3): 153–163. https://doi.org/10.3109/01612840.2015.1116030.

Bowman, Laura R., Carolyn McDougall, René Doucet, Brendon Pooran, Ying Xu, and Jeannette Campbell. 2024. Funding employment inclusion for ontario youth with disabilities: A theoretical cost-benefit model. *Frontiers in Sociology*. https://doi.org/10.3389/fsoc.2024.1281088.

Burge, Philip, Hélène Ouellette-Kuntz, and Nancy Hutchinson. 2008. A quarter century of inclusive education for children with intellectual disabilities in Ontario: Public perceptions. *Canadian Journal of Educational Administration and Policy* 87: 1–22. https://journalhosting.ucalgary.ca/index.php/cjeap/issue/view/2919.

Carman, Kristin L., Pam Dardess, Maureen Maurer, Shoshanna Sofaer, Karen Adams, Christine Bechtel, and Jennifer Sweeney. 2013. Patient and family engagement: a framework for understanding the elements and developing interventions and policies. *Health Affairs* 23 (2): 223–231. https://doi.org/10.1377/hlthaff.2012.1133.

Chapman, Rohhss, and Liz Tilley. 2013. Exploring the ethical underpinnings of self-advocacy support for intellectually disabled adults. *Ethics and Social Welfare* 7 (3): 257–271. https://doi.org/10.1080/17496535.2013.818160.

City of Toronto. n.d. *Accessible recreation*. https://www.toronto.ca/city-government/accessibility-human-rights/accessibility-at-the-city-of-toronto/city-services-for-people-with-disabilities/accessible-recreation/. Accessed 10 April 2024.

Complex Care at Home for Children. n.d. *Encourage autonomy: Suggestions by developmental stage.* https://complexcareathomeforchildren.com/prepare-your-child/encourage-autonomy/. Accessed 10 April 2024.

Daly-Cano, Meada, Annemarie Vaccaro, and Barbara Newman. 2015. College student narratives about learning and using self-advocacy skills. *Journal of Postsecondary Education and Disability* 28 (2): 213–227.

Ditchman, Nicole, Shirli Werner, Kristin Kosyluk, Nev Jones, Brianna Elg, and Patrick W. Corrigan. 2013. Stigma and intellectual disability: Potential application of mental illness research. *Rehabilitation Psychology* 58 (2): 206–216. https://doi.org/10.1037/a0032466.

Galer, Dustin. 2015. Disability rights movement in Canada. *The Canadian Encyclopedia.* Historica Canada: Canada. https://www.thecanadianencyclopedia.ca/en/article/disability-rights-movement. Accessed 10 April 2024.

Government of Ontario. 2005. *Accessibility for Ontarians with Disabilities Act, 2005, S.O. 2005, c. 11.*

Government of Ontario. 2009. Closing institutions for people with a developmental disability. *Ontario Newsroom*, March 31, 2009. https://news.ontario.ca/en/backgrounder/5213/closing-institutions-for-people-with-a-developmental-disability. Accessed 10 April 2024.

Holland Bloorview (Holland Bloorview Kids Rehabilitation Hospital). 2018. *Advocacy tool kit for parents of children and youth with autism spectrum disorder (ASD) and other neurodevelopmental disorders (NDDs).* https://hollandbloorview.ca/sites/default/files/2019-06/AdvocacyToolKit-July24-LINKED-LOCKED.pdf. Accessed 10 April 2024.

Holland Bloorview (Holland Bloorview Kids Rehabilitation Hospital). n.d.a. *Mission, vision & values.* https://hollandbloorview.ca/about-us/hospital/mission-vision-impact. Accessed 10 April 2024.

Holland Bloorview (Holland Bloorview Kids Rehabilitation Hospital). n.d.b. *Project inclusion: Enabling engagement in choices, goals, and plans.* https://deareverybody.hollandbloorview.ca/project-inclusion/. Accessed 10 April 2024.

Holland Bloorview (Holland Bloorview Kids Rehabilitation Hospital). n.d.c. *Timetable for growing up.* https://hollandbloorview.ca/services/programs-services/timetable-growing. Accessed 10 April 2024.

Holland Bloorview (Holland Bloorview Kids Rehabilitation Hospital). 2012. *Youth engagement.* https://hollandbloorview.ca/services/programs-services/youth-engagement. Accessed 10 April 2024.

King, Gillian, Eric Smart, Laura Bowman, and Madhu Pinto. 2023. Social participation interventions targeting relational outcomes for young people with physical and developmental disabilities: An umbrella review and narrative synthesis. *Disability and Rehabilitation* 45 (13): 2073–2086. https://doi.org/10.1080/09638288.2022.2085332.

Merriam-Webster.com Dictionary. n.d. *Advocacy.* https://www.merriam-webster.com/dictionary/advocacy. Accessed 10 April 2024.

Montreuil, Marjorie, and Franco A. Carnevale. 2016. A concept analysis of children's agency within the health literature. *Journal of Child Health Care* 20 (4): 503–511. https://doi.org/10.1177/1367493515620914.

Morris, S., G. Fawcett, L. Brisebois, and J.A Hughes. 2018. A demographic, employment and income profile of canadians with disabilities aged 15 years and over, 2017. *Statistics Canada.* https://www150.statcan.gc.ca/n1/en/catalogue/89-654-X2018002. Accessed 10 April 2024.

Pandey, Sudha, and Shalini Agarwal. 2013. Transition to adulthood for youth with disability: Issues for the disabled child and family. *IOSR Journal of Humanities and Social Science* 17 (3): 41–45.

Pearson, Charlotte, Nick Watson, Justine Gangneux, and Ida Norberg. 2021. Transition to where and to what? Exploring the experiences of transitions to adulthood for young disabled people. *Journal of Youth Studies* 24 (10): 1291–1307. https://doi.org/10.1080/13676261.2020.1820972.

PLAN (Planned Lifetime Advocacy Network). 2023. *Financial security.* https://plan.ca/what-we-do/planning/financial-security/. Accessed 10 April 2024.

Seko, Yukari, Anna Oh, Dolly Menna-Dack, C.J. Curran, Joanne Maxwell, and Gillian King. 2021. Integrating lived experience into clinical practice: a case study of young peer providers in pediatric rehabilitation. *Disability and Rehabilitation* 4 (23): 7246–7254. https://doi.org/10.1080/09638288.2021.1988735.

Statistics Canada. 2018. Canadians with disabilities: A demographic, employment and income profile, 2017 *The Daily*. https://www150.statcan.gc.ca/n1/daily-quotidien/181128/dq181128a-eng.htm. Accessed 10 April 2024.

Tompa, Emile, Amirabbas Mofidi, Arif Jetha, Pamela Lahey, and Alexis Buettgen. 2021. Development and implementation of a framework for estimating the economic benefits of an accessible and inclusive society. *Equality, Diversity and Inclusion: An International Journal* 41 (3): 318–339. https://doi.org/10.1108/EDI-07-2020-0186.

United Nations. 2017. *Become a UN youth volunteer*. https://www.unv.org/become-un-youth-volunteer. Accessed 10 April 2024.

United Nations. 1989. *Convention of the rights of the child*. https://www.ohchr.org/en/instruments-mechanisms/instruments/convention-rights-child. Accessed 10 April 2024.

# Part V
# Sexuality, Intimacy, Romance, and Parenting

# Chapter 19
# The Impact of Ableism on the Sexual, Reproductive, and Parenting Rights of People with Intellectual Disabilities

**Robyn M. Powell**

**Abstract** This chapter surveys the impact of ableism on the sexual, reproductive, and parenting rights of people with intellectual disabilities (ID). Entrenched within societal attitudes and structures, ableism creates barriers that hinder the full realization of fundamental human rights. First, the chapter explains ableism, its adverse effects on people with ID, and how ableism lays the groundwork for eugenic ideologies and practices. Then, it examines how ableism constrains the sexual lives of people with ID, exploring pervasive stigmas, stereotypes, and misconceptions that perpetuate their marginalization and disempowerment in matters of sexuality. Next, it explores ableism's influence on access to reproductive health services and decision-making autonomy. Thereafter, the chapter analyzes challenges and discriminatory practices confronted by people with ID in relation to parenting. Following a critical examination of inequities, the chapter concludes by offering recommendations to dismantle ableism's impact on the sexual, reproductive, and parenting rights of people with ID by transforming laws, policies, and practices.

**Keywords** Sexuality · Reproduction · Parenting · Intellectual disabilities

## 19.1 Introduction

Over time and across jurisdictions, ableism has exerted a pervasive and detrimental influence on the sexual, reproductive, and parenting rights of people with intellectual disabilities (ID). Through the perpetuation of damaging stereotypes and misconceptions, ableism has systematically marginalized and excluded this population, preventing them from fully engaging in matters of sexuality. Opportunities to express their sexuality and form meaningful relationships have been unjustly

R. M. Powell (✉)
University of Oklahoma College of Law, Norman, OK, USA
e-mail: rpowell@ou.edu

Lurie Institute for Disability Policy, Brandeis University, Waltham, MA, USA



A. Bianchi and J. A. Vogt (eds.), *Intellectual Disabilities and Autism: Ethics and Practice*, The International Library of Bioethics 108,
https://doi.org/10.1007/978-3-031-61565-8_19

denied. Additionally, their access to comprehensive sexual education has been deliberately constrained, compromising their ability to navigate consensual relationships.[1] Moreover, the reproductive rights of people with ID have been systematically undermined by denying their autonomy and decision-making power. Discriminatory practices (e.g., forced sterilizations) have been inflicted upon them based solely on their disability, resulting in gross violations of their reproductive and parenting rights and perpetuating the notion that their worth as parents is diminished. Despite incremental progress, ableism persists in present-day society, continuing to hinder access to inclusive sexual education, comprehensive reproductive health services, and supportive parenting systems. Recognizing and dismantling ableist beliefs, practices, and policies are imperative to fostering a more equitable society that upholds the sexual, reproductive, and parenting rights of people with ID.

This chapter explores the intersection of ableism and the sexual, reproductive, and parenting rights of people with ID. It begins by defining ableism and examining ableism's history, particularly its connection to eugenics. It then delves into the impact of ableism on sexual rights, addressing issues of consent and relationships. Next, it explores barriers faced by people with ID in regard to reproductive health services and decision-making autonomy. Moving on, the chapter analyzes the impact of ableism on parenting rights, discussing stereotypes, discrimination, and the need for support systems. The chapter concludes by presenting law, policy, and practice recommendations to dismantle the ableist barriers that influence the sexual, reproductive, and parenting rights of people with ID.

## 19.2 Understanding Ableism and Its Historical Impact

Ableism, as a form of discrimination, marginalizes people based on notions about disabilities, and people with ID face numerous challenges resulting from ableist attitudes and practices. The historical legacy of eugenics exacerbates these issues, leading to forced sterilizations and violations of reproductive autonomy. Additionally, stereotypes surrounding cognitive abilities and capacity to consent have hindered their sexual agency and access to comprehensive sexual health services. Parenting rights have also been undermined, with discriminatory assumptions leading to removing children or denying custody. This section defines ableism and its impact on people with ID. It then examines the convergence of ableism and eugenics.

---

[1] Sexual education is evidently important when thinking about sexual rights and people with ID. Extensive information about this topic can be found in Chap. 21 of this volume.

### *19.2.1 Defining Ableism*

Ableism is a system of prejudice and discrimination that devalues people with disabilities, perpetuating the perception of their inferiority in relation to those without disabilities (Campbell 2014). The concept is expounded through a range of nuanced interpretations. For instance, Ben-Moshe (2020, 16) defines ableism as "oppression faced due to disability/impairment (perceived or lived), which not only signals disability as a form of difference but constructs it as inferior." For Lewis (2022), ableism represents a form of discrimination that assesses people based on preconceived notions of ability, favoring those perceived as "normal" or nondisabled, while marginalizing those labeled as "abnormal."

Ableism encompasses the act of stereotyping, holding prejudiced beliefs, engaging in discriminatory actions, and enforcing social oppression against people with disabilities, and manifests in various ways (Bogart and Dunn 2019). Ambivalent ableism, particularly paternalistic ableism, has been identified as the most prevalent form of bias. Paternalistic ableism often involves expressions of pity, unsolicited assistance, infantilization, excessive protection, and disregard for the autonomy of people with disabilities (Nario-Redmond et al. 2019). Further, ableist perceptions encompass the notion that disability is "something that needs to be ameliorated, corrected or erased in order to come closer to the ideal able-bodied state" (Wu et al. 2019, 268).[2]

Like other forms of oppression, ableism functions across four interconnected levels: internalized, interpersonal, institutional, and structural. At the internalized level, people with disabilities may internalize negative beliefs and attitudes, resulting in self-doubt and diminished self-worth (Campbell 2014). Moreover, people, both with and without disabilities, may internalize a hierarchy of disabilities wherein certain disabilities, such as intellectual and psychiatric disabilities, are stigmatized more than physical or sensory disabilities (Thomas 2000). Conversely, interpersonal ableism encompasses prejudiced attitudes and behaviors exhibited between people, including acts of discrimination, exclusion, and differential treatment (Nario-Redmond et al. 2019). This form of ableism perpetuates through harmful stereotypes, microaggressions, and misplaced sympathy (Ibid.). Institutional ableism refers to discriminatory practices and policies embedded within organizations, creating systemic barriers and unequal treatment of people with disabilities (Fierros 2006). Finally, structural ableism shapes laws, social policies, and institutions that perpetuate disparities and unequal outcomes for people with disabilities (Keefe 2022; Powell 2021).

Critically, ableism intersects with other systems of oppression (Powell 2021). Ableism is closely interwoven with both white supremacy and patriarchy, and the identities of disabled people encompass a multitude of aspects, such as race, gender, and other social identities. These factors contribute to the challenge of distinguishing ableism from racism and sexism in certain situations. Furthermore, racism

[2] For more information about ableism, particularly in relation to health care services and autistic people, see Chap. 12 of this volume.

and sexism have historically operated by assigning various forms of deficiency to people belonging to marginalized groups (Clare 2017; Lewis 2022; Sins Invalid 2019; Tremain 2017).

People with ID face persistent discrimination (Dell'Armo and Tassé 2021). Some scholars use the term "cognitive ableism" to describe "prejudice or bias in favour of the interests of individuals who possess certain cognitive abilities" (Sandberg et al. 2021, 1422). The presence of ableist ideologies manifests in an overly protective environment that restricts people with ID's ability to engage in typical social exploration, including aspects such as sexuality, reproduction, and parenting (Gill 2015).

### *19.2.2 The Legacy of Ableism and Eugenics*

Throughout history, ableism and eugenics have been closely entwined, with ableist beliefs serving as the foundation for eugenic ideologies and actions (Powell 2021). The impact of ableism extends beyond people with disabilities, affecting those perceived as disabled or deemed inferior in various ways (Ibid.). Guided by ableist convictions, the eugenics movement led to the involuntary sterilization of countless women, and some men, with disabilities, to protect society from perceived moral decline and procreation from those labeled "socially inadequate" (Cohen 2017). This era was marked by a belief in the heritability of intellectual capacity, as exemplified by the infamous words of Justice Oliver Wendell Holmes of the United States Supreme Court in the case of *Buck v. Bell* (1927, § 207), where he remarked, "three generations of imbeciles are enough" when discussing the forced sterilization of Carrie Buck, a woman believed to have an intellectual disability. Laws that prevented people with disabilities from marrying were also prevalent during the eugenics era, along with institutionalization (Powell 2022). Today, people with disabilities, particularly those with ID, face structural and attitudinal obstacles in pursuit of sexual, reproductive, and parenting rights, which can be traced back to ideologies originating from the eugenics era (Lund et al. 2021; Powell 2022).

## 19.3 Ableism and Sexual Rights

Sexual rights are human rights that should be available to everyone. However, ableism frequently obstructs these rights for people with ID. Ableist attitudes and stereotypes tend to underestimate the autonomy and agency of this community, resulting in their exclusion from conversations about their own sexual experiences and desires. Additionally, insufficient support and limited access to sexual education perpetuate this exclusion, leaving people with ID ill-prepared to navigate healthy relationships and make informed choices. This section examines the impact of ableism on consent and capacity, and sexual expression and relationships.

### 19.3.1 Consent and Capacity

Everyone should be allowed to express their sexuality and engage in intimate relationships. However, ableist assumptions undermine the ability of people with ID to provide informed consent, which leads to their exclusion from discussions and decision-making processes concerning their sexual lives.[3] This exclusion arises from sexual ableism, a belief system that erroneously suggests people with ID lack sexual agency and disregards their rights as sexual beings (McCarthy et al. 2020). Societal portrayals often depict people with ID as either devoid of sexuality or, paradoxically, as possessing flawed and undesirable sexualities that necessitate social or medical intervention (Shakespeare 2000; Gill 2015). These misconceptions label people with ID as lacking the cognitive capacity for sexuality rather than acknowledging their diverse sexualities and recognizing them as active participants in their sexual experiences (Gill 2015).

Indeed, despite people with ID securing significant rights since the eugenics era, debates about their sexual decision-making capacity endure (Murphy and O'Callaghan 2004). Although international treaties like the United Nations Convention on the Rights of Persons with Disabilities state that capacity should not be restricted based on disability, domestic laws are often more restrictive (Powell and Stein 2016). For example, guardianship laws in the United States often restrict the sexual decision-making rights of people with ID (Ginn 2022).

Correspondingly, sexual consent capacity assessments have received increasing attention (Gill 2015; Lynden 2007; Onstot 2019). These are particularly popular in congregate care settings. In fact, some group homes adopt a general stance that assumes people with ID cannot consent (Gill 2015). This is all the more complicated because there is no one standard for determining the ability to consent (Ibid.). For example, some professionals and caregivers utilize "mental age" to determine the capacity to consent, even though the concept of mental age can be viewed as ableist (Ibid.).

Importantly, most people with ID, including those with guardians, have sexual decision-making capacity (Lyden 2007). Moreover, an adult's capacity to consent is often determined by the presence of requisite rationality, knowledge, and voluntariness, and with appropriate education, counseling, and exposure to diverse social situations, most people with ID can develop it (Lyden 2007; Murphy 2003).

### 19.3.2 Sexual Expression and Relationships

People with ID desire intimacy and can engage in sexual and romantic relationships (Friedman 2023).[4] However, their sexuality is often viewed through a lens of

---

[3] For more on sexual consent and people with ID, see Chap. 8 of this volume.

[4] For more information on people with ID's desire for romantic relationships and relevant experiences, see Chap. 20 of this volume.

deviance, leading to desexualizing them and believing they are not interested in or capable of such relationships (Ibid.). Women with ID are frequently portrayed as vulnerable victims, while men with ID are stereotyped as potential predators (Ibid.). As a result of these ableist perceptions, they are repeatedly denied opportunities to express their sexuality.

Caregivers, including family members, often act as gatekeepers, imposing restrictions on the sexual relationships of the people with ID they support (Abbott 2013; Azzopardi-Lane and Callus 2015; Bernert and Ogletree 2013). In fact, some caregivers discourage people with ID from pursuing sexual relationships, with some going as far as attempting to prevent them from masturbating (Powell et al. 2020). Moreover, people with ID residing in group homes often face limitations or denial of opportunities to engage in sexual relationships (Chin 2018). Critically, barriers to sexual expression and relationships are heightened for LGBTQ+ people with ID (Wilson et al. 2018).

Thus, sexual expression and relationships for people with ID are often hindered by societal barriers and misconceptions. Ableist assumptions limit their autonomy and participation in decision making regarding their own sexual lives. They are subjected to stereotypes that portray them as sexually unable or having abnormal sexualities, resulting in marginalization and exclusion. Additionally, people with ID often face restrictive policies and practices (e.g., institutional rules, guardianship arrangements) that further impede their sexual expression. Such policies are often based on ableist, paternalistic views that underestimate their capacity to engage in consensual relationships or make informed choices about their sexual lives.

## 19.4 Ableism and Reproductive Rights

Reproductive rights are fundamental, but people with ID face ongoing barriers to fully accessing them. Ableism plays a significant role in perpetuating these obstacles, limiting access to necessary services. Furthermore, there is a denial of decision making and self-determination regarding reproductive matters for people with ID. This section explores how ableism affects people with ID's access to reproductive health services and ability to make decisions.

### *19.4.1 Access to Reproductive Health Services*

People with ID face pervasive inequities in accessing various reproductive health services due to ableism. For example, women with ID face lower cancer screening rates (Parish et al. 2013; Swaine et al. 2013a, b)[5] and are less likely to get routine Pap tests due to mistaken assumptions about their sexual activity and HPV risk

[5] For more about the ethics of cancer care and people with IDD see Chap. 25 in this volume.

(Agaronnik et al. 2020; Parish et al. 2013). Ableism also plays a role in shaping access to contraception for people with ID. One manifestation of ableism is the lack of attention given by reproductive health providers to the contraceptive needs of women with ID, as evidenced by a scarcity of inquiries about contraception (Greenwood and Wilkinson 2013). Consequently, research indicates that women with ID have lower rates of contraception knowledge and use than others (Horner-Johnson et al. 2019; Mosher et al. 2018; Wu et al. 2018). Conversely, some women with ID experience coercion from their caregivers to use contraception out of fear of potential pregnancies (Powell 2022; Powell et al. 2020). These examples highlight how ableism shapes both the provision and utilization of contraception among people with ID.

Furthermore, women with ID face challenges in accessing adequate preconception and perinatal care (Akobirshoev et al. 2017; Mitra et al. 2015; Parish et al. 2015). They have a heightened risk of experiencing complications during pregnancy, delivery, and postpartum, including gestational diabetes, pre-eclampsia, cesarean delivery, preterm birth, low-birth-weight infants, and stillbirth. These disparities stem from ableist attitudes and systemic barriers that hinder people with ID from receiving comprehensive reproductive health services. Additionally, it is notable that Black women with ID are more likely to experience adverse outcomes than their white counterparts. This discrepancy can be attributed to the compounded effects of ableism and racism (Akobirshoev et al. 2020).

The outcomes experienced by people with ID in the realm of reproductive health are linked to the presence of communication, programmatic, and attitudinal barriers rooted in ableism. These barriers impact their access to quality reproductive healthcare and contribute to health disparities. For instance, reproductive health providers often struggle with communication when interacting with people with ID, relying on caregivers rather than engaging with the patients (Agaronnik et al. 2020; Taouk et al. 2018). Additionally, many providers hold entrenched beliefs about providing reproductive healthcare for women with ID, reflecting historical attitudes that endorsed forced sterilization (Greenwood and Wilkinson 2013). Unfortunately, these attitudes persist, leading to negative views on parenting and the promotion of sterilization among this population (Agaronnik et al. 2020). Furthermore, the lack of training for reproductive health providers poses a barrier to achieving equitable reproductive health services for people with ID (Amir et al. 2022; Powell 2022; Taouk et al. 2018). Insufficient training and knowledge among reproductive health professionals regarding the unique needs and rights of people with ID contribute to perpetual disparities in accessing comprehensive reproductive health services.

### *19.4.2 Decision Making and Self-Determination*

Throughout history, people with ID have faced challenges maintaining control over their bodies and lives. Regrettably, the denial of autonomy and agency persists, impeding reproductive rights. Ableist assumptions continue to underestimate their decision-making abilities, resulting in restrictions on reproductive choices and perpetuating inequalities in accessing reproductive health services.

Despite the decline of forced sterilization practices since the eugenics era, such violations have not been completely eradicated. In the United States, for instance, thirty states and the District of Columbia still have involuntary sterilization laws, granting court-appointed guardians the authority to sterilize people with ID against their wishes (National Women's Law Center 2022). Similarly, globally, courts and guardians often justify compulsory sterilization as a supposed protective measure (Powell and Stein 2016). Thus, decisions regarding sterilization are primarily driven by caregivers who express concerns about assuming the responsibility of caring for offspring (Powell 2022).

In addition, the denial of reproductive decision making includes instances of guardians forcing people with ID to proceed with unintended pregnancies or undergo unwanted abortions (Powell 2022). Further, court approval is often required for people under guardianship to access abortion services. This approach disregards the autonomy of people with ID, undermining their competence to make decisions regarding their bodies and lives. Such practices further perpetuate the denial of agency and invalidate their right to reproductive health services.

These examples underscore the continued impact of ableism on the reproductive rights, health, and justice of people with ID, necessitating concerted efforts to challenge these practices, advocate for autonomy and self-determination, and ensure equal access to reproductive health services without coercion or discrimination.

## 19.5 Ableism and Parenting Rights

The right to create and maintain a family is cherished by many, but ableism poses an obstacle to the parenting rights of people with ID. Stereotypes often lead to the assumption that people with ID are unfit parents. Limited access to resources further hampers their ability to fulfill their parenting roles effectively. Consequently, involvement with the child welfare system can increase, with biases rather than an accurate evaluation determining the removal of children from parents with ID. This section describes the impact of ableism on parents with ID, exploring how it influences stereotypes, the lack of available support systems, and, briefly, involvement with the child welfare system.

### *19.5.1 Stereotypes and Discrimination*

Ableism presents a significant barrier to the parental rights of people with ID. The ingrained presumption that they are unfit parents perpetuates discrimination and undermines their ability to create a nurturing environment for their children (National Council on Disability 2012; Powell 2016). Research consistently highlights the belief that people with ID should not have the right to have children (Dell'Armo and Tassé 2021), and parents with ID frequently express that societal attitudes and discrimination represent their greatest challenges (Llewellyn 2019). These harmful stereotypes often result in heightened scrutiny and interventions by others, and in some cases, the child welfare system may remove their children. This pervasive ableist presumption hinders the full realization of parental rights for people with ID, perpetuating inequality and denying them the opportunity to raise their children in a loving and supportive environment.

Moreover, parents with ID face scrutiny based on the assumption that intelligence is directly linked to parenting capacity (Llewellyn 2019). Flawed assessments of parental capacity (e.g., IQ scores) contribute to alarming rates of child removal by the child welfare system for parents with ID (IASSIDD SIRG 2008).[6] Ableism perpetuates the presumption that they cannot care for their children or acquire parenting skills (Powell 2016). These unfounded beliefs lead to unjust questioning of their parenting abilities, creating additional challenges.

The pervasive stereotypes and discrimination toward parents with ID have profound implications. Strikingly, the stigma and discrimination faced by parents with ID begin in childhood and persist as they transition into parenthood (Llewellyn 2019). Once they are parents, negative attitudes from others (e.g., family, professionals) contribute to a climate of suspicion and doubt. Fear becomes a constant companion for these parents, as they are at frequent risk of being reported to the child welfare system due to unwarranted speculations (National Council on Disability 2012). The biased assumptions surrounding their ID perpetuate an environment where their parenting abilities are questioned. Further, negative societal attitudes lead to unjust child welfare system referrals, even without evidence of maltreatment, resulting in the termination of their parental rights (Powell 2016; IASSIDD SIRG 2008). Hostile attitudes within the child welfare system compound these challenges. The pervasive stereotypes and discrimination deny parents with ID their fundamental right to care for and raise their children.

### *19.5.2 Support Systems*

Ableism impacts the availability and adequacy of both formal and informal parenting support. Formal services, such as parenting education and technology, effectively support these parents (Lightfoot and LaLiberte 2011; Powell 2016; Powell et al. 2022). However, the limited availability and accessibility of these services perpetuate inequities and leave many parents with ID without the necessary support they require (IASSIDD SIRG 2008). The scarcity of formal supports reflects a systemic failure to address the unique needs and rights of parents with ID, perpetuating ableist attitudes that underestimate their parenting abilities.

---

[6] For information regarding competency assessments and child custody cases for people with ID, see Chap. 7 of this volume.

Moreover, informal supports, including family and friends, play a vital role in providing assistance and understanding to parents with ID (IASSIDD SIRG 2008; Powell et al. 2022). Unfortunately, ableism can lead to social isolation and limited access to informal support networks for these parents. Discrimination and negative attitudes toward disability can create barriers, hindering their ability to build robust support systems. This isolation not only deprives parents of the practical assistance they may need, but also denies them the emotional support that can be invaluable to navigate parenting.

Adequate formal and informal supports are essential in combating ableism. Through formal support systems, we can ensure that parents with ID have access to necessary resources, services, and education to fulfill their parenting responsibilities (Powell et al. 2022). Additionally, fostering inclusive communities that promote acceptance, understanding, and support can help combat social isolation and associated consequences.

Addressing ableism in the provision of support systems for parents with ID is crucial not only for the well-being and rights of these parents but also for the welfare of their children. Adequate support can prevent unnecessary interventions from the child welfare system, reduce the risk of family separation, and support family reunification when necessary (Powell et al. 2022).

## 19.6 Conclusion

Confronting the profound impact of ableism on the sexual, reproductive, and parenting rights of people with ID necessitates a comprehensive transformation of laws, policies, and practices. It requires a shift in societal attitudes and a commitment to dismantling the barriers perpetuating discrimination and marginalization. This transformation must start with revising legal frameworks to ensure equal protection and opportunities for people with ID, while actively involving them in decision making. Additionally, comprehensive education programs should be implemented to challenge stigmas and stereotypes, providing accurate information and empowering people to make informed choices about their sexual lives. Access to reproductive health services and decision-making autonomy must be expanded, and support systems should be established to assist people with ID in their journey toward parenthood. Through these collective efforts, we can lay the foundation for a more just and equitable future, where the sexual, reproductive, and parenting rights of people with ID are fully realized and respected.

## References

Abbott, David. 2013. Nudge, nudge, wink, wink: Love, sex and gay men with intellectual disabilities—A helping hand or a human right? *Journal of Intellectual Disability Research* 57: 1079–1087.

Agaronnik, Nicole, Elizabeth Pendo, Tara Lagu, Christene DeJong, Aixa Perez-Caraballo, and Lisa I. Iezzoni. 2020. Ensuring the reproductive rights of women with intellectual disability. *Journal of Intellectual & Developmental Disability* 45: 365–376.

Akobirshoev, Ilhom, Monika Mitra, Susan L. Parish, Anne Valentine, and Tiffany A. Moore Simas. 2020. Racial and ethnic disparities in birth outcomes and labor and delivery charges among Massachusetts women with intellectual and developmental disabilities. *Intellectual and Developmental Disabilities* 58: 126–138.

Akobirshoev, Ilhom, Susan L. Parish, Monika Mitra, and Eliana Rosenthal. 2017. Birth outcomes among US women with intellectual and developmental disabilities. *Disability and Health Journal* 10: 406–412.

Amir, Nili, Lauren D. Smith, Anne M. Valentine, Monika Mitra, Susan L. Parish, and Tiffany A. Moore Simas. 2022. Clinician perspectives on the need for training on caring for pregnant women with intellectual and developmental disabilities. *Disability and Health Journal* 15: 101262.

Azzopardi-Lane, Claire, and Anne-Marie. Callus. 2015. Constructing sexual identities: People with intellectual disability talking about sexuality. *British Journal of Learning Disabilities* 43: 32–37.

Ben-Moshe, Liat. 2020. *Decarcerating disability: Deinstitutionalization and prison abolition.* Minneapolis: University of Minnesota Press.

Bernert, Donna J., and R.J. Ogletree. 2013. Women with intellectual disabilities talk about their perceptions of sex. *Journal of Intellectual Disability Research* 57: 240–249.

Bogart, Kathleen R., and Dana S. Dunn. 2019. Ableism special issue introduction. *Journal of Social Issues* 75: 650–664.

Campbell, Fiona A. Kumari. 2014. Ableism as transformative practice. In *Rethinking anti-discriminatory and anti-oppressive theories for social work practice*, ed. Christine Cocker and Trish Hafford-Letchfield, 78–92. New York: Palgrave Macmillan.

Chin, Natalie M. 2018. Group homes as sex police and the role of the Olmstead integration mandate. *New York University Review of Law & Social Change* 42: 379–450.

Clare, Eli. 2017. *Brilliant imperfection: Grappling with cure.* Durham, NC: Duke University Press.

Cohen, Adam. 2017. *Imbeciles: The Supreme Court, American eugenics, and the sterilization of Carrie Buck.* New York: Penguin Books.

Dell'Armo, Kristin A., and Marc J. Tassé. 2021. Attitudes, stigma, and ableism toward people with intellectual disability. In *APA handbook of intellectual and developmental disabilities: foundations*, eds. Laraine Glidden, Leonard Abbeduto, Laura Lee McIntyre, and Marc J. Tassé, 473–497. Washington, DC: American Psychological Association.

Fierros, Edward Garcia. 2006. One size does not fit all: A response to institutionalizing inequity. *Disability Studies Quarterly* 26 (2). https://doi.org/10.18061/dsq.v26i2.683.

Friedman, Carli. 2023. Sexual health and parenting supports for people with intellectual and developmental disabilities. *Sexuality Research and Social Policy* 20 (1): 257–272.

Gill, Michael. 2015. *Already doing it: Intellectual disability and sexual agency.* Minneapolis: University of Minnesota Press.

Ginn, G. Hannah. 2022. Securing sexual justice for people with intellectual disability: A systematic review and critical appraisal of research recommendations. *Journal of Applied Research in Intellectual Disabilities* 35 (4): 921–934. https://doi.org/10.1111/jar.12867.

Greenwood, Nechama W., and Joanne Wilkinson. 2013. Sexual and reproductive health care for women with intellectual disabilities: A primary care perspective. *International Journal of Family Medicine* 2013 (1): 642472. https://doi.org/10.1155/2013/642472.

Horner-Johnson, Willi, Esther L. Moe, Ryan C. Stoner, Krystal A. Klein, Alison B. Edelman, Karen B. Eden, Elena M. Andresen, Aaron B. Caughey, and Jeanne-Marie. Guise. 2019. Contraceptive knowledge and use among women with intellectual, physical, or sensory disabilities: A systematic review. *Disability and Health Journal* 12: 139–154.

IASSIDD SIRG (International Association for the Scientific Study of Intellectual and Developmental Disabilities, Special Interest Research Group on Parents and Parenting with Intellectual Disabilities). 2008. Parents labelled with intellectual disability: Position of the IASSIDD SIRG on Parents and parenting with intellectual disabilities. *Journal of Applied Research in Intellectual Disabilities* 21: 296–307.

Keefe, Elizabeth Stringer. 2022. From detractive to democratic: The duty of teacher education to disrupt structural ableism and reimagine disability. *Teachers College Record* 124: 115–147.

Lewis, Talila A. 2022. Working definition of ableism—January 2022 update. *Talila Lewis*. https://www.talilalewis.com/blog/working-definition-of-ableism-january-2022-update. Accessed 10 April 2024.

Lightfoot, Elizabeth, and Traci LaLiberte. 2011. Parental supports for parents with intellectual and developmental disabilities. *Intellectual and Developmental Disabilities* 49: 388–391.

Llewellyn, Gwynnyth. 2019. Parents with disabilities. In *Handbook of parenting: Social conditions and applied parenting*, 3rd ed., ed. Marc H. Bornstein, 234–273. New York: Routledge.

Lund, Emily M., Anjali J. Forber-Pratt, and Erin E. Andrews. 2021. Combating old ideas and building identity: Sexual identity development in people with disabilities. In *The Routledge handbook of disability and sexuality*, ed. Russell Shuttleworth and Linda Mona, 97–105. New York: Routledge.

Lyden, Martin. 2007. Assessment of sexual consent capacity. *Sexuality and Disability* 25: 3–20.

McCarthy, Michelle, Karen Milne Skillman, Nicola Elson, Claire Bates, Rachel Forrester-Jones, and Siobhan Hunt. 2020. Making connections and building confidence: A study of specialist dating agencies for people with ID *Sexuality and Disability* 38: 3–18.

Mitra, Monika, Susan L. Parish, Karen M. Clements, Xiaohui Cui, and Hafsatou Diop. 2015. Pregnancy outcomes among women with intellectual and developmental disabilities. *American Journal of Preventive Medicine* 48: 300–308.

Mosher, William, Rosemary B. Hughes, Tina Bloom, Leah Horton, Ramin Mojtabai, and Jeanne L. Alhusen. 2018. Contraceptive use by disability status: New national estimates from the National Survey of Family Growth. *Contraception* 97: 552–558.

Murphy, Glynis H. 2003. Capacity to consent to sexual relationships in adults with learning disabilities. *BMJ Sexual & Reproductive Health* 29: 148–149.

Murphy, Glynis H., and Ali Li O'Callaghan. 2004. Capacity of adults with intellectual disabilities to consent to sexual relationships. *Psychological Medicine* 34: 1347–1357.

Nario-Redmond, Michelle R., Alexia A. Kemerling, and Arielle Silverman. 2019. Hostile, benevolent, and ambivalent ableism: Contemporary manifestations. *Journal of Social Issues* 75: 726–756.

National Council on Disability. 2012. *Rocking the cradle: Ensuring the rights of parents with disabilities and their children*. Washington, DC. https://beta.ncd.gov/report/rocking-the-cradle-ensuring-the-rights-of-parents-with-disabilities-and-their-children/. Accessed 10 April 2024.

National Women's Law Center. 2022. *Forced sterilization of people with disabilities in the United States*. National Women's Law Center. https://nwlc.org/wp-content/uploads/2022/01/*f*.NWLC_SterilizationReport_2021.pdf. Accessed 10 April 2024.

Onstot, Andrea. 2019. Capacity to consent: Policies and practices that limit sexual consent for people with intellectual/developmental disabilities. *Sexuality and Disability* 37: 633–644.

Parish, Susan L., Jamie G. Swaine, Esther Son, and Karen Luken. 2013. Determinants of cervical cancer screening among women with intellectual disabilities: Evidence from medical records. *Public Health Reports* 128: 519–526.

Parish, Susan L., Monika Mitra, Esther Son, Alexandra Bonardi, Paul T. Swoboda, and Leah Igdalsky. 2015. Pregnancy outcomes among US women with intellectual and developmental disabilities. *American Journal on Intellectual and Developmental Disabilities* 120: 433–443.

Powell, Robyn M. 2016. Safeguarding the rights of parents with intellectual disabilities in child welfare cases: The convergence of social science and law. *CUNY Law. Review* 20: 127–149.

Powell, Robyn M. 2021. Confronting eugenics means finally confronting its ableist roots. *William & Mary Journal of Race, Gender, and Social Justice* 27: 607–633.

Powell, Robyn M. 2022. Disability reproductive justice. *University of Pennsylvania Law Review* 170: 1851–1903.

Powell, Robyn M., and Michael Ashley Stein. 2016. Persons with disabilities and their sexual, reproductive, and parenting rights: An international and comparative analysis. *Frontiers of Law in China* 11: 53–85.

Powell, Robyn M., Susan L. Parish, Monika Mitra, and Eliana Rosenthal. 2020. Role of family caregivers regarding sexual and reproductive health for women and girls with intellectual disability: A scoping review. *Journal of Intellectual Disability Research* 64: 131–157.

Powell, Robyn M., Susan L. Parish, Monika Mitra, Michael Waterstone, and Stephen Fournier. 2022. Child welfare system inequities experienced by disabled parents: Towards a conceptual framework. *Disability & Society* 39 (2): 1–28.

Sandberg, Linn J., Hanna Bertilsdotter Rosqvist, and Alisa Grigorovich. 2021. Regulating, fostering and preserving: the production of sexual norms through cognitive ableism and cognitive othering. *Culture, Health & Sexuality* 23: 1421–1434.

Shakespeare, Tom. 2000. Disabled sexuality: Toward rights and recognition. *Sexuality and Disability* 18: 159–166.

Sins Invalid. 2019. *Skin, tooth, and bone: The basis of movement is our people: A disability justice primer*. Sins Invalid.

Swaine, Jamie G., Sarah Dababnah, Susan L. Parish, and Karen Luken. 2013a. Family caregivers' perspectives on barriers and facilitators of cervical and breast cancer screening for women with intellectual disability. *Intellectual and Developmental Disabilities* 51: 62–73.

Swaine, Jamie G., Susan L. Parish, and Karen Luken. 2013b. Breast and cervical cancer screening for women with intellectual disabilities. *Health & Social Work* 38: 183–186.

Taouk, Laura H., Michael F. Fialkow, and Jay A. Schulkin. 2018. Provision of reproductive healthcare to women with disabilities: A survey of obstetrician–gynecologists' training, practices, and perceived barriers. *Health Equity* 2 (1): 207–215.

Thomas, Adrian. 2000. Stability of Tringo's hierarchy of preference toward disability groups: 30 years later. *Psychological Reports* 86: 1155–1156.

Tremain, Shelley, 2017. Knowing disability, differently. In *The Routledge handbook of epistemic injustice*, ed. Ian James Kidd, José Medina, and Gaile Pohlhaus, Jr., 175–183. New York: Routledge.

Wilson, Nathan J., Jemima Macdonald, Brenda Hayman, Alexandra M. Bright, Patsie Frawley, and Gisselle Gallego. 2018. A narrative review of the literature about people with intellectual disability who identify as lesbian, gay, bisexual, transgender, intersex or questioning. *Journal of Intellectual Disabilities* 22: 171–196.

Wu, Justine, Jianying Zhang, Monika Mitra, Susan L. Parish, and Geeth Kavya Minama Reddy. 2018. Provision of moderately and highly effective reversible contraception to insured women with intellectual and developmental disabilities. *Obstetrics and Gynecology* 132: 565–574.

Wu, Justine, Yael Braunschweig, Lisa H. Harris, Willi Horner-Johnson, Susan D. Ernst, and Bethany Stevens. 2019. Looking back while moving forward: A justice-based, intersectional approach to research on contraception and disability. *Contraception* 99: 267–271.

# Chapter 20
# "Somebody You Love": Romantic Relationships and Mental Health Amongst People with Intellectual and Developmental Disabilities

**Marina Heifetz**

**Abstract** This chapter provides an up-to-date overview of romantic relationship conceptualization and experiences of people with intellectual and developmental disabilities. While it is clear that individuals with intellectual and developmental disabilities generally desire and often experience romantic relationships, there continues to be limited socio-sexual education. The ongoing tendency to "shelter" individuals with intellectual and developmental disabilities from open and informative discussions of romantic relationships and sexuality often also "shelters" these individuals from understanding the importance of healthy and safe romantic experiences. Thus, these individuals are at increased risk for unplanned pregnancies, sexually transmitted infection, and abuse. Based on research, important recommendations and future directions are offered.

**Keywords** Intellectual and developmental disabilities · Romantic relationships · Sexual education · Mental health

## 20.1 Introduction

"[A romantic partner is] who you love to spend time with. You like to kiss them. "It is somebody you love." These are examples of the many ways people with intellectual and developmental disabilities conceptualize romantic relationships. Many people with intellectual and developmental disabilities have a desire to form romantic relationships (Heifetz et al. 2020). At the same time, these individuals have different opportunities to develop these relationships, compared to individuals without intellectual and developmental disabilities. People with intellectual and developmental disabilities are often perceived as less socially competent than typically developing

M. Heifetz (✉)
York University, Toronto, Canada
e-mail: heifetzm@yorku.ca

 
A. Bianchi and J. A. Vogt (eds.), *Intellectual Disabilities and Autism: Ethics and Practice*, The International Library of Bioethics 108,
https://doi.org/10.1007/978-3-031-61565-8_20

individuals (Solish et al. 2010). Yet, individuals with intellectual and developmental disabilities are also less likely to be educated about intimate relationships and sexuality in general, leading them to be more vulnerable to some risky romantic behaviours in relationships, such as engaging in unprotected sexual intercourse (Isler et al. 2009). As such, people with intellectual and developmental disabilities tend to be more monitored by their caregivers and have less opportunities to develop intimate relationships, which often leads to isolation and loneliness (English et al. 2018; Emerson et al. 2021a).

Given the history of negative views and myths about people with intellectual and developmental disabilities and their intimate relationships, it is unsurprising that there continues to be limited knowledge and misguided beliefs about their intimate desires and experiences. This chapter addresses the nature of intimate relationships among people with intellectual and developmental disabilities by contextualizing these relationships within the existent research framework to gain a better and more accurate understanding of these relationships.

## 20.2 Historical Perspectives on Romantic Relationships

In line with the history of institutionalization of people with intellectual and developmental disabilities, development of intimate relationships has been a rather recent topic of discussion. Given that people with intellectual and developmental disabilities are often in positions of vulnerability within our society, both families and support services have generally discouraged and denied opportunities for these individuals to form and develop intimate relationships. When intimate relationships did occur, they were generally done so in secrecy (Mattinson 1975). During the period of institutionalization, the eugenics movement contributed to the prevention of procreation via involuntary sterilization, while individuals with milder forms of intellectual disability had legal restrictions that denied them official marriage (Walsh 1997). In his popular and innovative book, *The Principle of Normalization in Human Services,* Wolfensberger (1972) wrote that the idea of socio-sexual and marriage relationships for people with more severe intellectual and developmental disability tends to be rejected because such relationships are perceived as very different from those of typically developing individuals. Wolfensberger introduced the idea that societal perceptions of socio-sexual relationships and people with intellectual and developmental disabilities are culturally bound and there is a need to re-evaluate and reflect on the biases held toward these members of our society.

Even in the twenty-first century, there continue to exist various unfounded beliefs about the romantic development of people with intellectual and developmental disabilities. Given the vulnerability of people with intellectual and developmental disabilities to be abused and exploited, along with risky sexual behaviours (e.g., lack of contraceptive measures), some have taken the stance of denying the existence of sexuality or sexual feelings in these individuals altogether. For instance, some myths imply that people with intellectual and developmental disabilities are "eternal children," and hence, need to be protected from having sexual experiences. These widespread misbeliefs continue to influence how some people in society treat individuals with intellectual and developmental disabilities in regard to their sexuality (Griffiths 2007). Yet, research shows that people with intellectual and developmental disabilities are similar to individuals without intellectual and developmental disabilities in that they attach significant value to romantic relationships and in how they spend time together (Heifetz et al. 2020; Ward et al. 2010). Importantly, many individuals with intellectual and developmental disabilities express wanting to spend more time together, something that may be limited to more organized events (e.g., supervised and organized through an organization) with their romantic partners (Ward et al. 2010; Hebblewhite et al. 2022). While research has long acknowledged that social participation, friendships, and emotional well-being are core components in measuring quality of life, sexuality and romantic relationships have been largely overlooked. These issues need to be addressed to enhance the quality of life for individuals with intellectual and developmental disabilities.

## 20.3 Romantic Relationship Conceptualization

Characterizing relationships requires considering distinctive perceptions, expectancies, schemata, and attributions regarding oneself, the other person, and the relationship (Collins 2003). Romantic conceptualizations are important because they demonstrate the understanding that people have of romantic relationships. For instance, research on typically developing individuals shows that a clear distinction exists in their mind between friendships and romantic relationships (Connolly et al. 1999). Yet, it is known that individuals with intellectual and developmental disabilities do not consistently show a clear ability in distinguishing between friendships and romantic relationships. Ward and colleagues (2010), for instance, found that approximately one third of their adult participants with intellectual and developmental disabilities indicated that "feelings toward each other" are what differentiate whether someone is a romantic partner or a friend. However, many other participants focused on more concrete behavioural factors to distinguish a romantic partner and a friend, such as being asked out on a date, exchanging phone numbers, or simply asking the person whether they are their girlfriend or boyfriend. Other researchers have also found that the distinction between a romantic partner and a friend is not always clear to individuals with intellectual and developmental disabilities. For instance, some may blur the role of a romantic partner and a paid carer, contributing to confusion of boundaries (English et al. 2018), or they may perceive a romantic partner as exclusively a husband or a wife, with no in-between option available (Heifetz et al. 2020). It is important to consider the factors that impact this variability in understanding romantic relationships. These factors include having romantic experiences and accessing and receiving developmentally appropriate socio-sexual education, as discussed further in the sections below.

### 20.3.1 Autonomy and Romantic Experiences

Not surprisingly, romantic conceptualizations may affect, and be affected by, romantic experiences (Furman and Winkles 2012). Yet, despite the frequent desire for these relationships by individuals with intellectual and developmental disabilities (Heifetz et al. 2020), there are numerous hurdles to build intimate connections. During adolescence, for instance, dating experiences are considered developmentally appropriate explorations for typically developing youth in Western cultures. Irrespective of whether or not adolescents are dating, research on typically developing adolescents shows that they are preoccupied with romantic concerns and participate in mixed-sex interactions that typically lead to dating activities (O'Sullivan et al. 2007). In fact, typically developing adolescents' dating tends to be characterized more by activities like being together with their partner in a group, holding hands, and thinking of themselves and their partner as a couple, rather than sexual activities, which typically develop later on (O'Sullivan et al. 2007).

Caregivers of youth with intellectual and developmental disabilities are often conflicted between a desire to foster their child's independence and to protect them from adverse outcomes. The lack of independence that youth with intellectual and developmental disabilities typically experience limits some of the social opportunities that typically developing youth experience, such as going to the mall without parental supervision, e-mailing, and going to the movies (Walker-Hirsch 2007). These adolescent social experiences may not be easily accessible for adolescents with intellectual and developmental disabilities, although these are the types of experiences that provide opportunities to engage with friends and develop romantic relationships. When caregivers limit these social opportunities, youth with intellectual and developmental disabilities, in turn, may be left with limited understanding of romantic relationships, given that such awareness develops primarily through social experiences (Furman and Wehner 1994).

Notably, caregiver supervision and lack of autonomy continues to be a concern for individuals with intellectual and developmental disabilities during their transition to adulthood. Researchers consistently find that these individuals feel like they are never alone and are constantly monitored, which they experience as overbearing. The lack of opportunity for independence, autonomous decision making, and privacy has influenced many people with intellectual and developmental disabilities to pursue romantic experiences in secrecy. These secretive behaviours may help individuals avoid the disapproval of others and allow for some degree of autonomy, while preventing further prohibition or punishment (English et al. 2018).

### 20.3.2 Socio-Sexual Education

Socio-sexual education plays an essential role in building individuals' understanding of intimate relationships and sexuality. Unsurprisingly, socio-sexual education has been shown to help prevent sexual abuse and reduce the risk of sexually transmitted infections (STIs) (Gust et al. 2003). Importantly, societal views about sexuality and reproduction among people with intellectual and developmental disabilities may limit their access to socio-sexual education. Research has consistently shown that individuals with intellectual and developmental disabilities have limited exposure to socio-sexual education in school and at home, with the consequence being that their socio-sexual knowledge is largely incorrect (Cheng and Udry 2003; Isler et al. 2009). For instance, women with intellectual and developmental disabilities tend to be less knowledgeable about contraception than women without disabilities (Horner-Johnson et al. 2020). Even when some socio-sexual education is provided at school, it may not be offered in a way that is easily understood and digestible to individuals with intellectual and developmental disabilities. The socio-sexual education provided to individuals with intellectual and developmental disabilities tends to be incomplete and does not address the full range of topics to allow for a comprehensive understanding of sexuality. This education is also most often provided sporadically, or in response to a problem, as opposed to proactively (Griffiths 2007). Moreover, research shows that individuals with intellectual and developmental disabilities describe the current relationship education materials available to them as too broad and overwhelming. They identify learning about relationships and responsible sexual activity as their major concerns. For instance, they want to know "What's the difference between sex and rape?" and "How do you use a condom?" (Swango-Wilson 2009). Additionally, youth with intellectual and developmental disabilities in mainstream schools may not respond to sex education in the same way as their peers, which may be attributed to lack of opportunity to consolidate their existing knowledge with peers and/or family (Baines et al. 2018). These individuals' knowledge about sexuality and intimate relationships has often been gained through other sources, such as through the internet, television, or observation of peers (Heifetz et al. 2020; English et al. 2018).

Notably, the internet has become an important source for learning about sexuality and relationships for individuals with intellectual and developmental disabilities. Through sources like forum discussions and YouTube videos, these individuals have become increasingly informed about physical and mental health, where this self-education contributes to their positive sense of well-being. At the same time, conflict, bullying, and abuse are present online, which may influence internet users to experience psychological distress (Hebblewhite et al. 2022). Providing accessible education about safe internet use is important for people with intellectual and developmental disabilities since they may use the internet to learn about romance, sex, and other topics about which they want to be informed.

In addition, offering socio-sexual education at home may be uncomfortable for parents of children with intellectual and developmental disabilities. Parents often fear that socio-sexual education will initiate interest in romantic involvement and that this interest will lead to risky behaviours and corresponding consequences, such as pregnancy and STIs (Walker-Hirsch 2007). As well, compared to parents of typically developing youth, parents of youth with intellectual and developmental disabilities often consider themselves to be less knowledgeable in how to discuss sex and birth control with their children. They also tend to feel that discussing these topics will embarrass their child or encourage them to have sex (Cheng and Udry 2003). These attitudes often lead to restrictions in socio-sexual education (Isler et al. 2009).

As distinct from caregivers who do not provide socio-sexual education, there also exist caregivers who actively discourage physical intimacy through disapproval. As a result, there may be ongoing conflict and ambivalence for these caregivers' children with intellectual and developmental disabilities who have a desire for intimate relationships while struggling with the negative messages provided by families and staff. As a consequence of strong caregiver disapproval, some individuals with intellectual and developmental disabilities are of the impression that any physical intimacy is shameful, wrong, or dangerous. For instance, some individuals with intellectual and developmental disabilities describe sexual activity as doing "weird things on the bed" (Heifetz et al. 2020). Moreover, the tendency for caregivers to disapprove of physical intimacy may lead to these individuals' lack of awareness of the connection between pleasure and physical intimacy, which can potentially lead to unpleasant experiences given that these individuals may engage in physical intimacy to "look normal" rather than to have an enjoyable experience (English et al. 2018).

## 20.4 Risks and Vulnerabilities (Given Limited Education)

The clear lack of socio-sexual education presents various risks for individuals with intellectual and developmental disabilities. Research has shown that most young people with mild-to-moderate intellectual and developmental disabilities have had sexual intercourse by about 19–20 years of age (Baines et al. 2018). However, given the lack of socio-sexual education, youth with intellectual and developmental disabilities who have sex are less likely to use contraception and more likely to become infected with STIs than their peers. Girls with intellectual and developmental disabilities are also at a higher risk for teen pregnancy (Cheng and Udry 2003; Baines et al. 2018). Moreover, since many people with intellectual and developmental disabilities may feel like they have to explore physical intimacy in secrecy, their experiences can end up being risky because they occur quickly and without an opportunity or thought about safe sexual practices (English et al. 2018).

### *20.4.1 Sexually Transmitted Infections*

Despite efforts to shelter individuals with intellectual and developmental disabilities, current data shows that the prevalence of HIV among individuals with intellectual and developmental disabilities is similar to the general population (Lunsky et al. 2017). Yet, many perceive individuals with intellectual and developmental disabilities as being at lower risk for HIV and, as such, these individuals are less likely to be offered

HIV testing (Neri et al. 2007). At the same time, since they are offered limited education about HIV and its prevention, individuals with intellectual and developmental disabilities may not self-identify as being at risk for HIV and may not request testing. Researchers' calls for action involves preventative measures, including appropriate socio-sexual education for individuals with intellectual and developmental disabilities, along with more informed clinical care for better monitoring and intervention (Lunsky et al. 2017).

### 20.4.2 Unplanned Pregnancy

Following deinstitutionalization and repeal of sterilization laws, opportunities for childbearing in individuals with intellectual and developmental disabilities have increased (Servais 2006) and fertility rates among young women with and without intellectual and developmental disabilities are now similar (Brown et al. 2016). Yet, women with intellectual and developmental disabilities have a higher proportion of *unplanned* pregnancy rates than women without disabilities (53% vs. 36%; Horner-Johnson et al. 2020). These unintended pregnancies have a substantial impact on the physical and mental health of these women, both during and after pregnancy (Kavanaugh et al. 2017). These high rates of unplanned pregnancies for women with intellectual and developmental disabilities may be directly linked to their limited reproductive knowledge. It is clear that these women would benefit from appropriate sex education and reproductive planning conversations as part of their routine care.

### 20.4.3 Romantic Relationships and Violence

Dating violence is unfortunately prevalent among individuals with intellectual and developmental disabilities. Some researchers estimate that individuals with intellectual and developmental disabilities are two and a half times more likely to be victims of abuse than individuals without disabilities. Approximately 60% of those with intellectual and developmental disabilities who have been in a romantic relationship have also reported experiencing interpersonal violence, primarily emotional and physical abuse. While no significant gender difference in experiences of interpersonal violence has been found (Ward et al. 2010), women with intellectual and developmental disabilities are more likely to be physically and sexually abused than women without disabilities (Lunsky and Canrinus 2005). Risk factors for exploitation and abuse may include restricting a person's ability to make autonomous decisions about romantic relationships, which may influence them to engage in secretive, exploitative relationships (Rushbrooke et al. 2014) as well as lack of knowledge about healthy relationships (Bernet and Ogletree 2013). Importantly, individuals with intellectual and developmental disabilities respond rather passively to abuse; women may become withdrawn and depressed, while men may display externalizing behaviours (Lunsky and Canrinus 2005).

## 20.5 Relationships and Mental Health

Researchers consistently find that people with intellectual and developmental disabilities have poorer physical and mental health (Lin et al. 2019). Community-based studies suggest that approximately 35–40% of children and youth with intellectual and developmental disabilities have a diagnosable mental health disorder, compared to 8% within the general population. Relationships are essential in enriching quality of life and they are generally seen as important contributors to individuals' mental health. Romantic relationships, in particular, are known to provide people with intellectual and developmental disabilities with a sense of feeling accepted and a secure connection (May and Simpson 2000). Yet, as mentioned above, while people with intellectual and developmental disabilities have great interest and desire for romantic relationships (Heifetz et al. 2020), they often lack opportunities, education, and social skills to initiate and develop intimate relationships. These difficulties may contribute to feelings of loneliness and mental health difficulties. Indeed, researchers have found that many individuals with intellectual and developmental disabilities identify alleviation of feelings of loneliness as a primary reason for wanting an intimate relationship (English et al. 2018).

Given that pregnancy rates are rising for women with intellectual and developmental disabilities (Brown et al. 2020), these women may experience increased risk to their mental well-being during the perinatal period. Women with disabilities in general are considered at increased risk for various adverse outcomes during pregnancy, including pregnancy complications, postpartum depression, and perinatal hospitalization (Tarasof et al. 2020). Indeed, researchers have found women with intellectual and developmental disabilities to be particularly vulnerable to mental illness both during and following the pregnancy period. This research has clear implications for these women's children, given that maternal mental health and well-being impacts children's health and well-being (Brown et al. 2020). Some factors that may contribute to this increased vulnerability include the possibility of unplanned pregnancy resulting from lack of previously available, accessible education, as well as the experiences of social isolation and abuse (Lunsky and Canrinus 2005).

## 20.6 Conclusion

Romantic relationships are desired and often experienced by individuals with intellectual and developmental disabilities. However, a lack of available and accessible socio-sexual education, caregivers' desires to prevent youth with intellectual and developmental disabilities from experiencing undue harm, and a failure to enable autonomous decision making and relationship building have impinged on their rights to experience healthy and safe romantic relationships. Research has consistently demonstrated that social participation, friendships, and emotional well-being are core components in measuring quality of life. However, sexuality and romantic relationships have been largely overlooked in this type of research amongst people with intellectual and developmental disabilities. Many people with intellectual and developmental disabilities hold misconceptions about what healthy romantic relationships may look like and they are at increased risk for unplanned pregnancies, STIs, and abuse. Moving forward, it is important to normalize sexual expression, given the link between sexuality and the confusion and shame that many individuals with intellectual and developmental disabilities may feel. These feelings, misunderstandings, and lack of available support may also contribute to mental illness.

Although research on romantic experiences of individuals with intellectual and developmental disabilities has come a long way, there are still many unexplored areas. For instance, it has been reported that people with intellectual and developmental disabilities, compared to typically developing individuals, may have experienced increased levels of stress and conflict with their romantic partner during the COVID-19 pandemic (Emerson et al. 2021b). These increased challenges also heighten these individuals' risk for mental illness. Research is limited in terms of how life-altering, prolonged, and worldwide events (e.g., a pandemic) may influence romantic relationships amongst people with intellectual and developmental disabilities, though this research is needed insofar as our society plans for future pandemics.

Another important area that deserves further exploration is internet use and people with intellectual and developmental disabilities. Research shows that people, in general, increasingly use the internet to seek information about sexuality and romantic relationships (International Telecommunication Union 2019). The internet is appealing in that it may help individuals avoid stigma, rejection, and any embarrassment that may come from talking face-to-face with parents or healthcare professionals (Kendall 2012). Recent research on mothers with intellectual and developmental disabilities found that they use the internet frequently for help with parenting questions, and that having the internet available contributed to their feelings of empowerment and served as a protective factor in terms of their mental health (Heifetz et al. 2020). Given the prominence of social media and insofar as the internet may be a source of relationship education for people with intellectual and developmental disabilities, it is important to explore ways to provide appropriate socio-sexual education and support through this medium.

Overall, in order to address the romantic relationship needs and desires of people with intellectual and developmental disabilities, appropriate socio-sexual education and opportunities for relationship development are essential. When developing educational opportunities, it is essential that voices of people with intellectual and developmental disabilities are heard and that the population is meaningfully represented during educational program development (e.g., asking people with intellectual and developmental disabilities to sit on program development advisory committees). In addition to providing education on relationship safety and exploitation, it is also important to normalize desire, pleasure, and sexuality. By taking the step to develop and normalize these educational opportunities, it is anticipated that people with intellectual and developmental disabilities may experience more open, transparent, satisfying, and safe romantic relationships. These relationships can serve as a positive contribution to their health, well-being, and inclusion in society.

## References


Baines, Susannah, Eric Emerson, Janet Robertson, and Chris Hatton. 2018. Sexual activity and sexual health among young adults with and without mild/moderate intellectual disability. *BMC Public Health* 18: 667–679.

Bernet, D.J., and R.J. Ogletree. 2013. Women with intellectual disabilities talk about their perceptions of sex. *Journal of Intellectual Disability Research* 57 (3): 240–249. https://doi.org/10.1111/j.1365-2788.2011.01529.x.

Brown, Hilary K., Simon Chen, Astrid Guttmann, Susan M. Havercamp, Susan Parish, Joel G. Ray, Lesley A. Tarasoff, Simone N. Vigod, Adele Carty, and Yona Lunsky. 2020. Rates of recognized pregnancy in women with disabilities in Ontario, Canada. *American Journal of Obstetrics and Gynecology* 222 (2): 189–192.

Brown, Hilary K., Yona Lunsky, Andrew Wilton, Virginie Cobigo, and Simone N. Vigod. 2016. Pregnancy in women with intellectual and developmental disabilities. *Journal of Obstetrics and Gynaecology Canada* 38 (1): 9–16.

Cheng, Mariah M., and Richard J. Udry. 2003. How much do mentally disabled adolescents know about sex and birth control? *Adolescent & Family Health* 3: 28–38.

Collins, Andrew. 2003. More than myth: The developmental significance of romantic relationships during adolescence. *Journal of Research on Adolescence* 13: 1–24.

Connolly, Jennifer, Wendy Craig, Adele Goldber, and Debra Pepler. 1999. Conceptions of cross-sex friendships and romantic relationships in early adolescence. *Journal of Youth and Adolescence* 28: 481–494.

English, Brad, Anna Tickle, Roshan das Nair. 2018. Views and experiences of people with intellectual disabilities regarding intimate relationships: A qualitative metasynthesis. *Sex Disabil* 36: 149–173.

Emerson, Eric, Nicola Fortune, Gwynnyth Llewellyn, and Roger Stancliffe. 2021a. Loneliness, social support, social isolation and wellbeing among working age adults with and without disability: Cross-sectional study. *Disability and Health Journal* 14 (1): 100965.

Emerson, Eric, Zoe Aitken, Vaso Totsika, Tania King, Roger J. Stancliffe, Chris Hatton, Gwynnyth Llewellyn, Richard P. Hastings, and Anne Kavanagh. 2021b. The impact of the COVID pandemic on working age adults with disability: Meta-analysis of evidence from four national surveys. *Health and Social Care in the Community* 30: 1–12.

Furman, Wyndol, and Elizabeth Wehner. 1994. Romantic views: Toward a theory of adolescent romantic relationships. In *Personal relationships during adolescence*, vol.6, ed. Raymond Montemayor, Gerald R. Adams, and Thomas P. Gullotta. Thousand Oaks, CA: Sage Publications.

Furman, Wyndol, and Jessica K. Winkles. 2012. Transformations in heterosexual romantic relationships across the transition into adulthood: "Meet me at the bleachers… I mean the bar." In *Relationship pathways: From adolescence to young adulthood*, ed. Brett Laursen and W. Andrew Collins, 23–42. Thousand Oaks, CA: Sage Publications.

Griffiths, Dorothy. 2007. Sexuality and persons with intellectual disabilities. In *A Comprehensive guide to intellectual and developmental disabilities*, ed. Ivan Brown, and Maire Percy, 571–581. Baltimore, MD: Paul H. Brookes.

Gust, Deborah A., Susan A. Wang, James Grot, Ray Ransom, and William C. Levine. 2003. National survey of sexual behavior and sexual behavior policies in facilities for individuals with mental retardation/developmental disabilities. *Mental Retardation* 41 (5): 365–373.

Hebblewhite, Gillian, Nick Hutchinson, and Kathleen Galvin. 2022. Adults with intellectual disabilities' lived experiences of wellbeing and the internet: A descriptive phenomenological study. *Disability & Society* 37 (4): 567–590.

Heifetz, Marina, Johanna Lake, Jonathan Weiss, Barry Isaacs, and Jennifer Connolly. 2020. Dating and romantic relationships of adolescents with intellectual and developmental disabilities. *Journal of Adolescence* 79: 39–48.

Horner-Johnson, Willie, Mekhala Dissanayake, Justine P. Wu, Aaron B. Caughey, and Blair G. Darney. 2020. Pregnancy intendedness by maternal disability status and type in the United States. *Perspectives on Sexual and Reproductive Health* 52 (1): 31–38.

Isler, Aysegul, Fatma Tas, Dilek Beytut, and Zeynep Conk. 2009. Sexuality in adolescents with intellectual disabilities. *Sexuality and Disability* 27: 27–34.

International Telecommunication Union. 2019. *Measuring digital development, Facts and figures.* https://www.itu.int/en/ITU-D/Statistics/Documents/facts/ICTFactsFigures2017.pdf. Accessed 2 August 2023.

Kavanaugh, Megan L., Kathryn Kost, Lori Frohwirth, Isaac Maddow-Zimet, and Vivian Gor. 2017. Parents' experience of unintended childbearing: A qualitative study of factors that mitigate or exacerbate effects. *Social Science & Medicine* 174: 133–141.

Kendall, Nancy. 2012. *The sex education debates.* Chicago: University of Chicago.

Lin, Elizabeth, Robert S. Balogh, Anna Durbin, Laura Holder, Nancy Gupta, Tiziana Volpe, Barry J. Isaacs, Jonathan A. Weiss, and Yona Lunsky. 2019. *Addressing gaps in the health care services used by adults with developmental disabilities in Ontario.* ICES. https://www.camh.ca/-/media/files/professionals/hcardd/snapshots/hcardd-snapshot-addressing-gaps-health-care-adults-dd-ontario-pdf.pdf. Accessed 9 July 2024.

Lunsky, Yona, and Maaike Canrinus. 2005. Gender issues, mental retardation, and depression. In *Mood disorders in people with mental retardation*, ed. Peter Sturmey, 113–130. Kingston, New York: National Association for the Dually Diagnosed.

Lunsky, Yona, Anna Durbin, Hilary Brown, Symron Bansal, Marina Heifetz, and Tony Antoniou. 2017. Health profiles and associated service use among adults with HIV and intellectual and developmental disabilities. *AIDS* 31 (5): 697–705.

Mattinson, Janet. 1975. *Marriage and mental handicap.* London: Institute of Marital Studies.

May, David, and Murray K. Simpson. 2000. Marriage and Parenting. In *Transition and change in the lives of people with intellectual disabilities*, ed. D. May, London: Jessica Kingsley.

Neri, Steven V., Elizabeth H. Bradley, and Nora E. Groce. 2007. Frequency of HIV testing among persons with disabilities; Results from the national health interview survey, 2002. *AIDS Education and Prevention* 19 (6): 545–554.

O'Sullivan, Lucia F., Mariah Mantsun Cheng, Kathleen Mullan Harris, and Jeanne Brooks Gunn. 2007. I wanna hold your hand: The progression of social, romantic and sexual events in adolescent relationships. *Perspectives on Sexual and Reproductive Health* 39 (2): 100–107.

Rushbrooke, Elizabeth, Craig Murray, and Samantha Townsend. 2014. The experiences of intimate relationships by people with intellectual disabilities: A qualitative study. *Journal of Applied Research in Intellectual Disabilities* 27: 531–541.

Servais, Laurent. 2006. Sexual health care in persons with intellectual disabilities. *Mental Retardation and Developmental Disabilities Research Reviews* 12: 48–56.

Solish, Abbie, Adrienne Perry, and Patricia Minnes. 2010. Participation of children with and without disabilities in social, recreational and leisure activities. *Journal of Applied Research in Intellectual Disabilities* 23: 226–236.

Swango-Wilson, Amy. 2009. Perception of sex education for individuals with developmental and cognitive disability: A four cohort study. *Sexuality and Disability* 27 (4): 223–228.

Tarasof, Lesley A., Saranyah Ravindran, Hannan Malik, Dinara Salaeva, and Hilary K. Brown. 2020. Maternal disability and risk for pregnancy, delivery, and postpartum complications: A systematic review and meta-analysis. *American Journal of Obstetrics and Gynecology* 222 (1): 27.e1–27.e32.

Walker-Hirsch, Leslie. 2007. Sexuality education and intellectual disability across the lifespan. In *The facts of life…and more: Sexuality and intimacy for people with intellectual disabilities*, ed. Leslie Walker-Hirsch. Maryland: Brookes.

Walsh, Patricia Noonan. 1997. Old world – new territory: European perspectives on intellectual disability. *Journal of Intellectual Disability Research* 41 (2): 112–119.

Ward, Karen M., Rebecca L. Bosek, and Erin L. Trimble. 2010. Romantic relationships and interpersonal violence among adults with developmental disabilities. *Intellectual and Developmental Disabilities* 48: 89–98.

Wolfensberger, Wolf. 1972. *The principle of normalisation.* Toronto: National Institute on Mental Retardation.

# Chapter 21
# Sexual Education and Empowerment Amongst People with Intellectual and Developmental Disabilities

**Elizabeth K. Schmidt**

**Abstract** The United Nations Convention on the Rights of Persons with Disabilities (CRPD) supports several rights that relate directly to sexuality. In spite of this, those with intellectual and developmental disabilities (IDD) have poor sexual health outcomes, including increased risk for sexually transmitted infection, unplanned pregnancy, and sexual abuse. Further, those with IDD express a desire for intimate relationships. Accessible, inclusive, comprehensive sexual health education can improve these outcomes. This education can be provided by a number of people, including, but not limited to, parents or caregivers, peers, educators, and health providers. Strategies to support education include Universal Design for Learning, videos, visuals, modeling, role playing, and digital interventions. Educators should evaluate outcomes of sexual education in the areas of sexual knowledge, attitudes, and skills to ensure the strategies they utilize are effective for the specific individual(s) with whom they are working. Research suggests utilization of accessible, inclusive, comprehensive sexual health education, while incorporating the above-mentioned strategies, has the potential to improve sexual knowledge, attitudes, and skills, as well as to improve sexual self-advocacy and support the sexuality-related rights articulated in the CRPD.

**Keywords** Sexual education · Peer educator · Parent educator · Accessibility · Learning modalities

## 21.1 Introduction

The United Nations Convention on the Rights of Persons with Disabilities (CRPD) articulates several rights that support sexual autonomy for people with disabilities, including those with intellectual and developmental disabilities (IDD) (UN 2006).

E. K. Schmidt (✉)
Department of Occupational Therapy, Lincoln Memorial University, Knoxville, USA
e-mail: Elizabeth.schmidt@lmunet.edu



A. Bianchi and J. A. Vogt (eds.), *Intellectual Disabilities and Autism: Ethics and Practice*, The International Library of Bioethics 108,
https://doi.org/10.1007/978-3-031-61565-8_21

Article 16 supports the right to educational measures to protect them "from all forms of exploitation, violence and abuse, including their gender-based aspects" (UN 2006, Article 16, ¶ 1) and advocates for "the provision of information and education on how to avoid, recognize and report instances of exploitation, violence and abuse" (UN 2006, Article 16, ¶ 2). Article 25 supports the right to health care (including sexual health) which entails awareness amongst healthcare providers of how to accommodate the learning needs of people with IDD (UN 2006, Article 25, ¶ 1a). Furthermore, Article 23 calls for "access to age-appropriate information, reproductive and family planning education" (UN 2006, Article 23, ¶ 1a). Together, these articles support the argument for more accessible, inclusive, and comprehensive sexual health education for people with IDD.

In spite of the rights expressed in the CRPD, many individuals with IDD experience bias surrounding their sexuality and sexual experiences (Wilkinson et al. 2014; Dinwoodie et al. 2020). Specifically, caregivers, healthcare providers, and educators often dismiss these individuals' sexual identities (Wilkinson et al. 2014; Dinwoodie et al. 2020; Ballan 2012; Holmes and Himle 2014; Sinclair et al. 2015), which can negatively influence mental health.[1] Those with IDD who hold intersectional identities, such as those who are lesbian, gay, bisexual, transgender, queer, intersex, asexual or another marginalized gender or sexual orientation (LGBTQIA+), or are persons of color, may further experience dismissal of their sexual identities or hypersexualization based on the related biases of many in Western society (Armin-Arsala et al. 2023; Epstein et al. 2017).

People with IDD experience barriers to receiving the respect that they inherently deserve surrounding their identities. Additionally, many face barriers accessing sexual health information, which has substantial implications for their overall sexual and reproductive health. Students with disabilities are significantly less likely to receive sexual education compared to their peers without disabilities (Barnard-Brak et al. 2014). Furthermore, few physicians provide the recommended sexual health information to adolescent clients (Bruener and Matson 2016; Marcell et al. 2017); let alone adolescents with IDD. Educators and healthcare providers often assume that individuals' parents or caregivers should instead be responsible for providing sexual health information to them (Rooks-Ellis et al. 2020). However, studies have shown that parents also feel ill-equipped to take on this responsibility (Pownall et al. 2012; Rooks-Ellis et al. 2020). Although sexual health information is rarely provided, educators, healthcare providers, parents, and individuals with IDD commonly report it to be an important topic (Schmidt et al. 2021a, 2022; Rooks-Ellis et al. 2020). Further, individuals with IDD experience the same desires and physiological responses as their peers without disabilities (Hellemans et al. 2007, 2010; Tissot 2009).

Serious implications of the limited sexual health information or education provided to those with IDD includes higher risk for sexually transmitted infections (STIs) (Mandell et al. 2008), sexual abuse, assault or dating violence (Jones et al. 2012; Gil-Llario et al. 2019; Pearson et al. 2022), and unplanned pregnancies

[1] Rexha et al. (in preparation).

(Horner-Johnson et al. 2020; Alhusen et al. 2020). The significant impact of ableism on people with IDD's sexual, reproductive, and parenting rights, as described in Chap. 19, alongside the inflated rates of these negative outcomes, indicate the vital importance of providing accessible sexual education and empowerment for them. Articles 16 and 23 of the CRPD speak to the importance of preventative education measures related to these issues.

## 21.2 Sexuality Education

Comprehensive sexuality education should include, at a minimum, anatomy and physiology, puberty and adolescent development, identity, pregnancy and reproduction, STIs, healthy relationships, and personal safety (Son and Draws 2023). Comprehensive education is an effective approach for delaying initiation of sexual intercourse, reducing STIs, and preventing unplanned pregnancies (Ott and Santelli 2007; Carter 2012; Goldfarb and Lieberman 2021). However, research suggests the need to expand the range of topics to further benefit youth sexual health knowledge, perception, and attitudes (Son and Draws 2023). For example, experts suggest that topics should include "issues of self-awareness and body image, interpersonal social skills, personal care and hygiene, developing relationships, risk prevention, and sexual expression, as well as legal aspects of sexual activities" (Sinclair et al. 2015, as cited in Rooks-Ellis et al. 2020, 446). Research with individuals with IDD, as well as their parents or caregivers, health professionals, and educators, has revealed various recommendations for provision of comprehensive sexuality education that is accessible to the unique needs of this population.

### *21.2.1 Education Onset and Educators*

Research suggests that sexual health information be provided at early ages for all individuals (Schmidt et al. 2021b) as this allows for ample learning opportunities. Sexual education can be tailored to the individual's age, should be provided by multiple individuals involved in their care (e.g., parents or caregivers, health professionals, educators), and can be accomplished using a shared decision-making model (Roden et al. 2020). For example, parents or caregivers can provide information about anatomy, using medically accurate terminology, and educating them on boundaries, consent, etc., regarding their specific reproductive anatomy. These early interactions are foundational for an understanding of reproduction and can be built into everyday experiences, such as dressing or toilet training. Health providers can reaffirm this when they clearly explain, even to young children, what they are doing, why they are doing it, which body parts they need to look at or touch in their exam and ask for the child's permission prior to examining. Doing so from a young age will teach the child that they have the power and right to say "no" in the future. Sexuality education provided in schools at a later age can reaffirm these boundaries. The shared decision-making model naturally incorporates other components of evidence-based sexual health education for people with IDD, specifically repetition, an effective strategy to promote understanding and generalizability of sexual health information for those with IDD (Roden et al. 2020; Schmidt et al. 2020, 2021b).

Although research indicates a need for multiple educators to provide sexual education to youth with IDD, many potential educators report they lack the training necessary to do so (Schmidt et al. 2021a; Pownall et al., 2012). This is unsurprising given that the required skillset includes strong instructional skills, content-specific knowledge that they may not receive in their formal training, and universal skills. Firpo-Triplett and colleagues (2023) define instructional skills as those needed to provide effective instruction to students and include student engagement. Content-specific skills refer to sexual health content, something rarely covered in education programs for teachers, certain health professionals (e.g., therapists, case workers, psychologists), and parenting courses. Additionally, content-specific skills are required to facilitate the intervention being utilized, such as "managing role-plays" or "leading condom demonstrations" (Firpo-Triplett et al. 2023, 260). Lastly, universal skills are those needed to handle sensitive questions and integrate trauma-informed, culturally responsive, and inclusive education for all people (Firpo-Triplett et al. 2023).

The complexity of skills required coupled with the paucity of formal education opportunities, indicate a need for additional training options for those providing sexual health information to all youth, but especially youth with IDD. Barriers to accessing training opportunities include time constraints, expenses, and a lack of inclusivity around LGBTQIA+ and disability-specific content (Firpo-Triplett et al. 2023; Schmidt et al. 2021a). One recommendation is to use online training programs which have been found efficacious and can include learning modules specifically designed for various cultural groups who may require unique content-specific information (Firpo-Triplett et al. 2023). Another recommendation is to empower parents and caregivers to increase the communication they have with their child on sexual health topics (Rooks-Ellis et al. 2020; Holmes and Himle 2014; Tissot 2009).

### *21.2.2 Educational Settings*

Some youth receive sexual education in formal group settings, like school-based or community-based settings, and some receive it in formal one–on–one settings, like healthcare settings or with an educator, case worker, etc. All of these formal settings can deliver in-person or online education. There are opportunities for informal education, including learning from parents or caregivers, peers, siblings, and/or online by oneself. The setting likely needs to be decided based on an individual's unique needs. Some people may benefit from learning in classroom or group settings, where they can learn from others' questions, responses, etc. Alternatively, some may benefit from a one-on-one approach, where they can receive sexual health information very specific to them, their needs, and learning styles. Different educators may prefer different settings, likely based on their own circumstances [e.g., educators prefer group-based settings and healthcare providers preferred individualized settings (Schmidt et al. 2021b)]. In the same study, preferences regarding the use of digital interventions (e.g., mobile applications, social media, wholly online programs) were mixed for both participant groups, with about half of each group adamantly opposed to their use (Schmidt et al. 2021b).

Digital interventions for sexual health education are commonly utilized, recommended, and have positive results for youth without disabilities (Wen Lan et al. 2023; DeSmet et al. 2015). They include computer-based games, internet programs, videos, and/or virtual reality games. One systematic review and meta-analysis found seven serious games that used various levels of immersion for sexual health promotion. DeSmet et al. (2015) found that these games delay sexual initiation, which is believed to decrease STIs and HIV risk, increase relationship skills, improve resistance against sexual coercion, prevent unintended pregnancies, and increase knowledge, attitudes, and self-efficacy related to sexual health.

Effective interventions allow the player to control their learning, offer a "just-right" challenge with repeated practice, are motivating and enjoyable, and include task involvement, immediate feedback, active participation, and interaction (Cooney and Darcy 2020; DeSmet et al. 2015). Immersive virtual reality games can also allow tailoring and personalization of the characters and scene to better match the player and their environment, which may improve generalizability (DeSmet et al. 2015; Schrier 2019). Although research on the effects of digital interventions for individuals with IDD is limited, it is likely that these would be more effective given the common challenges with generalizing information learned in other settings (Lorenzo et al. 2019). Further, there is evidence supporting the use of digital interventions to improve social skills, commonly related to intimate partnerships, and to reduce sexual victimization for autistic adolescents and adults specifically (Lorenzo et al. 2019).

### *21.2.3 Educational Activities*

Research into the methods or strategies used when delivering sexual health information can provide insight into the effects of various strategies with individuals with IDD. Two comprehensive reviews of the literature found evidence to support the use of Universal Design for Learning strategies, including a multi-modal approach using pictures, videos, didactic teaching, and activity-based learning, such as role playing, modeling and video modeling (Schmidt et al. 2020; Schaafsma et al. 2015). Different approaches or strategies may be more appropriate for different educators and educational settings (e.g., caregivers may use natural opportunities for learning and humor, whereas healthcare providers may use infographics with pictures, facts and video links) (Schmidt et al. 2021b).

### 21.2.4 Parent, Caregiver and Peer Educator Training

Many parents or caregivers either hold reserved views or misconceptions about the sexuality of their child with a disability, or they may feel unprepared and ill-equipped to provide accessible sexual health information (Ballan 2012; Holmes and Himle 2014; Sinclair et al. 2015; Frank and Sandman 2021; Rooks-Ellis and Achey 2023). Parents or caregivers have indicated a need and desire for resources, recommendations, and even opportunities to learn from other parents or caregivers' experiences when providing this information to their child with IDD (Rooks-Ellis and Achey 2023; Schmidt et al. 2021b).

There is a paucity of tools available to support parents and caregivers in developing the skills necessary to provide sexual health information to their child. Two examples of such tools are Frank and Sandman's (2021) sexuality education parent training and the Home B.A.S.E. workshop tools (Rooks-Ellis and Achey 2023; Rooks-Ellis et al. 2020). Other resources include pre-developed programs to help establish boundaries to prevent sexual victimization, such as the Circles strategy (Walker-Hirsch and Champagne 1991) and the use of social narratives to facilitate choice making and self-care skills (Test et al. 2011). Parents should use plain, simple, and medically accurate language and avoid euphemisms like "the birds and the bees" (Rooks-Ellis and Achey 2023). Additionally, they should consider using visuals, videos, modeling behaviors or engaging in role play, and provide repeated practice (Rooks-Ellis et al. 2020; Frank and Sandman 2021).

Research suggests that training parents increases their knowledge and improves their attitudes, beliefs, and behaviors about sexual health, which is hypothesized to improve outcomes for their child with IDD as well (Rooks-Ellis and Achey 2023; Rooks-Ellis et al. 2020). Additionally, training parents improves generalization of information for individuals with IDD, as parents and caregivers can incorporate information into the natural environment and across everyday activities (Rooks-Ellis and Achey 2023; Rooks-Ellis et al. 2020).

Peer educators are widely used for people with IDD and more recently have been used to promote retention of sexual health information. Additionally, peer-educator sexual education programs have been used for adolescents with other disabilities and have been found effective for both the learners and the peer educator in improving their sexual health knowledge (Bullard and Wallace 1978). However, people with IDD report limited opportunities to learn through peers (Schmidt et al. 2022), and there is scarce research in this area. The available research indicates peer educators are beneficial for youth to learn sexual health information related to forming and maintaining relationships (Frawley and Bigby 2014; James et al. 2022). Additional studies recommend that peer-to-peer models be used and researched further (Gougeon 2009; Frawley and Wilson 2016; Brown et al. 2020). Failing this, formal sexual education programs should, at the very least, be co-designed with people with IDD to ensure they are appropriate for their specific needs and desires (Frawley and Bigby 2014; Schaafsma et al. 2015; Brown et al. 2020).

### *21.2.5 Educational Outcomes*

It is vital that the outcomes associated with sexual health education programs for people with IDD be evaluated to ensure they are effective. Relevant outcomes include sexual health knowledge, interpersonal boundaries, attitudes, or specific skills and many research studies evaluating the effects of these programs have used surveys and interviews as assessment tools (Schmidt et al. 2020; Schaafsma et al. 2015). Thompson and colleagues (2016a, b) conducted a review of readily available sexual health knowledge surveys and found the Socio-Sexual Knowledge and Attitudes Assessment Tool-Revised (SSKAAT-R) (Griffiths and Lunsky 2003) has the most established psychometric properties indicating its reliability and validity for use with individuals with intellectual disabilities, followed by the Assessment of Sexual Knowledge (ASK) (Thompson et al. 2016a; Butler et al. 2003), the General Sexual Knowledge Questionnaire (GSKQ) (Talbot and Langdon 2006), and lastly the Sexual Knowledge, Experience, and Needs Scale for people with Intellectual Disability (SexKen-ID) (McCabe 1993). However, clinicians report concerns with understanding how to implement each assessment tool, misinterpreting results, the relevance of the information (e.g., many are outdated or are not inclusive of LGBTQIA+ identities), and the overall comprehensiveness of the tools available (Thompson et al. 2016a). Thus, many create their own surveys of sexual health knowledge related to the content provided in their own sexual education programs. Although it is ideal to utilize assessment tools that have been tested with individuals with IDD, if clinicians do not have access to these tools or they do not fit all their needs, it is still preferred to evaluate outcomes and report changes before and after sexual education is provided.

## 21.3 Sexual Self-Advocacy

People with disabilities often experience limited opportunities to enact their basic sexual and reproductive health rights due to societal attitudes and practices (Friedman et al. 2014; UN 2006). Specifically, the common misconception that people with IDD lack the ability or capacity to consent to sexual relationships and that they are either hypersexual, deviant, or childlike and in need of protection, often leads to harmful practices, such as withholding sexual health education, limiting opportunities for intimate partnerships, and even forced sterilization (Friedman et al. 2014; Kempton and Kahn 1991; Tilley et al. 2012). Individuals with IDD have indicated a desire for more control over these decisions and additional support to form and maintain intimate relationships (Kelly et al. 2009; Friedman et al. 2014; Schmidt et al. 2022). When Friedman and colleagues (2014) interviewed individuals with IDD, they found that sexual self-advocacy includes a central component of understanding one's choices and having the ability to make one's own choices.

To feel empowered to make their own choices, individuals with IDD reported two areas in which they need support. First, they need support to communicate their choices. This includes support when using one's voice to stand up for their own choices and rights or others' choices and rights, getting information they need to make educated decisions, and developing communication skills to support healthy relationships. Participants suggested that all these skills should be developed interdependently, or with the support and collaboration of professionals and families (Friedman et al. 2014). This emphasizes the importance of comprehensive, inclusive sexual health education provided by multiple people, using a shared decision-making model. Further, it relates to the recommendation to move beyond traditional didactic education and to utilize modeling, role playing, and digital interventions to help support the generalization of skills needed for effective communication.

The second theme, respect, includes knowing and respecting oneself, others, and their rights. One needs sexual health education to understand one's own rights and those of others (Friedman et al. 2014). Additional reflection is needed for individuals to understand their own wants, needs, desires, and identity (Ibid). This is something professionals and families can also incorporate into sexual health education. Specifically, exposing individuals to sexual health facts and cultural differences around sexual practices in nonjudgmental, unbiased environments can help them learn more about others and make decisions for themselves. Professionals and families can also encourage reflection, so that individuals have a better understanding of their own wants, needs, and desires. Friedman and colleagues (2014) emphasize the importance of expanding access to sexual health education and information, educating professionals about the rights of individuals with IDD, increasing access to counseling, and developing opportunities for sexual expression, which includes opportunities to spend time alone with individuals or express themselves sexually. Current unpublished research by the author suggests that exposure to information, representation of people with different sexual orientations and genders, and providing opportunities for sexual expression can help autistic people understand their own identities.[2] Educators should also reiterate the importance of respecting others' rights and choices. This can be modeled for individuals with IDD by demonstrating acceptance in various settings.

## 21.4 Conclusion

There is a strong need to provide accessible, inclusive, and comprehensive sexual health education to individuals with IDD. This education can be provided by a number of people, including, but not limited to, parents and caregivers, peers, educators, and health providers. Strategies to support education include Universal Design for Learning, videos, visuals, modeling, role playing, and digital interventions. Access to appropriate sexual health education is a right protected under the CRPD and has

[2] Schmidt et al. (unpublished).

the potential to improve sexual knowledge, attitudes, skills, and safety, as well as to support sexual autonomy and improve sexual self-advocacy.

## References


Alhusen, Jeanne L., Tina Bloom, Jacqueline Anderson, and Rosemary B. Hughes. 2020. Intimate partner violence, reproductive coercion, and unintended pregnancy in women with disabilities. *Disability and Health Journal* 13: 100849. https://doi.org/10.1016/j.dhjo.2019.100849. Accessed 9 April 2024.

Armin-Arsala, Temor, Elizabeth K. Schmidt, and Joseph Christian Ungco. 2023. Sexual and intimate occupations across the lifespan. In *Sexuality and intimacy: An occupational therapy approach*, ed. Kathryn Ellis and Joseph Christian Ungco, 93–122. AOTA Press. ISBN: 978-1-56900-608-5.

Ballan, Michelle S. 2012. Parental perspectives of communication about sexuality in families of children with Autism spectrum disorders. *Journal of Autism and Developmental Disorders* 42 (5): 2964–2970.

Barnard-Brak, Lucy, Marcelo Schmidt, Steven Chesnut, Tianlan Wei, and David Richman. 2014. Predictors of access to sex education for children with intellectual disabilities in public schools. *Intellectual and Developmental Disabilities* 52 (2): 85–97. https://doi.org/10.1352/1934-9556-52.2.85.

Brown, Michael, Edward McCann, Maria Truesdale, Mark Linden, and Lynne Marsh. 2020. The design, content and delivery of relationship and sexuality education programmes for people with intellectual disabilities: A systematic review of the international evidence. *International Journal of Environmental Research and Public Health* 17: 7568. https://doi.org/10.3390/ijerph17207568.

Bruener, Cora C., and Gerri Mattson. 2016. Sexuality education for children and adolescents. *Pediatrics* 138 (2): e20161348. https://doi.org/10.1542/peds.2016-1348.

Bullard, David G., and Douglas H. Wallace. 1978. Peer educator-counselors in sexuality for the disabled. *Sexuality and Disability* 1 (2): 147–152.

Butler, Jenny, Daniel Leighton, and Jennifer Galea. 2003. *The assessment of sexual knowledge*. Victoria, Melbourne, Australia: Centre for Developmental Disability Health.

Carter, David. 2012. Comprehensive sex education for teens is more effective than abstinence. *American Journal of Nursing* 112 (3): 15.

Cooney, Adeline, and Eamon Darcy. 2020. 'It was fun': Exploring the pedagogical value of collaborative educational games. *Journal of University Teaching & Learning Practice* 17 (3). https://doi.org/10.53761/1.17.3.4. Accessed 9 April 2024.

DeSmet, Ann, Ross Shegog, Dimitri Van Ryckeghem, Geert Crombez, and Ilse De Bourdeaudhuj. 2015. A systematic review and meta-analysis of interventions for sexual health promotion involving serious digital games. *Games for Health Journal: Research, Development, and Clinical Applications* 4 (2): 78–90. https://doi.org/10.1089/g4h.2014.0110.

Dinwoodie, Robert, Beth Greenhill, and Alex Cookson. 2020. "Them two things are what collide together": Understanding the sexual identity experiences of lesbian, gay, bisexual, and trans people labelled with intellectual disability. *Journal of Applied Research in Intellectual Disabilities* 33: 3–16. https://doi.org/10.1111/jar.12252.

Epstein, Rebecca, Jamilia Blake, and Thalia Gonzalez. 2017. *Girlhood interrupted: The erasure of Black girls' childhood.* Georgetown Center on Poverty and Inequality. Available at SSRN 3000695. https://www.blendedandblack.com/wp-content/uploads/2017/08/girlhood-interrupted.pdf. Accessed 9 April 2024.

Firpo-Triplett, Regina, Tamara Kuhn, Lane Edwards, and Diana Dull Akers. 2023. Video-based microskills training for educators. In *Sex education research: A look between the sheets*, ed.

William J. Taverner, 260–272. New York, London: Routledge, Taylor and Francis Group. ISBN: 978-1-032-01205-6.

Frank, Katherine, and Linda Sandman. 2021. Parents as primary sexuality educators for adolescents and adults with Down syndrome: A mixed methods examination of the Home B.A.S.E. for Intellectual Disabilities workshop. *American Journal of Sexuality Education* 16 (3): 283–302.

Frawley, Patsie, and Christine Bigby. 2014. "I'm in their shoes": Experiences of peer educators in sexuality and relationship education. *Journal of Intellectual and Developmental Disability* 39 (2): 167–176. https://doi.org/10.3109/13668250.2014.890701.

Frawley, Patsie, and Nathan J. Wilson. 2016. Young people with intellectual disability talking about sexuality education and information. *Sexuality and Disability* 34 (4): 469–484. https://doi.org/10.1007/s11195-016-9460-x.

Friedman, Carli, Catherine K. Arnold, Aleksa L. Owen, and Linda Sandman. 2014. "Remember our voices are our tools:" Sexual self-advocacy as defined by people with intellectual and developmental disabilities. *Sexuality and Disability* 32: 515–532. https://doi.org/10.1007/s11195-014-9377-1.

Gil-Llario, Maria Dolores, Vicente Morell-Mengual, Irene Díaz-Rodríguez, and Rafael Ballester-Arnal. 2019. Prevalence and sequelae of self-reported and other-reported sexual abuse in adults with intellectual disability. *Journal of Intellectual Disability Research* 63 (2): 138–148. https://doi.org/10.1111/jir.125555.

Goldfarb, Eva S., and Lisa D. Lieberman. 2021. Three decades of research: The case for comprehensive sex education. *Journal of Adolescent Health* 68 (1): 13–27. https://doi.org/10.1016/j.jadohealth.2020.07.036. Accessed 9 April 2024.

Gougeon, Nathalie A. 2009. Sexuality education for students with intellectual disabilities, a critical pedagogical approach: Outing the ignored curriculum. *Sex Education* 9 (3): 277–291. https://doi.org/10.1080/14681810903059094. Accessed 9 April 2024.

Griffiths, Dorothy, and Yona Lunsky. 2003. *The socio-sexual knowledge and attitude assessment tool instruction manual.* Wood Dale, IL: Stoelting Co.

Hellemans, Hans, Kathy Colson, Christine Verbraeken, Robert Vermeiren, and Dirk Deboutte. 2007. Sexual behavior in high-functioning male adolescents and young adults with Autism spectrum disorder. *Journal of Autism and Developmental Disorders* 37 (2): 260–290. https://doi.org/10.1007/s10803-006-0159-1. Accessed 9 April 2024.

Hellemans, Hans, Herbert Roeyers, Wouter Leplae, Tina Dewaele, and Dirk Deboutte. 2010. Sexual behavior in male adolescents and young adults with Autism spectrum disorder and borderline/mild mental retardation. *Sexuality and Disability* 28 (2): 93–104. https://doi.org/10.1007/s11195-009-9145-9. Accessed 9 April 2024.

Holmes, Laura G., and Michael B. Himle. 2014. Brief report: Parent-child sexuality communication and autism spectrum disorders. *Journal of Autism and Developmental Disorders* 44: 2964–2970. https://doi.org/10.1007/s10803-014-2146-2.

Horner-Johnson, Willi, Mekhala Dissanayake, Justine P. Wu, Aaron B. Caughey, and Blair G. Darney. 2020. Pregnancy intendedness by maternal disability status and type in the United States. *Perspectives on Sexual and Reproductive Health* 52 (1): 31–38. https://doi.org/10.1363/psrh.12130.

James, Michelle H., Joanne E. Porter, Sumitra Kattel, Valerie Prokopiv, and Peter Hopwood. 2022. Peer educators in the facilitation of sexuality and respectful relationship education for people with an intellectual disability: A scoping review and narrative synthesis. *Sexuality and Disability* 40 (3): 487–502. https://doi.org/10.1007/s1195-022-09740-4.

Jones, Lisa, Mark A. Bellis, Sara Wood, Karen Hughes, Ellie McCoy, Lindsay Eckley, Geoff Bates, Christopher Mikton, Tom Shakespeare, and Alana Officer. 2012. Prevalence and risk of violence against children with disabilities: A systematic review and meta-analysis of observational studies. *The Lancet* 380 (9845): 899–907. https://doi.org/10.1016/S0140-6736(12)60692-8. Accessed 9 April 2024.

Kelly, Grace, Helen Crowley, and Carol Hamilton. 2009. Rights, sexuality and relationships in Ireland: 'It'd be nice to be kind of trusted.' *British Journal of Learning Disabilities* 37: 308–315. https://doi.org/10.1111/j.1468-3156.2009.00587.x.

Kempton, Winifred, and Emily Kahn. 1991. Sexuality and people with intellectual disabilities: A historical perspective. *Sexuality and Disability* 9 (2): 93–111.

Lorenzo, Gonzalo, Asunción Lledó, Graciela Arráez-Vera, and Alejandro Lorenzo-Lledó. 2019. The application of immersive virtual reality for students with ASD: A review between 1990–2017. *Educational Information Technologies* 24: 127–151. https://doi.org/10.1007/s10639-018-9766-7. Accessed 9 April 2024.

Mandell, David S., Catharine C. Eleey, Julie A. Cederbaum, M. Elizabeth Noll, Katherine Hutchinson, Loretta S. Jemmott, and Michael B. Blank. 2008. Sexually transmitted infection among adolescents receiving special education services. *Journal of School Health* 78 (7): 382–388.

Marcell, Arik V., Gale R. Burstein, and AAP Committee on Adolescence. 2017. Sexual and reproductive health care services in the pediatric setting. *Pediatrics* 140 (5): e20172858.

McCabe, M.P. 1993. *Sexual knowledge, experience and needs scale for people with intellectual disability (SexKen-ID)*, 3rd ed. Melbourne, Australia: Psychology Research Centre, Deakin University.

Ott, Mary A., and John S. Santelli. 2007. Abstinence and abstinence-only education. *Current Opinion in Obstetrics and Gynecology* 19 (5): 446–452.

Pearson, Amy, Jon Rees, and Samantha Forster. 2022. "This was just how this friendship worked": Experiences of interpersonal victimization among autistic adults. *Autism in Adulthood* 4 (2): 141–150. https://doi.org/10.1089/aut.2021.0035.

Pownall, Jaycee D., Andrew Jahoda, and Richard P. Hastings. 2012. Sexuality and sex education of adolescents with intellectual disability: Mothers' attitudes, experiences, and support needs. *Intellectual and Developmental Disabilities* 50 (2): 140–154. https://doi.org/10.1352/1934-9556-50.2.140.

Rexha, F., J. Bauza de Garcia, and Elizabeth K. Schmidt. in preparation. *Contributing factors to mental health challenges for LGBTQIA+ autistic adults in the United States.*

Roden, Rosemary Claire, Elizabeth K. Schmidt, and C. Holland-Hall. 2020. Sexual health education for adolescents and young adults with intellectual and developmental disabilities: Recommendations for accessible sexual health information. *The Lancet Child & Adolescent Health* 4 (9). https://doi.org/10.1016/S2352-4642(20)30098-5.

Rooks-Ellis, Deborah L., and Nicole Achey. 2023. Sexuality education for parents of children with intellectual and developmental disabilities. In *Sex education research: A look between the sheets*, ed. William J. Taverner, 73–83. New York, London: Routledge, Taylor and Francis Group. ISBN: 978-1-032-01205-6.

Rooks-Ellis, Deborah L., Brooklin Jones, Ella Sulinski, Sarah Howorth, and Nicole Achey. 2020. The effectiveness of a brief sexuality education intervention for parents of children with intellectual and developmental disabilities. *American Journal of Sexuality Education* 15 (4): 444–464. https://doi.org/10.1080/15546128.2020.1800542.

Schaafsma, Dilana, Gerjo Kok, Joke M.T.. Stoffelen, and Leopold M.G.. Curfs. 2015. Identifying effective methods for teaching sex education to individuals with intellectual disabilities: A systematic review. *Journal of Sex Research* 52 (4): 412–432. https://doi.org/10.1080/00224499.2014.919373.

Schmidt, Elizabeth K., Christopher Brown, and Amy Darragh. 2020. Scoping review of sexual health education interventions for adolescents and young adults with intellectual or developmental disabilities. *Sexuality and Disability* 38: 439–453. https://doi.org/10.1007/s11195-019-09593-4. Accessed 9 April 2024.

Schmidt, Elizabeth K., Brittany N. Hand, Susan Havercamp, Carolyn Sommerich, Lindy Weaver, and Amy Darragh. 2021a. Sex education practices for people with intellectual and developmental disabilities: A qualitative study. *The American Journal of Occupational Therapy* 75 (3): 7503180060.

Schmidt, Elizabeth K., Natalie Robek, Megan Dougherty, Brittany N. Hand, Susan Havercamp, Carolyn Sommerich, Lindy Weaver, and Amy Darragh. 2021b. Recommendations to improve accessibility of sexuality education for individuals with intellectual or developmental disabilities: A qualitative study. *American Journal of Sexuality Education* 16 (1): 38–56. https://doi.org/10.1080/15546128.2020.1860177.

Schmidt, Elizabeth K., Megan Dougherty, Natalie Robek, Lindy Weaver, and Amy Darragh. 2022. Sexual experiences and perspectives of adolescents with intellectual and developmental disabilities. *Journal of Adolescent Research* 37 (2). https://doi.org/10.1177/07435584211028225. Accessed 9 April 2024.

Schmidt, Elizabeth K., S.M. Espinosa, L. Clausen, and R. Hickman. unpublished. *Identity development for LGBTQIA+ autistic adults in the United States.*

Schrier, Karen. 2019. Designing games for moral learning and knowledge building. *Games and Culture* 14 (4): 306–343. https://doi.org/10.1177/1555412017711514.

Sinclair, James, Deanne Unruh, Lauren Linstrom, and David Scanlon. 2015. Barriers to sexuality for individuals with intellectual and developmental disabilities: A literature review. *Education and Training in Autism and Developmental Disabilities* 50 (1): 3–16. https://www.jstor.org/stable/24827497. Accessed 9 April 2024.

Son, Juyeon, and Cindy Draws. 2023. A call for an expansive comprehensive sexual health education model. In *Sex education research: A look between the sheets,* ed. William J. Taverner, 129–139. New York, London: Routledge, Taylor and Francis Group. ISBN: 978-1-032-01205-6.

Talbot, Tiffany J., and Peter E. Langdon. 2006. A revised sexual knowledge assessment tool for people with intellectual disabilities: Is sexual knowledge related to sexual offending behaviour? *Journal of Intellectual Disability Research* 50 (7): 523–531. https://doi.org/10.1111/j.1365-2788.2006.00801.x.

Test, David W., Sharon Richter, Victoria Knight, and Fred Spooner. 2011. A comprehensive review and meta-analysis of the social stories literature. *Focus on Autism and Other Developmental Disabilities* 26 (1): 49–62. https://doi.org/10.1177/1088357609351573. Accessed 9 April 2024.

Thompson, Vanessa R., Roger J. Stancliffe, Nathan J. Wilson, and Alex Broom. 2016a. The content, usefulness and usability of sexual knowledge assessment tools for people with intellectual disability. *Sexuality and Disability* 34: 495–512. https://doi.org/10.1007/s11195-016-9458-4.

Thompson, Vanessa R., Roger J. Stancliffe, Alex Broom, and Nathan J. Wilson. 2016b. Clinicians' use of sexual knowledge assessment tools for people with intellectual disability. *Journal of Intellectual and Developmental Disability* 41 (3): 243–254. https://doi.org/10.3109/13668250.2016.1164303.

Tilley, Elizabeth, Jan Walmsley, Sarah Earle, and Dorothy Atkinson. 2012. 'The silence is roaring': Sterilization, reproductive rights and women with intellectual disabilities. *Disability and Society* 27 (3): 413–426. https://doi.org/10.1080/09687599.2012.654991.

Tissot, Catherine. 2009. Establishing a sexual identity. Case studies of learners with autism and learning difficulties. *Autism: The International Journal of Research and Practice* 13 (6):551–566. https://doi.org/10.1177/1362361309338183. Accessed 9 April 2024.

UN (United Nations General Assembly). 2006. Convention on the rights of persons with disabilities. In *Treaty series*, vol. 2515, 3, G.A. Res. 61/106, 76th plen. Mtg., U.N. Doc A/Res/61/106. https://social.desa.un.org/issues/disability/crpd/convention-on-the-rights-of-persons-with-disabilities-crpd. Accessed 9 April 2024.

Walker-Hirsch, Leslie, and Marklyn P. Champagne. 1991. The circles concept: Social competence in special education. *Educational Leadership* 49 (1): 65–67. http://www.ascd.org/ASCD/pdf/journals/ed_lead/el_199109_walker-hirsch.pdf. Accessed 9 April 2024.

Wen Lan, Chiao, Alexandra F. Lightfoot, David Gere, Arianna Taboada, and Norweeta G. Milburn. 2023. Digital sexual health interventions for young people. In *Sex education research: A look between the sheets*, ed. William J. Taverner, 251–259. New York, London: Routledge, Taylor and Francis Group. ISBN: 978-1-032-01205-6.

Wilkinson, Verity J., Kate Theodore, and Roman Raczka. 2014. 'As normal as possible': Sexual identity development in people with intellectual disabilities transitioning to adulthood. *Sexuality and Disability* 33 (1): 93–105. https://doi.org/10.1007/s11195-014-9356-6. Accessed 9 April 2024.

# Chapter 22
# The Intersection of Mothering, Intellectual Disabilities, and Culture

**Laura Pacheco**

**Abstract** Mothering with a disability is socially and culturally constructed; however, there is little research that has explored the experiences of mothers with intellectual disabilities from ethnocultural communities. This chapter examines the ways in which dominant and community culture shape the life story of Maria, a mother with an intellectual disability from an ethnocultural community in Canada that emerged from a larger narrative study. Implications for ethical and socially just praxis are discussed.

**Keywords** Mothers · Mothering · Intellectual disability · Culture · Narratives

## 22.1 Introduction

Mothering with an intellectual disability[1] has been a contested social identity for many decades across cultures around the world (Elkhateeb and Peter 2019; Pacheco and McConnell 2017). Within a cultural model of disability, disability is socially and culturally constructed by the hegemony (or dominant culture) to maintain unequal power relations and interactions (Daniels 2019). It is well documented, particularly within the Global North, that larger social structures and ideologies emerging from the "majority" culture have created and maintained a grand narrative of incompetence, often constructing these mothers as unfit due to their subordinate social position (Pacheco et al. 2022; Sigurjónsdóttir and Rice 2017). The feminist scholarship on motherhood has also been criticized for ostensibly ignoring the experiences of women with disabilities, as Elshout (1994, 100) argues: "Feminists who criticise the

[1] In this chapter, person-first language is utilized but recognize that people with disabilities have varying perspectives on affirming language (APA 2022). This language is consistent within the field of parents and parenting with an intellectual disability.

L. Pacheco (✉)
Memorial University, St. John's, Canada
e-mail: lpacheco@mun.ca



A. Bianchi and J. A. Vogt (eds.), *Intellectual Disabilities and Autism: Ethics and Practice*, The International Library of Bioethics 108,
https://doi.org/10.1007/978-3-031-61565-8_22

traditional sex roles of wife and mother are not sensitive to the fact that women with disabilities are taught from a very early age that they are not fit to be lover or wife, let alone mother."

Cultural norms and values are deeply entrenched in social institutions and communities. Gender roles, such as mothering and/or motherhood, and the guidelines to perform these roles are dictated by society and cultural communities (Malacrida 2009). In other words, cultural norms impact the way in which women are regarded and treated, particularly as it relates to their performance of "ideal mothering" within and across social worlds (Ibid.). The voices and experiences of mothers with intellectual disabilities from ethnocultural communities have been rendered invisible and/or dismissed within multiple sectors of society, resulting in their social exclusion and denial of citizenship rights. Disability scholarship has also been critiqued for lacking an intersectional lens that explores how interfacing identities such as disability, gender, race, and ethnicity create interwoven levels of oppression (Pacheco and McConnell 2017; Collins 2019). The International Association on the Scientific Study for Intellectual Disabilities (IASSIDD) Parenting Special Interest Research Group (IASSID 2008) has deemed this a significant gap in research knowledge. The purpose of this chapter is to bring forward the narratives of mothers with intellectual disabilities from different ethnocultural communities.

## 22.2 Background

### *22.2.1 Meanings of Motherhood for Women with Intellectual Disabilities*

There is a growing body of literature, primarily situated in the Global North, that foregrounds the voices of mothers with intellectual disabilities. This literature has increased our understanding of the meaning of motherhood for women with intellectual disabilities and their stigmatic experiences within child welfare and other contexts.

Mothering and motherhood are often sought-after social positions by women across the globe (Bilyk 2022). Many women with intellectual disabilities have the same desires to become mothers; however, many will face negative attitudes, discouragement, and sometimes coercion related to their reproductive autonomy (McConnell and Phelan 2022). These ideologies are rooted in ableist discourses and macro level culture that, in turn, create benchmarks for identifying who is "fit" or "unfit"; for example, to have children (Bilyk 2022; Campbell 2009). Furthermore, if these women follow through with their pregnancy, their competence to care for their child is often called into question by people in their social milieu (family, neighbors, and professionals) due to their intellectual disability (Booth and Booth 2000; Ignagni and Fudge Schormans 2016; Llewellyn 1997; McConnell and Llewellyn 2005; Mayes et al. 2008).

Many women with intellectual disabilities report feeling empowered by their mother and/or mothering identity (Franklin et al. 2022; Mayes et al. 2008; Pacheco and McConnell 2017; Roets et al. 2008), particularly as this identity is regarded positively within society. Becoming and being a mother appears to allow women with intellectual disabilities to feel a sense of belonging and to provide a transition into mainstream community and adulthood; an experience that might not have been possible without having their own child (Mayes et al. 2006, 2008; More 2022).

### *22.2.2 Experiences Within the Child Welfare Context*

International research studies show that mothers with intellectual disabilities are over-represented in child welfare systems and that stigmatized constructions of disabled mothering and gaps in service provision underpin the disparities within the health and social care system (McConnell et al. 2021; Sigurjónsdóttir and Rice 2017). The source of this prejudicial response to these parents is ableist ideologies (rooted in eugenics) that understand them to fall outside of the ideal "good", "proper", and "able" parent (Malacrida 2009; McConnell and Phelan 2022; McConnell and Llewellyn 2005).

The majority of research exploring the experiences of mothers with intellectual disabilities having child welfare involvement, revealed that they felt their disability was held against them, and consequently they were "fighting a losing battle" (Booth and Booth 2000). They described having been "caught in the child welfare web" and having had little agency in enacting change (Booth and Booth 1994; Gould and Dodd 2014; McConnell and Sigurjónsdóttir 2010). Within an intersectionality framework, it is widely accepted that where there is oppression, there is also resistance (Collins 2019). Interpretive scholarship has illustrated how these mothers have used strategies of resistance within different parts of their motherhood journey to combat the injustice they have experienced. This includes, for example, speaking out and up about the oppression they are experiencing within their social milieu and within the public eye, gathering competence-promoting allies, joining forces with disabled mothers, and becoming co-researchers (Franklin et al. 2022; Pacheco and McConnell 2017; Traustadóttir and Sigurjónsdóttir 2010).

Interpretative research has increasingly given voice to the experiences of mothers with intellectual disabilities; however, there is a paucity of research that explores the role of cultural diversity in their life stories. More specifically, there is insufficient exploration of how culture can impact the experiences of these mothers (IASSID 2008). Moreover, the lack of diversity in the literature reinforces hegemonic understandings of caregiving and support (Collins 2019). To begin to address this gap in knowledge, the author undertook a narrative study to further understand the experiences of mothers with intellectual disabilities from different cultural communities.

## 22.3 The Research Study

This chapter is based on the author's doctorate research to explore and document the life stories of mothers with intellectual disabilities from minority-culture communities in the Canadian provinces of Quebec and Alberta (Pacheco 2015). Critical self-reflexivity was embedded within the narrative research process where the author explored her intersecting identities as a cis-gender disabled woman from an ethnocultural community who was involved in the co-construction of the narratives within the study.

Eight mothers who self-identified as having an intellectual disability and a sense of belonging to an ethnocultural community, each participated in between three and seven interviews in their own homes (their preferred location). The interviews ceased when a rich and thick narrative was produced. After constructing the narratives, the author met with each mother, presented their narrative, and invited them to both review it and give feedback to ensure that it accurately represented their story.

The narrative research method employed was underpinned by intersectionality theory. Narrative research methods seek to understand the lives of individuals within a socio-cultural context as told through participant stories; they explore the meaning behind experiences and human action (Riessman and Quinney 2005). Intersectionality theory examines how social identities such as gender, disability, race, and ethnicity depend on, and mutually construct, one another to form interlocking identities that may result in levels of oppression (Collins 2019).

In order to understand how cultural context influences the everyday lives of mothers with intellectual disabilities, Maria's (a pseudonym) story will be presented. Maria identifies as a member of the Portuguese community.[2] The following is a snapshot of her much larger narrative; however, the storylines in this excerpt can help us further understand the experiences of mothers with intellectual disabilities from cultural communities within hegemonic societies. Maria's story does not represent the experiences or life stories of all mothers with intellectual disabilities from a Portuguese community or any other ethnocultural community; however, it explores the complex interplay of systemic forces (including culture) that shape lived experience and identity (Collins 2019).

## 22.4 Maria's Story

Maria was born to a farming family in the Portuguese countryside. She described a happy childhood and coming to Canada with her family when she was an adolescent.

Maria had always dreamed about getting married and having children; something that was also important to her family. She further described:

[2] In order to respect Maria's privacy, the province in which she lives will not be included in the description.

> Even though I was a bit slow and I had problems with learning and sometimes working, my family still wanted me to get married and have children. This is what women do. And my mom told me that by getting married; maybe I would learn to do [the] things properly [sic].

Her family introduced her to her first boyfriend, a young man close in age to Maria. They chaperoned them on all of their dates—with one exception. When Maria visited her boyfriend and his family in the United States, the family went to the beach leaving Maria and her boyfriend alone at his home. What ensued was one of the most difficult events in Maria's life. Maria described how her boyfriend dragged her to the basement and brutally raped her.

Maria saw the rape as a violation and stated: "No one deserves to be treated like an animal", and that "he was an animal for doing that to me". The rape resulted in a pregnancy. Her family wanted Maria to abort her baby because it "was not made out of love" and would bring shame to the family. Maria wanted to keep the child as she "had always dreamed about being a mother". However, unable to endure the pressure from her family, Maria had the abortion. Nonetheless, Maria married this man at the request of her family. She explained how they believed, because of the rape, no other man would marry her.

After a year, Maria, gathering her courage and strength, decided to end their marriage. She had known the abuse would continue and asked her sister for help. A few months after the divorce, Maria's family found a man twelve years older than her that they thought would be a good match and strongly encouraged their marriage. Maria married him and had two children. Maria described her children's births as "the best day of my life". When she became a mother, she felt like she belonged in her family and community for one of the first times in her life.

Maria reported how her children brought her the love that her husband did not. He was an alcoholic who spent most of his time outside the home and abused Maria and the children, both physically and emotionally. The abuse seemed to be worse when he complained that the house was not as it should be or when the children were seen as misbehaving.

Maria tried to protect her children from this abuse but did not always succeed. One night, after her husband had been out drinking for many hours, he came home and their young daughter was crying. Maria tried to soothe her daughter. She described how her husband became really upset, screamed at her, and threw the baby such that her head hit the headboard. Their daughter suffered brain damage as a result.

Her family instructed her not to tell anybody what had actually happened; instead saying that the baby fell from Maria's arms. After years of abuse, Maria ended the marriage against her family's wishes:

> So my family was not happy about it about me leaving him because they said that marriage is like until death do you part you know what I mean.... You know the Portuguese way [sic].

Maria explained that she ended the marriage in order to provide herself and her children with a better life. After her divorce, Maria found herself alone. She asked her family for help, but did not receive any support from them. She interpreted this to be a consequence of not being a "good woman" according to the standards of her family and cultural community.

Once on her own, Maria was seen by child welfare as not being able to meet her children's needs. Her intellectual disability was emphasized as a risk factor by child welfare and her family. Shortly after her divorce, her children were placed in out-of-home care. Maria was overwhelmed with emotions; her family blamed her for her inability to raise her children, whose loss she lived with daily.

Maria discussed how she felt judged and abandoned by her family and community:

> They used to say to me, you are not a good mother, you have a mental handicap and had the children replaced [placed in foster care]. Now I am abandoned. Abandoned from [sic] my family and by my Portuguese culture.

She felt that her family didn't fully understand her disability and that they blamed her for the behaviors of others, including her husband's, and for having child welfare in her life. She described how she finally stood up to a family member and declared:

> And I had enough, you know, they would always talk about me and I just couldn't take it. So in the hospital in the hall I screamed, I am mentally retarded, I am fucking mentally retarded, what don't you understand about that! I was just tired of that. She didn't say anything else after that, she knew that I was fedded [sic] up.

Despite her hardships, Maria recounted how her children were what sustained her:

> They are the only thing that keeps me in this world. If it wasn't for them, I wouldn't always want to be there. I thank for [sic] God that I have them.

Most importantly, she declared, "No matter where they live, I am always their mother. I will always be their mother".

## 22.5 What Does Maria's Story Tell Us?

Echoed throughout Maria's story is the demonstration of strength and resilience in the face of interpersonal and systemic struggles. Though Maria's story is her own, it illustrates common threads that are woven across the mothers' narratives (Booth and Booth 1994). Along with the personal stories of the other mothers in the doctoral study, Maria's demonstrates how particular socio-cultural expectations can accrue from their intersecting social locations and how these mothers are both seen and treated based on their achievements and their failures to accomplish these socio-cultural norms (Chisale 2018; Pacheco and McConnell 2017).

### *22.5.1 Meeting Socio-cultural Expectations*

Maria's narrative illustrates how she received different messages about what it means to be a "good" and "bad" woman within and across the different socio-cultural contexts she found herself in: her own cultural community and the larger Canadian context. She was regarded as a good woman within three points in her life story. The first was when she got married and became a mother. Maria described many benefits of motherhood. Interestingly, becoming a mother seemed to erase the other parts of her (including her disability) that had been seen as unwanted by her family and cultural community. Becoming a mother increased Maria's self-worth and the value afforded to her by others and seemed to increase her sense of belonging and inclusion within her family and cultural community. This was also the case within the larger Canadian context as long as her intellectual disability remained hidden. The second point was when she took care of their household and her husband in a way that matched the cultural standards that were imposed on her. The third instance was when she kept up appearances and hid certain facets of family life: for instance, hiding her husband's abusive behaviours and the real cause of her daughter's "fall".

### 22.5.2 Failing to Meet Socio-cultural Expectations

Maria was constructed negatively when she did not uphold family and cultural norms and expectations. Her family and cultural community viewed disability as an undesirable social state and Maria's getting and staying married seemed to somewhat neutralize her disability. When she "acted" in ways that were deemed culturally inappropriate, Maria's disability was highlighted. A particular instance was when she divorced her husband and, as a result of a lack of supports (and untreated trauma), struggled to parent her children, leading to their apprehension by child welfare. At this point, her family, her cultural community, and Canadian child welfare services attributed her failures to her disability.

Maria was also seen as a "bad" woman when she opposed her family and her cultural community. This happened when she questioned their authority; for example, when her family instructed her to abort her first baby and when she got a divorce. Maria also described being caught between cultural, familial, and personal ideals. When Maria chose to do what she felt was best for herself and her children, she was regarded as a "bad woman", as she was not meeting the cultural expectations set out for her by her family and cultural community. Consequently, she experienced isolation from both and often felt as though she didn't belong in any of her social worlds.

### 22.5.3 Resistance and Agency

Though the mothers in this study experienced many forms of oppression (including control, abuse, divorce, poverty, and loss of children to the child welfare system) they were, at times, able to resist and fight against some of the unequal power relations in their lives. Maria did this in her conceptualization of the rape as an act of violence;

a violation of her personhood and her rights. She refused to internalize the perpetrator's actions and words and instead, used negative language to describe him and his behaviours. Maria questioned her family's judgment in relation to the abortion and their instruction that she marry her rapist. She stood up to a family member and used the words that once stigmatized her (e.g., mentally retarded) to reclaim herself and regain power. She opposed her family's and cultural community's expectations by divorcing her abusive husband, hoping to provide her children with a better life. She also resisted some of the negative judgments around her by asserting her identity as a mother in the face of her children's apprehension. By doing this, Maria was holding onto an identity that she had made her own.

## 22.6 Implications for Ethical and Socially Just Praxis with Women with Intellectual Disability from Ethnocultural Communities

Buttressing the research identified in Sect. 22.2 of this chapter, the narratives of these eight women contribute to a burgeoning collective narrative based on personal experience rather than on stereotypes and/or medicalized knowledge constructed by those considered as "experts" or "knowledge bearers" (e.g., clinicians, researchers). This is especially important given the historical exclusion of the voices, and denial of rights, of women with intellectual disabilities (Johnson and Traustadóttir 2000; McConnell and Phelan 2022; Traustadóttir and Sigurjónsdóttir 2010).

Despite the recognition that mothers with intellectual disabilities are constructed as 'recipients of care' within the health and social service system, there is a lack of scholarship exploring how practitioners can engage in ethical and socially just praxis (Pacheco and McConnell 2017). Cameron (2015) identifies critical praxis as conceptualizing and combatting disabling social interactions within and across communities and health and social service systems. Inspired by Hailes et al.'s (2021) ethical guidelines for social justice within the field of psychology, we will next explore key principles for developing critical praxis by incorporating critical self-reflexivity, spaces for interrogating socially and ethically just practices, and social justice action.

### 22.6.1 *Critical Reflexivity*

The longstanding history of systemic abuse toward women with intellectual disabilities extends far beyond the Eugenics movement (McConnell and Phelan 2022). Systems and their agents have, and continue to have, power that if left unchecked, can oppress mothers with intellectual disabilities from ethnocultural communities.

Given this reality, critical self-reflection needs to be used within praxis to: interrogate the systems that reinforce marginalization, examine personal social identities, and investigate unequal power relations that emerge from the client-worker relationship (Hailes et al. 2021). Within these institutional structures, it is paramount to consider the diverse ways in which culture(s) (e.g., macro level culture, cultural/ethnic community, child welfare culture, disability services culture) define(s) and construct(s) disability, identity, caregiving, and community engagement and the potential conflict with how the families view themselves and their role(s) (Campos and Kim 2017). This is particularly important when we consider what is in the best interest of the child (Choate and Engstrom 2014).

Critical reflexivity involves 'turning the gaze in', where the practitioner examines their personal and professional identities and how these intersecting identities can influence praxis (Pacheco et al. 2022). As Baines argues (2011, 7), "While addressing where one is placed on a multi-level continuum of privilege and oppression, it is equally important to understand, critique and improve how one uses that privilege to challenge oppression in everyday life." As practitioners within health and social services and within the community, we need to deconstruct our social location(s), our role, and our practice setting to explore how they create unequal power relations. This is primordial given that our assessments and decisions hold a lot of weight in these families' lives (Aunos and Pacheco 2020; Norlin and Randell 2022). Decisions about what constitutes "good enough parenting" can impact where the child will end up living—either in or out of home care, with kin or with the mother (or parents) (Choate and Tortelli 2022). Workers' own biases and ideologies can underpin the work, perpetuating the cycle of marginalization. Critical reflexivity requires the practitioner to center their work on social justice principles; committing themselves to lifelong learning and internal interrogation.

Relational approaches to praxis acknowledge the diverse sources of knowledge and place emphasis on experiential knowledge held by marginalized communities (Ruch 2005). As Scully (2018, 110) contends, "Living with various kinds of impairment results in the accumulation of a wealth of quite distinctive knowledge." Narrative epistemologies can help foreground the perspectives of mothers with intellectual disabilities from ethnocultural communities within praxis. The development of personal narratives (in both research and practice) can be transformative on individual and structural levels (Jirek 2017). For mothers with an intellectual disability and intersecting social identities, narrative spaces could create opportunities for meaningful interpretation of self and their life story by examining power and resilience. They also help practitioners develop a deeper understanding of these mothers as a person, considering their different environments; as opposed to solely emphasizing either disability or their role as a child carer (Davies et al. 2007). Therefore, it is important for workers to create a safe space to allow for narratives to emerge and promote individual and community empowerment. This requires an open and transparent dialogue about what could be shared if there was a legal request for information (e.g., subpoena of professional testimony).

Ethical and socially just praxis also requires practitioners to engage in socially just action-oriented praxis (Hailes et al. 2021). The first step is building critical consciousness; that is, identifying the ways in which these mothers can be discriminated against (due to their intersecting social identities) within and across different systems. This involves shifting the focus from personal deficits to wider systemic issues and structures. Within this context, social justice action could include taking action with, or advocating for, the mother within the health and social service system (i.e., being identified as having parenting deficits solely attributed to their disability in the child welfare context). On the meso level, workers need to take action against unjust policies; for example, child welfare policies that identify parental intellectual disability as a criterion of risk within the initial assessment. Building meaningful alliances within the disability community is also essential in building epistemological inclusion. Given the tensions that emerge out of praxis, critical and collective interrogation should be undertaken to explore and unpack power and privilege. This could include the creation of critical dialogue spaces that challenge workers' power and privilege and can encompass constructive confrontation and exploration of unequal power relations (Collins 2019) using either formal discussions, within organizations that have ethics boards and/or committees, or informal peer collaborations.

Community-based participatory research can also create transformational possibilities (Collins 2019). Within the field of parents and parenting with intellectual disability, some studies include parents with intellectual disabilities as co-researchers (Strike and McConnell 2002; Franklin et al. 2022; Roets et al. 2008). Of the many themes in Maria's story and those of the other women, the theme of intimate partner violence especially deserves further investigation (McConnell and Phelan 2022). All of the mothers interviewed disclosed abuse, either within their family of origin or within their relationships with their significant others. The research shows that violence against women with disabilities, especially women with intellectual disability, is widespread and long standing (Horner-Johnson and Drum 2006; McConnell and Phelan 2022). Some studies have explored abuse or upbringing in the lives of these families as part of a larger research study (Booth and Booth 1994; Sigurjónsdóttir and Traustadóttir 2010) but few have focused on the abuse that mothers with intellectual disabilities often encounter in their daily lives (McConnell and Phelan 2022). There is a need for research exploring the experience of abuse from the women's vantage point (as co-researchers) within an intersectional perspective. Further, the ways that mothers with intellectual disabilities deal with (and resist) abuse in the face of great adversity should be explored, along with the ways that the larger socio-cultural context responds to the abuse and the impacts of such. These studies could provide increased understanding of survival narratives, and which multi-pronged strategies can be employed to create social change in the lives of women with intellectual disabilities.

## 22.7 Concluding Thoughts

The experiences of mothers with intellectual disabilities from ethnocultural communities were previously hidden, misunderstood, and/or confined to the private sphere (Pacheco and McConnell 2017). By providing their narratives, Maria and the other mothers in this study challenge dominant social and cultural discourses and "contribute to an alternative social understanding of disability" (Björnsdóttir and Traustadóttir 2010, 60).

## References

American Psychological Association (APA). 2022. Disability. In *APA style*. https://apastyle.apa.org/style-grammar-guidelines/bias-free-language/disability. Accessed 9 August 2023.

Aunos, Marjorie, and Laura Pacheco. 2020. Able or unable: How do professionals determine the parenting capacity of mothers with intellectual disabilities. *Journal of Public Child Welfare* 15 (3): 357–383. https://doi.org/10.1080/15548732.2020.1729923.

Baines, Donna. 2011. *Doing anti-oppressive practice: Social justice social work*. Halifax, NS: Fernwood Publishing.

Bilyk, Dallas. 2022. 'Good' mother versus 'Bad' mother: How societal conceptions of motherhood in Canada and Ukraine impact a mother's access to opportunities. *Crossing Borders: Student Reflections on Global Social Issues* 4 (November). https://doi.org/10.31542/cb.v4i1.2476.

Björnsdóttir, Kristín, and Rannveig Traustadóttir. 2010. Stuck in the land of disability? The intersection of learning difficulties, class, gender and religion. *Disability & Society* 25 (1): 49–62.

Booth, Tim, and Wendy Booth. 1994. *Parenting under pressure mothers and fathers with learning difficulties*. Buckingham UK: Open University Press.

Booth, Tim, and Wendy Booth. 2000. The uncelebrated parent: Stories of mothers with learning difficulties caught in the child protection net. *British Journal of Learning Disabilities* 34: 94–102.

Cameron, Colin G. 2015. Turning experience into theory: The affirmation model as a tool for critical praxis. *Social Work & Social Sciences Review* 17 (3): 108–121. https://doi.org/10.1921/swssr.v17i3.802.

Campbell, Fiona Kumari. 2009. *Contours of ableism: The production of disability and abledness*. London, UK: Palgrave Macmillan.

Campos, Belinda, and Heejung Kim. 2017. Incorporating the cultural diversity of family and close relationships into the study of health. *American Psychologist* 72: 543–554.

Chisale, Sinenhlanhla S. 2018. 'Disabled motherhood in an African community': Towards an African women theology of disability. *In Die Skriflig* 52 (1): a2375. https://doi.org/10.4102/ids.v52i1.2375.

Choate, Peter W, and Sandra Engstrom. 2014. The 'Good enough' parent: Implications for child protection. *Child Care in Practice* 20 (4): 368–382. https://doi.org/10.1080/13575279.2014.915794.

Choate, Peter W, and Christina Tortorelli. 2022. Attachment theory: A barrier for indigenous children involved with child protection. *International Journal of Environmental Research and Public Health* 19 (14): 8754. https://doi.org/10.3390/ijerph19148754.

Collins, Patricia Hill. 2019. *Intersectionality as critical social theory*. Durham, NC: Duke University Press. https://doi.org/10.2307/j.ctv11hpkdj.

Daniels, Julia. 2019. Disabled mothering? Outlawed, overlooked and severely prohibited: Interrogating ableism in motherhood. *Social Inclusion* 7 (January): 114–123. https://doi.org/10.17645/si.v7i1.1551.

Davies, Linda, Julia Krane, Sara Collings, and Sharon Wexler. 2007. Developing mothering narratives in child protection practice. *Journal of Social Work Practice* 21 (March): 23–34. https://doi.org/10.1080/02650530601173516.

Elkhateeb, Itaf, and Dimity Peter. 2019. Negation of the right of women with disabilities in Palestine to Marry: Cultural considerations for disability. *Sexuality and Disability* 37 (4): 559–570. https://doi.org/10.1007/s11195-019-09588-1.

Elshout, Elly. 1994. Roundtable discussion: Women with disabilities: A challenge to feminist theology. *Journal of Feminist Studies in Religion* 10 (2): 99–104.

Franklin, Laura, Kate Theodore, Daniel Foulds, Mary Cooper, Lisa Mallaghan, Paul Wilshaw, Alison Colborne, Emilie Flower, Deborah Dickinson, and Joyce Nga Yu Lee. 2022. 'They don't think i can cope, because i have got a learning disability…': Experiences of stigma in the lives of parents with learning disabilities. *Journal of Applied Research in Intellectual Disabilities* 35 (4): 935–947. https://doi.org/10.1111/jar.12934.

Gould, Sheila, and Karen J. Dodd. 2014. 'Normal people can have a child but disability can't': The experiences of mothers with mild learning disabilities who have had their children removed. *British Journal of Learning Disabilities* 42 (1): 25–35. https://doi.org/10.1111/bld.12006.

Hailes, Helen P., Christopher J. Ceccolini, Ellen Gutowski, and Belle Liang. 2021. Ethical guidelines for social justice in psychology. *Professional Psychology: Research and Practice* 52 (1): 1–11. https://doi.org/10.1037/pro0000291.

Horner-Johnson, Willi, and Charles E. Drum. 2006. Prevalence of maltreatment of people with intellectual disabilities: A review of recently published research. *Mental Retardation and Developmental Disabilities Research Reviews* 12 (1): 57–69.

IASSID Special Interest Research Group on Parents and Parenting with Intellectual Disabilities. 2008. Parents labelled with intellectual disability: Position of the IASSID SIRG on parents and parenting with intellectual disabilities. *Journal of Applied Research in Intellectual Disabilities* 21 (4): 296–307.

Ignagni, Ester, and Ann Fudge Schormans. 2016. Reimagining parenting possibilities: The experiences of people labeled with intellectual and developmental disabilities. *International Journal of Birth and Parenting Education (IJBPE)* 3 (3): 9–12.

Jirek, Sarah L. 2017. Narrative reconstruction and post-traumatic growth among trauma survivors: The importance of narrative in social work research and practice. *Qualitative Social Work* 16 (2): 166–188. https://doi.org/10.1177/1473325016656046.

Johnson, Kelley, and Rannveig Traustadóttir, eds. 2000. *Women with intellectual disability: Finding a place in the world*. London: Jessica Kingsley Publishers.

Llewellyn, Gwynnyth. 1997. Parents with intellectual disability learning to parent: The role of experience and informal learning. *International Journal of Disability, Development and Education* 44 (3): 243–261. https://doi.org/10.1080/0156655970440306.

Malacrida, Claudia. 2009. Performing motherhood in a disablist world: Dilemmas of motherhood, femininity and disability. *International Journal of Qualitative Studies in Education* 22 (1): 99–117.

Mayes, Rachel, Gwynnyth Llewellyn, and David McConnell. 2006. Misconception: The experience of pregnancy for women with intellectual disabilities. *Scandinavian Journal of Disability Research* 8 (2–3): 120–131.

Mayes, Rachel, Gwynnyth Llewellyn, and David McConnell. 2008. Active negotiation: Mothers with intellectual disabilities creating their social support networks. *Journal of Applied Research in Intellectual Disabilities* 21 (4): 341–350.

McConnell, David, and Gwynnyth Llewellyn. 2005. Social inequality, 'The deviant parent' and child protection practice. *Australian Journal of Social Issues* 40 (4): 553–566.

McConnell, David, and Shanon K. Phelan. 2022. The devolution of eugenic practices: Sexual and reproductive health and oppression of people with intellectual disability. *Social Science & Medicine* 298 (April): 114877. https://doi.org/10.1016/j.socscimed.2022.114877.

McConnell, David, and Hanna Björg Sigurjónsdóttir. 2010. Caught in the child protection net. In *Parents with intellectual disability: Past, present and futures*, ed. Gwynnyth Llewellyn,

Rannveig Traustadóttir, David McConnell, and Hanna Björg Sigurjónsdóttir, 171–187. Malden, MA: Wiley-Blackwell. https://onlinelibrary.wiley.com/doi/10.1002/9780470660393.ch11.

McConnell, David, Marjorie Aunos, Laura Pacheco, and Maurice Feldman. 2021. Child maltreatment investigations in Canada: Main and moderating effects of primary caregiver cognitive impairment. *Child Maltreatment* 26 (1): 115–125. https://doi.org/10.1177/1077559520910806.

More, Rahel. 2023. Meanings of parenting and dis/ability for mothers and fathers with intellectual disabilities in the context of social work in Austria: Potentials of deconstruction for shifts in meaning. *Qualitative Social Work* 22 (5): 859–878. https://doi.org/10.1177/14733250221109765.

Norlin, Jerry, and Eva Randell. 2023. Social workers' perceptions of assessing the parental capacity of parents with intellectual disabilities in child protection investigations. *Journal of Intellectual Disabilities* 27 (4): 76–993. https://doi.org/10.1177/17446295221113717.

Pacheco, Laura. 2015. *Mothers with intellectual disability from ethnocultural communities in Canada: A narrative study*. Unpublished Ph.D. diss., University of Alberta. https://doi.org/10.7939/R3D21RS40.

Pacheco, Laura, and David McConnell. 2017. Love and resistance of mothers with intellectual disability from ethnocultural communities in Canada. *Journal of Applied Research in Intellectual Disabilities* 30 (3): 501–510. https://doi.org/10.1111/jar.12342.

Pacheco, Laura, Rahel More, Marjorie Aunos, and Rachelle Rose. 2022. Unfit and cast aside: Portrayals of mothering with intellectual disability in Québec court reports. *Critical Discourse Studies* 21 (3): 322-40.https://doi.org/10.1080/17405904.2022.2152063.

Riessman, Catherine Kohler, and Laura Quinney. 2005. Narrative in social work. *Qualitative Social Work* 4 (4): 391–412. https://doi.org/10.1177/1473325005058643.

Roets, Griet, Rosa Reinaart, and Geert Van Hove. 2008. Living between borderlands: Discovering a sense of nomadic subjectivity throughout Rosa's life story. *Journal of Gender Studies* 17 (2): 99–115. https://doi.org/10.1080/09589230802008865.

Ruch, Gillian. 2005. Relationship-based practice and reflective practice: Holistic approaches to contemporary child care social work. *Child and Family Social Work* 10 (2): 111–123.

Scully, Jackie Leach. 2018. From 'She would say that, wouldn't she?' to 'Does she take sugar?' Epistemic injustice and disability. *IJFAB: International Journal of Feminist Approaches to Bioethics* 11 (1): 106–124. https://doi.org/10.3138/ijfab.11.1.106.

Sigurjónsdóttir, Hanna Björg, and James G. Rice. 2017. 'Framed': Terminating the parenting rights of parents with intellectual disability in Iceland. *Journal of Applied Research in Intellectual Disabilities* 30 (3): 543–552. https://doi.org/10.1111/jar.12301.

Sigurjónsdóttir, Hanna Björg, and Rannveig Traustadóttir. 2010. Family within a family. In *Parents with intellectual disabilities: Past, present and futures*, ed. Gwynnyth Llewellyn, Rannveig Traustadóttir, David McConnell, and Hanna Björg Sigurjónsdóttir, 49–63. Malden, MA: Wiley-Blackwell. https://onlinelibrary.wiley.com/doi/10.1002/9780470660393.ch3.

Strike, Robert, and David McConnell. 2002. Look at me, i have something important to say. *Sexuality and Disability* 20: 53–63.

Traustadóttir, Rannveig, and Hanna Björg Sigurjónsdóttir. 2010. Parenting and resistance: Strategies in dealing with services and professionals. In *Parents with intellectual disabilities: Past, present and futures*, ed. Gwynnyth Llewellyn, Rannveig Traustadóttir, David McConnell, and Hanna Björg Sigurjónsdóttir, 107–118. Malden, MA: Wiley-Blackwell. https://doi.org/10.1002/9780470660393.ch7.

# Part VI
# Health Care

# Chapter 23
# Collaborating with People with Intellectual Disabilities and Autism in Healthcare Education

**Shay Dawson, M. Ariel Cascio, Meredith Wiles, and Neli Ragina**

**Abstract** This chapter provides an overview of the Healthcare Education Engaging Disability Studies Program (HEEDS). HEEDS is an innovative approach to a medical curriculum change that helps to diversify healthcare students' knowledge of disabled patient populations and fosters compassionate practice techniques, especially in working with autistic patients and those with intellectual disability. Many medical schools and other healthcare professional schools lack implementation of diversity education related to working with disabled patients, while at the same time, many healthcare disparities are present in modern day medical practices. The HEEDS program will be described in detail, including an overview of the work of people with intellectual disabilities as standardized patient actors to help train medical students on interaction skills, an advisory board specializing in disability, a written lived experience from an autistic person navigating the healthcare system, and an overview of a disability and medicine seminar training series. Feedback from patients with intellectual disabilities on the importance of the HEEDS program is highlighted.

**Keywords** Disability · Medical education · Patient lived experience

S. Dawson (✉) · M. Wiles · N. Ragina
Central Michigan University, Mount Pleasant, USA
e-mail: dawso2s@cmich.edu

M. Wiles
e-mail: wiles1mj@cmich.edu

N. Ragina
e-mail: ragin1n@cmich.edu

M. A. Cascio
Michigan State University, East Lansing, USA
e-mail: casciome@msu.edu



A. Bianchi and J. A. Vogt (eds.), *Intellectual Disabilities and Autism: Ethics and Practice*, The International Library of Bioethics 108,
https://doi.org/10.1007/978-3-031-61565-8_23

## 23.1 Introduction

This chapter addresses a key issue of justice in healthcare for people with Intellectual Disabilities (ID) and Autism. People with disabilities face substantial health inequities in the United States. The Americans with Disabilities Act created a major stride forward in prohibiting discrimination against any individual with a disability (Jackson et al. 2007). However, obstacles remain that limit equal access to various services, including in healthcare and medical education. This chapter describes the interprofessional healthcare education program called "Healthcare Education Engaging Disability Studies" (HEEDS) and discusses its approaches to increase justice for people with ID and Autism by (1) collaborating with people with ID and Autism throughout the program to create a patient centered and patient informed intervention, and (2) emphasizing and paying close attention to issues of representational justice, with the ultimate goal of reducing health disparities. In this chapter, we focus on the collaborative approach of our Community Advisory Board, the role of the autistic third author and former program participant in revising program content around Autism, and the role of standardized patients with intellectual disabilities in teaching and assessing first-year medical students.

The HEEDS program began with a needs assessment (Hamilton et al. 2022). Interviews and focus groups found that people with physical and developmental disabilities in our local communities had (1) a lack of patient-centeredness that impedes the quality of care, (2) inadequate communication that marginalizes patients within the clinical encounter, and (3) accessibility barriers that interfere with navigating the healthcare system. These findings reflect and nuance those in other studies. The World Health Organization identified common barriers that individuals with a disability may face when accessing healthcare, including negative stigmatization from healthcare providers (Scheer et al. 2003; World Health Organization 2023). Physicians have reported feeling inadequately trained in treating patients with disabilities, which may also contribute to negative stigmatization (McColl et al. 2008). Physicians are less likely to examine their patients with disabilities, or only do partial examinations, due to the difficulty of transferring patients onto an examining table, the amount of time needed for those patients to dress and undress, the possible need for dressing assistance, and lack of adequate equipment to assist with transportation (McColl et al. 2008). Many factors influence the care and treatment of patients with disabilities. Notably, physicians who treat more patients with ID provide more information to the patient and have better communication that allows the patient to be more autonomous in their treatment choice, when compared to physicians who treat fewer patients with ID (Werner et al. 2017). The HEEDS program aims to address these healthcare disparities and provides important opportunities for healthcare students to gain early exposure to key knowledge and skills in caring for patients with disabilities, as well as build meaningful connections with local disability communities.

## 23.2 Collaborating with People with Disabilities on a Community Advisory Board

The HEEDS program is guided by a Community Advisory Board which includes members who have personal or professional experiences with disability. This board consists of external academic advisors Lisa Iezzoni, MD (Harvard) and Raymond Curry, MD, FACP (University of Illinois Chicago College of Medicine); representatives from local disability organizations including Peckham, Inc. (vocational rehabilitation non-profit), Special Olympics, Michigan Career and Technical Institute, Mid-Michigan Industries, Michigan Department of Health and Human Services, Community Mental Health of Mid-Michigan, and Riley Children's Foundation in Indiana; and individuals with personal lived experience of disability. Many board members have both professional roles and lived experience.

The Advisory Board provides input on, reviews, and recommends appropriate standardized patient (SP) scripts, current issues in health care that should be a part of students' learning, and typical problems that people with disabilities encounter when interacting with medical professionals. The Advisory Board also works with the College of Medicine to recruit people with lived experience to be standardized patients (SPs) for the disability-specific patient encounters during the SP sessions.

We have taken several steps to increase the accessibility of the Advisory Board. To reduce barriers to attend and to recognize members' time and expertise, we offer stipends for committee attendance. We also maintain connections with Michigan Rehabilitation Services, our local government agency working in the area of disability and employment, to answer questions and ensure that payment of a stipend does not compromise anyone's disability benefits. To clearly communicate the purpose of the committee, we have prepared an Easy Read invitation (Fig. 23.1), following guidance from the Autistic Self Advocacy Network (Autistic Self Advocacy Network 2021a).

At the time of writing this chapter, no one who identifies as having an intellectual disability sits on the advisory board. However, several family members of people with ID do, as well as professionals who work primarily with people with ID. The advisory board works to ensure people with ID can contribute meaningfully to the project, as described below.

## 23.3 HEEDS Program Overview

The HEEDS program, which began in 2021, consists of several education activities across the first three years of medical school, guided by a community advisory board of disability scholars and experts with lived experience of disability. All students participate in an Initial Clinical Experience (ICE) project with community partners in their first year of medical school, with some students working with disability-focused community organizations. First and second-year students also have the option of attending a five-session disability seminar series lead by author Shay Dawson. All

**Fig. 23.1** Easy Read invitation describing the Advisory Board, select pages

medical students complete SP encounters at the end of first year, with standardized patients with lived experiences of physical disability and ID portraying scenarios of clinical encounters with a patient with those disabilities. Starting in 2023, all third-year medical students will undergo Objective Standardized Clinical Examination (OSCE) encounters to test their retention of skills learned during initial standardized patient encounters from their first year. Student OSCE performance will be analyzed to examine the influence of completing (or not completing) the optional seminars and ICE projects with disability-focused organizations.

### 23.3.1 Initial Clinical Experience (ICE) Volunteer Projects

The ICE projects allow students to interact directly with people who have a disability and to practice their clinical skills through hands-on experiences embedded within the medical school curriculum. The ICE learning objectives include: (a) increase students' knowledge of community organizations that can help people with disabilities, (b) improve awareness of social and access issues related to various types of disabilities, and (c) foster communication skills and compassion through peer-to-peer activities. For example, one group working with Special Olympics delivered cardiopulmonary resuscitation (CPR) training to athletes. The mission of the Special Olympics is to promote physical fitness and health through the competition of sports for athletes with ID. Training in CPR empowers these athletes to respond to emergencies they, like anyone else, might encounter in daily life.

### 23.3.2 Disability and Medicine Seminar Series

The Disability and Medicine Seminar Series aims to expose students to critical philosophies and approaches in working with disabled patients, including people with ID and Autistic people. The seminars consist of five one-hour modules related to: (1) problem areas in disability and medicine, (2) implicit biases in disability and medicine, (3) models of disability, (4) patients with physical disability, and (5) patients with ID and Autism. Each module consists of a videotaped interview with a patient or medical professional that has lived experience with a disability.

The seminar series introduces students to key topics that are important in working effectively with disabled patients in the healthcare setting. Although medical and other health professions students are trained in diagnostics and treatment, typically their curricula focus exclusively on the use of the medical model of disability. This model posits that patients/clients are outside of the normal bell curve and need to be "fixed" or "healed" to bring them as close to "normal" as possible. Although this may be helpful in providing a clear treatment goal within medicine and research, it also perpetuates a negative interaction approach with patients/clients by creating stigma and social marginalization. This approach is especially problematic for patients/clients who have different goals or do not wish to be cured, e.g., people who prefer Deaf culture or Autistic pride approaches. Students are encouraged to move beyond narrow, medical conceptions of disability as a disease to be cured or a problem to be fixed by offering alternative lenses to view disability through the social and biopsychosocial models of disability (Shakespeare 2006; World Health Organization

2023), both of which value human variation and place disability and health on a continuum in which all members of society are included.

Major discussions within the seminar series include: (1) disability interaction skills training, including person-first versus identity-first language as well as using empowerment, engagement, and empathy as specific approaches with patients/clients, (2) overview of diagnostic categories coupled with health disparities within each population, (3) psychosocial functioning aspects related to disability, stigma, and social marginalization, and (4) interviews with patients and parents with lived experiences that help highlight more didactic approaches. For example, one interview consists of a Physician Assistant who has an adult child who is Autistic with high support needs. This medical professional, with lived experience as a parent of an Autistic child, has become a go-to service provider for Autistic patients and offers insights into providing exceptional care, including specific approaches within the clinic. The seminar series session on ID also focuses on the nuanced nature of caring for patients who have ID without Autism, those who are Autistic with ID, and those who are Autistic without ID. The following section highlights some of the discussion centered around this complex topic of ID and Autism spectrum disorder.

### *23.3.3 Ethics and Representation: Patients with Intellectual Disabilities and/or Autism*

In the initial offering of the seminar series, there was a lack of breadth and depth regarding topics such as differences in identity-first and person-first language preferences by patients, similarities and differences in support needs for Autistic patients and patients with ID, as well as a lack of reflection on Applied Behavioral Analysis (ABA) therapy and organizations such as Autism Speaks. After presenting the seminar during the first year of the grant, the third author of this chapter, a graduate student at the time, presented the research team with valuable feedback. Based on this feedback, our program now includes discussions about these divisive issues in the disability world, many of which may be unknown to the medical community (Kras 2010; Sparrow 2016).

Our program directly discusses person-first language and identify-first language, to help inform students of possible differences in patient preference. Within the Autism community, many but not all prefer to be called Autistic, using identity-first language (Bottema-Beutel et al. 2021). Arguments for identity-first language include the idea that Autism is not a negative thing, rather a positive part of identity within a proud minority group. On the other hand, following the advocacy of People First and other self-advocacy movements in combatting the high stigma associated with intellectual disability, originating partially from medical terminology associated with IQ scales with words such as idiot, moron, imbecile, and retarded, a more person-first approach is used with ID (Carey 2009). Person-first language such as "John has an ID" or "Sally has Down syndrome" assists in countering negative stigma associated with historic language describing ID. Finally, a person-centered approach simply includes calling patients by what they prefer to be called, which takes some discussion and evaluation first for each patient and is taught as a best practice approach. Students are also taught that it is important to state that Autistic clients have high or low support needs rather than high or low functioning (Bottema-Beutel et al. 2021). This same approach may be used with patients that have ID.

During the ID seminar session, students learn that epidemiological studies have found that approximately one third of Autistic people have ID, which indicates that the majority of their Autistic patients will not have ID (Maenner et al. 2023). This statistic has changed so rapidly that this lecture has needed to be updated every six months to include the most up-to-date data. Students also learn about the nuances of applying a pan-disability rights lens related to their future Autistic patients, as some may identify with the disability rights movement (Autistic Self Advocacy Network 2021b) and others may not (Kenny et al. 2016; Lester et al. 2014). Students learn that many prominent approaches to Autism are not well respected in the Autistic community. For example, many in the Autistic community do not support the use of ABA therapy, as it is seen as attempting to change a core part of their positive identity using abusive practices (Sparrow 2016). Furthermore, for similar reasons, Autism Speaks is seen as a negative non-profit as the historical approach for this organization is deeply rooted in a tragedy model of Autism as well as the medical model of disability, i.e., that Autism is something that is bad and needs to be changed to be more like the societal norm (Kras 2010). On the other hand, the Autistic Self Advocacy Network (ASAN) may be supported by some Autistic patients. Discussing these divisive topics prepares the student for future Autistic patients who may or may not hold the same views as the medical model most commonly taught, implicitly or explicitly, in healthcare professions education.

Autistic patients and people with ID do share some common needs, e.g., social skill support, educational support, daily structure support in overcoming implicit and explicit biases, financial resources for medical care, and communication supports. Some differences exist as well. Autistic people may have more sensory needs in a physician's office than patients with ID only. Patients with ID may need more concrete language and concepts while Autistic patients with low support needs will likely have a strong capability to understand medical jargon and abstract concepts. It is therefore important to distinguish differences between Autistic patients and those with ID, while also recognizing the support the disability community can provide to all. The third author provided insights on some of these specific communication needs from the perspective of having a lived experience with Autism and navigating the healthcare system. The following section offers valuable lessons from this first-hand perspective.

### 23.3.4 Lived Experience Insights into Healthcare System Navigation

*As an Autistic person, I experience health care differently than my peers. I perceive/feel/experience my own body differently. My pain and emotions are felt differently than others. When this is combined with the daily reality of having a condition that causes me to communicate differently, I have a difficult time communicating with healthcare practitioners and getting the care I need. Even if I perceive my own symptoms correctly, I may not be able to communicate them to my doctor. I typically do not know when to go to the doctor and will end up going too often or not often enough. I often feel that I have too many things to tell the doctor and will leave some things out, in fear of being seen as annoying, a hypochondriac, or attempting to get medication I do not need.*

*I have always been someone who carries and uses a notebook to stay organized and remember things. I rely on this heavily at my doctor's office to be sure I remember each medication I have taken, family and surgical history, issues I am having, and anything else I need for my appointment. Some doctors respond very well to this organization, but not all. I have unfortunately overheard healthcare workers "warn" each other outside the exam room I am in that I have a notebook and may be a difficult patient. This perception of me is not only hurtful, but also makes me concerned that my questions and concerns are not going to be taken seriously. In the least, it is discouraging.*

*Communication is an issue for me outside of the exam room as well. Understanding when to arrive for an appointment can be very frustrating for me. As someone who works full time, I must manage my schedule carefully to attend doctor's appointments after or before work. When on the phone with a receptionist, we will schedule an appointment for 4:15, giving me 15 minutes to drive there after getting out of work at 4:00. After the appointment is scheduled, the receptionist will tell me to arrive 15, sometimes 30, minutes early to my appointment. We will then need to start all over to find a time that works for me. To me, a 4:15 appointment means I need to be there by 4:15. This is often a major frustration for me. Being able to check in electronically on MyHealth-type apps has reduced this issue.*

*Autism affects every part of my life and I require a lot of support, but externally I seem to live a very typical life. I am able to speak, advocate for myself for the most part, follow treatment plans, and I have a college degree. I can hold a job and have strong relationships. I have family, friends, and a partner who supports me enough to live a fairly normal life. For many autistic people, these things are not true. They struggle more with communication and sensory issues. They may not have the same abilities or education as me. Many autistic adults do not have the support they need to live normal lives.*

*The issues and barriers I speak of here are specific to myself and my struggles, which tend to be social and executive function struggles. Other autistic people may have more severe sensory, mobility, and communication struggles. Each autistic person will need different support to have a successful healthcare experience.*

*Treating autistic people as individuals is very important to give them correct support and ensure their healthcare experience is of quality.*

### 23.3.5 Hiring People with Intellectual Disability as Standardized Patients

Through the HEEDS program, we added two new SP encounters to the first-year medical student curriculum. In this chapter, we focus on the SP encounter featuring a patient with an intellectual disability, as follows.

The patient in this case is Jo [Smith]. Jo came to the doctor's office with their parent. Jo is in their late teens/early 20's, lives with their parent, and attends a day program during the week. Jo especially likes swimming. Jo is at the doctor's office today because their ear hurts. Jo's parent and day program won't let them go swimming until the ear pain is gone and the doctor says it is safe. The student doctors will need to do an ear exam. Jo does not like the ear exam because it hurts. The student doctors will likely recognize that Jo's ear pain is "swimmer's ear" and recommend ear drops. The student doctors should talk to Jo, not only to Jo's parent.

The learning objectives are:

- Describe the healthcare needs of people with ID.
- Demonstrate respect, sensitivity and direct communication and care to the semi-independent young adult rather than to the caregiver.
- Identify and address how the patient's symptoms affect lifestyle/activities.
- Identify differential diagnoses of ear pain in this case.

Due to the variations in Autistic patients with and without ID described above, as well as the fact that many people still confuse the two, the research team has discussed and been hesitant about having Autistic people without ID portray SPs with ID within the simulation lab for SP encounters. Considering the learning objectives, the team concluded that the key hiring goals were SPs with someone who can come with them as a support person, and preferably SPs who have experience going to doctors' appointments with a support person. Some potential SPs who identify as Autistic and expressed interest did not share this experience with Jo, and so we looked for those who did. Working with SPs with lived experience gives students accurate feedback, both in their specific scenario and in a large group debrief with actors and other volunteers after the sessions. It also allows students to build relationships with the actors themselves.

Attending to accessibility barriers and needs in recruitment of SPs with ID has been important to the success of this specific component of the HEEDS program. Initially, discussion with a potential SP takes place to answer any questions or concerns that may be present. This is followed by providing assistance in getting hired by the university human resource department, e.g., filling out an application or navigating the hiring process. Before the training takes place, a written script of the SP encounter is sent to the SP with an intellectual disability so that they and their caregiver have time to practice and begin to memorize the role. During training, many of the authors were present, along with the Simulation Lab staff, to introduce themselves face-to-face in preparation for their role in the sessions themselves, help answer questions, alleviate concerns, and to make sure material was presented in a concrete, rather than abstract, manner. Flexible scheduling was also utilized, as many SPs had busy schedules due to the Special Olympics State Games happening around the same time, or due to medical appointments. As needed, a one-on-one meeting was scheduled with each SP if group training was not conducive to their success.

### 23.3.6 Perspectives from Standardized Patients with ID

Across the two years of sessions, four SPs have portrayed Jo Smith. We asked the three SPs in 2023 to help us write this section of the chapter. Authors Dawson, Cascio, and Ragina met with SPs Nicole Huff, Jesse, Jared Pappas, and caregivers, Janis, Barb, and Jamee Pappas to discuss what they wanted to tell other people about the session. Nicole described the SP sessions as "a step forward" toward her career goals in theatre. She explained that the sessions were important so that the students can get good grades and become good doctors. Jared and Jamee Pappas verbalized and/or indicated the ear exam as the central part of their work as SPs. Jared Pappas wrote that he became an actor in the standardized patient lab because it was good to interact with the doctors, and it was important to show the doctors, i.e., positive ways to interact with patients that have ID. Nicole Huff stressed that the students were incredible: easy-going, and easy to talk to. Jared Pappas wrote that he likes to tell the students they did a good job. Aware of the importance of shared interests in building rapport, we asked the SPs if they found anything in common to talk about. Barb shared that Jesse and the students talked about a shared interest in swimming.

Many of our SPs receive high levels of support from their caregivers. Jamee Pappas reflected explicitly on her and Jared Pappas's typical experience in the doctor's office and goals for the SP sessions. Jamee Pappas explained:

> It was important for the students to be exposed to people who communicate differently. It was a huge struggle for me to sit there and not say anything because typically I'm his voice. But I thought, you know, the students aren't going to get anything from this if I'm his voice, so it was a good lesson for me, because I had to sit there and be quiet to the best of my ability and let him lead and let them figure out how to ask in a different way. Many of them were very good at responding and asking further questions to get what they needed.

Barb, a lifelong special education teacher and Special Olympics coach, shared a common conversation she has had with her friends in these settings. She explained that when they had doctors' visits, she would ask how it went, and her friends often said "I don't know. They were talking to mom, and I didn't know what they were talking about." Barb encouraged them to get clarification and reminded them they deserve to know what the doctors are talking about. These experiences motivated her to work with the SP sessions, to train doctors to have those conversations because, in her experience, many don't.

### 23.3.7 Navigating Healthcare Conversations with Caregivers

As Jamee Pappas highlights above, healthcare providers and students learning via SP sessions must navigate a three-way conversation between the healthcare team, the patient, and the patient's caregiver or support person. While the seminars and learning objectives do stress talking to the patient directly rather than the caregiver, caregivers of standardized patients, and advisory board members gave feedback that students and providers can over-apply this advice to the extent of ignoring the caregiver. In this section, the third author uses her lived experience to talk about the nuances of teaching about these interactions, which are inevitably going to be individualized, not only between patients, but also between visits and topics.

*I prefer to bring a support person with me to appointments, especially when I need help communicating my symptoms or advocating for myself. When I do this, I expect to be the main person the doctor communicates with. My support person is there to clarify things, communicate things I may forget about, and learn about my treatment plan to help me adhere to it. I do not want the doctor to speak to my support person as if they are in charge of my care. However, I do not want the doctor to discount my support person either. They are a part of my care and can help with my treatment plan—a resource, if you will. I want the presence of someone else to be an opportunity for the doctor to ask them questions about their perception of my symptoms and behavior, so I can have quality care and treatment. My caregiver/support person preferences are not the same as other autistic peoples'. Some autistic people may prefer or need the doctor to communicate directly with their caregiver. The caregiver may have a deeper understanding of their issue, or simply be able to communicate the issue more completely. This is especially true when the caregiver is family. However, the autistic patient should still be a part of the conversation. They should be given the opportunity to answer questions and communicate with the doctor, in whatever way works for them, and be given the opportunity to speak without their caregiver present. People with disabilities are often infantilized* (Ditchman et al. 2016). *Their sexuality and substance use may not be addressed. This is a major disservice to autistic people. Autistic patients need the opportunity to discuss their sexual activity, substance use, and anything else sensitive with their doctor without the presence of their caregiver. Patient-only communication moments with a doctor should be standard, with the option to opt-out by the patient. It should be highly recommended by the doctor at every appointment, and the caregiver should not be the one to opt-out on behalf of the patient.*

*This patient-only communication practice not only creates opportunity for conversations about the patient's sexual activity and substance use, but also opportunity for conversations about their emotional and mental health needs. Additionally, this is a protection measure for autistic people who may be abused or neglected by their caregivers. Everyone needs opportunities to speak up about abuses or neglect they are experiencing. In these situations, it is imperative that types of abuse or neglect are explained to the autistic patient, with clear examples. An autistic person may believe the abuse or neglect they experience is normal or experienced by everyone.*

*Caregiving is difficult. It takes a lot of energy and can be hard on the caregiver's health. Caregiver well-being should not be ignored, and caregivers should be given support and the opportunities to find support. However, caregiver support and frustrations should never be discussed in front of the patient. Doing so will take away from the experience of the patient. Additionally, this creates a uniquely dangerous situation for the patient, where the caregiver becomes the "victim" of the patient's disability. The caregiver may view the patient as a burden, and the patient may feel like a problem and burden to the people around them. This could cause the patient to withdraw from the support they need, or give the caregiver an "excuse" to give less or improper care to the patient. Patients should not be made to feel like a burden or problem for the sake of supporting the caregiver. Caregiver support should be discussed at a different time, privately. The struggles of the caregiver are unique and intense. It is exhausting yet rewarding, important, high stress, often constant work, and often unpaid or thankless. For caregivers to be good caregivers, they need support in the form of monetary help and resources, respite, education, and more. Caregiver support benefits the caregiver as well as the patient. Caregiver support should be considered as a connection or part of the patient's care, but it must be kept separate and communicated carefully.*

## 23.4 Conclusion

The HEEDS program aims to contribute to the development of just healthcare experiences and environments for people with ID and Autism. In order to achieve this aim, the program addresses potential causes of health disparities: provider knowledge, communication skills, and personal experiences, by incorporating them into an interprofessional healthcare education program (HEEDS).

Social constructs continue to impede how people view individuals with disabilities. Challenging these views and unconscious biases will help to promote diversity, equity, and belonging for these patients and ultimately improve the quality of health care they receive. This program takes seriously its commitment to justice and inclusivity by collaborating with people with ID and ensuring representation of complex issues of ID and Autism in the healthcare provider education that is patient-driven and informed.

Aspects of the HEEDS program will be replicable for most medical schools and other health professional schools. Local connections have been vital for the program's collaborative approach and for building community between healthcare education students and local disability communities. The HEEDS program provides a blueprint for how to integrate local connections into healthcare education. People with ID and

Autism teach the research team and students in various ways, leading to constant improvement and relationship building.

While healthcare disparities motivated the development of the HEEDS program, heeding disability studies helps all. The things that will help Autistic people and people with ID—barrier reduction, flexibility, clear communication, individualism—also help others. Many people struggle with illiteracy, poverty, and communicating their own health needs. Paving the way for Autistics and people with ID to have excellent healthcare experiences paves the way for everyone to have excellent healthcare experiences and improved quality of life. Everyone needs clear communication with their healthcare provider, easy navigation through the healthcare system, simple treatment plans, and caring physicians. Embracing the aforementioned as the focus on training the future healthcare professionals will be beneficial for everyone, not only for the Autistic people and people with ID.

## References


Autistic Self Advocacy Network. 2021a. One idea per line: A guide to making easy read resources. https://autisticadvocacy.org/resources/accessibility/easyread/. Accessed 30 April 2023.

Autistic Self Advocacy Network. 2021b. What we believe—Autistic self advocacy network. https://autisticadvocacy.org/about-asan/what-we-believe/. Accessed 30 April 2023.

Bottema-Beutel, Kristen, Steven K. Kapp, Jessica Nina Lester, Noah J. Sasson, and Brittany N. Hand. 2021. Avoiding ableist language: Suggestions for Autism researchers. *Autism in Adulthood* 3 (1): 18–29.

Carey, Allison C. 2009. *On the margins of citizenship: Intellectual disability and civil rights in twentieth-century America*. Philadelphia: Temple University Press.

Ditchman, Nicole, Kristin Kosyluk, Eun-Jeong Lee, and Nev Jones. 2016. How stigma affects the lives of people with intellectual disabilities: An overview. In *Intellectual disability and stigma: Stepping out from the margins*, 31–47. London: Palgrave Macmillan.

Hamilton, Natalie, Oluwatoyin Olumolade, Madelyn Aittama, et al. 2022. Access barriers to healthcare for people living with disabilities. *Journal of Public Health* 30 (5): 1–9.

Jackson, Patricia M., Peter Blanck, and James Schmeling. 2007. *The impact of the Americans with disabilities act: Assessing the progress toward achieving the goals of the ADA*, vol. 29, 42. Washington, DC: National Council on Disability.

Kenny, Lorcan, Caroline Hattersley, Bonnie Molins, Carole Buckley, Carol Povey, and Elizabeth Pellicano. 2016. Which terms should be used to describe Autism? Perspectives from the UK Autism community. *Autism* 20 (4): 442–462.

Kras, Joseph F. 2010. The "Ransom notes" affair: When the neurodiversity movement came of age. *Disability Studies Quarterly* 30 (1).

Lester, Jessica Nina, Khalid Karim, and Michelle O'Reilly. 2014. 'Autism itself actually isn't a disability': Negotiating a 'normal' versus' abnormal' autistic identity. *Communication & Medicine* 11 (2): 139.

Maenner, Matthew J., Zachary Warren, Ashley Robinson Williams, et al. 2023. Prevalence and characteristics of Autism spectrum disorder among children aged 8 years—Autism and developmental disabilities monitoring network, 11 Sites, United States, 2020. *MMWR. Surveillance Summaries* 72 (2): 1–14. https://doi.org/10.15585/mmwr.ss7202a1.

McColl, Mary Ann, Donna Forster, Sam ED. Shortt, Duncan Hunter, John Dorland, Marshall Godwin, and Walter Rosser. 2008. Physician experiences providing primary care to people with disabilities. *Healthcare Policy* 4 (1): e129.

Scheer, Jessica, Thilo Kroll, Melinda T. Neri, and Phillip Beatty. 2003. Access barriers for persons with disabilities: The consumer's perspective. *Journal of Disability Policy Studies* 13 (4): 221–230.

Shakespeare, Tom. 2006. The social model of disability. *The Disability Studies Reader* 2: 197–204.

Sparrow, Max. 2016. ABA. https://unstrangemind.com/aba/. Accessed 30 April 2023.

Werner, S., S. Yalon-Chamovitz, M. Tenne Rinde, and A.D. Heymann. 2017. Principles of effective communication with patients who have intellectual disability among primary care physicians. *Patient Education and Counseling* 100 (7): 1314–1321.

World Health Organization. 2023. Disability. https://www.who.int/news-room/fact-sheets/detail/disability-and-health. Accessed 9 April 2024.

# Chapter 24
# Reproductive Health and Pregnancy Outcomes

**Hilary K. Brown**

**Abstract** Reproductive health in women with intellectual and developmental disabilities (IDD) is an area characterized by neglect and prejudice. For much of the twentieth century, involuntary sterilization of women with IDD was common. The United Nations Convention on the Rights of Persons with Disabilities now protects the reproductive rights of people with IDD, but eugenic ideologies remain entrenched in the health system. These issues have resulted in pervasive reproductive health inequities for women with IDD related to reproductive cancer screening, contraception, preconception health, fertility, pregnancy outcomes, and menopause. Significant changes to reproductive health care are required to address these disparities, including creation of tailored resources that support reproductive health care access and informed decision making in women with IDD; development of healthcare provider curricula that address knowledge, attitudes, and skills required to deliver high-quality reproductive health care for these women; and integration of greater flexibility and collaboration in the health system.

**Keywords** Developmental disabilities · Intellectual disability · Health equity · Pregnant women · Reproductive health

## 24.1 Introduction

Intellectual and developmental disabilities (IDD) have a prevalence of one to three percent (Anderson et al. 2019), thus impacting a significant proportion of the population. Adults with IDD experience important disparities in health outcomes and health care access (Krahn and Fox 2013). However, reproductive health in women with IDD is an area characterized by "neglect, prejudice, disapproval, and misunderstanding" (Richards et al. 2008, 184).

H. K. Brown (✉)
University of Toronto Scarborough and University of Toronto, Toronto, ON, Canada
e-mail: hk.brown@utoronto.ca

 
A. Bianchi and J. A. Vogt (eds.), *Intellectual Disabilities and Autism: Ethics and Practice*, The International Library of Bioethics 108,
https://doi.org/10.1007/978-3-031-61565-8_24

For much of the twentieth century, women with IDD experienced high rates of involuntary sterilization. Sexual sterilization acts introduced in the 1920s allowed involuntary sterilization of people with IDD with the aim of preventing reproduction in individuals deemed to have "undesirable" genetic traits (Richards et al. 2008). In the United States, 65,000 Americans with IDD and psychiatric disabilities were sterilized between the 1920s and 1970s, and Belgium, Canada, Germany, Sweden, and numerous other nations saw similarly high numbers of sterilizations during this period (Roy et al. 2012). Most regions repealed their sterilization acts in the 1970s. Yet, there continue to be examples of "new eugenics", including genome editing and screening of fetuses for disabilities such as Down Syndrome, perpetuating ideas of disability as a negative characteristic (McConnell and Phelan 2022).

Major international developments have begun to impact the landscape of reproductive rights for people with IDD. In 2006, the United Nations Convention on the Rights of Persons with Disabilities (CRPD) became the first international treaty on disability, promoting the view that people with disabilities are active members of society who have the same human rights as non-disabled people and are capable of exercising those rights and making decisions of their own free will. Article 23 ensures the right of persons with disabilities to marry, found a family, decide on the number and spacing of their children, have access to family planning education, and retain their fertility (United Nations 2006), and Article 25 states people with disabilities have the right to quality and affordable reproductive health care. To date, one hundred and eighty-one states and the European Union have ratified the CRPD. This legally obligates them to ensure their laws, policies, and programs uphold the CRPD's values. The World Health Organization (WHO) and United Nations Population Fund (2009) developed a guidance note to support the CRPD's implementation in the context of reproductive care. Its recommendations include the need to develop policies and programs in partnership with people with disabilities, improve accessibility within existing services rather than creating separate or parallel services, and create laws, policies, and budgets that consider the needs of people with disabilities.

However, despite these milestones, eugenic ideologies remain entrenched in the health and social service systems (McConnell and Phelan 2022). For example, many pregnant women with IDD report experiencing pressure from healthcare providers to terminate their pregnancies (Björnsdóttir et al. 2017). Parents with IDD are also over-represented in child protection cases and are more likely than parents without IDD to lose custody, even in the absence of evidence of child abuse or neglect (Tøssebro et al. 2017).

These ongoing negative experiences occur against a backdrop of broader social and health disparities among people with IDD. Indeed, women with IDD disproportionately experience many of the known risk factors for adverse reproductive health outcomes. They are more likely than women without IDD to experience barriers to employment and education (Garrels and Sigstad 2021); poverty, unstable housing, and food insecurity (Brucker and Nord 2016; Roebuck 2021); and physical and mental illness (Krahn and Fox 2013). They also experience barriers to health care, including ableist provider attitudes and lack of knowledge of IDD, communication difficulties with providers, problems with transportation, and lack of coordination between health and social services (Balogh et al. 2016).

The repercussions of these ongoing issues are reflected in pervasive reproductive health inequities in women with IDD related to reproductive cancer screening, contraception, preconception health, fertility, pregnancy outcomes, and menopause. Significant changes to reproductive health care are needed to address these disparities. Such changes require a shift in thinking, from a "roaring silence" (Tilley et al. 2012, 413) on IDD and reproductive health, to strengths-based, person-centred care approaches that honour the reproductive rights of women with IDD.

## 24.2 Reproductive Health Disparities

### *24.2.1 Reproductive Cancer Screening*

Breast and cervical cancer are significant causes of morbidity and mortality in women (Sung et al. 2021). Reproductive cancer screening has therefore emerged as a key component of preventive care, shown to be effective in reducing the incidence of, and mortality associated with, these cancers (WHO 2006). In most regions, routine cervical cancer screening using a Pap test is recommended for average-risk women in their mid-twenties to mid-seventies (Fontham et al. 2020), while breast cancer screening using mammography is recommended for average-risk women in their mid-fifties to mid-seventies (Oeffinger et al. 2015). Yet, studies from diverse health systems have shown women with IDD are less likely than their peers to receive cervical and breast cancer screening (Horsbøl et al. 2023; Reidy et al. 2014).

Several factors have been explored as reasons for these disparities, including individual-level factors such as inadequate knowledge about screening and its importance, fear and embarrassment related to screening processes, and, in some cases, inability to provide consent for the procedures. Provider and system-level factors include negative provider assumptions about the need for reproductive cancer screening in women with IDD, and concerns about the invasiveness of the tests, which might lead to patient trauma and jeopardize the patient-provider relationship (Chan et al. 2022). For cervical cancer screening, one hypothesis has been that women with IDD are less likely to be offered Pap tests because of the assumption they are not sexually active and are therefore less likely to be exposed to the human papillomavirus, the major cause of cervical cancer. However, one study found that even in individuals with a history of pregnancy, women with IDD were less likely than their peers to receive a Pap test (Brown et al. 2015), suggesting disparities persist even in women who are clearly eligible. These disparities are concerning given similar rates of breast cancer, for example, in women with and without IDD, as well as growing evidence of later stage cancer diagnoses in women with IDD (Batchelor 2017).

### 24.2.2 Contraception and Family Planning

Among sexually active, reproductive-aged women, contraception use is an important indicator of reproductive health care access. Contraception prevents unintended pregnancy and provides an opportunity for individuals to benefit from preconception health promotion efforts to address risk factors for adverse pregnancy outcomes, such as smoking, nutrition deficits, and chronic illness (WHO 2012). Systematic reviews show women with IDD have lower contraceptive knowledge and use a narrower range of methods compared to their peers (Horner-Johnson et al. 2019a). Women with IDD are less likely to use highly effective long-acting reversible contraception such as intrauterine devices (Wu et al. 2018). On the other hand, they are more likely to experience sterilization (Brown et al. 2018a) and use the progestin shot (Brown et al. 2018a; Wu et al. 2018), which has side effects related to mood swings, weight gain, and bone density loss (Bigrigg et al. 1999).

Several hypotheses have been posited for these patterns. Adolescents with IDD are less likely than their peers to receive sexual education in school (Grove et al. 2018), and healthcare providers are less likely to initiate family planning conversations with women with IDD (Marquis et al. 2023). Often, contraception is provided to women with IDD to prevent pregnancy resulting from sexual abuse, rather than as a means of reproductive life planning, as it would be for other women (Greenwood and Wilkinson 2013). Contraception choices may also relate to health issues. For example, women with IDD have elevated rates of dysmenorrhea and cyclic mood and behaviour alterations, so menstrual suppression may be a means of addressing such concerns. Further, oral contraception may be contraindicated for women who experience other conditions such as epilepsy or cardiovascular disease, because of interference with medication effectiveness or concerns about risks of side effects (Quint 2016). Yet, research also suggests providers make assumptions about the capacity of women with IDD to reliably take oral contraception (Taouk et al. 2018). Higher rates of sterilization in women with IDD are also concerning. While laws now prohibit involuntary sterilization and outline consent procedures, studies have raised questions about whether women with IDD are given accessible information on, and understand the ramifications of, sterilization for their health and reproductive options (Rowlands and Amy 2019). Thus, there remain concerns about the autonomy of women with IDD and equitable contraception access.

### 24.2.3 Fertility

Historical eugenic practices have led to an assumption among healthcare providers that pregnancy is uncommon in women with IDD (Tilley et al. 2012). For many years, there was little information on pregnancy in women with IDD, with most studies being case series (Chamberlain et al. 1984). Recently, availability of large surveys and population-wide administrative datasets have enabled more comprehensive analyses.

Studies from different jurisdictions have reported conflicting findings on fertility rates in women with IDD, with some studies showing a lower rate compared to women without IDD (Brown et al. 2016; Horner-Johnson et al. 2016) and others reporting similar rates (Rubenstein et al. 2022). However, overall, these data provide important evidence that pregnancy is not uncommon in women with IDD.

Research also suggests important patterns in pregnancies that reflect the need for improved family planning services. For example, Horner-Johnson et al. (2020) showed women with cognitive disabilities were at elevated risk of reporting an unwanted or mistimed pregnancy. Women with IDD also have elevated rates of rapid repeat pregnancy within twelve months of a previous delivery (Brown et al. 2018b), with short inter-pregnancy intervals carrying risks of pregnancy complications (Wendt et al. 2012). Women with IDD are also more likely to have an induced abortion (Schuengel et al. 2022). These studies suggest the need for improved, patient-centered family planning services for these women.

### 24.2.4 Preconception Health

Preconception health describes the health of all reproductive-aged individuals (WHO 2012). In the general population, nearly fifty percent of all pregnancies are unintended (Sedgh et al. 2014), and prenatal care is often initiated well into or after the critical first trimester when many behavioural and environmental exposures that carry risks for adverse perinatal outcomes have already occurred (WHO 2012). As such, preconception health has emerged as an important concept for health promotion in all individuals who could become pregnant, regardless of their intentions (WHO 2012). While numerous studies have documented health disparities experienced by adults with IDD (Krahn and Fox 2013), fewer have examined health in reproductive-aged women with IDD. However, emerging evidence from population-based studies and large-scale surveys suggests reproductive-aged women with IDD have higher rates of chronic physical conditions such as diabetes mellitus, mental illness, substance use, teratogenic medication use, smoking, and physical inactivity compared to their peers (Mitra et al. 2016; Tarasoff et al. 2020a). These disparities have implications for the development of tailored preconception health promotion efforts for women with IDD that address these risks.

### 24.2.5 Pregnancy Outcomes and Care

Despite advances in obstetric and neonatal medicine, rates of maternal mortality and preterm birth are increasing or plateauing in industrialized countries rather than decreasing (Creanga et al. 2014; Richter et al. 2019). Identification of high-risk groups is thus important for developing interventions to reduce adverse outcomes. Meta-analyses suggest women with IDD have elevated risks of common pregnancy complications, such as gestational hypertension and caesarean section (Tarasoff et al. 2020b). Recent population-based studies also show they are at elevated risk of rare but serious outcomes, such as severe maternal morbidity and mortality (Brown et al. 2021; Mitra et al. 2021). Women with IDD are also more likely to access acute care in pregnancy and postpartum, with higher rates of emergency department visits and hospitalizations for obstetric, other medical, and psychiatric reasons (Brown et al. 2017). They also experience disparities in largely preventable newborn complications, with meta-analyses showing elevated risk of preterm birth and low birth weight (Tarasoff et al. 2020c), and smaller numbers of studies suggest increased risk of rare outcomes such as severe neonatal morbidity (Brown et al. 2022; Gleason et al. 2023). These studies suggest there is considerable room for improvement in obstetric and newborn care for women with IDD.

Indeed, studies have found gaps in pregnancy care for women with IDD. Population-based studies in Canada (Nishat et al. 2022) and the United States (Horner-Johnson et al. 2019b) suggest women with IDD are more likely to enter prenatal care later and receive fewer than the recommended number of visits. They are also more likely to be dissatisfied with their prenatal, delivery, and postpartum care (Redshaw et al. 2013). Moreover, qualitative studies from multiple jurisdictions provide insight into the barriers to perinatal care experienced by women with IDD. For example, they report negative provider attitudes about their pregnancies; lack of accessible information about pregnancy; communication barriers such as provider over-use of medical jargon; lack of involvement in decision making about labour and birth; fear of child protection involvement; and lack of coordination between health and social services (Höglund and Larsson 2013; Khanna et al. 2022; Saeed et al. 2022). These barriers have implications for maternal health: for example, the 2006–2008 Confidential Enquiries into Maternal Deaths in the United Kingdom reported how a breakdown in care for a woman with IDD with preeclampsia ultimately led to her death (Cantwell et al. 2011). These data suggest the need for tailored, supportive perinatal care for women with IDD.

### *24.2.6 Menopause and Ageing*

The menopausal transition is an important milestone for women. In addition to menopause-related symptoms, biological, psychosocial, and behavioural changes associated with menopause have implications for women's health (El Khoudary et al. 2019). Compared to other life stages, less research has focused on the menopausal transition in women with IDD. Some researchers have posited that this lack of data may be driven by a historically lower life expectancy in women with IDD, leading to assumptions about the need for research on ageing (Willis et al. 2010). However, given increases in life expectancy in people with IDD over the last several decades (Coppus 2013), there is a growing need for data to support development of tailored supports. Existing research suggests women with IDD have poor knowledge about menopause (Willis et al. 2010). Little research has examined the timing or complications of menopause in this population, but women with some types of IDD, such as Down Syndrome, experience menopause at earlier ages (Seltzer et al. 2008). Given the relation between early menopause and ovarian failure with health complications such as cardiovascular disease (El Khoudary et al. 2019), there is a need to explore risks of post-menopausal health disorders in women with IDD.

## 24.3 Recommendations

The CRPD supports the reproductive rights of women with IDD (United Nations 2006). However, strong evidence from multiple jurisdictions shows women with IDD experience significant disparities related to their reproductive outcomes and health care access across the lifespan. One can argue that continued failure to ameliorate the barriers that contribute to these disparities, in the face of the evidence presented, violates the principle of non-maleficence. This presents an urgent need, from a human rights perspective, for changes to the healthcare system to improve reproductive health care for women with IDD.

Improvements are required related to availability of tailored resources and services for women with IDD that support reproductive health care access, autonomy, and decision making. Most resources for women on reproductive cancer screening, contraception, preconception health, pregnancy, and menopause are not created with the specific learning and social needs of women with IDD in mind, and healthcare resources for adults with IDD do not often address reproductive healthcare needs beyond basic topics related to menstruation and contraception (Greenwood and Wilkinson 2013). Women with IDD may require plain language resources and resources in other formats that do not require reading (e.g., videos). To ensure equitable access to high-quality reproductive health care, there is a need for development of tailored resources.

There is a need for development of provider curricula that address the knowledge, skills, and attitudes required to deliver high-quality reproductive care to women with IDD. In the general population, healthcare providers play a pivotal role in sharing reproductive health information with patients to encourage reproductive cancer screening, promote healthy relationships, prevent unintended pregnancy, and support healthy pregnancies. The effectiveness of this patient-provider interaction is crucial to support the individual's autonomy. However, most providers receive little to no training on IDD (Amir et al. 2022). Providers also report an absence of evidence-based resources to support delivery of reproductive health care to this group. For example, there is a paucity of clinical guidelines on IDD and pregnancy, and broader guidelines on health care for adults with IDD tend to briefly mention reproductive health, if at all (Sullivan et al. 2018). The creation of disability competency training for providers in the United States holds promise (Havercamp et al. 2021), as does the growing focus on equity and diversity in medical education (Janz 2019). Such efforts must include training on IDD and reproductive health.

There is also a need for greater flexibility and collaboration within the healthcare system, given considerations related to communication, memory, and social determinants of health that may result in the need for longer healthcare appointments for women with IDD. However, health insurance policies are often built around visits of a specified length (e.g., ten minutes), making funding and time a barrier to care for this group (Thompson et al. 2014). Likewise, there may be times when women might benefit from more frequent visits than recommended for the general population. One example is the postpartum period, when women with IDD are at elevated risk of mortality even up to one year postpartum (Brown et al. 2021), well after the standard last postpartum visit, at around six weeks. These examples show the need for more flexibility in the health system through insurance policies that support more care for women with IDD when needed, and collaboration with other services that can address their needs.

## 24.4 Conclusion

Evidence from multiple diverse healthcare systems shows women with IDD experience important reproductive health disparities that are preventable through the provision of better care. To uphold the reproductive rights of women with IDD, there must be a will to make the necessary changes to our reproductive healthcare systems.

## References

Amir, Nili, Lauren D. Smith, Anne M. Valentine, Monika Mitra, Susan L. Parish, and Tiffany A. M Simas. 2022. Clinician perspectives on the need for training on caring for pregnant women with intellectual and developmental disabilities. *Disability and Health Journal* 15 (2): 101262. https://doi.org/10.1016/j.dhjo.2021.101262.

Anderson, Lynda Lahti, Sheryl A. Larson, Sarah Mapel Lentz, and Jennifer Hall-Lande. 2019. A systematic review of U.S. studies on the prevalence of intellectual or developmental disabilities since 2000. *Intellectual and Developmental Disabilities* 57 (5): 421–38. 10.1352/1934-9556-57.5.421.

Balogh, Robert, Carly A. McMorris, Yona Lunsky, Helene Ouellette-Kuntz, Laurie Bourne, Angela Colantonio, Daniela C. Gonçalves-Bradley. 2016. Organizing healthcare services for persons with an intellectual disability. *Cochrane Database of Systematic Reviews* 4 (4): CD007492. Chichester, UK: John Wiley & Sons.

Batchelor, Natasha C. 2017. *Breast cancer incidence and stage at diagnosis in Ontarians with and without intellectual disabilities.* Master's thesis, University of Ontario Institute of Technology, 2017. ProQuest 10758516. https://www.proquest.com/openview/f7a8d2661298a082b6f725ae7fffaf90/1?pq-origsite=gscholar&cbl=18750&diss=y.

Bigrigg, Alison, Marc Evans, Babatunde A. Gbolade, Julia Newton, Laura Pollard, Anne Szarewski, C Thomas, and Martyn Walling. 1999. Depo Provera. Position paper on clinical use, effectiveness and side effects. *The British Journal of Family Planning* 25 (2): 69–76.

Björnsdóttir, Kristín, Ástríður. Stefánsdóttir, and Guðrún. V. Stefánsdóttir. 2017. People with intellectual disabilities negotiate autonomy, gender and sexuality. *Sexuality and Disability* 35 (3): 295–311. https://doi.org/10.1007/s11195-017-9492-x.

Brown, Hilary K., Natasha Plourde, Helene Ouellette-Kuntz, Simone Vigod, and Virginie Cobigo. 2015. Brief report: Cervical cancer screening in women with intellectual and developmental disabilities who have had a pregnancy. *Journal of Intellectual Disability Research* 60 (1): 22–27. https://doi.org/10.1111/jir.12225.

Brown, Hilary K., Yona Lunsky, Andrew S. Wilton, Virginie Cobigo, and Simone N. Vigod. 2016. Pregnancy in women with intellectual and developmental disabilities. *Journal of Obstetrics and Gynaecology Canada* 38 (1): 9–16. https://doi.org/10.1016/j.jogc.2015.10.004.

Brown, Hilary K., Virginie Cobigo, Yona Lunsky, and Simone Vigod. 2017. Postpartum acute care utilization among women with intellectual and developmental disabilities. *Journal of Women's Health* 26 (1): 329–337. https://doi.org/10.1089/jwh.2016.5979.

Brown, Hilary K., Yolanda A. Kirkham, Yona Lunsky, Virginie Cobigo, and Simone N. Vigod. 2018a. Contraceptive provision to postpartum women with intellectual and developmental disabilities: A population-based cohort study. *Perspectives on Sexual and Reproductive Health* 50 (3): 93–99. https://doi.org/10.1363/psrh.12060.

Brown, Hilary K., Joel G. Ray, Simon Chen, Astrid Guttmann, Susan M. Havercamp, Susan Parish, Simone N. Vigod, Lesley A. Tarasoff, and Yona Lunsky. 2021. Association of preexisting disability with severe maternal morbidity or mortality in Ontario, Canada. *JAMA Network Open* 4 (2): e2034993. https://doi.org/10.1001/jamanetworkopen.2020.34993.

Brown, Hilary K., Simon Chen, Astrid Guttmann, Susan M. Havercamp, Susan L. Parish, Joel G. Ray, Simone N. Vigod, Lesley A. Tarasoff, and Yona Lunsky. 2022. Neonatal outcomes of mothers with a disability. *Pediatrics* 150 (3): e2021055318. https://doi.org/10.1542/peds.2021-055318.

Brown, Hilary K., Joel G. Ray, Ning Liu, Yona Lunsky, and Simone N. Vigod. 2018b. Rapid repeat pregnancy among women with intellectual and developmental disabilities: A population-based cohort study. *CMAJ: Canadian Medical Association Journal* 190 (32): E949–56. https://doi.org/10.1503/cmaj.170932.

Brucker, Debra L., and Derek Nord. 2016. Food insecurity among young adults with intellectual and developmental disabilities in the United States: Evidence from the national health interview survey. *American Journal on Intellectual and Developmental Disabilities* 121 (6): 520–532. https://doi.org/10.1352/1944-7558-121.6.520.

Cantwell, Roch, Thomas Clutton-Brock, Griselda Cooper, Andrew Dawson, James Drife, Debbie Garrod, Ann Harper et al. 2011. Saving Mothers' lives: reviewing maternal deaths to make motherhood safer: 2006–2008. The Eighth report of the confidential enquiries into maternal deaths in the United Kingdom. *BJOG : An International Journal of Obstetrics & Gynaecology* 118 (S1): 1–203. 10.1111/j.1471-0528.2010.02847.x.

Chamberlain, Allan, Joseph Rauh, Alice Passer, Marianne McGrath, and Robert Burket. 1984. Issues in fertility control for mentally retarded female adolescents: Sexual activity, sexual abuse, and contraception. *Pediatrics* 73 (4): 445–450.

Chan, Dorothy N. S., Bernard M. H. Law, Doreen W. H. Au, Winnie K. W. So, and Ning Fan. 2022. A systematic review of the barriers and facilitators influencing the cancer screening behaviour among people with intellectual disabilities. *Cancer Epidemiology* 76: 102084. https://doi.org/10.1016/j.canep.2021.102084.

Coppus, Antonia M. W. 2013. People with intellectual disability: What do we know about adulthood and life expectancy? *Developmental Disabilities Research Reviews* 18 (1): 6–16. https://doi.org/10.1002/ddrr.1123.

Creanga, Andreea A., Cynthia J. Berg, Jean Y. Ko, Sherry L. Farr, Van T. Tong, F. Carol Bruce, and William M. Callaghan. 2014. Maternal mortality and morbidity in the United States: Where are we now? *Journal of Women's Health* 23 (1): 3–9. https://doi.org/10.1089/jwh.2013.4617.

El Khoudary, Samar R., Gail Greendale, Sybil L. Crawford, Nancy E. Avis, Maria M. Brooks, Rebecca C. Thurston, Carrie Karvonen-Gutierrez, L. Elaine Waetjen, and Karen Matthews.

2019. The menopause transition and women's health at midlife: a progress report from the study of women's health across the nation (SWAN). *Menopause* 26 (10): 1213–27. https://doi.org/10.1097/GME.0000000000001424.

Fontham, Elizabeth T. H., Andrew M. D. Wolf, Timothy R. Church, Ruth Etzioni, Christopher R. Flowers, Abbe Herzig, Carmen E. Guerra, et al. 2020. Cervical cancer screening for individuals at average risk: 2020 guideline update from the American Cancer Society. *CA: A Cancer Journal for Clinicians* 70 (5): 321–46. https://doi.org/10.3322/caac.21628.

Garrels, Veerle, and Hanne Marie Høybråten. Sigstad. 2021. Employment for persons with intellectual disability in the Nordic countries: A scoping review. *Journal of Applied Research in Intellectual Disabilities* 34 (4): 993–1007. https://doi.org/10.1111/jar.12880.

Gleason, Jessica L., Jagteshwar Grewal, Zhen Chen, Alison N. Cernich, and Katherine L. Grantz. 2023. Risk of adverse neonatal outcomes among pregnant women with disabilities. *International Journal of Epidemiology* 52 (1): 203–213. https://doi.org/10.1093/ije/dyac183.

Greenwood, Nechama W., and Joanne Wilkinson. 2013. Sexual and reproductive health care for women with intellectual disabilities: A primary care perspective. *International Journal of Family Medicine*: 642472. https://doi.org/10.1155/2013/642472.

Grove, Linsey, Dianne Morrison-Beedy, Russel Kirby, and Janet Hess. 2018. The birds, bees, and special needs: Making evidence-based sex education accessible for adolescents with intellectual disabilities. *Sexuality and Disability* 36 (4): 313–329. https://doi.org/10.1007/s11195-018-9547-7.

Havercamp, Susan M., Wesley R. Barnhart, Ann C. Robinson, and Cara N. Whalen Smith. 2021. What should we teach about disability? National consensus on disability competencies for health care education. *Disability and Health Journal* 14 (2): 100989. https://doi.org/10.1016/j.dhjo.2020.100989.

Höglund, Berit, and Margareta Larsson. 2013. Struggling for motherhood with an intellectual disability—a qualitative study of women's experiences in Sweden. *Midwifery* 29 (6): 698–704. https://doi.org/10.1016/j.midw.2012.06.014.

Horner-Johnson, Willi, Blair G. Darney, Sheetal Kulkarni-Rajasekhara, Brian Quigley, and Aaron B. Caughey. 2016. Pregnancy among US women: Differences by presence, type, and complexity of disability. *American Journal of Obstetrics & Gynecology* 214 (4): 529.e1–9. https://doi.org/10.1016/j.ajog.2015.10.929.

Horner-Johnson, Willi, Frances M. Biel, Aaron B. Caughey, and Blair G. Darney. 2019a. Differences in prenatal care by presence and type of maternal disability. *American Journal of Preventive Medicine* 56 (3): 376–382. https://doi.org/10.1016/j.amepre.2018.10.021.

Horner-Johnson, Willi, Esther L. Moe, Ryan C. Stoner, Krystal A. Klein, Alison B. Edelman, Karen B. Eden, Elena M. Andresen, Aaron B. Caughey, and Jeanne-Marie. Guise. 2019b. Contraceptive knowledge and use among women with intellectual, physical, or sensory disabilities: A systematic review. *Disability and Health Journal* 12 (2): 139–154. https://doi.org/10.1016/j.dhjo.2018.11.006.

Horner-Johnson, Willi, Mekhala Dissanayake, Justine P. Wu, Aaron B. Caughey, and Blair G. Darney. 2020. Pregnancy intendedness by maternal disability status and type in the United States. *Perspectives on Sexual and Reproductive Health* 52 (1): 31–38. https://doi.org/10.1363/psrh.12130.

Horsbøl, Trine Allerslev, Susan Ishøy Michelsen, Tina Harmer Lassen, Knud Juel, Janne Bigaard, Christina Engel Hoei-Hansen, Ilse Vejborg, and Lau Caspar Thygesen. 2023. Breast cancer screening among women with intellectual disability in Denmark. *JAMA Network Open* 6 (1): e2248980. https://doi.org/10.1001/jamanetworkopen.2022.48980.

Janz, Heidi L. 2019. Ableism: The undiagnosed malady afflicting medicine. *Canadian Medical Association Journal* 191 (17): E478–E479. https://doi.org/10.1503/cmaj.180903.

Khanna, Aishwarya, Lauren D. Smith, Susan L. Parish, and Monika Mitra. 2022. Pregnancy recommendations from women with intellectual and developmental disabilities to their peers. *Disability and Health Journal* 15 (4): 101343. https://doi.org/10.1016/j.dhjo.2022.101343.

Krahn, Gloria L., and Michael H. Fox. 2013. Health disparities of adults with intellectual disabilities: What do we know? What do we do? *Journal of Applied Research in Intellectual Disabilities* 27 (5): 431–446. https://doi.org/10.1111/jar.12067.

Marquis, Sandra, Yona Lunsky, Kimberlyn M. McGrail, and Jennifer Baumbusch. 2023. Population-level use of gynecological health services by female youth with intellectual/developmental disabilities in British Columbia Canada. *Disability and Health Journal:* 101478. https://doi.org/10.1016/j.dhjo.2023.101478.

McConnell, David, and Shanon Phelan. 2022. The devolution of eugenic practices: Sexual and reproductive health and oppression of people with intellectual disability. *Social Science & Medicine* 298: 114877. https://doi.org/10.1016/j.socscimed.2022.114877.

Mitra, Monika, Karen M. Clements, Jianying Zhang, and Lauren D. Smith. 2016. Disparities in adverse preconception risk factors between women with and without disabilities. *Maternal and Child Health Journal* 20 (3): 507–515. https://doi.org/10.1007/s10995-015-1848-1.

Mitra, Monika, Ilhom Akobirshoev, Anne Valentine, Hilary K. Brown, and Tiffany A. Moore Simas. 2021. Severe maternal morbidity and maternal mortality in women with intellectual and developmental disabilities. *American Journal of Preventive Medicine* 61 (6): 872–81. https://doi.org/10.1016/j.amepre.2021.05.041.

Nishat, Fareha, Yona Lunsky, Lesley A. Tarasoff, and Hilary K. Brown. 2022. Prenatal care adequacy among women with disabilities: A population-based study. *American Journal of Preventive Medicine* 62 (1): 39–49. https://doi.org/10.1016/j.amepre.2021.05.037.

Oeffinger, Kevin C., Elizabeth T. H. Fontham, Ruth Etzioni, Abbe Herzig, James S. Michaelson, Ya-Chen Tina Shih, Louise C. Walter, et al. 2015. Breast cancer screening for women at average risk: 2015 guideline update from the American Cancer Society. *JAMA* 314 (15): 1599–1614. https://doi.org/10.1001/jama.2015.12783.

Quint, Elisabeth H. 2016. Adolescents with special needs: Clinical challenges in reproductive health care. *Journal of Pediatric and Adolescent Gynecology* 29 (1): 2–6. https://doi.org/10.1016/j.jpag.2015.05.003.

Redshaw, Maggie, Reem Malouf, Haiyan Gao, and Ron Gray. 2013. Women with disability: The experience of maternity care during pregnancy, labour and birth and the postnatal period. *BMC Pregnancy and Childbirth* 13 (1): 174. https://doi.org/10.1186/1471-2393-13-174.

Reidy, Mary, Suzanne Denieffe, and Sinéad Foran. 2014. Cancer screening in women with intellectual disabilities: An Irish perspective. *Journal of Intellectual Disabilities* 18 (1): 51–60. https://doi.org/10.1177/1744629514521067.

Richards, Deborah, Nancy Miodrag, Shelley L. Watson, Maurice Feldman, Marjorie Aunos, Diane Cox-Lindenbaum, et al. 2008. Sexuality and human rights of persons with intellectual disabilities. In *Challenges to the human rights of people with intellectual disabilities*, ed. Frances Owen and Dorothy Griffiths, 184–218. London: Jessica Kingsley Publishers.

Richter, Lindsay L., Joseph Ting, Giulia M. Muraca, Amélie Boutin, Qi Wen, Janet Lyons, Anne Synnes, and Sarka Lisonkova. 2019. Temporal trends in preterm birth, neonatal mortality, and neonatal morbidity following spontaneous and clinician-initiated delivery in Canada, 2009-2016. *Journal of Obstetrics and Gynaecology Canada* 41 (12): 1742–1751. https://doi.org/10.1016/j.jogc.2019.02.151.

Roebuck, Maryann. 2021. Housing for people with intellectual disabilities: A scoping review. *Journal on Developmental Disabilities* 26 (2): 1–25.

Rowlands, Sam, and Jean-Jacques Amy. 2019. Sterilization of those with intellectual disability: Evolution from non-consensual interventions to strict safeguards. *Journal of Intellectual Disabilities* 23 (2): 233–249. https://doi.org/10.1177/1744629517747162.

Roy, Ashwin, Ameeta Roy, and Meera Roy. 2012. The human rights of women with intellectual disability. *Journal of the Royal Society of Medicine* 105 (9): 384–389. https://doi.org/10.1258/jrsm.2012.110303.

Rubenstein, Eric, Deborah B. Ehrenthal, Jenna Nobles, David C. Mallinson, Lauren Bishop, Marina C. Jenkins, Hsiang-Hui. Kuo, and Maureen S. Durkin. 2022. Fertility rates in women with intellectual and developmental disabilities in Wisconsin Medicaid. *Disability and Health Journal* 15 (3): 101321. https://doi.org/10.1016/j.dhjo.2022.101321.

Saeed, Gul, Hilary K. Brown, Yona Lunsky, Kate Welsh, Laurie Proulx, Susan Havercamp, and Lesley A. Tarasoff. 2022. Barriers to and facilitators of effective communication in perinatal care: A qualitative study of the experiences of birthing people with sensory, intellectual, and/or developmental disabilities. *BMC Pregnancy and Childbirth* 22 (1): 364. https://doi.org/10.1186/s12884-022-04691-2.

Schuengel, Carlo, Maarten Cuypers, Lianne Bakkum, and Geraline L. Leusink. 2023. Reproductive health of women with intellectual disability: Antenatal care, pregnancies and outcomes in the Dutch population. *Journal of Intellectual Disability Research* 67 (12):1306-1316. https://doi.org/10.1111/jir.12982.

Sedgh, Gilda, Susheela Singh, and Rubina Hussain. 2014. Intended and unintended pregnancies worldwide in 2012 and recent trends. *Studies in Family Planning* 45 (3): 301–314. https://doi.org/10.1111/j.1728-4465.2014.00393.x.

Seltzer, Gary B., Nicole Schupf, and Hseuh-Sheng Wu. 2008. A Prospective study of menopause in women with Down's syndrome. *Journal of Intellectual Disability Research* 45: 1–7. 10.1046/j.1365-2788.2001.00286.x.

Sullivan, William F., Heidi Diepstra, John Heng, Shara Ally, Elspeth Bradley, Ian Casson, Brian Hennen, et al. 2018. Primary care of adults with intellectual and developmental disabilities: 2018 Canadian consensus guidelines. *Canadian Family Physician* 64 (4): 254–279.

Sung, Hyuna, Jacques Ferlay, Rebecca L. Siegel, Mathieu Laversanne, Isabelle Soerjomataram, Ahmedin Jemal, and Freddie Bray. 2021. Global cancer statistics 2020: GLOBOCAN estimates of incidence and mortality worldwide for 36 cancers in 185 countries. *CA: A Cancer Journal for Clinicians* 71 (3): 209–49. https://doi.org/10.3322/caac.21660.

Taouk, Laura H., Michael F. Fialkow, and Jay A. Schulkin. 2018. Provision of reproductive healthcare to women with disabilities: A survey of obstetrician-gynecologists' training, practices, and perceived barriers. *Health Equity* 2 (1): 207–215. https://doi.org/10.1089/heq.2018.0014.

Tarasoff, Lesley A., Yona Lunsky, Simon Chen, Astrid Guttmann, Susan M. Havercamp, Susan L. Parish, Simone N. Vigod, Adele Carty, and Hilary K. Brown. 2020a. Preconception health characteristics of women with disabilities in Ontario: A population-based, cross-sectional study. *Journal of Women's Health* 29 (12): 1564–1575. https://doi.org/10.1089/jwh.2019.8273.

Tarasoff, Lesley A., Fahmeeda Murtaza, Adele Carty, Dinara Salaeva, Angela D. Hamilton, and Hilary K. Brown. 2020b. Health of newborns and infants born to women with disabilities: A meta-analysis. *Pediatrics* 146 (6): e20201635. https://doi.org/10.1542/peds.2020-1635.

Tarasoff, Lesley A., Saranyah Ravindran, Hannan Malik, Dinara Salaeva, and Hilary K. Brown. 2020c. Maternal disability and risk for pregnancy, delivery, and postpartum complications: A systematic review and meta-analysis. *American Journal of Obstetrics & Gynecology* 222 (1): e1-27. https://doi.org/10.1016/j.ajog.2019.07.015.

Thompson, Vanessa R., Roger J. Stancliffe, Alex Broom, and Nathan J. Wilson. 2014. Barriers to sexual health provision for people with intellectual disability: A disability service provider and clinician perspective. *Journal of Intellectual & Developmental Disability* 39 (2): 137–146. https://doi.org/10.3109/13668250.2014.898742.

Tilley, Elizabeth, Jan Walmsley, Sarah Earle, and Dorothy Atkinson. 2012. 'The silence is roaring': Sterilization, reproductive rights and women with intellectual disabilities. *Disability & Society* 27 (3): 413–426. https://doi.org/10.1080/09687599.2012.654991.

Tøssebro, Jan, Turid Midjo, Veronika Paulsen, and Berit Berg. 2017. Prevalence, trends and custody among children of parents with intellectual disabilities in Norway. *Journal of Applied Research in Intellectual Disabilities* 30 (3): 533–542. https://doi.org/10.1111/jar.12304.

United Nations. 2006. *Convention on the rights of persons with disabilities*. Treaty series, 2515, 3. New York: United Nations General Assembly.

Wendt, Amanda, Cassandra M. Gibbs, Stacey Peters, and Carol J. Hogue. 2012. Impact of increasing inter-pregnancy interval on maternal and infant health. *Paediatric and Perinatal Epidemiology* 26 (S1): 239–258. https://doi.org/10.1111/j.1365-3016.2012.01285.x.

Willis, D.S., J.G. Wishart, and W.J. Muir. 2010. Menopausal experiences of women with intellectual disabilities. *Journal of Applied Research in Intellectual Disabilities* 24 (1): 74–85. https://doi.org/10.1111/j.1468-3148.2010.00566.x.

World Health Organization (WHO). 2006. *Comprehensive cervical cancer control: A guide to essential practice*. Geneva: World Health Organization.

World Health Organization (WHO). 2012. *Meeting to develop a global consensus on preconception care to reduce maternal and childhood mortality and morbidity*. Geneva: World Health Organization.

World Health Organization and United Nations Population Fund. 2009. *Promoting sexual and reproductive health for persons with disabilities—WHO/UNFPA guidance note*. World Health Organization. https://apps.who.int/iris/bitstream/handle/10665/44207/9789241598682_eng.pdf?sequence=1.

Wu, Justine, Jianying Zhang, Monika Mitra, Susan L. Parish, and Geeth Kavya Minama. Reddy. 2018. Provision of moderately and highly effective reversible contraception to insured women with intellectual and developmental disabilities. *Obstetrics & Gynecology* 132 (3): 565–574. https://doi.org/10.1097/AOG.0000000000002777.

# Chapter 25
# The Ethics of Cancer Care for People with Intellectual and/or Developmental Disabilities

**Julie S. Armin and Heather J. Williamson**

**Abstract** In this chapter, we examine how case reports of cancer treatment for people with intellectual and/or developmental disabilities describe their inclusion in research and medical decision making and bring forward assumptions regarding their capacity for engagement in their cancer care. Looking at case reports as "social practices" that generate and sustain ideas about deservingness and configure resources for patients with intellectual and/or developmental disabilities in particular ways, we bring forth several key points about cancer treatment for people with intellectual and/or developmental disabilities and offer recommendations for ethical treatment practices. We find examples in the case reports of people with intellectual and/or developmental disabilities, of their not receiving standard-of-cancer-care, their exclusion from cancer research, and their limited engagement in decision making about their care. Several cases, however, highlight recommendations for ethical care, including interprofessional team-based care that facilitates patient engagement in decision making and support for the oncology team.

**Keywords** Cancer treatment · Decision making · Intellectual disability · Developmental disability

J. S. Armin (✉)
The University of Arizona, Tucson, AZ, USA
e-mail: jarmin@arizona.edu

H. J. Williamson
Northern Arizona University, Flagstaff, AZ, USA
e-mail: heather.williamson@nau.edu



A. Bianchi and J. A. Vogt (eds.), *Intellectual Disabilities and Autism: Ethics and Practice*, The International Library of Bioethics 108,
https://doi.org/10.1007/978-3-031-61565-8_25

## 25.1 Introduction

While certain groups of people with intellectual and/or developmental disabilities[1] are more likely to develop cancer (e.g., leukemias or other cancers among people with Down syndrome or intellectual disability) (de Volo et al. 2012; Hasle et al. 2000), describing biological risk is not our purpose in this paper. Instead, we examine social processes that may influence the administration of guideline-concordant cancer treatment for people with intellectual and/or developmental disabilities. We examine how case reports of cancer treatment for people with intellectual and/or developmental disabilities describe their inclusion in research and medical decision making, and bring forward assumptions regarding their capacity for engagement in their cancer care. We start this analysis using the citations from a scoping literature review published in *Lancet Oncology* (Boonman et al. 2022), which outlines the published evidence regarding cancer treatment and decision making among people with intellectual disabilities.

Cancer has been termed a "dread disease", though developments in oncology have led to greater life expectancy after diagnosis (Mukherjee 2010; Patterson 1987). However, it is misleading to call cancer a single disease, since the treatments used to rid the body of it or send it into remission vary according to the type of disease, the biological characteristics of the patient and tumor, the spread of disease throughout the body, and if the disease is recurring after successful treatment (Burke et al. 2019; Mukherjee 2010). Treatment of varied cancers may involve surgeries, chemotherapies, hormone therapies, immunotherapies, radiation therapy, or stem cell transplantation, many of which can result in short-term or long-term side effects (National Cancer Institute 2023). Further, the success of cancer treatment can be influenced by patients' ability to access services (Stalford 2019), trust in and expectations of the cancer care systems (Mullins et al. 2019), and the availability of care (Dyer 2019). Embedded in cancer care are also the decisions that must be made by the cancer patient regarding treatment options offered by health professionals (Gattellari et al. 2001). For people with intellectual and/or developmental disabilities, ableism in health care may create assumptions about their quality of life, which may, in turn, influence what treatment options are made available to them (Agaronnik et al. 2021; Iezzoni et al. 2021).

The moral concern of "deservingness" (Willen 2012a; b) provides insight into how medical practices can configure and reproduce inequality and social exclusion. Although initially used to describe the healthcare rights of immigrants, we translate its use to consider the rights of people with intellectual and/or developmental disabilities to receive adequate health care. Rather than focusing on universal or legally (or politically) secured rights, we examine deservingness in "situationally specific,

[1] The authors use person-first language unless citing a resource that uses identity-first language. They also refer to language about people with disabilities that is currently viewed by advocacy communities as biased to examine its use among the oncology cases cited in this article. When reporting results of previous research, the authors use the diagnostic language used by the original authors.

vernacular moral arguments" by examining "social practices" or the "interpretation and enactment of [ethical] commitments in everyday bureaucratic, clinical and advocacy interactions" (Willen 2012b, 806). We examine how the cases discussed in this chapter generate and sustain ideas about deservingness and configure resources for patients with intellectual and/or developmental disabilities in particular ways, determining the "proper" allocation of valuable cancer resources. During the height of the COVID-19 pandemic in the United States, the question of deservingness was highlighted as people with disabilities became central in decisions regarding healthcare rationing and vaccine priority populations (Lund and Ayers 2022).

A number of the cases described here acknowledge the extended life expectancy for people with intellectual and/or developmental disabilities (Kurkjian et al. 2006) and the need to document cancer cases (Satgé et al. 2013) or particular cancer treatment approaches in adults with intellectual and/or developmental disabilities (Enomoto et al. 2015; Takagi et al. 2003). However, they also highlight the disconnection of medicine from self-advocate communities that could provide guidance regarding their concerns, including preferred language. For instance, several articles use the term "mental retardation" in their reports (Engel et al. 2014; Enomoto et al. 2015), which is a term that has been replaced by intellectual disability (US Congress 2010). Further, the case reports foreground two main ethical issues for consideration: (1) people with intellectual and/or developmental disabilities contributing to the evidence base for cancer care; and (2) engaging people with intellectual and/or developmental disabilities in their cancer care.

## 25.2 People with Intellectual and/or Developmental Disabilities Contributing to the Evidence Base for Cancer Care

Multiple case reports articulate the patient's successful completion of the standard of care for a particular malignancy, highlighting the person's treatment tolerance. An example is the description of a patient with Nijmegen Breakage Syndrome who developed Stage IVB Hodgkin's Lymphoma and was treated with four cycles of chemotherapy. In regard to the treatment of Hodgkin's Lymphoma, the authors note that many patients with intellectual and/or developmental disability are untreated. However, the four cycles of chemotherapy in this case resulted in remission, and the patient with Nijmegen Breakage Syndrome returned to his job as a cook (Engel et al. 2014).

Satgé and colleagues (2013) note that standard cancer treatment may be difficult for people with intellectual disabilities, though they describe the care of a sixty-four-year-old man with "moderate intellectual disability", two years beyond diagnosis, who received "the usual treatment" for breast cancer. In the case, the authors note that the man, who resided in an institutional setting, found the lump himself and agreed to the recommended cancer therapy after "the disease and treatment were explained to him". His therapy included a right mastectomy and lymph node dissection, with radiation and chemotherapy. Additionally, the authors highlight the need to report cases of breast cancer in men with intellectual disability, noting that theirs is the second reported case in the literature.

Case reports also highlight the need for patients with intellectual and/or developmental disabilities to participate in oncology research studies. For instance, Wadhwa et al. (2018) describe the case of a twenty-one-year-old man with Down syndrome diagnosed with high-risk B-precursor acute lymphoblastic leukemia. The authors note that the patient had multiple comorbidities including obesity, obstructive sleep apnea, and a history of skin infections. The patient received brief inpatient chemotherapy infusion of blinatumomab, followed by twenty-eight days of outpatient infusions. He attended the clinic twice a week "for physical examination, laboratory studies, and replacement of his blinatumomab infusion bag through a home health company that provides the drug. He experienced no significant toxicity or complications during the outpatient treatment phase" (Wadhwa et al. 2018, 2). Of note, patients with Down syndrome were excluded from a clinical trial of the medication (blinatumomab) that successfully treated their patient, and the authors conclude that patients with Down syndrome "should be included in upcoming clinical trials testing these modalities" (Wadhwa et al. 2018, 2).

Another case study describes a twenty-two-year-old man with Down syndrome who participated in a supportive care intervention for people receiving cancer treatment (Bühl et al. 2019). The supportive care intervention involved a mixed modality of physical activities. The authors acknowledge that the patient was excluded from the clinical trial to test this exercise program due to his disability, but they reference the United Nations Convention on the Rights of Persons with Disabilities as a motivation for offering the program to him outside of the trial. This case report highlighted the acceptability and feasibility of the program for the patient, noting one hundred percent participant adherence to the intervention and no adverse events (Bühl et al. 2019).

## 25.3 Engaging People with Intellectual and/or Developmental Disabilities in Their Cancer Care

The case reports described below introduce two main areas related to the engagement of people with intellectual and/or developmental disabilities in their cancer care: (1) their involvement in (or exclusion from) decision making, particularly as it relates to informed consent; and (2) assumptions about patients' ability to participate in certain types of cancer care and/or the providers' inability to provide care.

### 25.3.1 Decision Making with a Focus on Informed Consent

Many articles describe an inability to obtain informed consent for recommended cancer treatments due to the patient's disabilities. Articles discussing cancer care and people with intellectual and/or developmental disability also ignore the topic of assent entirely, noting that the family provided consent. For adults with guardians, assent involves explaining a procedure or research activities to a patient or participant and gaining their agreement to participate, while the guardian provides consent (Bittick and Holliman 2023). Best practices for consent and assent dictate that these are ongoing processes, rather than a one-time decision (Wheeler et al. 2020). In several cases, the reasons for determining the patient could not provide consent are poorly defined, such as "not competent", with little explanation about what is meant by "competency" and how a person's incompetency was determined. For instance, in one case report involving nine patients with testicular cancer (Hafeez et al. 2007), the authors note that none of the patients were able to make an informed decision about their care or were deemed competent enough to provide informed consent. In general, the consent processes described for many of the case reports neither recognize nor accommodate the needs of people with intellectual and/or developmental disabilities to empower informed decision making (Boonman et al. 2022). Even cases in which an individual's autonomy and well-being are seemingly honored, such as the above-mentioned case of the person enrolled in a supportive care intervention, the description of consent includes only physician approval and parental consent (Bühl et al. 2019). The case report does not mention assent or describe why they did not seek assent from the individual himself.

In some articles, consent is not expressly discussed. The authors of these articles note that cancer treatment decisions were made for the person with intellectual and/or developmental disability, though they do not provide any contextual information about the decision-making process. For instance, in a case involving a twenty-three-year-old with hepatic tumors who died four months after diagnosis, the authors stated that they could not operate on the tumor. Furthermore, systemic chemotherapy was discontinued "due to the patient's mental retardation" (Horiguchi et al. 2015, 2869). Another case involving a person with stage III squamous cell carcinoma of the tongue, reports that a general anesthesia was used for chemotherapy infusion because such anesthesia is often used for routine dental treatments among people with intellectual and/or developmental disabilities (Enomoto et al. 2015). No further justification regarding this individual's particular need for general anesthesia for cancer treatment was provided. The same case also deemed daily chemotherapy appointments to be "unacceptable and difficult" without offering any further explanation.

In several cases, family members or other caregivers are mentioned as part of decision-making processes, with some families declining treatment altogether. Articles seldom discuss the process of decision making between the clinical team and patients' families. Instead, the details regarding the reasons for complex treatment decisions are represented by statements such as "family and caregivers opted not to treat the patient with chemotherapy" (Ji et al. 2017, 1391). For example, one article describes a "58-year-old Hispanic male with Down syndrome" (Kwok-Oleksy et al. 2012, 341) who is diagnosed with Stage IIIC melanoma in his big toe. After surgery, the authors note, "The patient's caregiver (sister) refused further treatment with interferon-alpha in the adjuvant setting because of toxicity associated with the regimen in conjunction with *the patient's altered mental status*" (339). The authors note that interferon-alpha is the standard of care. The authors also note that there were no melanoma vaccine trials open to him "given his unique status", (341) suggesting to the reader that the patient's intellectual and/or developmental disability is the barrier to his research participation.

### 25.3.2 Assumptions About Disability Status and Cancer Care

Case reports also indicate that clinicians may question the ability for people with intellectual and/or developmental disabilities to handle or adhere to diagnostic procedures or treatment options (Boonman et al. 2022; Hafeez et al. 2007). A case exemplifying these assumptions and corresponding treatment involves a twenty-nine-year-old woman with Rubinstein-Taybi syndrome who presented to her primary care doctor with irregular menstrual symptoms. A pelvic ultrasound showed tumors and the patient was diagnosed with low-grade serous carcinoma of the ovary and an endometrioid adenocarcinoma of the uterus. The patient elected to have a hysterectomy (surgery to remove the uterus) and bilateral salpingo-oophorectomy (surgery to remove both ovaries and fallopian tubes) for treatment, completing chemotherapy after the surgery. However, her treating clinicians opted not to provide radiation "because the patient would have been unable to cooperate with the procedures involved" (Johannesen et al. 2015, 133). The authors provide no explanation as to why the patient would be "unable" to "cooperate" with radiation procedures, which potentially suggests that assumptions were made about the woman's disability, and the way it may influence her ability to handle all recommended treatments.

In other cases, disability status itself was the reason that clinicians opted not to treat the cancer more aggressively. One of these cases involves a forty-four-year-old man with Down syndrome, who was diagnosed with esophageal cancer at an advanced state (Moreels et al. 2007). The authors of this report state that medical complications, in combination with the patient's "impaired mental condition", made him inoperable and they subsequently referred the patient to palliative care. The patient died six months later. Hsiung et al. (2003) review a similar case involving a thirty-nine-year-old man with Down syndrome who was diagnosed with testicular cancer. In response to his cancer, health providers offered three management approaches: (1) radiation; (2) chemotherapy; and (3) observation with close follow-up CT scans due to his "mental disability", which they assumed would make it difficult for him to tolerate side effects. In this case, however, the patient's mother selected radiation and the patient was cancer free four years later.

Finally, in a case outlined by Wei et al. (2019), a poor outcome experienced by the cancer patient was attributed to the patient's Down syndrome and corresponding non-adherence. In this circumstance, a forty-three-year-old woman with Down syndrome had a neck growth and was diagnosed with Hashimoto thyroiditis and primary diffuse large B-cell lymphoma of the thyroid, which is a rare occurrence. Because of the disease process, the clinical team opted for surgery (palliative unilateral thyroidectomy) as a first step, along with a recommendation for combination targeted therapy and chemotherapy. The patient did not undergo the recommended chemotherapy, and due to uncontrolled growth of the tumor, she died three weeks after surgery. In their discussion regarding the poor outcome, the authors state that the patient's Down syndrome "contributed to the lack of follow-up and lack of compliance" (Wei et al. 2019, 4).

## 25.4 Conceptualizing Best Practices for Ethical Cancer Care

The data presented here—case reports of cancer treatment for people with intellectual and/or developmental disabilities—distills what is a complex process of interactions between the patient, their supporters, and the health care team, as well as negotiations about cancer treatment and other health care, into a multi-page report. Based on our review of these reports, the primary ethical concerns are that people with intellectual and/or developmental disabilities are not always adequately included as part of consent processes, provided the opportunity to indicate assent, included in clinical research opportunities, or offered standard-of-care treatment. These concerns are noteworthy given that patients with intellectual and/or developmental disabilities are rights-bearing individuals who deserve to have equitable access to cancer care treatment and research opportunities. Another ethical concern relates to *the reporting* of individuals' cancer care experiences. Several cases note the need to report cases in order to highlight particular tumor types which were thought to be rare in people with intellectual and/or developmental disabilities (Kwok-Oleksy et al. 2012) or to provide guidance to other practitioners who may treat a person with intellectual and/or developmental disabilities diagnosed with a similar cancer (Hafeez et al. 2007). While laudable goals, as we note above, the lack of detail regarding treatment decisions limits our understanding of a complex process, thus providing a limited model for others.

In terms of the oncology care of people with intellectual and/or developmental disabilities, we ask: How do we configure cancer care to demonstrate and enact the deservingness of people with intellectual and/or developmental disabilities? Some cases reviewed in this chapter, while missing important details about their specific approach to cancer care decision making, acknowledge the challenges of cancer treatment for this population which can be translated into recommendations for better care. Lesage and colleagues (2017) report that a thirty-two-year-old patient with melanoma was not offered interferon, which might have successfully treated the cancer. They note the challenges of treating patients with "cognitive impairments" due to challenges in communication, including the patients' ability to communicate pain or symptoms or offer a fully informed consent, the physician's inability to do a full workup with a patient, and the prevalence of comorbid conditions among people with intellectual and/or developmental disabilities. They call for interdisciplinary team-based care and better training for oncology providers in working with patients with intellectual and/or developmental disabilities, which follows recommendations made elsewhere in the literature as well (Samtani et al. 2021).

Two cases written by colorectal cancer nurses examine the ethics of decision making and interprofessional care (Black and Hyde 2004). In one case, a man in his mid-fifties with mental health disabilities was diagnosed with colon cancer. The authors describe the interprofessional process for supported decision making with the patient, support professionals, a psychiatrist, and a "learning disability nurse".[2] After he initially refused surgery for the cancer, the multidisciplinary team assessed the patient's ability to consent versus simply acquiesce to surgery. The authors acknowledge that conversations with the patient were "disconcerting and difficult", but eventually, the patient consented to receive the surgery. After the patient's death, the learning disability team leader worked with the article's authors to provide a debrief session with healthcare staff who were distraught by the experience of this patient's care. In this debrief, they processed staff's concerns about the patient's care, including his consent for surgery and the staff's "lack of rapport" with the patient (2004, 975).

Black and Hyde's second case focused on supportive care for a twenty-nine-year-old woman with colorectal cancer, who received a stoma (an incision that allows feces to exit the body into a bag). The authors describe the patient as "mostly non-verbal". After struggling with her stoma bag, her support professionals joined her in the hospital to help the healthcare staff. One of the nurses used Makaton (a mode of communication that uses signs, symbols, and verbal language) to explain why the patient needed to use the stoma. The authors also describe (in less detail) working with support staff and the patient to plan end-of-life care after a recurrence of cancer.[3]

In another case reported in the UK, a sixty-one-year-old man with a learning disability was diagnosed with testicular cancer (Mott et al. 2019). The patient first had access to nursing staff trained in learning disabilities who advised the cancer care specialists to provide a longer appointment time and to include his primary support person (sister) in the consultation. They also took time to have the patient repeat back key points in order to assess his capacity to understand his diagnosis and the proposed treatment.

[2] Learning disability is a term used in the UK to indicate intellectual and/or developmental disabilities.

[3] While an important topic, end-of-life planning is not addressed in detail in any of the cases discussed. Others have provided insights into end-of-life planning for cancer patients with intellectual and/or developmental disabilities (e.g. Tuffrey-Wijne et al. 2013).

## 25.5 Conclusions

In the "social practice" of case reports, we can see how the consolidation of a complicated cancer treatment process into a brief paper may serve to minimize the deservingness of cancer patients with intellectual and/or developmental disabilities. Many of the cases discussed in this chapter omit critical details about patient care that would support the deservingness of people with intellectual and/or developmental disabilities to receive standard-of-cancer-care, to enroll in cancer research, and to engage in decision making about their care. In contrast, there are case report examples that provide critical details from which we can draw recommendations. Black and Hyde (2004), for instance, use two cases to provide recommendations for care of the patient and the oncology team. They highlight the importance of respecting patient autonomy in decision making and, relatedly, the need for accessible information about cancer for patients with intellectual and/or developmental disabilities. Importantly, they emphasize that treatment decisions should be based on the patient's medical situation and not a provider's interpretation of the patient's quality of life. Further, they emphasize the importance of interprofessional teams in addressing the care of patients with intellectual and/or developmental disabilities, a call echoed by Mott and colleagues (2019) and Lesage and co-authors (2017). In these cases, we can see the benefits of interprofessional care for addressing consent issues, patient-provider communication, and oncology provider support.

**Acknowledgement** Julie Armin's work on this chapter was partly supported by the Administration on Community Living Award # 90DDUC0090-01- 00.

## References

Agaronnik, Nicole, Areej El-Jawahri, Kristi Kirschner, and Lisa Iezzoni. 2021. Exploring cancer treatment experiences for patients with pre-existing mobility disability. *American Journal of Physical Medicine and Rehabilitation* 100 (2): 113.

Bittick, Amy, and Ryan Holliman. 2023. Informed consent with people judged incapable of legally consenting. *Advances in Mental Health and Intellectual Disabilities* 17 (2): 95–108. https://doi.org/10.1108/AMHID-10-2022-0040.

Black, Pat, and Christine Hyde. 2004. Caring for people with a learning disability, colorectal cancer and stoma. *British Journal of Nursing* 13 (16): 970–975.

Boonman, Anne J., Maarten Cuypers, Geraline L. Leusink, Jenneken Naaldenberg, and Haiko J. Bloemendal. 2022. Cancer treatment and decision making in individuals with intellectual disabilities: A scoping literature review. *The Lancet Oncology* 23 (4): e174–e183.

Bühl, Linda, Thomas Abel, Florian Wolf, Max Oberste, Wilhelm Bloch, Michael Hallek, Thomas Elter, and Philipp Zimmer. 2019. Feasibility and potential benefits of an exercise intervention in a male with Down syndrome undergoing high-dose chemotherapy for acute lymphoblastic leukemia: A case report. *Integrative Cancer Therapies* 18: 1534735419832358.

Burke, Nancy, Julie Armin, and Laura Eichelberger. 2019. Introduction: framing cancer and structural vulnerability. In *Negotiating Structural Vulnerability in Cancer Control*, edited by Julie

Armin, Nancy Burke and Laura Eichelberger. Santa Fe and Albuquerque: School for Advanced Research Press and University of New Mexico Press.
de Volo, Chiara Palka Bayard, Melissa Alfonsi, Valentina Gatta, Antonio Novelli, Laura Bernardini, Donatella Fantasia, Ivana Antonucci, Domenico Angelucci, Robert Zori, and Liborio Stuppia. 2012. 16q22. 1 microdeletion detected by array-CGH in a family with mental retardation and lobular breast cancer. *Gene* 498 (2): 328–331.
Dyer, Karen. 2019. Colonial legacies: population panics, reproductive control, and cancer-related fertility care in Puerto Rico. In *Negotiating Structural Vulnerability in Cancer Control*, ed. Julie Armin, Nancy Burke and Laura Eichelberger. Santa Fe and Albuquerque: School for Advanced Research Press and University of New Mexico Press.
Engel, Katharina, Martina Rudelius, Felix G. Meinel, Christian Peschel, and Ulrich Keller. 2014. An adult patient with Nijmegen Breakage Syndrome and Hodgkin's Lymphoma. *BMC Hematology* 14 (1): 2. https://doi.org/10.1186/2052-1839-14-2.
Enomoto, Akifumi, Takanori Nakatani, Eri Morikage, Takeshi Shimoide, and Suguru Hamada. 2015. Tongue cancer with mental retardation due to microcephaly: A clinical report. *World Journal of Surgical Oncology* 13 (1): 216. https://doi.org/10.1186/s12957-015-0634-2.
Gattellari, Melina, Phyllis N. Butow, and Martin H N. Tattersall. 2001. Sharing decisions in cancer care. *Social Science and Medicine* 52 (12): 1865–1878. https://doi.org/10.1016/S0277-9536(00)00303-8.
Hafeez, S., R.A. Sharma, R.A. Huddart, D.P. Dearnaley, and A. Horwich. 2007. Challenges in treating patients with down's syndrome and testicular cancer with chemotherapy and radiotherapy: The royal marsden experience. *Clinical Oncology* 19 (2): 135–142. https://doi.org/10.1016/j.clon.2006.10.006.
Hasle, Henrik, Inge Haunstrup Clemmensen, and Margareta Mikkelsen. 2000. Risks of leukaemia and solid tumours in individuals with down's syndrome. *The Lancet* 355 (9199): 165–169.
Horiguchi, Norio, Satoru Kakizaki, Keisuke Iizuka, Satoshi Hagiwara, Tatsuya Ohyama, Yuichi Yamazaki, Nobuyuki Shibusawa, Ken Sato, Motoyasu Kusano, and Yoichi Nakazato. 2015. Hepatic angiosarcoma with dyskeratosis congenita. *Internal Medicine* 54 (22): 2867–2872.
Hsiung, Stripp, C. Diana, David Vaughn, Keith Van Arsdalen, and Richard Whittington. 2003. Three cases of advanced seminoma and down's syndrome: A possible association. *American Journal of Clinical Oncology* 26 (2): 197–199.
Iezzoni, Lisa I., Sowmya R. Rao, Julie Ressalam, Dragana Bolcic-Jankovic, Nicole D. Agaronnik, Karen Donelan, Tara Lagu, and Eric G. Campbell. 2021. Physicians perceptions of people with disability and their health care: Study reports the results of a survey of physicians perceptions of people with disability. *Health Affairs* 40 (2): 297–306.
Ji, Jianling, Catherine Quindipan, David Parham, Lishuang Shen, David Ruble, Moiz Bootwalla, Dennis T. Maglinte, Xiaowu Gai, Sulagna C. Saitta, and Jaclyn A. Biegel. 2017. Inherited germline ATRX mutation in two brothers with ATR-X syndrome and osteosarcoma. *American Journal of Medical Genetics Part A* 173 (5): 1390–1395.
Johannesen, Eric J., Todd Williams, Douglas C. Miller, and Erin Tuller. 2015. Synchronous ovarian and endometrial carcinomas in a patient with Rubinstein-Taybi syndrome: A case report and literature review. *International Journal of Gynecological Pathology* 34 (2): 132–135. https://doi.org/10.1097/pgp.0000000000000125.
Kurkjian, Carla, Sunil Patel, Rammurti Kamble, S. Terrence Dunn, William Kern, and Mohamed A. Kharfan-Dabaja. 2006. Acute promyelocytic leukemia and constitutional trisomy 21. *Cancer Genetics and Cytogenetics* 165 (2): 176–179. https://doi.org/10.1016/j.cancergencyto.2005.08.014.
Kwok-Oleksy, Christina, Manijeh Berenji, Nicholas G. Argerakis, Michael Trepal, and Marc K. Wallack. 2012. Subungual melanoma of the great toe in a Hispanic male with down syndrome: A case report and review of the literature. *The Journal of Foot and Ankle Surgery* 51 (3): 337–341. https://doi.org/10.1053/j.jfas.2012.01.013.

Lesage, Candice, Isabelle Coupier, and Bernard Guillot. 2017. A family with two cases of melanocytic tumors and fragile X syndrome. *Melanoma Research* 27 (6): 645–648. https://doi.org/10.1097/cmr.0000000000000400.

Lund, Emily M., and Kara B. Ayers. 2022. Ever-changing but always constant: Waves of disability discrimination during the COVID-19 pandemic in the United States. *Disability and Health Journal* 15 (4): 101374.

Moreels, T.G., E.P.M. Van Vliet, H.W. Tilanus, T.C.K. Tran, E.J. Kuipers, and P.D. Siersema. 2007. Down syndrome and esophageal cancer. *Diseases of the Esophagus* 20 (2): 183–186. https://doi.org/10.1111/j.1442-2050.2007.00667.x.

Mott, Charlotte Emily, Kenrick Ng, and Constantine Alifrangis. 2019. The management of patients with learning difficulties and testicular cancer: overcoming barriers to improve care. *BMJ Case Reports* 12 (11): e227330. https://doi.org/10.1136/bcr-2018-227330. https://casereports.bmj.com/content/bmjcr/12/11/e227330.full.pdf.

Mukherjee, Siddhartha. 2010. *The Emperor of all Maladies: A Biography of Cancer*. Vol. Book, Whole. New York: Scribner.

Mullins, Megan A., Lauren C. Peres, Anthony J. Alberg, Elisa V. Bandera, Jill S. Barnholtz-Sloan, Melissa L. Bondy, Ellen Funkhouser, Patricia G. Moorman, Edward S. Peters, and Paul D. Terry. 2019. Perceived discrimination, trust in physicians, and prolonged symptom duration before ovarian cancer diagnosis in the African American Cancer epidemiology study. *Cancer* 125 (24): 4442–4451.

National Cancer Institute. 2023. *Cancer Treatment*. https://www.cancer.gov/about-cancer/treatment/types. Accessed 9 April 2024.

Patterson, James T. 1987. *The Dread Disease: Cancer and Modern American Culture*. Cambridge, MA: Harvard University Press.

Samtani, Grace, Tamsen L. Bassford, Heather J. Williamson, and Julie S. Armin. 2021. Are researchers addressing cancer treatment and survivorship among people with intellectual and developmental disabilities in the US? A scoping review. *Intellectual and Developmental Disabilities* 59 (2): 141–154.

Satgé, Daniel, Bernard Leduc, Fernand Raffi, and Etienne Roux. 2013. Adenocarcinoma of the right breast in a man with intellectual disability. *Case Reports in Oncological Medicine* 2013: 968035.

Stalford, M. 2019. Connecting rural patients with urban hospitals across the cancer care continuum: A view from Vietnam on a global problem. In *Negotiating Structural Vulnerability in Cancer Control*, ed. Julie Armin, Nancy Burke and Laura Eichelberger. Santa Fe and Albuquerque: School for Advanced Research Press and University of New Mexico Press.

Takagi, K., A. Yoshida, K. Kinoshita, D. Iwashima, S. Imamaura, H. Iwasaki, H. Tsutani, and T. Ueda. 2003. Unrelated donor bone marrow transplantation for acute mixed lineage (myeloid and B-lymphoid lineage) leukemia in an adult with down syndrome. *Annals of Hematology* 82 (4): 236–240. https://doi.org/10.1007/s00277-003-0617-1.

Tuffrey-Wijne, Irene, Nikoletta Giatras, Gary Butler, Amanda Cresswell, Paula Manners, and Jane Bernal. 2013. Developing guidelines for disclosure or non-disclosure of bad news around life-limiting illness and death to people with intellectual disabilities. *Journal of Applied Research in Intellectual Disabilities* 26 (3): 231–242. https://doi.org/10.1111/jar.12026.

US Congress. 2010. S.2781–111th Congress (2009–2010): Rosa's Law. October 5, 2010. https://www.congress.gov/bill/111th-congress/senate-bill/2781/text. Accessed 9 April 2024.

Wadhwa, Aman, Matthew A. Kutny, and Ana C. Xavier. 2018. Blinatumomab activity in a patient with down syndrome B-precursor acute lymphoblastic leukemia. *Pediatric Blood and Cancer* 65 (2): e26824.

Wei, Luyang, Ning Zhao, Yalun Li, Xiaodan Zheng, Huaxin Zhang, and Zhongtao Zhang. 2019. Down syndrome with primary thyroid diffuse large B-cell lymphoma and Hashimoto thyroiditis: A CARE compliant case report. *Medicine* 98 (35): e16994.

Wheeler, Anne C., Amanda Wylie, Melissa Raspa, Adrienne Villagomez, Kylee Miller, Anne Edwards, Margaret DeRamus, Paul S. Appelbaum, and Donald B. Bailey. 2020. Decisional capacity for informed consent in males and females with fragile X syndrome. *Journal of Autism and Developmental Disorders* 50 (5): 1725–1747. https://doi.org/10.1007/s10803-019-03930-4.

Willen, Sarah S. 2012a. How is health-related deservingness reckoned? Perspectives from unauthorized im/migrants in Tel Aviv. *Social Science and Medicine* 74 (6): 812–821. https://doi.org/10.1016/j.socscimed.2011.06.033. http://www.sciencedirect.com/science/article/pii/S0277953611003984.

Willen, Sarah S. 2012b. Migration, illegality, and health: Mapping embodied vulnerability and debating health-related deservingness. *Social Science and Medicine* 74 (6): 805–811. https://doi.org/10.1016/j.socscimed.2011.10.041.

# Chapter 26
# Intellectual Disability, Euthanasia, and Assisted Suicide

**Tim Stainton and Trudo Lemmens**

**Abstract** This chapter elucidates areas of concern regarding euthanasia and people with intellectual and developmental disabilities (IDD). It broadly asks, "does the expansion of euthanasia and assisted suicide (EAS) regimes threaten the rise of a quiet eugenics?" The chapter examines some of the moral and ethical rationales that may support this trajectory. It considers parallel actions such as selective termination and neonatal euthanasia as indicators of how people with IDD may fare with broad access to EAS. A key consideration is the question of consent and whether current practice is sufficient. Finally, the chapter considers what evidence we have regarding IDD and euthanasia before considering what may lie ahead.

**Keywords** Euthanasia · Assisted suicide · Intellectual disability · Vulnerability · Medical assistance in dying

## 26.1 Introduction

On July 26, 2016, a former employee of the Tsukui Yamayuri-en care home for people with developmental disabilities in Japan murdered nineteen residents, injuring twenty-six others. Satoshi Uematsu, consistently maintained he was acting for the good of society and that "people with disabilities who were unable to communicate well had no human rights" (BBC 2020). Prior to his rampage he wrote "I envision a world where a person with multiple disabilities can be euthanized, with an agreement from the guardians, when it is difficult for the person to carry out household and social activities. [...] Now is the time to carry out a revolution and make an inevitable but

T. Stainton (✉)
Canadian Institute for Inclusion and Citizenship, University of British Columbia, Vancouver, BC, Canada
e-mail: timst@mail.ubc.ca

T. Lemmens
Faculty of Law, Dalla Lana School of Public Health, University of Toronto, Toronto ON, Canada
e-mail: trudo.lemmens@utoronto.ca

 
A. Bianchi and J. A. Vogt (eds.), *Intellectual Disabilities and Autism: Ethics and Practice*, The International Library of Bioethics 108,
https://doi.org/10.1007/978-3-031-61565-8_26

tough decision for the sake of all mankind… I believe there is still no answer about the way of life for individuals with multiple disabilities. The disabled can only create misery" (BBC 2020). He subsequently noted that killing the disabled would stimulate the world economy and maybe prevent a World War III (BBC 2020; Yamamoto 2016).

Uematsu's views bare a chilling similarity to those expressed in the first half of the twentieth century and align with the theory of eugenics, widely held throughout the western world (De Bruin and Robertson 2006). Eugenic measures impacted a range of persons, but fell most heavily and tenaciously on those with intellectual and developmental disabilities (IDD) (Hawkins 1997). The eugenics movement reached its nadir with the Nazi *Aktion T4* program, a precursor of the wider holocaust to come. *Aktion T4* led to state sanctioned, physician led murder of some 300,000 people with disabilities (Hohendorf 2016).

The current rapid expansion of euthanasia and assisted suicide (EAS) begs the question: are we heading into an era of quiet, more "subtle" eugenics, where those considered to be suffering or to have a poor quality of life will be assisted to die rather than assisted to live? This chapter elucidates some reasons that we may be facing a new era of elimination of people with IDD, not through the coercive and state mandated programs of the eugenics era, but rather a quieter, more "civilized" approach of state sponsored EAS, centered around the concept of choice and using subtle encouragement to opt for death.

In considering the vulnerability of people with IDD to EAS for reasons other than terminal illness, this chapter examines some of the moral and ethical rationales that may support this elimination. We will also consider what parallel actions can tell us about how people with IDD may fare in a world of broadly available EAS. Additionally, the chapter will consider what evidence we have regarding IDD and euthanasia before considering where it may be headed.

## 26.2 Sites of Precarity

### *26.2.1 Moral and Legal Status and IDD*

An overarching vulnerability of persons with IDD is the belief that they lack the necessary type of reason to qualify for moral status and by extension, legal status. Indeed, their humanity has been questioned throughout modern western history. To offer just one example: John Locke, in the *Essay*, contemplates the status of entities that look human but lack sufficient reason and concludes they are some sort of interstitial species "between man and beast" (Locke 1975, 416). This view was reflected in his political writings, making it clear that people with IDD are not counted as rights-bearing citizens of emergent liberal democracies. The idea that people with IDD lack moral status and the basic requisites for being human, creates an existential threat in a context where EAS is widely accepted. More fundamental is the association between rationality and moral personhood, which is axiomatic in much of Western moral and political philosophy. For instance, twentieth century philosopher Gilbert Ryle noted that "specifically human behavior" is that "which is unachieved by animals, idiots and infants". Anthony Quinton noted that "defective human beings who look and are physically constructed like men... are only marginally or by a sort of prudent and humane courtesy fully human beings" (cited in Goodey 1992, 28fn). While this view is not universal, it remains a powerful foundation to contemporary moral philosophy and applied ethics. It is not hard to see how the elimination of persons with IDD may be justified and tolerated when their humanity is in question.

### 26.2.2 Contemporary Bioethics

The idea that IDD leads to a life of suffering and poor quality underpins the view that providing access to EAS is compassionate and, in many cases, morally correct. This is most evident in certain strains of utilitarian ethics. Perhaps best known are Peter Singer's views about disabled persons' moral status and the ethics of euthanasia, wherein he has consistently expressed his views supporting the morality, in some cases, of killing disabled persons and the relative moral status of some disabled persons and certain animals (Singer 1993, 175–217; Singer 1995). Tremain goes as far as to say that eugenic ableism is '*a fundamental condition of possibility for* the field of bioethics' (Tremain 2023, 22).

As Tuffrey-Wijne et al. (2019, 116) note: "the fact that the disability itself, rather than an acquired medical condition, can be accepted as a cause of suffering that justifies euthanasia is deeply worrying". Where this becomes most concerning is when it is operationalized through approaches like Quality Adjusted Life Years (QALYs) to determine what, if any, interventions offer the best cost benefit outcome. Some bioethicists have endorsed the view that utilitarianism requires discrimination against the disabled in the allocation of health care resources based on the maximization of quality adjusted life years (McKie et al. 1998; Stein 2002). In the context of EAS, using QALYs or similar methods raises serious concerns. As Barrie (2015) notes, "problems… [with QALYs] …relate closely to the debate over euthanasia and assisted suicide because negative QALY scores can be taken to mean that patients would be "better off dead"" (635).

### 26.2.3 Selective Termination and Neonatal Euthanasia

It is helpful to consider how disabled persons have fared in related contexts, notably neonatal euthanasia and selective termination after prenatal diagnosis of Down syndrome.[1] Current trends in prenatal testing indicate a negative view toward having

[1] For more on selective termination and Down syndrome, see Chap. 5 of this volume.

a child with a disability (Parens and Asch 2003; Skotko 2009). The potential to "eliminate Down syndrome" through prenatal testing and selective termination is discussed as a positive development with regards to elimination itself and potential cost savings (Quinones and Lajka 2017). While the issue of selective termination is complex, it lends credence to the view that the lives of people with IDD are not worth living.

Another area which suggests this negative valuation of disabled persons in health care[2] is neonatal euthanasia. Legal in Belgium and the Netherlands, evidence suggests it is practiced elsewhere despite being illegal. In 2002, the Groningen Protocol for neonatal euthanasia was developed in the Netherlands. Significant numbers of these cases involve neonates with non-life threatening, medically treatable conditions and disabilities (American College of Pediatricians 2004; Verhagen and Sauer 2005; Vizcarrondo 2014). The American College of Pediatricians (2004) note that there is room for parental, physician, personal, social, and economic bias. In their review of every case reported from 1998 to 2005, Verhagen and Sauer (2005) found that all involved spina bifida. They report that the considerations used to decide on euthanizing included: *expectation* (emphasis added) of extremely poor quality of life (suffering) in terms of functional disability, pain, discomfort, poor prognosis, and hopelessness; predicted lack of self-sufficiency; expected hospital dependency; and long-life expectancy. What is striking is that none of the individuals in these cases were terminal, nor apparently experiencing significant physical pain. In all cases these were third party, subjective determinations of perceived future quality of life and a determination that nothing else could make their life tolerable.

Both of these examples suggest that, when given the opportunity and sufficient justification, there is support for the elimination of disabled persons. It is difficult to see why a similar scenario would not play out with EAS. In Canada, the head of the Quebec College of Physicians and Surgeons already recommended before a Parliamentary Committee that was reviewing the practice, that future legislation should permit euthanasia of severely disabled newborns (Cullen and Zabjek 2022).

### 26.2.4 Inadequate Supports

A primary objection to EAS by disabled persons is the risk that disabled people will seek EAS not to end *physical suffering* related to a medical condition, but rather due to suffering caused by inadequate social or healthcare supports or forced institutionalization (Stainton 2023). In Canada, and elsewhere, socio-economic data paints a negative picture of life with a disability (Morris et al. 2018). People with IDD also "suffer" from serious health inequities (Sullivan 2023).

The lack of appropriate social supports also creates an increased reliance on family. In Canada's 2022 report on medical assistance in dying (MAiD), 35.7% of

[2] For more on the negative and ableist attitudes that exist within the health care system, see Chap. 12 this volume.

people seeking MAiD reported "perceived burden on family, friends, or caregiver" as a component of the intolerable suffering leading to their decision (Health Canada 2022). Feelings of burdensomeness observed in persons with disabilities have been associated with suicidal ideation or attempts (Khazem et al. 2015; Russell et al. 2009). This is largely a function of the adequacy of disability supports.

While one might think that accessing MAiD for reasons related to inadequate supports would be cause for concern, it has been argued that this should not trump people's rights to access MAiD, even if their death could be prevented by provision of appropriate supports (Wiebe and Mullin 2023). In fact, the authors frame it even as "harm reduction" to offer access to death as a tool to escape suffering at least in part created by social injustice. This produces a dangerous cycle: people are not given appropriate supports—this causes suffering—ending suffering is good—therefore, EAS is not only "good" but may in fact be ethically required. The problem is that it is impossible to escape the eugenic outcomes this promotes.

### *26.2.5 Negative Perception of Disability in Health Care*

People with IDD and their families often encounter negative or ableist attitudes within the health care system (Gill 2010). A recent study which surveyed seven hundred fourteen physicians in the United States found that 82.4% reported that people with significant disability have worse quality of life than nondisabled people. The authors note "these findings about physicians' perceptions of this population raise questions about ensuring equitable care to people with disability. Potentially biased views among physicians could contribute to persistent health care disparities affecting people with disability" (Iezzoni et al. 2021, 297). Heidi Janz discusses how medical professionals with "undiagnosed ableism" risk signing off on EAS requests based on how they perceive disability as "the greatest source of suffering" (Janz 2023). This suggests that unbiased practice of EAS in regard to people with IDD is highly questionable.

## 26.3 EAS and Persons with IDD

There is limited information on IDD and EAS. Some may assume they will not be eligible as they would lack the capacity to consent. This is, of course, an inherently ableist assumption since most people with IDD are more than capable of giving consent for care decisions. Article twelve of the United Nations Convention on the Rights of Persons with Disabilities (CRPD) implies all persons with IDD are capable of providing consent and have a right to the support required to do so (Stainton 2016; UN 2006).

Both a recent report of a Canadian "Expert Panel on MAID and Mental Illness" (Health Canada 2022) and a professional psychiatric association in Québec (AMPQ

2020) state that persons with IDD should not be excluded from accessing MAiD. The reports recognize that there may be heightened concerns about voluntariness, structural vulnerabilities, and decision-making capacity, but fail to discuss how this may play out. The report from Québec explicitly mentions the relevance of the history of eugenics in relation to IDD, but also emphasizes the importance of "ensuring full access to their rights as citizens, including self-determination" (AMPQ 2020, 44–45). They thus embrace a rights rhetoric which assumes, in our view wrongly, that there has been an explicit recognition of a constitutional right to EAS (Lemmens et al. 2018) for a broad range of illness and disability-related suffering.

Currently, there are no official accounts of people with IDD accessing EAS in Canada, and no data is collected which provide this information; although one recent injunction case, discussed further, dealt with a request of a 27-year-old woman with autism. In Belgium there have been several cases of people with higher end autism spectrum disorder (ASD) being approved for euthanasia based on ASD (Day 2018). One notorious case involved Tine Nys, a 38-year-old woman with a history of mental illness. She was diagnosed with ASD after requesting EAS and died by euthanasia shortly thereafter. Tine's death resulted in the only court case against the three physicians involved. Although they were found not guilty of "poisoning" in the jury trial (Lemmens 2023), a recent article informed by the court proceedings documents failures in the care she received (De Hert et al. 2022).

The most comprehensive information we have comes from the Netherlands. The Dutch RTE[3] system requires a number of case reports, narratives of the individuals' situations, and the process of assessment and approval. Tuffrey-Wijne et al. (2018; 2019; 2023) used the online RTE database of nine hundred twenty-seven EAS case reports (2012–2021) to search for individuals with IDD and/or ASD and identified 39 cases.[4] They note "19 patients (49%) had intellectual disability and 24 (62%) had ASD; of these, 4 (10%) had both intellectual disability and ASD. All age groups were represented, with 18 patients (46%) younger than 50 when they died" (Tuffrey-Wijne et al. 2023, 2).

They found that in twenty-one percent of cases, factors directly associated with IDD or ASD were the cause of suffering described and a contributing factor in a further forty-two percent of cases. In a third of cases "physicians noted there was 'no prospect of improvement' as ASD and intellectual disability are not treatable" (Tuffrey-Wijne 2023, 1). In other words, these are not end of life cases or with other significant factors associated with their suffering (e.g., cancer diagnosis). Suffering is, in all jurisdictions, a key eligibility criterion for EAS, and so to attribute the suffering to the person's IDD creates a very vulnerable situation.

---

[3] *Regionale Toetsingscommissie Euthanasie.*

[4] It should be noted that the narrative reports represent only a fraction of all EAS cases in the Netherland. Hence, the 39 cases identified would not represent the total number of cases of people with IDD euthanized in the Netherlands which potentially could be exponentially higher.

In their analysis of reasons cited for EAS requests, the preventable social factors associated with the suffering are evident. These include social isolation and loneliness, lack of resilience or coping strategies, lack of flexibility (rigid thinking, difficulty adapting to change), and oversensitivity to stimuli (Tuffrey-Wijne 2023). Based on their analysis, it seems that people with IDD are likely to be offered EAS based largely on their disability and/or suffering caused by socially determined factors. Additionally, the authors note in their first study "We found no evidence of safeguards against the influence of the physicians' own subjective value judgements when considering EAS decision, nor of processes designed to guard against transference of the physicians' own values and prejudices" (Tuffrey-Wijne 2018, 19).

## 26.4 Capacity and Consent

People with IDD have the same rights as anyone to access publicly available services. This creates a dilemma if an individual with IDD is deemed capable of consenting to EAS, but as noted above, there are concerns regarding how to assess their request, as well as how to ensure their request is sufficiently voluntary and informed. There are inherent concerns about relying strongly on capacity and informed consent assessments as a tool to determine access to life-ending proceedings (Charland et al. 2016).

Concerns associated with consent and capacity assessments for EAS become even more prominent when considering the impact of Article twelve of the CRPD, which has been interpreted as assuming everyone is capable of making any decision if the appropriate supported decision-making arrangements are in place. This usually takes the form of either more intensive and/or alternative means of communication with the individual by a trusted other to help them grasp the nature and potential consequences of the decision.

There are various questions about how the CRPD provision would play out in the context of EAS. Firstly, in cases where alternative means of communication may be required, there is no legal framework in any of the jurisdictions which allow EAS for those not at the end of life which would compel or require this type of intervention. As the Tuffrey-Wijne et al. (2018) findings show, the consent process can be questionable when IDD is involved. Even where oversight and reporting are required, the absence of data or exceptional due process in cases of IDD suggests there is a risk that persons with IDD may be subject to a flawed consent or capacity determination process. In most cases, we simply would never know if there were people with IDD being euthanized or if appropriate capacity and consent procedures were followed.

A related concern is the tendency for people with IDD to be more credulous than the general population. Greenspan et al. (2001) have argued that "it is credulity and gullibility, as much or more than deficiencies in daily living skills, which cause people with …[IDD] (and, for that matter, other forms of cognitive impairment) to be viewed as "disabled"" (101–102). They note that people with IDD "may be particularly gullible because they lack the content knowledge or cognitive processing ability to evaluate false claims and, also, because they typically have made personality adaptations (e.g., an "external" motivational orientation) which cause them, even more than is typically the case, to base their behavior on the cues provided by others" (Ibid., 102). While much of their concern focused on victimization, the argument applies equally well in the case of IDD and EAS, particularly in the consent phase.

In regard to consent, a strict interpretation of Article twelve of the CRPD implies that "trusted others" can act as the representative of the person's will and preferences. In a strict sense, there is no reason why this would not extend to EAS. This is of course easily prevented by simply excluding EAS from the scope of the supported decision maker's authority. For example, the British Columbia Representation Agreement Act (1996) articulates areas where the representative cannot make decisions.

Under section nine, which requires a more traditional test of capacity, and requires the individual to authorize the scope of decisions the representative can make, there is the power to "give or refuse consent to health care for the adult, including giving or refusing consent, in the circumstances specified in the agreement, to specified kinds of health care, even though the adult refuses to give consent at the time the health care is provided" (Ibid., § 9, ¶ 1, vii).This then would potentially extend to decisions regarding EAS, as (1) they are not otherwise excluded from the scope, and (2) the person may indicate they wish this to be within scope. In essence, it would essentially be a form of advance directive (AD). This leads to the second concern around consent. Advance directives are allowed for EAS in both the Netherlands and Belgium, even though they are restricted in Belgium to situations where the person is irreversibly unconscious (Lemmens 2018). In Canada, only a 'waiver of consent' is allowed for situations where the person loses capacity after being approved for MAiD. However, a recent parliamentary committee recommended the legalization of a broad AD regime, for example for persons developing dementia (Parliament of Canada 2023). Noteworthy is that the Canadian Association of MAID Assessors and Providers argues that a person diagnosed with early onset dementia can already sign a 'waiver of consent' after being approved for MAiD. Their mere diagnosis would make their death reasonably foreseeable, which is a circumstance in which a waiver of consent can be signed (CAMAP n.d.).

This discussion regarding EAS and IDD is relevant because ADs are a form of substitute consent (Council of Canadian Academies 2018). While the person may have indicated their wish for EAS in advance, they are no longer capable of consenting at the time of the procedure, meaning that a substitute makes the final decision to proceed. In a broader sense what this does is weaken the core safeguard of direct contemporaneous consent by a capable individual. People with IDD have long been subject to forms of substitute decision making and it is not unreasonable to see this extend to decisions regarding EAS.

## 26.5 The Final Frontier?

It did not take long after EAS was legalized in Canada for the first case to surface of a physician recommending MAiD to a mother of a young women with multiple disabilities.

> His words were 'assisted suicide death was legal in Canada,'" she told CBC. "I was shocked, and said, 'Well, I'm not really interested,' and he told me I was being selfish." (Bartlett 2017).

This is not particularly shocking given the general attitude of healthcare providers toward disability, and it is likely the case that many people would agree with the doctor given what we know about public reaction to so called "mercy killings" of disabled people.

The reaction of the courts and the public to cases of filicide of disabled children also suggests a sympathetic view toward ending disabled people's lives. In a study examining over 200 news reports from North America between 2011 and 2015 about the murder of people with disabilities by their caregivers, the motive routinely given was "hardship" (Perry 2017). The "hardship narrative"—describing the life of the disabled person and their family as unbearably difficult—appears to be acceptable to news media with minimal questioning. More significantly, the same appears to be true for the courts before which these killings are tried. The law does not excuse them, but does seem to see these murders as excusable given the fact that imprisonment for the killers is relatively short or not ordered. Looking at these reports it seems that "explaining" these acts as "mercy killings" produces killer-oriented rather than victim-oriented stories, in which the actual lives of the disabled persons disappear (Perry 2017). Of course, the obvious disparity with EAS is the absence of a medical context in these murders, but the verdict of unbearable suffering in the medical cases is not dissimilar to the "hardship narrative".

In a high profile 1993 case in Canada, Robert Latimer murdered his disabled twelve-year-old daughter, Tracy. He characterized his act as "mercy killing", citing her pain, quality of life, and disability as motivating factors. While the act itself was shocking, equally so was the support he received from the public and the lower court. The trial court initially gave a sentence well below the mandatory minimum sentence for second degree murder, invoking an exemption based on the concept of 'necessity'. The Court of Appeal confirmed the conviction but imposed the mandatory minimum sentence of life imprisonment without parole eligibility for ten years, which the Supreme Court eventually upheld.[5]

In 2012, Latimer appeared on a prominent public affairs program in support of Annette Corriveau, a mother of two disabled adult children seeking the right to "euthanize" them. The case, as presented in the program, indicated the bias of what the above study calls a *killer-oriented* rather than a *victim-oriented* story. It was announced as "a mother's plea for mercy" and "a father who has been down this road before" for which both have paid a price (Tryon et al. 2012). Subsequently, Corriveau

[5] R. v Latimer, [2001] 1 S.C.R. 3.

appeared on the popular Dr. Phil Show to make her case. His audience voted ninety percent in support of her plea for "mercy" (Tada 2012).

A recent request for an injunction by a father, aimed at preventing the death by EAS of his 27-year daughter reveals how we already have EAS requests in Canada that raise concerns about the impact of broad legalization on persons with IDD.[6] The father alleged that she did not qualify for EAS, and that her autism and mental illness drove her request for EAS (mental health is currently excluded as a sole basis for requesting EAS in Canada). He also invoked that there were procedural concerns about doctor shopping since two out of four physicians found her ineligible, and the doctor who approved the MAID request the first time was used as 'tiebreaker' to gain approval. The judge agreed that the father had a reasonable cause to go to court, but that in the interim, EAS could go ahead since the harm of delaying the procedure outweighed the harm of grief to the parents. Harm of the woman's premature death was not even considered.

## 26.6 Conclusion

It may seem alarmist to suggest that the expansion of EAS may herald a new era of eugenics. The evidence in this chapter suggests, however, a real possibility of a slow and uncritical movement toward this end. While some will argue that people with IDD have a "right" to EAS, we argue that this is a naïve and uncritical position. Failure to critically engage with the reality persons with IDD face and the subtle pressures that exist in the context of EAS may result in the advent of a quiet eugenics.

## References

American College of Pediatricians. 2004. *Neonatal Euthanasia: The Groningen Protocol*.

AMPQ (Association des médecins psychiatres du Québec). 2020. *Access to Medical Assistance in Dying for People with Mental Disorders: Discussion Paper.* https://ampq.org/wp-content/uploads/2020/12/mpqdocreflexionammenfinal.pdf. Accessed 10 April 2024.

Barrie, Stephen. 2015. QALYS, euthanasia and the puzzle of death. *Journal of Medical Ethics* 41: 635–638.

Bartlett, Geoff. 2017. Mother says doctor brought up assisted suicide option as sick daughter was within earshot. *CBC*, July 25. https://www.cbc.ca/news/canada/newfoundland-labrador/doctor-suggested-assisted-suicide-daughter-mother-elson-1.4218669. Accessed 10 April 2024.

BBC. 2020. Satoshi Uematsu: Japanese man who killed 19 disabled people sentenced to death. BBC News, March 16. https://www.bbc.com/news/world-asia-51903289. Accessed 10 April 2024.

British Columbia Representation Agreement Act. 1996. Chapter 405, Part 2, Sections seven and nine. https://www.bclaws.gov.bc.ca/civix/document/id/complete/statreg/96405_01#section7. Accessed 10 April 2024.

[6] *WV v MV*, [2024] ABKB 174.

CAMAP (Canadian Association of MAiD Assessors and Providers). n.d. Medical assistance in dying (MAiD) and Dementia. https://camapcanada.ca/wp-content/uploads/2022/02/Assessing-MAiD-in-Dementia-FINAL-Formatted.pdf. Accessed 7 April 2024.

Charland, Louis C., Trudo Lemmens, and Kyoko Wada. 2016. Decision-making capacity to consent to medical assistance in dying for persons with mental disorders. *Journal of Ethics in Mental Health* Open volume: 1–14. https://ssrn.com/abstract=2784291. Accessed 10 April 2024.

Council of Canadian Academies. 2018. *State of Knowledge on Medical Assistance in Dying for Mature Minors, Advance Requests, and Where a Mental Disorder Is the Sole Underlying Medical Condition: Summary of Reports*. Ottawa, ON: The Expert Panel Working Group on MAiD Where a Mental Disorder Is the Sole Underlying Medical Condition. Available at https://cca-reports.ca/reports/medical-assistance-in-dying/. Accessed 7 April 2024.

Cullen, Catherine, and Alexandra Zabjek. 2022. Federal minister says she's shocked by suggested of assisted deaths for some babies. *CBC*, October 22. https://www.cbc.ca/news/politics/assisted-dying-carla-qualtrough-1.6625412#:~:text=Canada%27s%20minister%20of%20disability%20inclusion%20says%20she%27s%20offended,issues.%20%22I%20find%20that%20completely%20shocking%20and%20unacceptable. Accessed 7 April 2024.

Day, Michael. 2018. Three Belgian doctors are investigated over euthanasia of woman with Asperger's. *BMJ* 363: k5106. https://doi.org/10.1136/bmj.k5106.

De Bruin, Tabitha, and Gerald Robertson. 2006. Eugenics in Canada. *The Canadian Encyclopedia*. Historica Canada. https://www.thecanadianencyclopedia.ca/en/article/eugenics#:~:text=Support%20for%20Eugenics%20in%20Canada&text=Many%20prominent%20Canadians%20of%20that,Helen%20MacMurchy. Accessed 7 April 2024.

De Hert, Marc, Sien Loos, Sigrid Sterckx, Erik Thys, and Kristof Van Assche. 2022. Improving control over euthanasia of persons with psychiatric illness: Lessons from the first Belgian criminal court case concerning euthanasia. *Frontiers in Psychiatry* 13: 933748. https://doi.org/10.3389/fpsyt.2022.933748.

Gill, Carol J. 2010. No we don't think our doctors are out to get us: Responding to the straw man distortions of disability rights arguments against assisted suicide. *Disability and Health Journal* 3: 31–38.

Goodey, C.F. 1992. Mental disabilities and human values in Plato's late dialogues. *Archiv Für Geschichte Der Philosophie* 74 (1): 26–42. https://doi.org/10.1515/agph.1992.74.1.26.

Greenspan, Stephen, Gail Loughlin, and Rhonda S. Black. 2001. Credulity and gullibility in people with developmental disorders: A framework for future research. In *International Review of Research in Mental Retardation*, vol. 24, 101–135. ed. Laraine Masters Glidden. Academic Press.

Hawkins, Mike. 1997. *Social Darwinism in European and American thought 1860–1945: Nature as Model and Nature as Threat*. London: Cambridge University Press.

Health Canada. 2022. *Expert Panel on MAiD and Mental Illness: Final Report. Government of Canada.* https://www.canada.ca/en/health-canada/corporate/about-health-canada/public-engagement/external-advisory-bodies/expert-panel-maid-mental-illness/final-report-expert-panel-maid-mental-illness.html. Accessed 10 April 2024.

Hohendorf, Gerrit. 2016. Death as a release from suffering—The history and ethics of assisted dying in Germany since the end of the 19th century. *Neurology, Psychiatry and Brain Research* 22 (2): 56–62. https://doi.org/10.1016/j.npbr.2016.01.003.

Iezzoni, Lisa I., Sowmya R. Rao, Julie Ressalam, Dragana Bolcic-Jankovic, Nicole D. Agaronnik, Karen Donelan, Tara Lagu, and Eric G. Campbell. 2021. Physicians' perceptions of people with disability and their health care. *Health Affairs* 40 (2): 297–306. https://doi.org/10.1377/hlthaff.2020.01452.

Janz, Heidi. 2023. MAID to die by medical and systemic ableism. In *Medical Assistance in Dying (MAID) in Canada: Key Multidisciplinary Perspectives*, ed. J. Kotalik and D.W. Shannon, 299–308. Cham: Springer.

Khazem, Lauren R., Danielle R. Jahn, Kelly C. Cukrowicz, and Michael D. Anestis. 2015. Physical disability and the interpersonal theory of suicide. *Death Studies* 39 (10): 641–646. https://doi.org/10.1080/07481187.2015.1047061.

Lemmens, Trudo, Heesoo Kim, and Elizabeth Kurz. 2018. Why Canada's medical assistance in dying legislation should be *C(h)arter* compliant and what it may help to avoid. *McGill Journal of Law and Health* 11 (1): S61–S148.

Lemmens, Trudo. 2018. Charter scrutiny of Canada's medical assistance in dying law and the shifting landscape of Belgian and Dutch euthanasia practice. *Supreme Court Law Review* 85: 459–544.

Lemmens, Trudo. 2023. *What Counts as Evidence in the Polarized Euthanasia/Assisted Dying Debate: Lessons from a Belgian Criminal Case.* University of Toronto Faculty of Law blog, May 5. https://www.law.utoronto.ca/blog/faculty/what-counts-evidence-in-polarized-euthanasiaassisted-dying-debate-lessons-belgian. Accessed 10 April 2024.

Locke, John. 1975. *An Essay Concerning Human Understanding.* Clarendon edition. ed. Peter H. Nidditch. Oxford: Oxford University Press.

Mckie, John, Jeff Richardson, Peter D. Singer, and Helga Kuhse. 1998. *The Allocation of Health Care Resources: An Ethical Evaluation of the QALY Approach.* Aldershot: Ashgate Publishing Ltd.

Morris, Stuart, Gail Fawcett, Laurent Brisebois, and Jeffrey Hughes. 2018. *A Demographic, Employment and Income Profile of Canadians with Disabilities Aged 15 Years and Over, 2017*. Government of Canada, Statistics Canada. https://www150.statcan.gc.ca/n1/en/catalogue/89-654-X. Accessed 7 April 2024.

Parens, Erik, and Adrienne Asch. 2003. Disability rights critique of prenatal genetic testing: Reflections and recommendations. *Mental Retardation and Developmental Disabilities Research Reviews* 9 (1): 40–47.

Parliament of Canada. 2023. *Medical Assistance In Dying In Canada: Choices For Canadians, Report of the Special Joint Committee, 44th Parliament, 1st Session.* https://publications.gc.ca/collections/collection_2023/sen/yc3/YC3-441-0-2-2-eng.pdf. Accessed 8 April 2024.

Perry, David. 2017. *The Ruderman White Paper on Media Coverage of the Murder of People with Disabilities by their Caregivers.* https://issuu.com/rudermanfoundation/docs/murders_by_caregivers_wp_final_fina_ac379254b115d2. Accessed 10 April 2024.

Quinones, Julian, and Arijeta Lajka. 2017. What kind of society do you want to live in?: Inside the country where Down syndrome is disappearing. *CBS News*, August 15. https://www.cbsnews.com/news/down-syndrome-iceland/. Accessed 10 April 2024.

Russell, David, R. Jay Turner, and Thomas E. Joiner. 2009. Physical disability and suicidal ideation: A community-based study of risk/protective factors for suicidal thoughts. *Suicide and Life-Threatening Behavior* 39 (4): 440–451.

Russell, David, R. Jay Turner, and Thomas E. Joiner. 1995. *Animal Liberation.* New York, New York.

Singer, Peter. 1993. *Practical Ethics*, 2nd ed. Cambridge: Cambridge University Press.

Singer, Peter. 1995. *Animal liberation.* New York, N.Y.

Skotko, Brian G. 2009. With new prenatal testing, will babies with down syndrome slowly disappear? *Archives of Disease in Childhood* 94 (11): 823–826. https://doi.org/10.1080/23297018.2015.1063447.

Stainton, Tim. 2023. Assisted life before assisted death: Disability discomfort regarding MAiD. In *Medical Assistance in Dying (MAID) in Canada: Key Multidisciplinary Perspectives,* ed. Jaro Kotalik and David W. Shannon. Switzerland: Springer Nature.

Stainton, Tim. 2016. Supported decision-making in Canada: Principles, policy, and practice. *Research and Practice in Intellectual and Developmental Disabilities* 3 (1): 1–11.

Stein, Mark S. 2002. Utilitarianism and the disabled: Distribution of resources. *Bioethics* 16 (1): 1–19.

Sullivan, William. 2023. Implementing medical assistance in dying in Canada: Implications for people with intellectual and developmental disabilities. In *Medical Assistance in Dying (MAID) in Canada: Key Multidisciplinary Perspectives,* ed. Jaro Kotalik and David W. Shannon. Switzerland: Springer Nature.

Tada, Joni Eareckson. 2012. *Disappointed in Dr. Phil. HuffPost, July 25.* https://www.huffingtonpost.com/joni-earecksontada/disappointed-in-dr-phil_b_1546354.html. Accessed 10 April 2024.

Tremain, Shelly Lynn. 2023. Disaster Ableism, Epistemologies of Crisis, and the Mystique of Bioethics. In *The Bloomsbury Guide to Philosophy of Disability: Radical Resistances and Intersectional Imaginings*, ed. Shelly Lynn Tremain, 19–47. London: Bloomsbury.

Tryon, Jennifer, Hannah James, and Megan Rowney. 2012. *Taking Mercy.* Global News, March 8 (updated March 18). https://globalnews.ca/news/220653/takingmercy/. Accessed 10 April 2024.

Tuffrey-Wijne, Irene, Leopold Curfs, Ilora Finlay, and Sheila Hollins. 2018. Euthanasia and assisted suicide for people with an intellectual disability and/or autism spectrum disorder: An examination of nine relevant euthanasia cases in the Netherlands (2012–2016). *BMC Medical Ethics* 19 (1): 17. https://doi.org/10.1186/s12910-018-0257-6.

Tuffrey-Wijne, Irene, Leopold Curfs, Ilora Finlay, and Sheila Hollins. 2019. Because of his intellectual disability, he couldn't cope. Is euthanasia the answer? *Journal of Policy and Practice in Intellectual and Developmental Disabilities* 16 (2): 113–116. https://doi.org/10.1111/jppi.12307.

Tuffrey-Wijne, Irene, Leopold Curfs, Sheila Hollins, and Ilora Finlay. 2023. Euthanasia and physician-assisted suicide in people with intellectual disabilities and/or autism spectrum disorders: Investigation of 39 Dutch case reports (2012–2021). *British Journal of Psychology Open* 9 (3): e87. https://doi.org/10.1192/bjo.2023.69.

UN (The United Nations). 2006. (CRPD) *Convention on the Rights of Persons with Disabilities.* Treaty Series, 2515, 3, G.A. Res. 61/106, 76th plen. Mtg., U.N. Doc A/Res/61/106. https://social.desa.un.org/issues/disability/crpd/convention-on-the-rights-of-persons-with-disabilities-crpd. Accessed 10 April 2024.

Verhagen, Eduard, and Pieter J.J. Sauer. 2005. The Groningen Protocol—Euthanasia in severely ill newborns. *New England Journal of Medicine* 352 (10): 959–962.

Vizcarrondo, Felipe E. 2014. Neonatal Euthanasia: The Groningen protocol. *The Linacre Quarterly* 81 (4): 388–392. https://doi.org/10.1179/0024363914Z.000000000.

Wiebe, Kayla, and Amy Mullin. 2023. Choosing death in unjust conditions: Hope, autonomy and harm reduction. *Journal of Medical Ethics.* https://doi.org/10.1136/jme-2022-108871.

Yamamoto, Arata. 2016. Japan alleged mass murderer: All disabled should cease to exist. *NBC News*, July 26. https://www.nbcnews.com/news/world/japan-alleged-mass-murderer-all-disabled-should-cease-exist-n616791. Accessed 10 April 2024.

# Part VII
# Ethical Responses to 'Behaviours'

# Chapter 27
# Engendering a Feeling of Safety as an Ethical Imperative in Preventing Emotional Distress and Behaviours that Challenge Services

**Elspeth Bradley and Janet A. Vogt**

**Abstract** Safety is a common organizational value, generally understood as the obligation to ensure the physical safety of service recipients, staff, and visitors. Missing from this definition is the service recipient's right to *feel safe*. Cultural safety theory recognizes the inherent imbalance between a service provider and recipient and recommends shifting the power to the recipient to decide whether they feel safe with the service. This chapter focuses on people with intellectual and developmental disabilities, with or without co-existing autism (IDD/ASD), especially those engaging in behaviours that challenge services. We offer an ethical argument for engendering a feeling of safety in the lives of people with IDD/ASD and discuss the neurobiology underpinning human feelings and responses to safety, danger, and life threat; the latter responses to danger and life threat being similar to emotional distress and behaviours that challenge services observed in people with IDD/ASD. In these contexts, we highlight the importance of practices such as co-regulation and advocate that organizations broaden their understanding of "patient safety" to include recipients' Inside-Out perspectives and their *felt sense of being safe*.

**Keywords** Feeling safe · Polyvagal · Co-regulation · Cultural safety · Challenging behaviours · Inside-out perspective · IDD/ASD

E. Bradley (✉)
University of Toronto, Toronto, ON, Canada
e-mail: e.bradley@utoronto.ca

J. A. Vogt
Surrey Place, Toronto, ON, Canada
e-mail: j.vogt@alumni.utoronto.ca; janet.vogt@surreyplace.ca



A. Bianchi and J. A. Vogt (eds.), *Intellectual Disabilities and Autism: Ethics and Practice*, The International Library of Bioethics 108,
https://doi.org/10.1007/978-3-031-61565-8_27

## 27.1 Introduction

In line with emerging neurobiological understanding of our human responses to danger and life threat, and the observation that these responses are similar to behaviours that challenge services (BTCS) (Bradley and Korossy 2022), we advocate that people with intellectual and developmental disabilities with or without co-existing autism (IDD/ASD) are the authority on what feels safe to them. The challenge is how to access this felt sense in those who may have little or no verbal language.

The practice of *co-regulation* (see Sect. 27.5), along with trying to imagine the perspective of the service recipients[1] (*Inside-Out* perspective—see Sect. 27.3.2), offers promise as a therapeutic approach in determining whether, and ensuring that, service recipients with IDD/ASD feel safe. We recommend that organizations expand their understanding of safety to include the *felt sense of being safe*. This is especially important for people who rely on others, and the structures and predictability of their daily environments and supports, to prevent and/or contain their anxieties.

In this chapter, we propose an ethical argument for engendering a feeling of safety in the lives of people with IDD/ASD; outline the neurobiology underpinning our feelings and responses to safety, danger, and life threat; describe the context wherein emotional distress and BTCS arise; explore the practice of co-regulation; and discuss the implications of redefining patient safety for organizational policy and practice.

## 27.2 An Ethical Argument for Ensuring Patients' Felt Sense of Being Safe

Patient safety is an ethical imperative connected to principles of beneficence and non-maleficence (Kangasniemi et al. 2013). Beneficence refers to the moral obligation to act for the benefit of others (Sharpe 2003). Non-maleficence entails a negative obligation to refrain from inflicting harm or risk of harm to others (Beauchamp and Childress 2013). Therefore, acts of commission or omission by the healthcare provider which harm the patient, are a violation of the patient's right to receive safe care and a failure of the provider's duty to deliver this (Kangasniemi et al. 2013).

The concept of *cultural safety* emphasizes the importance of practitioner self-awareness in establishing trust such that the patient feels culturally safe (Ramsden 2002). The understanding of culture may include, for example, considerations of gender, sexual orientation, socioeconomic status, and disability (McEldowney and Connor 2011). The practitioner must be aware of their own biases toward patients who are "culturally" different from themselves. In this context, safety is deliberately

[1] The arguments in this chapter apply to encounters in any environment where a person with IDD/ASD receives care and support. The terms *patient* and *practitioner* are used when healthcare literature is cited. Otherwise, the terms *service recipient* and *service provider* are used.

subjective and describes the patient's perception of trust (which is central to feeling safe). Trust can be understood as one's "acceptance of being, to some degree and in some respects, in another's power" (Baier 2007, 137). The *trust moment* shifts the balance of authority in the clinical encounter toward the patient; it is their *felt sense* that ultimately determines if the service and/or practitioner is safe (Ramsden 2002).

Any person may self-identify as belonging to a specific culture, including disability culture (Flemmer et al. 2014). Therefore, people with IDD/ASD have the right to feel safe in their service encounters. Whether they do is informed by their lived experience (e.g., how they process external sensory information, whether they receive necessary accommodations, and their past experiences in these regards). For many, feeling connected to their practitioner is an essential step in care delivery (O'Brien 2023)—and healing. However, if the practitioner does not recognize that a person feels unsafe, they may mistakenly regard them as "difficult". Cultural safety requires the practitioner to consider those things which make the patient unique (Papps and Ramsden 1996) and, if they identify things that make the person feel unsafe, they are obliged to try to mitigate these things (O'Brien 2023).

In 2000, the Institute of Medicine released *To Err is Human* (Kohn et al. 2000) which led to discussions of patient safety focused on avoiding harmful medical errors. Consequently, patient safety as the right to *feel safe* is not typically articulated in organizational safety plans. Feeling *unsafe* is associated with stress, trauma, and adversity which, in turn, especially when experienced in childhood, are linked to poor mental and physical health through the life span (Felitti et al. 1998). Thus, feeling safe in services is not just an ethical consideration, but a public health concern, especially for those with little capacity for self-advocacy.

To support the argument for redefining patient safety, one needs to have a basic understanding of the neurobiology underpinning the experience of safety, the origins of emotional and behavioural distress observed in people with IDD/ASD, and the importance of practices such as co-regulation.

## 27.3 The Neurobiology of Feeling Safe

*Feeling* safe is a subjective experience and we must embrace each person's circumstances to understand their experience of what it means to feel safe, or *not* safe. In this section, we explore differences in the personal sense of safety from what appears to be the same objective circumstance. First, we will explore the neurobiology of feelings.

### *27.3.1 Feelings and Interoception*

Neurobiological research (Craig 2014; Damasio 2018; Davis and Panksepp 2018; Porges 2011) is helping us understand our sensory bodies and providing a language that connects our physical body to our feelings. Feelings are mental experiences related to the body's sensory detection and physiology, through which we regulate our behaviours to keep our bodies in balance (homeostasis). Feelings can only be described by the person having them (*Inside-Out perspective*),[2] whereas emotions can be observed by others (*Outside-In perspective*). Craig describes two types of feelings: feelings that come from our bodies, e.g., "my fingers feel cold" and affective feelings that relate to our moods, e.g., "I feel anxious, I do not feel safe" (Craig 2014, xvii). Both types reflect our physiological state. We express our feelings through words, body language (e.g., facial expression, gestures) and the tone of what we say, our posture, and how we move. When we have limited or no verbal language (as for many with IDD/ASD), others make inferences (often incorrectly) from our non-verbal expressions.

Feeling and being tuned-in to what is happening in our sensory and physiological body (e.g., the urge to urinate) arises through *interoception.* Interoception describes how our feelings link to the experienced physiology of our bodies, and consequently how we act from moment to moment. Interoception draws our attention to how our body's physiological state impacts our emotional state (and vice versa); influences our motivations, perceptions, and thoughts; and consequently, impacts our physical and mental well-being (Mahler 2015). Feeling is a human experience, personally felt, that can be shared with others. Human feelings are independent of cognitive and functional capacities, though our sharing of feelings may be profoundly affected if others are insensitive to the non-verbal expression of our feelings and are therefore unable to "hear" us.

Our sensory system continuously reports what is going on outside (exteroception) and inside (interoception) our body to our brain. Such feedback has been referred to as *the material me* (Craig 2002). Functional brain imaging studies demonstrate subjective feelings (e.g., emotions, moods) correlated with interoceptive cortical activity, i.e., subjective awareness—*the feeling me* (Craig 2003). *The material me* and *the feeling me* contribute to the *sense of self*.

### 27.3.2 Inside-Out and Outside-In

The "Inside-Out" concept originated in the writings of Donna Williams, a multi-talented Australian artist and person with autism, who wrote "from the time someone came up with the word "autism", the condition has been judged from the outside, by its appearances and not from the inside according to how it is experienced" (Williams 1996, 14). Bradley and Caldwell (2013) describe how diagnostic criteria in psychiatry make assumptions about the intentions behind behaviours (Outside-In perspective) that differ from reports by individuals engaging in these behaviours (Inside-Out perspective). In *Inside-Out not Outside-In*, Elbard (2015, 9) writes:

[2] The terms *Inside-Out* and *Outside-In* are explained in Sect. 27.3.2.

> I want to share my life experiences with you and help you understand how my learning and developmental disability has shaped my life. These are the thoughts and feelings I have *inside*. I would like you to understand how people's perceptions and attitudes about this disability can be positive and negative. These are the thoughts and feelings *outside* of me which belong to other people. Sometimes these beliefs and attitudes do not help me and often they are not how I think and feel.

The *Inside-Out* perspective includes bodily and emotional *feelings* reported by the individual; others observe these as "affect" and emotional states (*Outside-In* perspective). One might add the *Outside-In me* perspective to refer to the inferences others, e.g., healthcare providers, make about *me*, based on *their Inside-Out* perspective, i.e., their own personal experience informing their perception. The mismatch between how *I* feel, and how *I* experience validation from others about how I feel, can undermine a secure sense of self.

### *27.3.3 Sense of Self*

Developmental psychology (e.g., Kohut 1991; Stern 1985), relational neurobiology (Siegel 2012), theories of attunement (e.g., Schore 2003; Trevarthen 1979), and attachment (Bowlby et al. 1965) draw our attention to the sense of self also being embedded in relationships; the first relationship(s) being with our primary caregiver(s).[3] We are wired to connect (Dana 2018; Porges 2011), seek connection, and when connection is broken, our emotional response is palpable (Tronick and Gold 2020) as evidenced in Tronick's "Still Face" experiment[4] (Tronick 2007). Healthy emotional development relies on our primary caregiver(s) being attuned to our affective and emotional states, helping us link different states with what is happening; for example, internally (e.g., upset over an injury) and externally (e.g., sound of a siren), as well as modulate an appropriate response (*co-regulation*).

These interactions set the stage for trust in interpersonal exchanges that occur even in the absence of verbal language. Through these reciprocal exchanges and co-regulation (Dana 2018), first with our primary caregiver(s), and later with others (e.g., teachers), we learn to self-regulate independently. Even so, co-regulation remains a stabilising resource throughout life—especially when circumstances may leave us dysregulated—and is an ongoing part of our sense of self. Disruption in opportunities for co-regulation alters our sense of self and, if experienced early in life, leaves our nervous system easily triggered into dysregulation; a state that can become chronic without healing interventions. Individuals with IDD/ASD are especially vulnerable to absence of, and disruptions in, co-regulatory experiences appropriate to their developmental needs; additionally, chronological age and adult status often masks

[3] The term *caregiver* refers to someone who provides personal care (and often has an emotional relationship with the person) such as, a family member, friend, or paid caregiver (e.g., a primary support worker). A *care provider* is any health or social service provider (e.g., family doctor or social worker).

[4] The "still face" experiment starts at 1′ 56″.

their stage of emotional development and therefore what emotional supports are needed. They rely on family, paid caregivers, and care providers to recognise these vulnerabilities and provide sensitive co-regulation, according to past experiences and stage of emotional development. Such interactions are critical to establishing *trust* and a *felt sense of being safe.*

### 27.3.4 Feeling Safe

Feeling safe is an emergent property of nervous system activity consequent to physical and mental health, healthy relationships, access to attuned supports and expectations according to one's developmental needs, along with capacity and flexibility to manage everyday challenges with ease. *Dis-ease* can occur when any of these are disrupted. Porges's polyvagal theory provides a cohesive account of autonomic nervous system (ANS) and vagal nerve activation helpful in understanding the neurobiological underpinnings of feeling safe and what happens when this felt sense is threatened (Porges 2017).

Porges introduces the concept of *neuroception* to describe the automatic process by which the nervous system and brain evaluate risk such as safety, danger, and life threat without awareness (Porges 2017). Once detected, body physiology adjusts to optimise survival (homeostasis). Ventral vagal activity ("social engagement system") mobilises social behaviours (e.g., smiling, vocalisation, and listening) that attempt to avoid or avert the danger. If this does not work, or the neuroception of danger is too great, the sympathetic nervous system is mobilised into *fight* (fight off the threat), *flight* (run away), or *freeze* (be still). When a lethal threat is perceived, the body defaults to an even earlier phylogenetic survival strategy, that of "feigning death" and *shutdown* (immobilisation).

Bradley and Korossy (2022) point to similarities between the survival behaviours of *fight-flight-freeze* and *shutdown* and BTCS seen in some with IDD/ASD (see Sect. 27.4). For example, *fight* may present as self-injurious behaviours, or hitting others barring the way; *flight* may present as running and be misinterpreted as *deliberately* "trying to escape", or "avoiding", while *freeze* may be interpreted as *consciously* "refusing" or "non-compliance". In humans, "feigning death" has been likened to a severe trauma response which includes behavioural collapse, fainting, urination, defecation, and dissociation (being detached from physical and emotional experiences). These behaviours are biologically triggered by the perceived degree of threat, prior trauma, coping capacities, and other physiological states that may compromise resilience and coping (e.g., physical illness).

Polyvagal understanding is a paradigm shift that "normalises" the plight of individuals with IDD/ASD experiencing emotional and behavioural distress. Rather than "challenging or problem behaviours" these might equally well be viewed as a healthy ANS working around the clock, being propelled into ANS survival states consequent to *overwhelm.* Such overwhelm may occur when the individual has limited resources to manage difficult situations (Bradley and Korossy 2022).

## 27.4 The Context in Which Emotional Distress and BTCS Arise

High rates of emotional and behavioural distress in people with intellectual and developmental disabilities (IDD) are well documented, most often being reported as increased prevalence rates of psychiatric disorder or "challenging behaviours" (Bradley and Korossy 2015). People with IDD represent a diverse population with up to one third having co-existing autism. The overlap between symptoms and signs of "psychiatric disorder" and observations of emotional and behavioural distress is due, in part, to the heterogeneity of IDD (e.g., mild to profound) and of autism (e.g., verbal to non-verbal). Some with IDD/ASD can communicate verbally what may be underlying their distress; direct information that is crucial to psychiatric assessment and diagnosis. Others instead communicate distress through their behaviours which, if causing harm to self, others, or the environment, may be labelled as "challenging" (Bradley and Korossy 2016). In this chapter, we discuss *behaviours that challenge services* (BTCS) and their implicit co-construction; that is, behaviours that challenge arise in the context of whether services are effective in supporting unique developmental needs (Mansell 2007).

Behaviours that challenge services remain difficult and distressing for all involved. Historically, psychotropic medication and behavioural supports have been mainstay interventions, raising concerns about these being restraining practices rather than treatment. In fact, emotional distress and BTCS can be considered through several lenses, including those of trauma, pain and medical conditions, unmet developmental needs, as well as adaptive survival responses to feeling unsafe.

### *27.4.1 The Lens of Trauma*

Any situation that leaves someone feeling overwhelmed and alone can be traumatic. It is the person's subjective emotional experience that determines whether an event is traumatic; the more frightened and helpless they feel, the more likely the event will be traumatizing (Ford et al. 2011). Because adults without IDD often do not fully appreciate the impact of living with limited knowledge, verbal, social, practical, and life skills, people with IDD are more vulnerable to being traumatized. Without adequate supports, they may live in a chronic state of overwhelm, being easily triggered into ANS survival responses of *fight—flight—freeze* or *shutdown* which may be labelled as BTCS.

People with IDD are subject to increased incidences of adverse life events, exposure to abuse, neglect, exploitation, victimization, and hate crimes (Sobsey 1994). Shockingly, studies amongst disability services have documented increased risk of abuse (up to 78%), with 14% of direct care staff admitting to acts of violence against adults with IDD and 35% saying they had witnessed such violence (Strand et al. 2004). People with IDD also experience greater exposure to painful medical procedures, particularly in early and late stages of life as well as greater rates of discontinuities in care related to hospital admissions, respite, group and institutional living, and staff turnover. Trauma associated with physical restraint has been reported (Wilton 2020), as well as negative consequences of psychotropic medication for BTCS (Scheifes et al. 2016). Trauma also occurs due to environmental factors (e.g., distress from sensory hypersensitivities) (Bradley and Caldwell 2013).

Post-traumatic stress disorder (PTSD) is an outcome of single trauma experiences. Chronic, repeated, and prolonged traumas (e.g., childhood abuse; bullying; care disruptions) tend to result in more complex reactions (Wikipedia 2023). Complex PTSD affects three key domains: emotion regulation, self-identity, and relational capacities (Cloitre 2020). Symptoms include prolonged feelings of terror, emotional numbness, worthlessness, helplessness, distortions in identity or sense of self, and feeling on edge; behaviours overlap with PTSD and include avoidance, changes in mood, physical and emotional reactions, and hyper-vigilance (NHS 2022).

### 27.4.2 The Lens of Developmental Needs and Appropriate Supports and Expectations

The developmental profiles of individuals with IDD are unique to each individual consequent to the cause of their disabilities (e.g., specific genetic syndromes, pre- or post-natal injury) and lived experiences, e.g., *enabling* supports optimizing growth and development, while environments not suited to developmental needs adding further to their disability and emotional distress. Emotional and behavioural distress occurs when their coping skills, abilities, and personal supports are insufficient to manage their experience of being overwhelmed; being overwhelmed is a subjective feeling (Inside-Out) observed by others (Outside-In) as dysregulated behaviours and displays of emotions.

In general, BTCS are associated with greater severity of IDD, and when IDD and autism co-exist (as occurs in 20 to 30% of the IDD population) (Bradley et al. 2014). Research has shown that this greater prevalence of BTCS in autism is associated with differences in emotional regulation (Licence et al. 2020; Nicholls et al. 2020). Different stages in emotional development and functioning may predict BTCS; a greater prevalence of BTCS in individuals with more severe IDD and autism correlating with less well-developed emotional functioning (Sappok et al. 2014).

Syndrome-specific patterns of behaviour can be challenging for services; for example, self-injurious behaviour in Cornelia de Lange syndrome, temper outbursts in Prader-Willi Syndrome, and displays of anxiety in fragile X. These patterns in Prader-Willi Syndrome and fragile X are differentially elicited by changes in routine or expectations (Woodcock et al. 2009). Attention to differences in sensory sensitivities and to communication barriers, provision of low arousal environments and emotional engagement with the individual all diminish distress (Bradley and Caldwell 2013; Caldwell et al. 2019).

To understand the individual's developmental needs, and whether these are being met, requires observing them and *how* they are being supported (e.g., engagement with caregivers), in the different environments where they receive support. Emotional engagement with the individual (e.g., through co-regulation) allows greater access to their *Inside-Out* experience. Without this engagement or being able to observe the patient in their familiar environments, care providers will only have *Outside-In* perspectives, based on their non-IDD experiences. This commonly results in ascribing intentions such as "non-compliant", "avoiding", "escaping", even "malingering" to the individual; interpretations which are often at odds with their felt internal state. This "knee jerk" ascribing of intention reflects the subjective experience of the observer (*Outside-In*) which is unhelpful and can cause harm (Bradley and Caldwell 2013; Bradley et al. 2014).

### *27.4.3 The Lens of Adaptive "Survival" Responses to Feeling Unsafe*

Bradley and Korossy (2022) provide evidence for considering emotional and behavioural distress in terms of responses to cues of safety, danger, and life threat (from the individual's perspective). Cues of danger come from inside our bodies (e.g., associated with pain, medical conditions); from outside our bodies (e.g., sight, sound, difficulties communicating); and "in between" such as through distressing interpersonal relationships (e.g., due to inappropriate expectations, absence of attunement with the individual's perspective, state and felt experience).

Cues of safety, and the *felt sense of safety*, arise from safe and *enabling environments* (Banks et al. 2007), with communication and emotional needs being met through safe connection to others and availability of co-regulation, according to emotional and developmental needs. Social connection, curiosity, and learning occur when an individual's neuroception detects cues of safety. Neuroception of danger triggers a hypervigilant, dysregulated nervous system, feeling unsafe, and survival responses (BTCS), which in turn inhibit social connection and learning.

## 27.5 Co-Regulation and Practice

Co-regulation can be understood from several perspectives, including neurobiological, the felt sense, and the practice. Neuroscience describes how the human brain and mammalian ANS are wired for connection (Dana 2018; Porges 2011). In engaging with others, we are continually, below awareness, sending and receiving cues of safety and danger through what we are saying and our behaviour (see Sect. 27.3). The felt sense of co-regulation refers to the experience, in a safe place, of a calming presence and soothing support to manage distress, difficult emotions and sensations, along with the availability of behaviours we can model to modulate our arousal. This felt sense engendered in the other is experienced through the tone of our words and behaviour. Responsive care providers pay close attention to shifting emotional and physiological cues in the individuals they support, and in themselves, while also regulating their own emotional state. Our ability to be regulated, and offer co-regulation to others, is essential to the co-regulation process (Dana 2021).

Porges (2017) describes *feeling safe* as an important moderator influencing the effectiveness of medical procedures, psychological interventions, and psychoeducation. Activating the Social Engagement System supports health, growth, and restoration of body function; in this state the ANS is not easily triggered into defensive responses. He promotes the principle of feeling safe as a precursor to treatment and advises vetting the treatment environment for danger cues (e.g., sensory triggers). Feeling safe also necessitates cues of safety (e.g., familiarity, predictability, safe support people). Porges describes our *tuned in* capacity to derive the state and intentions of others from their vocal tone, facial expressions, gestures, and postures. We rely on social supports and predictable reliable environments to self-regulate. Withdrawal or absence of supports, or loss of predictability and trust, are sufficient to trigger survival responses (Porges 2017, Chap. 2).

Several approaches to care implicitly embracing these principles are available: for example, Intensive Interaction (Caldwell 2013), Responsive Communication (Caldwell et al. 2019), Validation Therapy (Feil 1992), Adaptive Interaction (Ellis and Astell 2017), Circle of Security (Powell et al. 2014), and Roots of Empathy (Gordon 2005). Such attunement and supportive, consistent responses to the individual's dysregulated states, permit a growing capacity for self-regulation. Some with IDD may have limited developmental capacity for independent self-regulation and require ongoing co-regulation; however, their capacity will be enhanced by the provision of structured environments they perceive as safe.

## 27.6 Implications for Organizational Policy and Practice

If we accept safety as an ethical imperative, and that cultural safety requires service providers to establish a therapeutic context in which the recipient feels safe, what does this mean for organizations that serve people with IDD/ASD? In organizational ethics, the purposeful use of values guides the decisions of a system or organization (Potter 1996); thus offering a critical opportunity to reshape practice. First, we need to advocate for an expanded definition of service recipient safety that prioritizes their *felt sense of being safe,* meaning the person's Inside-Out perspective. Second, we must identify *service recipient safety* as a key organizational value. An organization's values are strongly aligned with its ethical commitment (Kotalik et al. 2014) and the new understanding of service recipient safety articulated above ought to inform relevant policies. This has implications for the accessibility of health care and other services (Ramsden 2002; O'Brien 2023) and treatment effectiveness (Porges 2017).

This proposal has clear implications for organizational practices. For instance, many people with IDD/ASD require accommodations, whether to ameliorate the impact of certain environmental factors (Bradley and Korossy 2016) or to support their capability to communicate (McNeil et al. 2018) and participate in decision making (Sullivan and Heng 2018). In order to feel safe, a person's attempts to communicate must not be thwarted. They must be protected from an assault on their senses if they are to connect with a service provider and engage in a meaningful way. It follows that service providers must make every attempt to identify and provide the necessary and reasonable accommodations and supports, while being mindful of their own behaviour and demeanor so as to avoid contributing to the person's emotional distress. Furthermore, the deliberate practice of *co-regulation*, along with attention to imagining the *Inside-Out* perspective, offers promise as a therapeutic approach for helping service recipients contain their anxieties and fears and thereby *feel safe*. It follows that co-regulation should be a core competency for those spending significant time in direct contact with individuals with IDD/ASD. This includes various caregivers (e.g., personal support workers) and providers (e.g., healthcare). To support this, colleges, universities, and other professional programs should include such training for people destined for clinical careers in the developmental services, social services, or health care.

## 27.7 Conclusion

The circumstances and lived experiences of people with IDD/ASD are diverse. Many have been profoundly traumatized related to their developmental vulnerabilities, not having their voices heard or perspectives considered, and experience a daily reality of *overwhelm* when not adequately supported. Neuroception of overwhelm triggers ANS survival responses which are similar to the emotional distress and BTCS observed in people with IDD/ASD. Alarmingly, people with IDD/ASD experience trauma often in the very systems charged with their support, care, and safety. Therefore, healing and prevention through system change is needed. Safety is a key principle in *trauma informed care* (Menschner and Maul 2016). Complex trauma seen in people with IDD/ASD has often happened in relationships and through environmental and social adversity; treatment and healing must necessarily occur through environments and relationships that are experienced as safe (Herman 2015), including the availability of safe co-regulation. There has been a paradigm shift in understanding trauma that should encourage us to ask *what happened to you* rather than *what is wrong with you* (Winfrey and Perry 2021). What has happened to people with IDD/ASD has happened often and sometimes even intentionally within their systems of care. We must ensure that care systems are experienced as safe from their perspective.

## References

Baier, Annette. 2007. Trust, suffering, and the Aesculapian virtues. In *Working Virtue: Virtue Ethics and Contemporary Moral Problems*, ed. Rebecca L. Walker and Philip J. Ivanhoe, 135–153. Oxford: Clarendon.

Banks, Roger, Alick Bush, Peter Baker, Jill Bradshaw, Peter Carpenter, Shoumitro Deb, Theresa Joyce, Jim Mansell, and Kiriakos Xenitidis. 2007. *Challenging Behaviour: A Unified Approach. Royal College Report CR144.* https://www.rcpsych.ac.uk/docs/default-source/improving-care/better-mh-policy/college-reports/college-report-cr144.pdf?sfvrsn=73e437e8_2. Accessed 21 May 2024.

Beauchamp, Tom L., and James F. Childress. 2013. *Principles of Biomedical Ethics* 7th ed. New York: Oxford University Press.

Bowlby, John, Margery Fry, and Mary D. Ainsworth. 1965. *Child Care and the Growth of Love.* Baltimore, MD: Penguin Books.

Bradley, Elspeth, and Phoebe Caldwell. 2013. Mental health and autism: Promoting autism favourable environments (PAVE). *Journal on Developmental Disabilities* 19 (1): 8–23.

Bradley, Elspeth, Phoebe Caldwell, and Lisa Underwood. 2014. Autism Spectrum Disorder. In *Handbook of Psychopathology in Intellectual Disability: Research, Practice, and Policy*, ed. Elias Tsakanikos and Jane McCarthy, 237–264. New York: Springer. https://doi.org/10.1007/978-1-4614-8250-5.

Bradley, Elspeth, and Marika Korossy. 2015. Behaviour problems. In *Clinical Topics in Disorders of Intellectual Development*, 72–112. London: The Royal College of Psychiatrists.

Bradley, Elspeth, and Marika Korossy. 2016. Help with Behaviours that Challenge. *Journal on Developmental Disabilities* 22 (2): 101–120.

Bradley, Elspeth, and Marika Korossy. 2022. Are difficult behaviours described in intellectual and developmental disabilities and autism actually adaptive responses to feeling unsafe? *Journal on Developmental Disabilities* 27 (1): 1–26.

Caldwell, Phoebe. 2013. Intensive interaction: Using body language to communicate. *Journal on Developmental Disabilities* 19 (1): 33–39.

Caldwell, Phoebe, Elspeth Bradley, Janet Gurney, Jennifer Heath, Hope Lightowler, Kate Richardson, and Jemma Swales. 2019. *Responsive Communication: Combining Attention to Sensory Issues with Using Body Language (Intensive Interaction) to Interact with Autistic Adults and Children.* Shoreham-by-Sea, UK: Pavilion Publishing and Media Ltd.

Cloitre, Marylène. 2020. ICD-11 complex post-traumatic stress disorder: Simplifying diagnosis in trauma populations. *British Journal of Psychiatry* 216 (3): 129–131. https://doi.org/10.1192/bjp.2020.43.

Craig, A. D. (Bud). 2002. How do you feel? Interoception: The sense of the physiological condition of the body. *Nature Reviews Neuroscience* 3 (8): 655–666. https://doi.org/10.1038/nrn894.

Craig, A. D. (Bud). 2003. Interoception: The Sense of the Physiological Condition of the Body. *Current Opinion in Neurobiology* 13 (4): 500–505. https://doi.org/10.1016/S0959-4388(03)00090-4.

Craig, A. D. (Bud). 2014. *How Do You Feel? An Interoceptive Moment with Your Neurobiological Self*. Princeton, NJ: Princeton University Press. https://doi.org/10.23943/princeton/9780691156767.001.0001.

Damasio, Antonio R. 2018. *The Strange Order of Things: Life, Feeling, and the Making of Cultures.* New York: Pantheon Books.

Dana, Deborah A. 2018. *The Polyvagal Theory in Therapy: Engaging the Rhythm of Regulation.* New York: W.W. Norton.

Dana, Deborah A. 2021. *Anchored: How to Befriend Your Nervous System using Polyvagal Theory.* Louisville, CO: Sounds True.

Davis, Kenneth L., and Jaak Panksepp. 2018. *The Emotional Foundations of Human Personality: A Neurobiological and Evolutionary Approach.* New York: W.W. Norton.

Elbard, Kareem. 2015. Inside out not outside in—A change of attitude for all: A self-advocate's vision for those with intellectual and developmental disabilities. *Journal of Religion and Society* supplement 12: 8–12. https://cdr.creighton.edu/server/api/core/bitstreams/5636d3b3-04e4-4764-bb23-10c2fdd38cf2/content. Accessed 21 May 2024.

Ellis, Maggie, and Arlene Astell. 2017. *Adaptive Interaction and Dementia: How to Communicate Without Speech*. Philadelphia, PA: Jessica Kingsley Publishers.

Feil, Naomi. 1992. Validation therapy. *Geriatric Nursing* 13 (3): 129–133. https://doi.org/10.1016/S0197-4572(07)81021-4.

Felitti, Vincent J., Robert F. Anda, Dale Nordenberg, David F. Williamson, Alison M. Spitz, Valerie Edwards, Mary P. Koss, and James S. Marks. 1998. Relationship of childhood abuse and household dysfunction to many of the leading causes of death in adults: The adverse childhood experiences (ACE) study. *American Journal of Preventive Medicine* 14 (4): 245–258. https://doi.org/10.1016/j.amepre.2019.04.001.

Flemmer, Nicole, Lida Dekker, and Dawn Doutrich. 2014. Empathetic partnership: An interdisciplinary framework for primary care practice. *The Journal for Nurse Practitioners* 10 (8): 545–551. https://doi.org/10.1016/j.nurpra.2014.04.009.

Ford, Gail, Nora Baladerian, and Janice Schaad. 2011. *How Trauma Affects People with Intellectual Disabilities*. Washington, DC: The Arc http://www.thearc.org/wp-content/uploads/forchapters/Trauma.pdf. Accessed 21 May 2024.

Gordon, Mary. 2005. *Roots of Empathy: Changing the World, Child by Child*. Toronto: Thomas Allen and Son.

Green, Laurie, Karen McNeil, Marika Korossy, Kerry Boyd, Elizabeth Grier, MacKenzie Ketchell, Alvin Loh, et al. 2018. Help for behaviours that challenge in adults with intellectual and developmental disabilities. *Canadian Family Physician* 64 (Suppl 2): S23–S31.

Herman, Judith Lewis. 2015. Trauma and Recovery: The Aftermath of Violence—From Domestic Abuse to Political Terror. New York: Basic Books.

Kangasniemi, Mari, Mojtaba Vaismoradi, Melanie Jasper, and Hannele Turunen. 2013. Ethical issues in patient safety: Implications for nursing management. *Nursing Ethics* 20 (8): 904–916. https://doi.org/10.1177/0969733013484488.

Kohn, Linda T., Janet M. Corrigan, and Molla S. Donaldson (eds). 2000. *To Err is Human: Building a Safer Health System. Report of the Institute of Medicine*. Washington, D.C.: National Academy Press.

Kohut, Heinz. 1991. *The Search for the Self: Selected Writings of Heinz Kohut: 1978–1981*, vol. 4 ed. Paul H. Ornstein. Madison, CT: International Universities Press, Inc.

Kotalik, Jaro, Cathy Covino, Nadine Doucette, Steve Henderson, Michelle Langlois, Karen McDaid, and Louisa M. Pedri. 2014. Framework for ethical decision-making based on mission, vision and values of the institution. *HEC Forum* 26 (2): 125–133. https://doi.org/10.1007/s10730-014-9235-7.

Licence, Lucy, Chris Oliver, Jo Moss, and Caroline Richards. 2020. Prevalence and risk-markers of self-harm in autistic children and adults. *Journal of Autism and Developmental Disorders* 50 (10): 3561–3574. https://doi.org/10.1007/s10803-019-04260-1.

Mahler, Kelly J. 2015. *Interoception: The Eigth Sensory System: Practical Solutions for Improving Self-regulation, Self-awareness and Social Understanding of Individuals with Autism Spectrum and Related Disorders*. Shawnee Mission: AAPC Publishing.

Mansell, Jim L. 2007. *Services for People with Learning Disabilities and Challenging Behaviour or Mental Health Needs. October*. London, UK: Department of Health. https://research.kent.ac.uk/tizard/wp-content/uploads/sites/2302/2019/01/dh2007mansellreport.pdf. Accessed 21 May 2024.

McEldowney, Rose, and Margaret J. Connor. 2011. Cultural safety as an ethic of care: A praxiological process. *Journal of Transcultural Nursing* 22 (4): 342–349. https://doi.org/10.1177/1043659611414139.

McNeil, Karen, Meg Gemmill, Dara Abells, Samantha Sacks, Terry Broda, Catherine R. Morris, and Cynthia Forster-Gibson. 2018. Circles of care for people with intellectual and developmental disabilities: Communication, collaboration, and coordination. *Canadian Family Physician* 64 (Suppl 2): S51–S56.

Menschner, Christopher, and Alexandra Maul. 2016. *Key Ingredients for Successful Trauma-informed Care Implementation.* Princeton, N.J.: Robert Wood Johnson Foundation. https://www.samhsa.gov/sites/default/files/programs_campaigns/childrens_mental_health/atc-whitepaper-040616.pdf. Accessed 21 May 2024.

NHS (National Health Service). 2022. *Symptoms: Post-traumatic Stress Disorder*. https://www.nhs.uk/mental-health/conditions/post-traumatic-stress-disorder-ptsd/symptoms/. Accessed 21 May 2024.

Nicholls, Gemma, Richard P. Hastings, and Corinna Grindle. 2020. Prevalence and correlates of challenging behaviour in children and young people in a special school setting. *European Journal of Special Needs Education* 35 (1): 40–54. https://doi.org/10.1080/08856257.2019.1607659.

O'Brien, Katherine. 2023. The importance of feeling safe. *eDialogue: A Publication for Ontario Doctors.* Toronto: College of Physicians and Surgeons of Ontario. https://dialogue.cpso.on.ca/2023/03/the-importance-of-feeling-safe/. Accessed 21 May 2024.

Papps, Elaine, and Irihapeti Ramsden. 1996. Cultural safety in nursing: The New Zealand experience. *International Journal for Quality in Health Care* 8 (5): 491–497.

Porges, Stephen W. 2011. *The Polyvagal Theory: Neurophysiological Foundations of Emotion, Attachment, Communication and Self-Regulation.* New York: W.W. Norton.

Porges, Stephen W. 2017. *The Pocket Guide to the Polyvagal Theory: The Transformative Power of Feeling Safe.* New York: W.W. Norton.

Porges, Stephen W. 2022. Polyvagal theory: A science of safety. *Frontiers in Integrative Neuroscience* 16: 1–15. https://doi.org/10.3389/fnint.2022.871227.

Potter, Robert L. 1996. From clinical ethics to organizational ethics: The second stage of the evolution of bioethics. *Bioethics Forum* 12 (2): 3–12.

Powell, Bert, Glen Cooper, Kent Hoffman, and Bob Marvin. 2014. *The Circle of Security Intervention: Enhancing Attachment in Early Parent-child Relationships.* New York: Guildford Press.

Ramsden, Irihapeti Merenia. 2002. Cultural safety and nursing education in Aotearoa and Te Wai Pounamu. *Doctor of Philosophy in Nursing.* Victoria: University of Wellington.

Sappok, Tanja, Jan Budczies, Isabel Dziobek, Sven Bölte, Anton Dosen, and Albert Diefenbacher. 2014. The missing link: Delayed emotional development predicts challenging behavior in adults with intellectual disability. *Journal of Autism and Developmental Disorders* 44: 786–800. https://doi.org/10.1007/s10803-013-1933-5.

Scheifes, Arlette, Sanne Walraven, Joost Jan Stolker, Henk L. I. Nijman, Toine C. G. Egberts, and Eibert R. Heerdink. 2016. Adverse events and the relation with quality of life in adults with intellectual disability and challenging behaviour using psychotropic drugs. *Research in Developmental Disabilities* 49–50: 13–21. https://doi.org/10.1016/j.ridd.2015.11.017.

Schore, Allan N. 2003. *Affect Regulation and the Repair of the Self*. New York: W.W. Norton.

Sharpe, Virginia A. 2003. Promoting patient safety: An ethical basis for policy deliberation. *The Hastings Center Report* 33 (5): S1–S20. https://doi.org/10.2307/3528637.

Siegel, Daniel J. 2012. *Pocket Guide to Interpersonal Neurobiology: An Integrative Handbook of the Mind*. New York: W.W. Norton.

Sobsey, Richard. 1994. *Violence and Abuse in the Lives of People with Disabilities: The end of Silent Acceptance?* Baltimore, MD: Paul H. Brookes.

Stern, Daniel N. 1985. *The Interpersonal World of the Infant: A View from Psychoanalysis and Developmental Psychology*. New York: Basic Books. https://doi.org/10.4324/9780429482137.

Strand, Marie Louise, Eva Benzein, and Britt-Inger Saveman. 2004. Violence in the care of adult persons with intellectual disabilities. *Journal of Clinical Nursing* 13 (4): 506–514. https://doi.org/10.1046/j.1365-2702.2003.00848.x.

Sullivan, William F., and John Heng. 2018. Supporting adults with intellectual and developmental disabilities to participate in health care decision making. *Canadian Family Physician* 64 (Suppl 2): S32–S36.

Sullivan, William F., Michael Bach, John Heng, Megan Henze, Lana Kerzner, Karen McNeil, Andrea Perry, and Janet Vogt. 2020. *Decision Making in Health Care of Adults with Intellectual and Developmental Disabilities: Promoting Capabilities*. Toronto, ON: Developmental Disabilities Primary Care Program of Surrey Place. https://ddprimarycare.surreyplace.ca/wp-content/uploads/2020/11/Decision-Making-Approaches.pdf. Accessed 21 May 2024.

Trevarthen, Colwyn. 1979. Communication and cooperation in early infancy: A description of primary intersubjectivity. In *Before Speech: The Beginnings of Interpersonal Communication*, ed. Margaret Bullowa, 321–348. Cambridge: Cambridge University Press.

Tronick, Edward. 2007. *A Vision for the Developmental Sciences*. Filmed in Boston, MA. Video for Developmental Sciences at University of Massachusetts. https://www.youtube.com/watch?v=vmE3NfB_HhE&t=116s. Accessed 21 May 2024.

Tronick, Edward, and Claudia M. Gold. 2020. *The Power of Discord: Why the Ups and Downs of Relationships are the Secret to Building Intimacy, Resilience, and Trust*. New York: Little, Brown Spark.

Wikipedia. 2023. *Complex Post-traumatic Stress Disorder*. Wikimedia Foundation. https://en.wikipedia.org/wiki/Complex_post-traumatic_stress_disorder. Accessed 21 May 2024.

Williams, Donna. 1996. *Autism: An Inside-Out Approach*. London: Jessica Kingsley.

Wilton, Jo. 2020. *Briefing 54: Trauma, Challenging Behaviour and Restrictive Interventions in Schools*. Centre for Mental Health (UK). https://www.centreformentalhealth.org.uk/publications/trauma-challenging-behaviour-and-restrictive-interventions-schools#:~:text=Briefing%2054,-How%20do%20restrictive&text=Trauma%2C%20challenging%20behaviour%20and%20restrictive%20interventions%20in%20schools%20is%20a,between%20trauma%20and%20challenging%20behaviour. Accessed 21 May 2024.

Winfrey, Oprah, and Bruce D. Perry. 2021. *What Happened to You? Conversations on Trauma, Resilience, and Healing*. New York: Flatiron Books.

Woodcock, Kate A., Chris Oliver, and Glyn Humphreys. 2009. Task-switching deficits and repetitive behaviour in genetic neurodevelopmental disorders: Data from children with Prader-Willi syndrome chromosome 15 q11–q13 deletion and boys with Fragile X syndrome. *Cognitive Neuropsychology* 26 (2): 172–194. https://doi.org/10.1080/02643290802685921.

# Chapter 28
# Self-Determination and Alcohol Use: Exploring the Intersection of Ethics and Harm Reduction for People with Intellectual Disabilities

**Brooke Magel and Daniel Z. Buchman**

**Abstract** The closure of government-run institutions for people with intellectual disabilities (ID) has led to the advancement of equal rights, status, and community integration. As a result of increased community integration, people with ID now have more access and choice to consume substances such as alcohol. This becomes a significant concern insofar as alcohol use (AU) is on the rise in this population, and people with ID face many barriers to accessing AU/Alcohol Use Disorder (AUD) services. In this chapter, we explore concepts such as "Nothing About Us Without Us", dignity of risk, choice, autonomy, and self-determination, and highlight their alignment with the principles of harm reduction. We argue that adopting a harm reduction approach is ethically justifiable for people with ID related to AU/AUD. It is crucial to have people with ID lead or be involved in harm reduction strategies to ensure the strategies are meaningful and accessible.

**Keywords** Harm reduction · Alcohol · Intellectual disabilities · Ethics

---

B. Magel · D. Z. Buchman (✉)
Education, Centre for Addiction and Mental Health, Toronto, ON, Canada
e-mail: daniel.buchman@camh.ca

B. Magel
e-mail: brooke.magel@camh.ca

D. Z. Buchman
Bioethics, Centre for Addiction and Mental Health; Clinical Public Health, Dalla Lana School of Public Health, University of Toronto, Toronto, ON, Canada

Joint Centre for Bioethics, University of Toronto, Toronto, ON, Canada

 
A. Bianchi and J. A. Vogt (eds.), *Intellectual Disabilities and Autism: Ethics and Practice*, The International Library of Bioethics 108,
https://doi.org/10.1007/978-3-031-61565-8_28

## 28.1 Introduction

The closure of government-run institutions for people with intellectual disabilities (ID)[1] was prompted by the disability rights movement focused on greater independence and community integration (Bhatt and Gentile 2021). When institutionalization was typical, many people perceived people with ID as dependent, lacking capacity, burdensome, and incapable of obligations associated with citizenship. While there is still a long way to go, attitudes have since evolved with greater understanding and advocacy. Governments have acknowledged that people with ID deserve to be recognized as full members of society and should be extended equal rights, liberty, and status to redress past injustices (Altermark 2017). Such injustices include denying rights for people with ID (e.g., the right to vote) and segregation from family and mainstream society (Johnson and Bagatell 2020).

One outcome of community integration and deinstitutionalization is an increased ability to access psychoactive substances, such as alcohol (Gee et al. 2020; Huxley et al. 2019). In institutions, people with ID were closely monitored and had little to no ability to access substances without others' facilitation (Huxley et al. 2005). Community integration allows for greater autonomy for people with ID to access alcohol, which may increase their potential risk for developing conditions such as Alcohol Use Disorder (AUD) (Bhatt and Gentile 2021).

In this chapter, we argue for the ethical justification of adopting a harm reduction approach for alcohol use (AU)/AUD within the ID population. This justification is established upon an exploration of ethics and harm reduction. We argue that AU/AUD treatment must ensure the meaningful inclusion and full participation of people with ID, thus promoting their autonomy and dignity through empowered choice.

## 28.2 Alcohol and Alcohol Use Disorder Among People with ID

There is a misconception that people with ID do not consume alcohol or use substances (Lamanna et al. 2020). This misconception may be due to assumptions about their capacity to make choices and express preferences (Bhatt and Gentile 2021). Alcohol is the most used substance (Chaplin et al. 2011; Didden et al. 2020; van Duijvenbode and Van Der Nagel 2019) followed by cannabis (van Duijvenbode and Van Der Nagel 2019). The age of initiation for substance use (e.g., AU) is often later in life for people with ID compared to people without ID[2] (Gee et al. 2020), and AU patterns can differ significantly based on access, support, and cultural factors.

[1] We adopt person-first language 'people with intellectual disabilities' to honor the preferences of many people with ID. However, people may prefer identity-first language and preferences will continue to evolve.

[2] 'People without intellectual disabilities (ID)' refers to both non-disabled people and people with disabilities that do not have impairments falling under ID.

Alcohol use is often considered a milestone in transitioning from youth to adulthood for all, which is particularly precarious for people with ID.

People with ID use substances similarly to people without ID, such as to celebrate (Gee et al. 2020; Simpson 1998, 2012) or to cope with poor life experiences (Taggart et al. 2007). Alcohol can be a pleasurable activity that serves as a central component of social inclusion for people with ID (Ibid.). As Simpson (2012, 189) highlights:

> [A]lcohol use and nonuse by adults with intellectual disabilities is more than a simple matter of the management of risky behaviour, or even individual rights; neither of which acknowledges the social and cultural context and function of drinking. Drinking serves as a definite marker of the genuine 'cultural participation' that people with intellectual disabilities enjoy, or, perhaps, do not.

There are inconsistencies in both the collection and reporting on the prevalence of substance use and substance use disorders (e.g., AUD) in the ID population (Didden et al. 2020; Van Der Nagel et al. 2017). Some inconsistencies can be attributed to varying definitions of ID (Didden et al. 2020), denial of substance use (Didden et al. 2020), lack of formal assessments (Bhatt and Gentile 2021), and data collection instruments that are not inclusive for people with ID (Williams et al. 2018).

Some research indicates that the prevalence of substance use disorder may be lower in adults with ID (Carroll Chapman and Wu 2012; Chaplin et al. 2011; Huxley et al. 2005) and many may abstain from using substances (Simpson 2012; Van Der Nagel et al. 2011). Other research suggests that substance use disorders may be higher among people with ID (Bhatt and Gentile 2021) and may also be increasing due to deinstitutionalization (Day et al. 2016; Didden et al. 2020).

The literature focuses on the potential harms of excess AU and emphasizes the adverse effects of AU/AUD (Gee et al. 2020; Simpson 1998; Slayter 2007). Simpson (2012) argues that pathologizing and framing AU within the discourse of risk and as a "problem" negates the cultural importance and social inclusivity of AU. Instead, Simpson (2012, 184) suggests that discussions should be reoriented toward the "socially constructive aspects of drinking".

## 28.3 Alcohol and Alcohol Use Disorder Services for People with ID

It is often unclear whether people with ID should seek mainstream AU/AUD services or community ID services for AU/AUD (Slayter and Steenrod 2009) and there is a lack of evidence to suggest which services are more suitable (Huxley et al. 2005). Many community ID service employees do not feel adequately trained nor have the expertise to treat AUD (Kiewik 2018; Lamanna et al. 2020; Williams et al. 2018) and seek advice from mainstream AU/AUD service programs on an ad hoc basis (McLaughlin et al. 2007). In one study from the Netherlands, seventy-nine percent of community ID service employees cited a need for more knowledge and felt ill-equipped to treat AUD in people with ID (Van Der Nagel et al. 2011). Despite this, another study found that community ID service employees were able to provide valuable education about the perils of AU to people with ID (Taggart et al. 2007).

Often limited AU/AUD services are available for people with ID, so generic mainstream services are used (Huxley et al. 2019). However, employees at these services often lack knowledge about ID (Van Der Nagel et al. 2011; Lamanna et al. 2020). Bhatt and Gentile (2021) found that people with ID were denied AU/AUD services due to their intellectual impairments, as mainstream services cited that they could not meet their needs. Further, when people with ID *did* attend mainstream services, they often reported a negative experience and discontinued treatment (Taggart et al. 2007). It ought to be noted that AU/AUD services are likely overburdened in providing services which may make them hesitant to offer tailored services to people with ID unless there are established best practice frameworks. Unfortunately, such frameworks are absent (Taggart et al. 2007) and access to treatment programs continues to be a barrier for people with ID (Didden et al. 2020; Van Der Nagel et al. 2018). Given this, integrated strategies and service reconfiguration that bridge the gaps between ID care and mainstream AU/AUD services are strongly encouraged (Taggart et al. 2007) and literature suggests that it may improve treatment outcomes (Huxley et al. 2019; McLaughlin et al. 2007). Bridging ID care and AU/AUD treatments could offer the ability to reduce potential barriers to treatment and improving access to treatment should be prioritized (Lamanna et al. 2020).

## 28.4 Harm Reduction

AUD is a chronic condition and treatment should prioritize reducing harm over time (i.e., harm reduction approach) (Constantino et al. 2020). At its core, harm reduction is a social justice movement based on respecting people who use substances and encompasses various strategies to minimize the negative effects associated with substance use (National Harm Reduction Coalition n.d.). Presently, there is no universal definition for harm reduction (National Harm Reduction Coalition n.d.; Pineau 2021) or universal approach to implementing harm reduction principles (National Harm Reduction Coalition n.d.).

The origins of harm reduction can be traced back to grassroots advocates who used drugs (Marlatt 1996). Harm reduction measures (e.g., sterile needle and syringe distribution) began to gain acceptance in the early 1990s, fueled by the emerging HIV epidemic (Ball 2007). Harm reduction continues to expand beyond its roots, and the philosophy of harm reduction has been applied within medical contexts, such as anorexia nervosa (Bianchi et al. 2021; Tsiandoulas et al. 2023) and AU (Marlatt and Witkiewitz 2002). Everyday examples of harm reduction include condoms to reduce pregnancy and sexually transmitted infections, supervised injection sites, and naloxone distribution (Denning and Little 2017). However, harm reduction does come with its critiques. Some have argued that harm reduction may unintentionally reinforce a prescriptive moralistic stance by emphasizing that people are individually responsible for maintaining their health which diverts attention from broader systemic issues (e.g., poverty, ableism, and racism) that can shape health outcomes (Miller 2001).

## 28.4.1 Harm Reduction and Ethics

Harm reduction is often described as value or morally neutral (Fry et al. 2005; Pauly 2008; Rieder 2023; Stoljar 2020). For example, harm reduction commonly emphasizes that the ethical focus should not be on the question of whether people should have the choice to consume or not consume substances, including alcohol (Keane 2003), rather the emphasis should be on minimizing harm (Christie et al. 2008; Keane 2003; Pineau 2021). By not focusing on the question of morality, harm reduction encourages others to refrain from passing judgment based on their choices (Pauly 2008; Rieder 2023). However, as Fry et al. (2005), Keane (2003), and Pauly (2008) remark, harm reduction cannot truly be "value-neutral", as the "overarching goal of reducing harm is in itself a value notion" (Fry et al. 2005, 450).

### 28.4.1.1 Values of Harm Reduction

**Rights**

At the core of the tension surrounding harm reduction lies the conflicting perspectives on whether a person has the right to consume alcohol and the responsibility of institutions (e.g., the government or community ID services) to prevent, support, and intervene (Fry et al. 2005). According to Keane (2003), it is unclear whether a human rights framework would offer the optimal guidance on ethical issues in the realm of harm reduction. Many policies, regulations, and practices threaten the freedom of people with ID. Such policies and practices categorize individuals such as people with ID as marginalized and are often excluded from the "normal" definitions of citizenship (Ibid.). Adopting a human rights perspective is one way of operating against the status quo by making strong claims for the interests of people with ID who consume alcohol (Ibid.). Ultimately, a rights-based approach may confront the stigma of AU among people with ID, which is critical to enable people with ID to lead self-determined lives.

Hathaway's (2001) perspective further supports this argument and suggests that rights lie at the core of harm reduction and can establish a foundation for equality and necessary policy change. Additionally, Ezard (2001) identified that focusing on rights is paramount because it underscores the government's responsibility to minimize the adverse effects associated with substance use (e.g., AU). In this rights-based view, the government is obligated to offer services that can mitigate the harms of substance use. However, a rights-based approach may increase political barriers encountered in harm reduction as it places existing policies, regulations, and laws at the center of discussion (Keane 2003).

**Dignity of Risk**

Honoring and respecting the rights of people with ID relates to the concept of dignity of risk. The concept was coined by Perske (1972) in response to the overprotection of people with ID. Perske notes that the overprotection of people with ID denies them the opportunity to grow and take part in the risk-taking of everyday "normal" life. Dignity of risk emphasizes that a person with ID should be at the center of their own decision making, recognizing that life for all people, regardless of whether they have a disability, contains risk (CoughDrop 2021; Woolford et al. 2019). Dignity of risk suggests that people with ID should have the autonomy to make decisions that affect their lives and the right to take risks while leveraging desired supports (Bumble et al. 2022; CoughDrop 2021; Woolford et al. 2019). Although dignity of risk is an important concept that highlights the need to respect individuals' autonomous decision making, risk-taking for people with ID is often denied due to a misconception that they cannot calculate risk independently (CoughDrop 2021).

**Autonomy, Self-Determination, and Choice**

Autonomy means self-governance; the right to live according to one's values, interests, and priorities. Choice is tied to the notion of autonomy, and social norms regarding what people with ID should or should not do can perpetuate infantilization and marginalization. Social norms may hinder the ability of people with ID to exercise autonomy in terms of deciding whether to consume alcohol (Johnson and Bagatell 2020).

Historically, rather than focusing on a person's rights and choices, the exercise of rights for a person with ID was firmly controlled by societal norms and whether society deemed people with ID "competent" (Wolfensberger and Thomas 2007). To this day, self-determination is often diminished for people with ID when they make choices that are perceived as "socially unacceptable" thus denying people with ID the ability to make decisions about their lives based on their values and preferences (Johnson and Bagatell 2020).

The restriction of choice for people with ID that many people without ID take for granted may be due to a deficit-focused view of people with ID fueled by prejudice and discrimination (Johnson and Bagatell 2020). This deficit-focused view assumes that people with ID have functional limitations which may affect choice making (Devi 2013). This perspective can directly impact how people without ID understand and facilitate choice making in people with ID (Johnson and Bagatell 2020). Nonetheless, people with ID can express unique preferences and desires, and become self-determined if adequate social support is accessible (Devi 2013). In a small qualitative study by Caton et al. (2012), people with ID were asked about their understanding of health and healthy choices, including AU. The study found that people with ID were extremely knowledgeable about the adverse health effects of AU and capable of choosing to limit AU when necessary. Similarly, Taggart et al. (2006) found that people with ID could articulate many of the outcomes associated with excess AU to their overall physical and mental well-being.

To that end, the facilitation of choice making related to AU is a key ethical dilemma in which community ID services must negotiate a balance between their obligation to uphold the right of people with ID to self-determination while ensuring that people with ID aren't subjected to "unreasonable" risk (Pezzoni and Kouimtsidis 2015). For instance, suppose an adult with ID receives support in their home with daily activities. On one occasion, the adult with ID asks their caregiver to drive them to the liquor store because they want to purchase alcohol. The caregiver is unsure of what to do. Should the caregiver support them in fulfilling their desire to purchase alcohol by driving them to the store, or should they refuse? In this example, tension exists between providing optimal caregiver support and honoring the rights of people with ID to exert their preferences, even if they may be considered risky or inappropriate from some perspectives. This prompts a much-needed discussion about the misconception that all people with ID are inherently vulnerable and in need of protection. We ought to recognize that people with ID can make choices and formulate opinions that should be respected and honored (Savage and Bowers 2022).

**"Nothing About Us Without Us"**

"Nothing About Us Without Us" is a slogan from the disability rights movement, which acts as a reminder that decisions about the lives of people with ID should not be made without their full, meaningful inclusion. With respect to harm reduction, this slogan acknowledges the importance of prioritizing harm reduction goals identified by the people with ID themselves. It emphasizes the active involvement of people with ID or, ideally, leading the development of programs on AU/AUD (Pauly 2008). Furthermore, by prioritizing the voices of people with ID and co-occurring AU/AUD, it can offer opportunities to assist both community ID services and mainstream AU/AUD services in identifying and developing specific support systems needed to meet the population's needs. By developing AU/AUD interventions led by people with ID, we prioritize their expertise, perspectives, and agency, which can hopefully lead to interventions that are more likely to be meaningful.

## 28.5 Harm Reduction and Alcohol

Harm reduction provides practical means to decreasing AU/AUD centered around three tenets:

(1) to reduce harmful consequences associated with alcohol use;
(2) to provide an alternative to zero-tolerance approaches by incorporating drinking goals (abstinence or moderation) that are compatible with the needs of the individual; and
(3) to promote access to services by offering low-threshold alternatives to traditional alcohol prevention and treatment (Marlatt and Witkiewitz 2002, 868).

Some harm reduction approaches to AU/AUD emphasize the person's ability to choose their desired outcome(s), such as managed alcohol programs and wet shelters, which are low-barrier housing programs that do not require people who consume alcohol or have AUD to abstain. Rather, these housing programs permit the consumption of personal alcohol or offer regularly supplied alcoholic drinks to residents (Smith-Bernardin et al. 2017). This non-judgmental approach will likely provide a safe environment for AU that respects people's dignity. Other examples include counseling and evaluations to help people assess harm associated with their AU without mandating any particular intervention (van Wormer and Davis 2008) or allowing people to define their own goals regarding AU without requiring that they adhere to any pre-determined treatment (e.g., moderate AU or abstinence) (Neighbors et al. 2006). Goal-oriented approaches recognize that people have different objectives when it comes to their AU and allow them to personalize and choose their path.

Many existing AU harm reduction strategies lack comprehensive treatments specifically targeting marginalized populations, such as people with AUD and co-existing structural vulnerability (Ivsins et al. 2019), in this case, people with ID. Most importantly, as mentioned by Measham (2006), the challenges associated with consuming alcohol are not solely rooted in individual AU or the policies surrounding it but also from wider historical, structural, and environmental influences.

### *28.5.1 Harm Reduction Approaches Adapted to People with ID*

Resources that follow a harm reduction approach for substance use (such as AU) within the ID population are limited. One resource is a harm reduction policy from the Burnaby Association for Community Inclusion (BACI) which offers services to people with ID in Canada (Burnaby Association for Community Inclusion, n.d.). The BACI policy is not exclusive to AU/AUD and encompasses guidelines to help decrease the negative effects associated with substance use. BACI holds the belief that a harm reduction approach could be impactful and valuable in providing support for people with ID. BACI firmly believes that all substance treatment must "respect human rights and dignity by adhering to basic ethical principles such as fairness, respect for autonomy, and beneficence" (Burnaby Association for Community Inclusion 2019, 1). BACI aims to mitigate adverse consequences arising from "risky behavior," recognizing that such behavior exists on a continuum. The organization adheres to a harm reduction model in its policy, encompassing prevention, treatment, and the "dignity of risk" for people with ID (Ibid.).

Another example of adapting a harm reduction approach to the needs of people with ID is The Shift. The Shift is a Canadian evidence-based resource focused on holistic and intersectional harm reduction services for people with ID (Thomason et al. 2022) and acknowledges that people with ID have "will, preferences, desires, talents, shortcomings, barriers, and gifts" (Pegg 2022, 14). Its objective is to reduce barriers and address the insufficient services faced by people with ID while acknowledging intersecting social factors and systems of inequality (e.g., AUD, socioeconomic status, trauma). The Shift operates based on the "Nothing about Us, Without Us" principle from the disability rights movement, and is "focused on non-judgement and inclusive care that is culturally sensitive, evidence-based, accessible and more than anything, individual-led and based on incremental gains" (Ibid.). Most importantly, The Shift acknowledges that people with ID have "will, preferences, desires, talents, shortcomings, barriers, and gifts" (Pegg 2022, 14).

## 28.6 Conclusion

The concepts of dignity of risk, self-determination, choice, right to citizenship, and "Nothing About Us, Without Us" are fundamental ethics principles that align with the core principles of harm reduction. To effectively address the needs of people with ID who consume alcohol or may have AUD, we argue that a harm reduction approach is an ethically defensible approach to supporting this population around their AU. It is crucial to approach harm reduction with sensitivity. Ideally, the creation and implementation of AU/AUD harm reduction strategies for people with ID should be led or at the very least, include the involvement of people with ID to ensure that programs and treatments are accessible and meaningful.

## References

Altermark, Niklas. 2017. The post-institutional era: Visions of history in research on intellectual disability. *Disability and Society* 32 (9): 1315–1332. https://doi.org/10.1080/09687599.2017.1322497.

Ball, Andrew Lee. 2007. HIV, injecting drug use and harm reduction: A public health response. *Addiction* 102 (5): 684–690. https://doi.org/10.1111/j.1360-0443.2007.01761.x.

Bhatt, Nita V., and Julie P. Gentile. 2021. Co-occurring intellectual disability and substance use disorders. *AIMS Public Health* 8 (3): 479–484. https://doi.org/10.3934/publichealth.2021037.

Bianchi, Andria, Katherine Stanley, and Kalam Sutandar. 2021. The ethical defensibility of harm reduction and eating disorders. *The American Journal of Bioethics* 21 (7): 46–56. https://doi.org/10.1080/15265161.2020.1863509.

Bumble, Jennifer L., Christopher R. J. Worth, Lindsay S. Athamanah, Magen Rooney-Kron, April Regester, and Jonathan Lidgus. 2022. Messy inclusion: A call for dignity of risk in inclusive postsecondary education. *Inclusive Practices* 1 (2): 64–69. https://doi.org/10.1177/27324745211050023.

Burnaby Association for Community Inclusion. n.d. About BACI. https://gobaci.com/about/. Accessed 10 April 2024.

Burnaby Association for Community Inclusion. 2019. Harm reduction policy. https://gobaci.com/wp-content/uploads/2019/10/7.8-Harm-Reduction-Policy.pdf. Accessed 10 April 2024.

Carroll Chapman, Shawna L., and Li-Tzy Wu. 2012. Substance abuse among individuals with intellectual disabilities. *Research in Developmental Disabilities* 33 (4): 1147–1156. https://doi.org/10.1016/j.ridd.2012.02.009.

Caton, Sue, Darren Chadwick, Melanie Chapman, Sue Turnbull, Duncan Mitchell, and Jois Stansfield. 2012. Healthy lifestyles for adults with intellectual disability: Knowledge, barriers, and facilitators. *Journal of Intellectual and Developmental Disability* 37 (3): 248–259. https://doi.org/10.3109/13668250.2012.703645.

Chaplin, Eddie, Catherine Gilvarry, and Elias Tsakanikos. 2011. Recreational substance use patterns and co-morbid psychopathology in adults with intellectual disability. *Research in Developmental Disabilities* 32 (6): 2981–2986. https://doi.org/10.1016/j.ridd.2011.05.002.

Christie, Timothy, Louis Groarke, and William Sweet. 2008. Virtue ethics as an alternative to deontological and consequential reasoning in the harm reduction debate. *International Journal of Drug Policy* 19 (1): 52–58. https://doi.org/10.1016/j.drugpo.2007.11.020.

Constantino, John N., Shae Strom, Michael Bunis, Cy Nadler, Teresa Rodgers, Julia LePage, Connie Cahalan, Amber Stockreef, Lucas Evans, Rachel Jones, and Alyssa Wilson. 2020. Toward actionable practice parameters for Dual diagnosis: Principles of assessment and management for co-occurring psychiatric and intellectual/developmental disability. *Current Psychiatry Report* 22 (9): 7–13. https://doi.org/10.1007/s11920-020-1127-8.

CoughDrop. 2021. *Cole Sorensen and Donnie TC Denome: Autonomy, Self-Determination, Dignity of Risk, and Harm Reduction.* https://www.youtube.com/watch?v=BFU0iVDmSME. Accessed 10 April 2–24.

Day, Christine, Alexandra Lampraki, Dean Ridings, and Karen Currell. 2016. Intellectual disability and substance use/misuse: A narrative review. *Journal of Intellectual Disabilities and Offending Behaviour* 7 (1): 25–34. https://doi.org/10.1108/JIDOB-10-2015-0041.

Denning, Patt, and Jeannie Little. 2017. *Over the Influence: The Harm Reduction Guide to Controlling Your Drug and Alcohol Use*, 2nd ed. New York: The Guilford Press.

Devi, Nandini. 2013. Supported decision-making and personal autonomy for persons with intellectual disabilities: Article 12 of the UN convention on the rights of persons with disabilities. *Journal of Law, Medicine and Ethics* 41 (4): 792–806. https://doi.org/10.1111/jlme.12090.

Didden, Robert, Joanne Van Der Nagel, Monique Delforterie, and Neomi van Duijvenbode. 2020. Substance use disorders in people with intellectual disability. *Current Opinion in Psychiatry* 33 (2): 124–129. https://doi.org/10.1097/YCO.0000000000000569.

Ezard, Nadine. 2001. Public health, human rights and the harm reduction paradigm: From risk reduction to vulnerability reduction. *International Journal of Drug Policy* 12 (3): 207–219. https://doi.org/10.1016/S0955-3959(01)00093-7.

Fry, Craig L., Carla Treloar, and Lisa Maher. 2005. Ethical challenges and responses in harm reduction research: Promoting applied communitarian ethics. *Drug and Alcohol Review* 24 (5): 449–459. https://doi.org/10.1080/09595230500263905.

Gee, Grahame, Jenny Conder, and Simon Adamson. 2020. Alcohol use by people with an intellectual disability in Aotearoa/New Zealand: The protective influence of family and social and support networks. *British Journal of Learning Disabilities* 48 (3): 216–223. https://doi.org/10.1111/bld.12326.

Hathaway, Andrew D. 2001. Shortcomings of harm reduction: Toward a morally invested drug reform strategy. *International Journal of Drug Policy* 12 (2): 125–137. https://doi.org/10.1016/S0955-3959(01)00085-8.

Huxley, Adam, Alex Copello, and Ed Day. 2005. Substance misuse and the need for integrated services. *Learning Disability Practice* 8 (6): 14–17. https://doi.org/10.7748/ldp2005.07.8.6.14.c1633.

Huxley, Adam, Martha Dalton, Yvonne Y. Y. Tsui, and Karen P. Hayhurst. 2019. Prevalence of alcohol, smoking, and illicit drug use amongst people with intellectual disabilities: Review. *Drugs: Education, Prevention and Policy* 26 (5): 365–384. https://doi.org/10.1080/09687637.2018.1488949.

Ivsins, Andrew, Bernie Pauly, Meaghan Brown, Joshua Evans, Erin Gray, Rebecca Schiff, Bonnie Krysowaty, Kate Vallance, and Tim Stockwell. 2019. On the outside looking in: Finding a place for managed alcohol programs in the harm reduction movement. *International Journal of Drug Policy* 67: 58–62. https://doi.org/10.1016/j.drugpo.2019.02.004.

Johnson, Khalilah R., and Nancy Bagatell. 2020. No! you can't have it: Problematizing choice in institutionalized adults with intellectual disabilities. *Journal of Intellectual Disabilities* 24 (1): 69–84. https://doi.org/10.1177/1744629518766121.

Keane, Helen. 2003. Critiques of harm reduction, morality and the promise of human rights. *International Journal of Drug Policy* 14 (3): 227–232. https://doi.org/10.1016/S0955-3959(02)00151-2.

Kiewik, Marion. 2018. Prevention and intervention of substance use and misuse among persons with intellectual disabilities. https://repository.ubn.ru.nl/bitstream/handle/2066/200654/200654.pdf?sequence=1. Accessed 10 April 2024.

Lamanna, Denise, Yona Lunsky, Sophia Wen, Denise Dubois, and Vicky Stergiopoulos. 2020. Supporting efforts by intellectually disabled adults to exit homelessness: Key ingredients of a cross-sector partnership. *Psychiatric Services* 71 (1): 96–99. https://doi.org/10.1176/appi.ps.201800590.

Marlatt, G. Alan. 1996. Harm reduction: Come as you are. *Addictive Behaviors* 21 (6): 779–788. https://doi.org/10.1016/0306-4603(96)00042-1.

Marlatt, G. Alan., and Katie Witkiewitz. 2002. Harm reduction approaches to alcohol use. *Addictive Behaviors* 27 (6): 867–886. https://doi.org/10.1016/S0306-4603(02)00294-0.

McLaughlin, D.F., L. Taggart, B. Quinn, and V. Milligan. 2007. The experiences of professionals who care for people with intellectual disability who have substance-related problems. *Journal of Substance Use* 12 (2): 133–143. https://doi.org/10.1080/14659890701237041.

Measham, Fiona. 2006. The new policy mix: Alcohol, harm minimisation, and determined drunkenness in contemporary society. *International Journal of Drug Policy* 17 (4): 258–268. https://doi.org/10.1016/j.drugpo.2006.02.013.

Miller, Peter G. 2001. A critical review of the harm minimization ideology in Australia. *Critical Public Health* 11 (2): 167–178. https://doi.org/10.1080/09581590110039865.

National Harm Reduction Coalition. n.d. *Principles of Harm Reduction*. https://harmreduction.org/about-us/principles-of-harm-reduction/. Accessed 10 April 2024.

Neighbors, Clayton, Mary E. Larimer, Ty W. Lostutter, and Briana A. Woods. 2006. Harm reduction and individually focused alcohol prevention. *International Journal of Drug Policy* 17 (4): 304–309. https://doi.org/10.1016/j.drugpo.2006.05.004.

Pauly, Bernadette. 2008. Harm reduction through a social justice lens. *International Journal of Drug Policy* 19 (1): 4–10. https://doi.org/10.1016/j.drugpo.2007.11.005.

Pegg, Shawn. 2022. Innovations in housing for people who have an intellectual disability. Community Living Ontario. https://communitylivingontario.ca/wp-content/uploads/2022/11/Innovations-in-Housing.pdf. Accessed 10 April 2024.

Perske, Robert. 1972. Dignity of risk and the mentally retarded. *Mental Retardation* 10 (1): 24–27.

Pezzoni, V., and C. Kouimtsidis. 2015. Screening for alcohol misuse within people attending a psychiatric intellectual disability community service. *Journal of Intellectual Disability Research* 59 (4): 353–359. https://doi.org/10.1111/jir.12168.

Pineau, Nicholas. 2021. *Harm Reduction and Human Rights: Emphasizing the Dignity and Capabilities of People Who Use Drugs*. McGill human rights internships working paper series. https://www.mcgill.ca/humanrights/files/humanrights/pineau_nicholas_-_harm_reduction_and_human_rights_emphasizing_the_dignity_and_capabilities_of_people_who_use_drugs.pdf. Accessed 10 April 2024.

Rieder, Travis. 2023. The future of health care must be harm reductionist—To bring it about, we need moral philosophy. *Saint Louis University Journal of Health Law and Policy* 16 (2): 199–220.

Savage, Teresa A., and Amy Bowers. 2022. Dignity of risk, intellectual/developmental disabilities, and living in the community. *Perspectives in Biology and Medicine* 65 (2): 262–273. https://doi.org/10.1353/pbm.2022.0022.

Simpson, Murray K. 1998. Just say 'no'? Alcohol and people with learning difficulties. *Disability and Society* 13 (4): 541–555. https://doi.org/10.1080/09687599826597.

Simpson, Murray K. 2012. Alcohol and intellectual disability: Personal problem or cultural exclusion? *Journal of Intellectual Disabilities* 16 (3): 183–192. https://doi.org/10.1177/1744629512455595.

Slayter, Elspeth M. 2007. Substance abuse and mental retardation: Balancing risk management with the dignity of risk. *Families in Society: The Journal of Contemporary Social Services* 88 (4): 651–659. https://doi.org/10.1606/1044-3894.3688.

Slayter, Elspeth, and Shelley A. Steenrod. 2009. Addressing alcohol and drug addiction among people with mental retardation in nonaddiction settings: A need for cross-system collaboration. *Journal of Social Work Practice in the Addictions* 9 (1): 71–90. https://doi.org/10.1080/15332560802646547.

Smith-Bernardin, Shannon, Adam Carrico, Wendy Max, and Susan Chapman. 2017. Utilization of a sobering center for acute alcohol intoxication. *Academic Emergency Medicine* 24 (9): 1060–1071. https://doi.org/10.1111/acem.13219.

Stoljar, Natalie. 2020. Disgust or dignity? The moral basis of harm reduction. *Health Care Analysis* 28 (4): 343–351. https://doi.org/10.1007/s10728-020-00412-y.

Taggart, L., D. McLaughlin, B. Quinn, and V. Milligan. 2006. An exploration of substance misuse in people with intellectual disabilities. *Journal of Intellectual Disability Research* 50 (8): 588–597. https://doi.org/10.1111/j.1365-2788.2006.00820.x.

Taggart, Laurence, D. McLaughlin, B. Quinn, and C. McFarlane. 2007. Listening to people with intellectual disabilities who misuse alcohol and drugs. *Health and Social Care in the Community* 15 (4): 360–368. https://doi.org/10.1111/j.1365-2524.2007.00691.x.

Thomason, Tina, Tony Rachwalski, Suzanne Willett, and Jill Faber. 2022. SHIFT-holistic and intersectional harm reduction services for people with intellectual disabilities. Ontario Association on Developmental Disabilities. https://oadd.org/conferences/schedule/. Accessed 10 April 2024.

Tsiandoulas, Kate, Gordon McSheffrey, Lindsay Fleming, Vandana Rawal, Marc P. Fadel, Debra K. Katzman, and Melissa D. McCradden. 2023. Ethical tensions in the treatment of youth with severe anorexia nervosa. *The Lancet Child and Adolescent Health* 7(1): 69–76. https://doi.org/10.1016/S2352-4642(22)00236-X.

van Duijvenbode, Neomi, and Joanne E. L. Van Der Nagel. 2019. A systematic review of substance use (disorder) in individuals with mild to borderline intellectual disability. *European Addiction Research* 25 (6): 263–282. https://doi.org/10.1159/000501679.

van Wormer, Katherine S., and Diane Rae Davis. 2008. *Addiction Treatment: A Strengths Perspective*, 2nd edn. California: Thomson Brooks/Cole.

Van Der Nagel, Joanne E. L., Neomi van Duijvenbode, Stephen Ruedrich, Astri P. Ayu, and Arnt F. A. Schellekens. 2018. The perception of substance use disorder among clinicians, caregivers and family members of individuals with intellectual and developmental disabilities. *Journal of Mental Health Research in Intellectual Disabilities* 11 (1): 54–68. https://doi.org/10.1080/19315864.2017.1390712.

Van Der Nagel, Joanneke E. L., Marion Kiewik, Robert Didden, Hubert P. L. M. Korzilius, Marike van Dijk, Job van der Palen, Jan K. Buitelaar, and Cor A. J. De Jong. 2017. Substance use in individuals with mild to borderline intellectual disability: An exploration of rates and risks in the Netherlands. *Advances in Neurodevelopmental Disorders* 1 (4): 283–293. https://doi.org/10.1007/s41252-017-0035-3.

Van Der Nagel, Joanneke, Marion Kiewik, Jan Buitelaar, and Cor DeJong. 2011. Staff perspectives of substance use and misuse among adults with intellectual disabilities enrolled in Dutch disability services. *Journal of Policy and Practice in Intellectual Disabilities* 8 (3): 143–149. https://doi.org/10.1111/j.1741-1130.2011.00304.x.

Williams, Fionnuala, Christos Kouimtsidis, and Alexander Baldacchino. 2018. Alcohol use disorders in people with intellectual disability. *Bjpsych Advances* 24 (4): 264–272. https://doi.org/10.1192/bja.2017.37.

Wolfensberger, Wolf, and Susan Thomas. 2007. *PASSING: A Tool for Analyzing Service Quality According to Social Role Valorization Criteria: Ratings Manual*, 3rd edn. Syracuse, New York: Training Institute for Human Service Planning, Leadership and Change Agentry.

Woolford, Marta H., Cassandra de-Lacy-Vawdon, Lyndal Bugeja, Carolina Weller, and Joseph E. Ibrahim. 2019. Applying dignity of risk principles to improve quality of life for vulnerable persons. *International Journal of Geriatric Psychiatry* 35 (1): 122–130. https://doi.org/10.1002/gps.5228.

# Chapter 29
# Reducing Restraint and Restrictive Behaviour Management Practices: Ethics and Practice

**Peter Sturmey**

**Abstract** Restraint encompasses a variety of procedures that reduce movement. Their use raises numerous ethical questions for human services. Their use can safely be reduced through behavioral treatment of the problem behavior that occasions their use and organization-wide interventions using honest and valid measurement, goal setting, feedback, and staff training and support.

**Keywords** Autism · Intellectual disabilities · Ethics · Restraint · Seclusion · PRN medication · Evidence-based practice · Applied behavior analysis

## 29.1 Introduction

Restraints and related restrictive behavior management practices continue to be used in contemporary human services with children and adults with intellectual disabilities and autism. They are often concentrated in a relatively few settings and used with relatively few individuals. Restraints present numerous ethical challenges to individuals, practitioners, and organizations. Their continued use is problematic as alternative, effective, and less restrictive practices often exist. This chapter reviews some of these issues and potential solutions to the problems of restraints in human services.

P. Sturmey (✉)
The Graduate Center and Department of Psychology, Queens College, City University of New York, New York, USA
e-mail: Peter.Sturmey@qc.cuny.edu



A. Bianchi and J. A. Vogt (eds.), *Intellectual Disabilities and Autism: Ethics and Practice*, The International Library of Bioethics 108,
https://doi.org/10.1007/978-3-031-61565-8_29

## 29.2 Definitions

"Restraint" refers to restriction in movement. We restrain one another commonly and without objection by saying "stop" and holding our hands up; through traffic signs; straps and breaks on strollers; seat belts and child car seats; and medical and dental devices such as splints, gloves, and slings to promote healing. Such restraints are considered acceptable, if sometimes inconvenient, because they are conventional, usually unharmful or minimally harmful, and sometimes have benefits such as an orderly society or faster healing of injuries. We even consider some forms of restraint as ethical, such as preventing a child running onto the road to prevent injury; failure to restrain would result in penalties for the negligent caretaker. Even when such restraints are distressing to the person restrained, many find them acceptable because the risks may not be so great, their benefits appear to outweigh the risks and costs, and there appears to be no available effective alternatives. Restraint is even used as a rehabilitative strategy after stroke in constraint-induced movement therapy during which the non-affected limb is restrained in order to induce movement in the affected limb (Grotta et al. 2004).

On other occasions, restraining people's movement is seen as unacceptable. This often occurs when the harm is great, such as restraint-related death or severe injury; when the procedure unnecessarily restricts movement; when the procedure is stigmatizing and disrespectful; and when alternative better strategies exist. For example, locking a client in a group home in a basement room with no emergency fire egress, mechanical restraints that rub the skin and injure, personal restraints that are dangerous, and globally sedating *pro re nata (PRN)* drugs are often seen as unacceptable because of the injury and stigma associated with the procedures and the availability of equally or more effective alternative strategies.

Definitions of restraint often distinguish different methods of restraint. Personal restraint can include holding someone's hands down for various durations; prompting procedures; prevention of egress from a teaching situation; cross-chest restraints, sometimes referred to as "basket holds"; personal escort procedures; personal take-down procedures; and personal procedures to maintain the person on the ground or in a chair. Mechanical restraints include a variety of physical devices to restrict movement such as restraint boards; mitts; handcuffs; helmets; various belts, such as gait and wheelchair belts, and that may be used to restrict walking and related movement; and furniture that prevents movement, such as chairs that are tipped back to prevent egress or immobilizing a client's wheelchair. Devices to prevent egress from a room or building include, for example, baffle locks that clients cannot operate, placing door handles out of reach, locking doors, and blocking egress with furniture or mats. Finally, there are two kinds of chemical restraints. PRN, or as needed, psychotropic medication is sometimes used to calm and/or protect someone who is agitated, self-injurious, aggressive, or scared, for example, prior to a medical or dental procedure or during a severe behavioral episode and often suppresses all behavior rather than a specific problem behavior. Whereas infrequent use of PRN medication may be of only modest concern since it is often relatively harmless, its repeated and routine use often represents programmatic failure. For instance, when an individual is routinely sedated for dental and medical procedures, this may be due to the person not being taught how to manage their own behavior effectively. There is professional disagreement as to whether routine psychotropic medication should be considered restraint. For example, when accompanied by sedation that interferes with ongoing adaptive behavior some authorities consider it a form of pharmacological restraint. Finally, people sometimes restrain their own behavior, such as when we fold our hands in our laps to stop fidgeting. An interesting example of this is when a client is given a choice to have a PRN medication that results in movement restriction. Restraining one's own behavior is almost always seen as non-problematic and, indeed, is often seen as a useful self-management skill to be learned. It is sometimes a part of interventions to manage self-injury and can be part of the picture of chronic self-injurious behavior (SIB) when the person self-restraints to avoid engaging in SIB. One exception to this generalization is when someone is extremely inhibited, as in some anxiety disorders, and is too scared to talk or move.

Applied Behavior Analysis International (ABAI)'s *Statement on Restraint and Seclusion* (Vollmer et al. 2011) distinguished three related concepts: restraint, seclusion, and time out from reinforcement. ABAI defined restraint as "physically holding or securing the individual, either: (a) for a brief period of time to interrupt and intervene with severe problem behavior or (b) for an extended period of time using mechanical devices to prevent otherwise uncontrollable problem behavior (e.g., self-injurious behavior)" (Ibid., 104–105). Seclusion was defined as "isolating an individual from others to interrupt and intervene with problem behavior that places the individual or others at risk of harm" (Ibid., 105). Finally, time out from reinforcement as "an evidence-based treatment intervention that involves reducing or limiting the amount of reinforcement that is available to an individual for a brief period of time. It can entail removing an individual from his or her environment, or it may entail changes to the existing environment itself" (Ibid., 105). These distinctions are quite useful. Sometimes a similar procedure—locking someone in an isolation room—may function in different ways depending upon the context and the purpose of the procedure.

Services often distinguish non-emergency (or planned) and emergency restraints. Planned restraints are usually written into a behavior support plan with specified criteria, such as contingent restraint or a written restraint reduction plan. Emergency restraints are usually permitted as part of service, state policy, or standard operating procedure to prevent imminent harm to self or others and may also be permitted for other circumstances, such as property destruction. Tracking their frequency can indicate service deficiencies either for individuals or service units. Some services now only allow emergency restraint in an effort to reduce restraint. However, there is little evidence that such practices are effective in reducing restraint; in fact, such policies and legislation may result in an *increase* in restraint (Søndenaa et al. 2015).

Practitioners and services may use euphemisms for restraint procedures or engage in other practices that appear to deny or minimize their use. Some examples include referring to a mechanical restraint as a "lap lizard" or as a physical therapy device, or naming a seclusion room as a "calming room". Sometimes restraint is deliberately mislabeled to avoid providing intervention or to cheat on data returns. For example, if their state only requires that restraints lasting over five minutes be reported, staff may be trained to restrain their clients for four minutes and 59 seconds, let go for one second, and then reapply the restraint. Additionally, restraints may be maintained for healing with a medical order long after an injury has healed. Such euphemistic "word magic" and misrepresentation is both dishonest—a breach of basic professional ethics—and harmful to the client, as its effect is to deny services rather than removing restraint and promoting ethical alternatives.

## 29.3 Ethics

### *29.3.1 Ethics for Individual Clients*

As stated earlier, restraints are neither inherently good nor bad, but their appropriateness depends on the context and other factors. Some restraints are viewed as ethical and are required by law to ensure safety, whereas some are viewed as heinous, as in the case of preventable deaths.

Concern over unethical restraints dates back to at least the eighteenth century. In William Hogarth's series of engravings of *A Rake's Progress*, Tom's final tragic state is portrayed by an image of him in restraints in Bedlam. Freeing people with mental illness from restraint is also reflected in the nineteenth century movement toward restraint-free care (Bart 1870). This imagery is echoed in contemporary images of excessive use of restraint in psychiatric institutions in Nigeria (Human Rights Watch 2019). These examples of mechanical restraints are seen as unethical because they are highly restrictive and dangerous; alternatives exist, and often have not been attempted in any serious manner; and there is good evidence that large scale reduction in restraint can readily be achieved in some circumstances (Sturmey 2015, 2018).

Today, international rights statements for people with disabilities state or imply that individuals with disabilities have the right to movement, safety, community participation, and citizenship; rights that are often reflected in individual national and state laws in various countries. These rights are incompatible with many restrictive behavior management practices, including restraint. Thus, rights statements and national and state laws often imply elimination of almost all restraints (Weissbrodt et al. 2012). Professional standards too require honesty; benevolence; competent, timely, efficient, and effective least restrictive treatment; and informed consent and also imply elimination of most restraints and competent, evidence-based treatment of the underlying problem behavior that occasions the use of restraints (see below).

### 29.3.2 *Ethics for Individual Service Providers and Professionals*

Service providers and professionals usually work under contract with a service agency in which they promise to provide services, and for which they receive money and benefits. This money usually comes from the taxpayer, with health insurance paid by the family's employer, or from the family themselves directly. Such contracts are explicit promises between the employer and employee; ethical conduct requires honesty on the part of both. The employee is being dishonest if they do not provide the services as they agreed; the employer is also being dishonest when they do not ensure that contracted services are delivered faithfully. There are serious temptations for both parties and both may fail to meet their promises. In so doing, both parties engage in unethical conduct because failing to meet their promised contractual obligations is both dishonest and cheats the client, their families, and the taxpayer out of what was not delivered.

Individual staff and family members are often highly motivated to ensure the safety of the individuals they support. Often with no alternative behavioral repertoire and/or because of specific training or instruction to do so, they use restraint to keep the individual safe from self-harm or to prevent harm to others. In so doing, a behavioral trap may evolve. Their use of restraint may be negatively reinforced by the termination of aversive client behavior and avoidance of injuries to others and/or peers, and supervisors may thank them for keeping everyone safe. The application of restraint may sometimes reinforce client problem behavior (Favell et al. 1978). Failure to apply restraints may be punished through continuation of aversive client behavior, injuries, informally through peer censure, or formally through disciplinary action for neglect. Hence, staff and family members have a dilemma. They can achieve an ethically important outcome—minimizing harm in the short term—but at the cost of various long-term harms. Absent any alternative strategy or motivation to act in another manner, they may continue to use restraint, believing their actions are the best option available. Of course, not all staff and family use of restraint is benevolent. There are ample contemporary examples of abuse in community services involving many problematic practices, among them abuse of restraint (Flynn and Citarella 2013).

The contracts that service providers, professionals, and administrators enter into have implications for ethical use of restraint. Requirements to provide services, promote personal independence, ensure safety and dignity, etc., all imply avoiding and/or reducing restraint in most circumstances. Sometimes this requirement is spelled out in job contracts. Reducing excessive restraint in this type of situation requires changing the behaviors of staff, professionals, and administrators. The most obvious strategy is the use of effective interventions to prevent and/or treat problem behavior and avoidance of provocative practices that occasion the use of restraint. To do this effectively, all parties must be knowledgeable, skilled, and motivated to change. Services have a responsibility to provide a work environment that motivates employees at all organizational levels to learn about and engage in appropriate practices. Service auditing and professional licensure requirements can also sometimes support this function.

### *29.3.3 Ethics for Organizations*

Many contemporary human services are challenged by restraint use. Often understaffed and underfunded, they may easily focus on dealing with each day's emergencies, staffing issues, and ensuring funding and accreditation is maintained; the latter being essential for the maintenance of an organization's viability and providing immediate relief from aversive consequences for them. Unfortunately, although human services may publicly espouse values of client autonomy and respect, they may find it very difficult to address restraint use proactively due to competing organizational goals and dealing with emergencies. Thus, some organizations ignore or tolerate the use of restraint, offshore clients to other services, use police, or admit clients to restrictive psychiatric units to remove the problem of restraint, at least temporarily. It is often tempting for such organizations to ignore or deny the problem or use tokenistic and ineffective solutions, such as policies without effective follow-up, or once-off training to reduce restraint use.

The process of accrediting such organizations is also often deficient, focusing on paperwork, the process of care, and negative client outcomes such as injuries, rather than the process and outcomes of restraint reduction and promotion of positive client outcomes. They often rely on progressively increasing punishment—from warnings through to threats of large sanctions—the least effective use of punishment procedures. This may encourage organizations and professions to focus on avoidance or minimization of negative client outcomes at the expense of prevention and treatment, knowing that the worst sanctions are rarely delivered.

Organizations and their administrators can act ethically with regard to safe restraint reduction in a number of ways. They can become educated about effective practices, adopt these practices, and evaluate their own efforts to do so empirically and honestly (See Sect. 29.4.2).

### *29.3.4 Safety Data*

One aspect of acting ethically with regard to restraints is to monitor and reduce client and staff restraint-related injuries and deaths. Such data can include simple frequency or rate of restraint-related injuries per 1000 client days. Although useful to evaluate intervention efforts, these metrics do not capture the severity of injuries; however, for clients, simple rating scales might capture this information. For staff, it might be captured by number of sick days and workman's compensation. Data on client deaths can be useful for investigating sentinel events and systems failures but, due to very low frequency, are relatively insensitive to intervention efforts.

Despite the importance of safety data, there is surprisingly little research for individuals with intellectual disability (ID) and/or autism. Spreat et al (1986) reported that injuries occurred in 2.9% of applications of restraint; 91% of which were classified as non-serious injuries, whereas nine percent were serious injuries (involving sutures and loss of teeth). Personal restraints also resulted in a higher rate of injuries (4.8%) than mechanical restraint (1.5%), and emergency personal restraints resulted in the highest rate of injuries (7.1%) compared to planned personal restraints (3.2%). Williams (2009) reported that in a sample of 209 individuals in an institutional setting over twelve months, injuries occurred at a rate of 0.54 injuries per 100 restraints. Rates of restraint-related injuries, however, varied depending on whether or not they were emergency restraints. Injuries during emergency personal and mechanical restraints occurred in 5.6% and 3.5% of applications, respectively, whereas injuries during planned personal and mechanical restraints occurred during only 0.3% and 0.4% of applications, respectively. Other studies have found that effective interventions to reduce restraint result in very large and sustained reductions in both client and staff injuries, workman's compensation costs (Craig and Sanders 2018), and reduction in staff restraint-related tasks (LeBel and Goldstein 2005). These data suggest that effective organization-wide interventions to reduce restraint decrease both the number and severity of worker injuries.

### *29.3.5 Prevalence of Restraints*

Despite widespread repugnance for the use of restraints, they continue to be used widely in educational and residential settings for people with disabilities. Data from the US Department of Education was analyzed by Propublica and found at least 267,000 incidents of restraints and seclusion in 2012 (Vogell 2014). These data may be a significant undercount due to school districts failing to report required data and the US Department of Education failing to require accurate and complete data collection. Children with disabilities are seven times more likely than other students to be restrained or secluded in American schools (Gage et al. 2022). Restraints are also used disproportionally by the American school system with non-White students (Scheuermann et al. 2016). Restraints are often concentrated in special education and schools for students with ID and/or autism due in part to school systems placing students with autism out of mainstream and out of district education.

Large-scale studies have also found widespread use of restraints and other restrictive behavior management practices. Luiselli et al. (2011) reported restraints in two samples, totaling 448 adults living in community services, consisting of 52 homes and four day habilitation settings. Personal restraint was used with 5.2% of clients in one sample and with 7.9% of clients in the other. Almost all restraints were planned and most lasted for less than nine minutes. A very small proportion of each sample accounted for a very large proportion of restraints. In 509 residential units for 3902 instances of service use in England, approximately 89% of service units used PRN psychotropic medication and over half used personal and mechanical restraints; a small number having very high rates of restraints (Sturmey 2009). In a Swedish sample of 556 residents in 118 group homes, 17.8% had been subjected to restraint in the previous week, almost all having experienced multiple types of restraint, most commonly (73.7%) a belt (Lundström et al. 2011). The most common risk factors were being non-ambulatory, being non-verbal, externalizing behavior problems, seizures, and cerebral palsy. Thus, restraints are commonly used in contemporary community services for adults with disabilities and are concentrated in certain settings and individuals.

## 29.4 Restraint Reduction

There are several approaches to safe restraint reduction. This can take place at the level of an individual service user, an organizational unit, and even a region or national entity.

### *29.4.1 Individuals*

As noted above, restraints are distributed very unevenly, with a small proportion of individuals accounting for most restraints and sometimes only one or two individuals accounting for the majority of restraints. Recognition of this enables services to identify and focus resources on high-risk individuals. Restraints are most commonly used for aggression, other externalizing behavior problems, SIB, and hand mouthing. Hence, the adoption of effective, evidence-based interventions for these problems is a key component of restraint reduction.

For aggression and other externalizing behavior disorders, the intervention with the most evidence is function-based applied behavior analysis (ABA) including differential reinforcement, functional communication training, and antecedent interventions (Brosnan and Healy 2011). In some circumstances, anger management may also be effective (Hamelin et al. 2013) and mindfulness (Singh et al. 2006; 2015) may be relevant. Evidence-based treatment for SIB is primarily ABA (Gregori et al. 2018); this also holding true for hand mouthing which sometimes results in restraint in mittens (Cannella et al. 2006).

There are many good examples of behavioral interventions that can eliminate or greatly reduce restraints for individuals. For instance, Busch et al. (2018) successfully treated pica in a 19-year-old man with autism and severe ID and faded a helmet restraint using functional analysis, preference assessment, competing stimuli, manipulation of effort for pica, restraint fading, and programming generalization from treatment to community settings. Jensen et al. (2012) reduced the use of mechanical restraint in a 28-year-old woman with autism and moderate ID by conducting a functional assessment and giving her the choice of her preferred staff. She engaged in self-biting and asked for restraints, which were applied during baseline. Providing the choice of preferred staff members resulted in large reductions in both the frequency and duration of restraints.

In 2013, Gaskin, McVilly, and McGillivray conducted a systematic review and meta-analysis of fourteen small-sample experiments involving a wide range of behavioral interventions. Effect sizes were large with an average reduction of 79% in frequency of restraint use and 45% for duration; and, in studies focused on its use to manage agitation and/or aggression, an average reduction of 79% and 28% in frequency and duration, respectively. Thus, behavioral interventions can result in large reductions in restraint. In a second systematic review and meta-analysis of 59 small-sample experiments to reduce restraint, with 94 participants, mostly with severe or profound ID, Heyvaert et al. (2014) found that the intervention resulted in a reduction of 3.16 standard deviations in restraint use. They concluded that restraint interventions were highly effective.

There is also evidence that other interventions can reduce restraint. For example, several studies have indicated that the use of mindfulness practices by caregivers and clients may reduce restraint by residential staff (e.g., Singh et al. 2006). Most of this research comes from one research group and hence more work on independent replication is needed. To date there are no recent systematic and comprehensive reviews of the effectiveness of non-behavioral interventions on restraint reduction in individuals with ID and/or autism.

In addition to evidence-based interventions for the problem behavior that seemingly requires restraint, there are also specific restraint-reduction strategies for some individuals. These strategies include restraint fading protocols (Foxx 1990; Foxx and Duphrense 1984) which are often quite idiosyncratic and require frequent ongoing modification, and the use of restraint as a positive reinforcer for the absence of the problem behavior or for alternate behavior in some unusual circumstances (Favell et al. 1978; Kahng et al. 2008).

There are also periodically problematic caregiver strategies related to a particular caregiver. These range from inadvertently provoking a problem behavior, such as ignoring the person whose problem behavior is maintained by attention and thereby invoking an extinction burst, to more or less deliberately provoking the client; for example, by saying "if you do that you won't go home on the weekend", etc. An important component of restraint reduction for individuals can be to identify these problematic practices and to set clear expectations, provide training, and monitor the use of effective alternative strategies. Such practices can sometimes be dealt with through retraining, but sometimes other strategies such as reassigning staff and supervisors may be necessary if training of current caregivers is ineffective or not possible.

### 29.4.2 Organizations

Several factors suggest that organization-wide restraint reduction is possible and desirable. First, some studies have found that client individual differences, even including the presence or absence of problem behaviors that may occasion the use of restraint in some settings, are not strongly related to the use of restraint (Sturmey 1999). This suggests that other factors, such as organizational features and staff behavior, might be more important than the problem behavior per se in predicting restraint use. Second, as noted earlier in the chapter, organization-wide restraint reduction was apparently highly successful in the nineteenth century (Sturmey 2015). Third, organization-wide restraint reduction has been successfully practiced in other contexts, such as in nursing homes for seniors and psychiatric institutions, thereby suggesting that it is very possible. This literature includes some randomized controlled trials (e.g., Testad et al. 2010). Finally, there are numerous case examples of effective, large-scale restraint reduction in services for individuals with neurodevelopmental disabilities (Sturmey 2018). Although organization-wide restraint reduction appears to be effective, intervention methods are quite varied. However, common features across different papers include accurate data collection, effective data use including goal setting and feedback, and various forms of staff support including teaching staff alternative strategies to restraint (Sturmey 2018).

Despite the recognized desirability of organization-wide restraint reduction, and some well-known strategies, most human services organizations have not adopted appropriate and effective strategies for this purpose, leading LeBel et al. (2010) to ask, "Why are best practices ignored?" In response, one might ask why some organizations adopt restraint reduction programs. Obviously, some are motivated to reduce restraints in response to accreditation, external threats to funding, and/or public shaming in the press. Others appear driven by the motivation of a single heroic clinician or administrator who assigns the task of restraint reduction to a few key players and holds them accountable for client outcomes over an extended period of time (Sturmey 2018). Sometimes organizations try to reduce restraints but fail. They may fail because: they adopt ineffective practices, such as writing a policy with no effective follow through; they omit one of the four essential components of effective organization-wide restraint reduction listed above; they may not have the clinical and administrative skills to effectively address severe problem behavior or alter practices in high restraint program areas; or they simply fail to persist over the several months and years needed to address this issue effectively.

### 29.4.3 Regions and Countries

Periodically, restraint and restraint-related injuries and deaths become an issue in some regions or countries. In response, some, including national organizations, have attempted to reduce restraint on a large scale. There are precedents elsewhere in the literature, such as state-wide restraint reduction in forensic settings in Pennsylvania between 2001 and 2010 (Smith et al. 2015). LeBel et al. (2004) reported that over a 22-month period, state-wide adoption of data analysis, quality assurance, regulatory oversight, and technical assistance resulted in a large reduction in the use of restrictive behavior management practices, including restraint and seclusion reductions of 73% in child units, 47% in adolescent units, and 59% in mixed child and adolescent units. Other reports have been mixed or clearly negative. For example, in Victoria, Australia, the establishment of a senior practitioner to coordinate state-wide restraint reduction was associated with no change in restraint use during the first twelve months. In contrast, introduction of legislation to reduce restraint in Norway resulted in a very large and sustained *increase* in restraint use over an eleven-year period (Søndenaa et al. 2015), whereas others have reported mixed outcomes (Swett and Noones 1989).

## 29.5 Summary

Restraint refers to a wide variety of procedures that restrict movement. Their use raises numerous ethical questions for caretakers, practitioners, and administrators in contemporary services. Fortunately, there are a wide range of interventions that can safely reduce their use with individuals and organizations. Services are encouraged to adopt these measures.

## References


Bart, James Clark. 1870. Comprising a Sketch of the Treatment of the Insane in Europe and America. *American Journal of Psychiatry* 26 (4): 470–485.

Brosnan, Julie, and Olive Healy. 2011. A Review of Behavioral Interventions for the Treatment of Aggression in Individuals with Developmental Disabilities. *Research in Developmental Disabilities* 32 (2): 437–446.

Busch, Louis P.A.., Valdeep Saini, Carobeth Zorzos, and Lanre Duyile. 2018. Treatment of Life-Threatening Pica with 5-year Follow-up. *Advances in Neurodevelopmental Disorders* 2: 335–343.

Cannella, Helen I., Mark F. O'Reilly, and Giulio E. Lancioni. 2006. Treatment of Hand Mouthing in Individuals with Severe to Profound Developmental Disabilities: A review of the Literature. *Research in Developmental Disabilities* 27 (5): 529–544.

Craig, Jason H., and Kimberly L. Sanders. 2018. Evaluation of a Program Model for Minimizing Restraint and Seclusion. *Advances in Neurodevelopmental Disorders* 2 (4): 344–352.

Favell, Judith E., James F. McGimsey, and Michael L. Jones. 1978. The Use of Physical Restraint in the Treatment of Self-Injury And As Positive Reinforcement. *Journal of Applied Behavior Analysis* 11 (2): 225–241.

Flynn, Margaret, and Vic Citarella. 2013. Winterbourne View Hospital: A Glimpse of the Legacy. *The Journal of Adult Protection* 15 (4): 173–181.

Foxx, Richard M. 1990. "Harry": A Ten-year Follow-up of the Successful Treatment of a Self-injurious Man. *Research in Developmental Disabilities* 11 (1): 67–76.

Foxx, Richard M., and Derrick Dufrense. 1984. "Harry": The Use of Physical Restraint as a Reinforcer, Timeout from Restraint, and Fading Restraint in Treating a Self-injurious Man. *Analysis and Intervention in Developmental Disabilities* 4 (1): 1–13.

Gage, Nicholas A., Danielle L. Pico, and Lauren Evanovich. 2022. National Trends and School-Level Predictors of Restraint and Seclusion for Students with Disabilities. *Exceptionality* 30 (1): 1–13.

Gaskin, Cadeyrn J., Keith R. McVilly, and Jane A. McGillivray. 2013. Initiatives to Reduce the Use of Seclusion and Restraints on People with Developmental Disabilities: A Systematic Review and Quantitative Synthesis". *Research in Developmental Disabilities* 34 (11): 3946–3961.

Gregori, Emily, Mandy Rispoli, Stephanie Gerow, and Catharine Lory. 2018. Treatment of Self-Injurious Behavior in Adults with Intellectual and Developmental Disabilities: A Systematic Review. *Journal of Developmental and Physical Disabilities* 30: 111–139.

Grotta, James C., Elizabeth A. Noser, Tony Ro, Corwin Boake, Harvey Levin, Jarek Aronowski, and Timothy Schallert. 2004. Constraint-Induced Movement Therapy. *Stroke* 35 (11 suppl 1): 2699–2701.

Hamelin, Jeffery, Robert Travis, and Peter Sturmey. 2013. Anger Management and Intellectual Disabilities: A Systematic Review. *Journal of Mental Health Research in Intellectual Disabilities* 6 (1): 60–70.

Heyvaert, Mieke, Lore Saenen, Bea Maes, and Patrick Onghena. 2014. Systematic Review of Restraint Interventions for Challenging Behaviour Among Persons with Intellectual Disabilities: Focus on Effectiveness in Single-case Experiments. *Journal of Applied Research in Intellectual Disabilities* 27 (6): 493–510.

Human Rights Watch 2019. *Nigeria: People with Mental Health Conditions Chained, Abused.* https://www.hrw.org/sites/default/files/media_2020/11/Nigeria%20Presser%20Plus%20Brochure_0.pdf. Accessed 9 Apr 2024.

Jensen, Craig C., Tore Lydersen, Paul R. Johnson, Shannon R. Weiss, Michael R. Marconi, Mary L. Cleave, and Patricia Weber. 2012. Choosing Staff Members Reduces Time in Mechanical Restraint Due to Self-Injurious Behaviour and Requesting Restraint" *Journal of Applied Research in Intellectual Disabilities* 25 (93): 282–287.

Kahng, SungWoo, Jean-Marie Leak, Chau Vu, and Brian Mishler. 2008. Mechanical restraints as positive reinforcers for aggression. *Behavioral Interventions: Theory & Practice in Residential & Community-Based Clinical Programs* 23 (2): 137–142.

LeBel, Janice, and Robert Goldstein. 2005. Special Section on Seclusion and Restraint: The Economic Cost of Using Restraint and the Value Added by Restraint Reduction or Elimination. *Psychiatric Services* 56 (9): 1109–1114.

LeBel, Janice, Nan Stromberg, Ken Duckworth, Joan Kerzner, Robert Goldstein, Michael Weeks, Gordon Harper, and Marylou Sudders. 2004. Child and Adolescent Inpatient Restraint Reduction: A State Initiative to Promote Strength-based Care. *Journal of the American Academy of Child & Adolescent Psychiatry* 43 (1): 37–45.

LeBel, Janice, Kevin Ann Huckshorn, and Beth Caldwell. 2010. Restraint Use in Residential Programs: Why are Best Practices Ignored? *Child Welfare* 89 (2): 169–187.

Luiselli, James K., James M. Sperry, and Christine Magee. 2011. Descriptive Analysis of Physical Restraint (Protective Holding) Among Community Living Adults with intellectual Disability. *Journal of Intellectual Disabilities* 15 (2): 93–99.

Lundström, Mats O., Helena Antonsson, Stig Karlsson, and Ulla H. Graneheim. 2011. Use of Physical Restraints with People with Intellectual Disabilities Living in Sweden's Group Homes. *Journal of Policy and Practice in Intellectual Disabilities* 8 (1): 36–41.

Scheuermann, Brenda, Reece Peterson, Joseph B. Ryan, and Glenna Billingsley. 2016. Professional Practice and Ethical Issues Related to Physical Restraint and Seclusion in Schools. *Journal of Disability Policy Studies* 27 (2): 86–95.

Singh, Nirbhay N., Giulio E. Lancioni, Alan S.W.. Winton, W. John Curtis, Robert G. Wahler, Mohamed Sabaawi, Judy Singh, and Kristen McAleavey. 2006. Mindful Staff Increase Learning and Reduce Aggression in Adults with Developmental Disabilities. *Research in Developmental Disabilities* 27 (5): 545–558.

Singh, Nirbhay N., Giulio E. Lancioni, Bryan T. Karazsia, Rachel E. Myers, Alan SW. Winton, Larry L. Latham, and Kristen Nugent. 2015. Effects of Training Staff in MBPBS on the Use of Physical Restraints, Staff Stress and Turnover, Staff and Peer Injuries, and Cost Effectiveness in Developmental Disabilities. *Mindfulness* 6 (4): 926–937.

Smith, Gregory M., Donna M. Ashbridge, Aidan Altenor, William Steinmetz, Robert H. Davis, Phil Mader, and Dale K. Adair. 2015. Relationship Between Seclusion and Restraint reduction and Assaults in Pennsylvania's Forensic Services Centers: 2001–2010. *Psychiatric Services* 66 (12): 1326–1332.

Søndenaa, Erik, Frode Dragsten, and Richard Whittington. 2015. Practitioner Explanations for the Increasing Use of Restraint Measures in the Care of People With Intellectual Disabilities in Norway 2000–11. *Journal of Policy and Practice in Intellectual Disabilities* 12 (1): 58–63.

Spreat, Scott, David Lipinski, Jane Hill, and Mary Ellen Halpin. 1986. Safety Indices Associated with the Use of Contingent Restraint Procedures. *Applied Research in Mental Retardation* 7 (4): 475–481.

Sturmey, Peter. 1999. Correlates of Restraint Use in an Institutional Population. *Research in Developmental Disabilities* 20 (5): 339–346.

Sturmey, Peter. 2009. Restraint, Seclusion and PRN Medication in English Services for People with Learning Disabilities Administered by the National Health Service: An analysis of the 2007 National Audit Survey. *Journal of Applied Research in Intellectual Disabilities* 22 (2): 140–144.

Sturmey, Peter. 2018. Reducing Restraint in Individuals with Intellectual Disabilities and Autism Spectrum Disorders: A Systematic Review Group Interventions. *Advances in Neurodevelopmental Disorders* 2 (4): 375–390.

Sturmey, Peter. 2015. *Reducing Restraint and Restrictive Behavior Management Practices*. Cham: Springer International Publishing: Imprint: Springer.

Swett, Chester, and Jerry Noones. 1989. Factors Associated with Premature Termination from Outpatient Treatment. *Hospital and Community Psychiatry* 40 (9): 947–951.

Testad, Ingelin, Clive Ballard, Kolbjørn Brønnick, and Dag Aarsland. 2010. The Effect of Staff Training on Agitation and Use of Restraint in Nursing Home Residents with Dementia: A Single-Blind, Randomized Controlled Trial. *The Journal of Clinical Psychiatry* 71 (1): 5897.

Vogell, Heather. 2014. Violent and Legal: The Shocking Ways School Kids Are Being Pinned Down, Isolated Against Their Will. Propublica. https://www.propublica.org/article/schools-restraints-seclusions. Accessed 9 Apr 2024.

Vollmer, Timothy R., Louis P. Hagopian, Jon S. Bailey, Michael F. Dorsey, Gregory P. Hanley, David Lennox, Mary M. Riordan, and Scott Spreat. 2011. The Association for Behavior Analysis International Position Statement on Restraint and Seclusion. *The Behavior Analyst* 34 (1): 103–110.

Weissbrodt, David, Willy Madeira, Daniel Stewart, and William Dikel. 2012. Applying International Human Rights Standards to the Restraint and Seclusion of Students with Disabilities". *Law & Inequality* 30: 287–307.

Williams, Don E. 2009. Restraint Safety: An Analysis of Injuries Related to Restraint of People with Intellectual Disabilities. *Journal of Applied Research in Intellectual Disabilities* 22 (2): 135–139.

# Chapter 30
# Ethical Considerations Arising in the Assessment, Treatment, and Support of Individuals and Families Living with FASD

**Kiruthiha Vimalakanthan and Valerie Temple**

**Abstract** Fetal Alcohol Spectrum Disorder (FASD) is the disability associated with prenatal alcohol exposure (PAE). Because it is linked specifically to maternal alcohol consumption, FASD presents several unique ethical challenges for professionals (e.g., clinicians), families, and those living with FASD. Foremost of these is balancing the need for strong prevention messaging regarding the dangers of PAE with the stigma and blame that may be placed on mothers who have consumed alcohol in their pregnancy. Given the negative stereotypes associated with FASD, there is controversy about the value, advantages, and disadvantages of conferring and receiving this diagnosis for individuals and families. This chapter discusses the ethical challenges associated with crafting effective and non-stigmatizing public health messaging about FASD, issues of mother-blame and the villainizing of women who consume alcohol in pregnancy, and an overview of some of the ethical considerations in the assessment, diagnosis, treatment, and support of individuals and families living with FASD. The chapter also offers recommendations for improving our approach and support of these individuals and their families and creating programs and services which are both welcoming and effective.

**Keywords** Fetal alcohol spectrum disorder · Fetal alcohol syndrome · Prenatal alcohol exposure · Developmental disability · Ethical issues

K. Vimalakanthan (✉) · V. Temple
Surrey Place, Toronto, ON, Canada
e-mail: kira.vimalakanthan@surreyplace.ca

V. Temple
e-mail: valerie.temple@surreyplace.ca

 
A. Bianchi and J. A. Vogt (eds.), *Intellectual Disabilities and Autism: Ethics and Practice*, The International Library of Bioethics 108,
https://doi.org/10.1007/978-3-031-61565-8_30

## 30.1 Introduction

Fetal Alcohol Spectrum Disorder (FASD) is a diagnostic term which describes the effects that may be associated with prenatal alcohol exposure (PAE) (Cook et al. 2016). Canada has consensus guidelines (Ibid.) describing the levels of PAE required for diagnosis of FASD, as well as the associated physical, cognitive, and behavioural symptoms. Current estimates suggest prevalence rates in Canada of two to three percent, translating to over 785,000 people (Popova et al. 2019).

Receiving a diagnosis of FASD implies that an individual will experience substantial lifelong challenges across multiple domains of functioning. Ten such domains are assessed in FASD with significant deficits in three or more required for diagnosis (Cook et al. 2016). FASD does not require the presence of low intelligence; however, approximately sixty percent of individuals will have an overall IQ score below the Average range and about half will qualify for a diagnosis of Intellectual Disability by adulthood (McLachlan et al. 2020). Clinically significant challenges with attention, communication, learning and academics, memory, executive functioning, motor skills, affect regulation, and adaptive functioning may also be present (McLachlan et al. 2020).

FASD assessment is generally carried out by a multi-disciplinary team including service coordinators or social workers, medical experts, and allied health professionals such as psychologists, occupational therapists, and speech-language pathologists. Diagnosis involves direct assessment of the individual's strengths and weaknesses in ten domains, medical examination, extensive review of health and social history, interviews with family members and other care providers, and recommendations for future development and support.

The diagnosis and support of families living with FASD presents several unique ethical challenges. These include understanding and working to mitigate the effects of negative stereotypes against both the family members and individuals with FASD; balancing the risks and benefits of an FASD diagnosis to individuals and the broader society; and addressing the psycho-social, mental health, relational, and emotional issues often associated with the disorder. This chapter (1) discusses issues of stigma, mother-blame, and ethical issues inherent in assessment, diagnosis, treatment, and support; and (2) offers considerations for clinicians, professionals, and policy makers around these issues and recommendations for the ethical support of these individuals and families.

## 30.2 Preventative Public Health Messaging and Mother-Blame

North American society has a complex relationship with alcohol. It is legal, widely available to adults, and an accepted part of many social activities. It is also known to cause a variety of harmful effects if overused or consumed at high rates. Aside from its addictive properties and the behavioural problems which can result from intoxication, its chronic consumption in larger doses also leads to other medical problems including high blood pressure, liver damage, and heart disease (Husain et al. 2014; Osna et al. 2017). Regardless, over seventy-five percent of Canadians report consuming alcohol (Government of Canada 2021).

Alcohol use during pregnancy is a significant public health issue that has been met with a range of preventative strategies garnering mixed results (Bell et al. 2015; Drabble et al. 2014). In 2017, roughly four percent of Canadian women who had given birth since 2012 reported alcohol consumption during their pregnancy (Varin et al. 2021). Given that PAE is a necessary condition for its development, FASD is often described as "preventable"; theoretically, we know that if maternal alcohol consumption is eliminated, the disability cannot occur.

Public health messaging regarding alcohol consumption during pregnancy is generally consistent—there is no safe amount and no safe time to drink alcohol (Health Canada 2022). The bulk of research underlying these prevention strategies is also clear, concluding that: (1) alcohol consumption in pregnancy is associated with permanent disability in children; and (2) the level of alcohol required for disability to occur is indeterminate due to the complexity of intrapersonal factors for each woman and child (Cook et al. 2016). Nevertheless, controversy regarding messaging remains fueled by research emphasizing "scientific uncertainty" regarding the effects of low to moderate alcohol consumption and characterizing concern around FASD as a "moral panic" (Armstrong and Abel 2000; Zizzo and Racine 2017).

Some prevention efforts have faced criticism for being culturally insensitive, paternalistic, and stigmatizing rather than supportive of women (Wolfson et al. 2022). Although many disabilities are associated with stigmatization, FASD is unique in being connected to a specific behaviour—maternal alcohol use during pregnancy. Compounded by broader stereotypes and gendered expectations around motherhood (Armstrong 2003; Ganong and Coleman 1995) wherein mothers are held most responsible for their children's well-being, this well-intentioned messaging may place undue focus on maternal behaviour, inadvertently characterizing people with FASD as *victims* while villainizing their birth mothers. Indeed, these mothers are perceived as more blameworthy relative to other "bad" women (e.g., incarcerated women) (Corrigan et al. 2017). Promoting the mother-blame narrative (Bell et al. 2016) obscures the role of partners, extended family, and the broader community in supporting maternal health and well-being. It also discounts other contributing factors, such as childhood trauma, culture, intimate partner violence, pre-existing physical or mental health concerns, and the effects of addiction, among others (Flannigan et al. 2020; Popova et al. 2022). Finally, it ignores multigenerational risks, whereby individuals with PAE are themselves at increased risk of several challenges, including experiencing alcohol abuse problems and having alcohol-exposed pregnancies (Flannigan et al. 2020; Paley and O'Connor 2009).

Paradoxically, public health messaging and strategies that target maternal alcohol use may create even more barriers to care. Pregnant women who have consumed alcohol and/or struggle with alcohol abuse may be afraid to disclose this or access prenatal care altogether for fear of being discriminated against or reported to child protection agencies (Gonzales et al. 2021). Stigma from healthcare practitioners is a common barrier for pregnant women with substance use challenges (Eggertson 2013). These fears and the failure to access care may result in even greater harm to mother and child due to additional factors (e.g., poor nutrition, continued smoking, or a lack of prenatal education) (Zizzo and Racine 2017).

The biological parents of individuals with FASD may continue to experience alcohol misuse problems throughout their lifetime and may encounter judgment and stigmatization by service providers who may view past/present alcohol misuse as evidence of inability to care for children. They may also rightly fear blame and shame from the child and/or other family members. Custodial battles, loss of the opportunity to parent, and/or damage to familial relationships are all possible concerns. It is therefore unsurprising that families at-risk of FASD may be less likely to seek assessment, treatment, and other support services.

To move forward with FASD prevention and screening, the preponderance of evidence supports the provision of better education and training for healthcare practitioners and community service providers. There is a need to improve awareness and understanding of the complex factors contributing to maternal alcohol use during pregnancy, the impact of PAE, the signs and symptoms of FASD, and available interventions (Popova et al. 2022). Cultivating a safe, supportive, and empathetic alliance early in prenatal care, with open discussions about FASD, screening procedures, and possible outcomes, may encourage women to share their prenatal history in primary care settings (Yan et al. 2014). Zizzo and Racine (2017, p. 416) call for a balance of "respect for autonomy and the principle of beneficence", proactively discussing alcohol use with patients of all genders and backgrounds and providing compassionate patient-centered care and support in decision making. Awareness campaigns should target specific audiences, include culturally based materials, foster hope, avoid stigma and blame/shame, and increase understanding of how families, social networks, and the community can support prevention (Burgoyne 2006; Wolfson et al. 2022).

## 30.3 Diagnosing FASD: Risks and Benefits for Individuals, Families, and the Community

There is much ethical debate regarding the value of receiving an FASD diagnosis (Helgesson et al. 2018). Advocates for conferring the diagnosis argue that important advantages include gaining access to appropriate supports, FASD-informed interventions for the affected individual, opportunities for FASD prevention, and improved understanding of FASD through research (Cook et al. 2016; Joly et al. 2022; Temple et al. 2020). Opponents argue that supports should not depend on diagnosis, that such a label may stigmatize parent and child, and that a label of disability can contribute to otherness and promotion of marginalization and de-valuing difference

(Gonzales et al. 2021; Hardy et al. 2023). Furthermore, evidence for the effectiveness of FASD-specific interventions for symptoms is limited, making diagnosis less useful for choosing appropriate treatment modalities (Helgesson et al. 2018; Pei et al. 2017).

Publicly funded healthcare systems often use diagnoses as qualification criteria to receive supports—an oft-criticized practice. From an equity perspective, all individuals who need/benefit from support should receive it and eligibility should be needs- or function-based. Nobody should be excluded from receiving support because they do not meet criteria for a specific disorder. Among rural, remote, or underserved communities, access to medical professionals who can impart specific diagnoses can be limited; consequently, residents' ability to access supports may be impaired (Anderson et al. 2019). A society ought to create structures and programs for promoting healthy outcomes, positive identities, and personal growth for all its members regardless of diagnosis (Joly et al. 2022). A more function-based perspective would obviate the need to diagnose FASD or attribute a child's disability to their mother's alcohol consumption, thus reducing the opportunity for stigma and blame.

Conversely, not using diagnoses as qualification criteria is also challenging. When funds or available placements are limited, debate may ensue regarding how much disability is enough to warrant supports, and who is most in need or deserving of the supports. Developing appropriate and inclusive function-based criteria can also be difficult. For example, a combination of low intelligence and deficits in daily living skills is often used as criteria for functional challenges. This method often excludes autistic individuals who may struggle with social communication but have a high IQ. Similarly, individuals with FASD frequently have challenges in the executive functioning and emotion regulation domains—neither category being captured in many functionally-based deficit guidelines (Anderson et al. 2019). Ultimately, diagnosis identifies a constellation of impairments and challenges across various situations which can be used to group individuals based on function, need, and severity of disability. Thus, diagnosis is a valuable tool for identifying individuals in need of supports and services.

If we accept that conferring an FASD diagnosis is in itself useful, one could consider doing this without discussing PAE, thus avoiding the stigmatizing issue of maternal alcohol consumption. There are many precedents for this approach, such as the *Diagnostic and Statistical Manual of Mental Disorders* (DSM-5-TR), which outlines sets of symptoms that commonly occur together based on research evidence and clinical experience (APA 2022). Disorders are generally discussed in terms of correlations and risk factors rather than causal agents, thereby eliminating blame and stigma. Advantages include the ability to designate profiles of disability and severity as well as allowing researchers to group similar individuals for the purpose of developing and administering treatments. However, this approach would still disadvantage those who cannot access medical diagnoses. Also, grouping individuals for the purposes of intervention using only symptom profiles may hide important differences in intervention effectiveness related to etiology.

The most fundamental problem with not discussing PAE, however, relates to prevention. There is ample evidence that alcohol is associated with birth defects, damage to the central nervous system, and a broad array of adverse outcomes (Jacobson et al. 1993; O'Leary et al. 2010). Thus, we are fully aware of the teratogen and have a clear path to prevention. However, more and better research is required to understand the impact of dose, timing, multi-substance exposure, and prenatal care on the clinical picture. Better public health messaging and more effective prevention and outreach programs are needed to limit future FASD prevalence (Wolfson et al. 2022). Of note, between 2015 and 2021, FASD research involving evidence-based prevention efforts supporting positive health outcomes for both women and their children increased (Ibid.) suggesting that focusing on the dangers of PAE and FASD prevention is leading to more evidence-based research and positive programming for families.

Given that FASD diagnosis in Canada currently involves ascertaining PAE, how do we mitigate the potential negative outcomes of this discussion? First, clinicians need to consider the ethical and practical challenges that may arise before, during, and after the assessment, and proactively engage families and individuals in discussion to enable individuals' autonomous decision making; identifying the preferences of both parties for collection and dissemination of assessment and diagnostic information. Follow-up consultation and psychosocial support should be provided, such as helping clients understand fully the implications of diagnosis, advocating with other service providers and healthcare supports, and supporting conflict resolution within families (McFarlane 2011). Clinicians will require increased empathy, patience, and understanding when working with family members and they ought to be transparent about who will participate in the diagnostic process, the questions that will be asked, and the limits to confidentiality.

Clinicians can aim to promote a positive identity, autonomy, and healthy outcomes in the course of the assessment. Using a balanced, strength-based approach, assessments should emphasize that all individuals, regardless of ability level, have strengths and challenges (Joly et al. 2022). Allowing individuals to craft their own narrative and identity, and discussing what diagnosis means to them and others, will assist in creating greater autonomy and respect for their rights (Hardy et al. 2023). Some may embrace the diagnosis as a way of explaining and understanding their struggles, while others may wish to focus on specific abilities or challenges associated with the diagnosis that better fit with their life and experiences (Temple et al. 2020). The diagnostic team can also educate and support families, empowering them to make healthy choices and supporting positive outcomes for future pregnancies.

## 30.4 Ethical Considerations for Assessing and Diagnosing FASD in Children

Compared to typically developing children, children with PAE may be more likely to be judged incapable of providing informed consent. Due to the wide range of neurodevelopmental deficits with which they present, their capacity to understand information relevant to the assessment and the consequences of their decision regarding consent, may be limited.[1] Therefore, the clinical team must determine capacity to consent on a case-by-case basis prior to assessment. If the child is deemed incapable, consent must be obtained from their legal guardian, which may involve tactful navigation to determine who is lawfully entitled to give consent. Informed consent should entail full awareness that this assessment may result in an FASD diagnosis (Todorow et al. 2012).

Regardless of the child's capacity to consent, it is important to explain the purpose of the assessment in language they understand. In so doing, clinicians face the unique risk of disclosing information that may negatively impact the child's relationship with their biological mother, their emotional well-being, and self-concept. Does telling a child that their mother consumed alcohol while pregnant, and that this may be associated with their disability, bring about more irremediable harm than good? This might be especially likely when the assessment does not support the diagnosis of FASD. Ultimately, it may be best to explain reasons for testing without mentioning PAE. For example, the team might explain that the assessment is to help understand how the child thinks and learns, and to identify their areas of strength and weakness. This decision may be informed by parent or caregiver report on what the child has been told regarding the nature and purpose of the assessment. Such collaboration with family is crucial in setting the tone around the potential communication of an FASD diagnosis (or lack thereof) and supporting child and family in integrating this new information into their personal and/or family identity and self-concept.

If the child meets criteria for an FASD diagnosis, then the clinical team must weigh the potential benefits versus risks of disclosing this diagnosis to the child. Relevant factors to consider include the child's age and developmental level, their profile of impairments, current mental health, and their family's stability and living situation (Todorow et al. 2012). The legal guardian's concerns regarding this disclosure must also be considered and discussed, and their wishes ultimately respected, especially if the child was deemed incapable of consenting. This may result in postponement of the diagnosis to minimize misunderstanding and/or harm—for example, for very young children, or for children exhibiting self-harm or suicidal behaviour. While it is important to communicate assessment findings honestly and openly (veracity), this must be done in a manner that maintains respect and dignity for the child and their family, with the primary goal of promoting the child's well-being (beneficence) while avoiding causing harm (non-maleficence). Thus, the clinical team might only disclose the diagnosis if the potential benefits outweigh the potential risks.

---

[1] In Ontario, Canada, where the authors practice, as per the *Health Care Consent Act* (1996), there is currently no specific age of consent for many personal care-related decisions.

Sensitive communication of the FASD diagnosis to a child carries several benefits. Labelling the child's challenges can help externalize them, improve understanding of self, and impart a sense of relief and optimistic outlook. Through psychoeducation about FASD, a child might learn that they are not alone in their struggles, normalizing their experiences and perhaps even helping them feel more connected to the broader FASD community. The diagnosis may also facilitate access to supports and can help the child and their family be more informed self-advocates as they navigate a variety of institutional systems. Finally, this may even help the child make healthier choices about their own consumption habits in future.

Communicating the FASD diagnosis to a child also carries risks; for example, that of negatively impacting their relationship with their biological mother. This may be especially salient when the child remains in her custody and care, or when there is an eventual plan for family reunification. Disclosure to the child may also lead to the development of distorted cognitive self- and other- schemas and negatively impact the child's self-worth, such that they begin to believe that they are worthless, inherently flawed, or unlovable; and expect others to view them the same. Finally, an FASD diagnosis may open up the non-negligible risk of experiencing stigma.

Thoughtful and sensitive communication can help the benefits outweigh the risks. Some suggest that the only way potential harms can be evaluated and mitigated in this context is for the clinical team to directly communicate the diagnosis to the child (Todorow et al. 2012). This may prevent the child from generating their own, potentially inaccurate, explanations for the challenges they face, which could end up being more damaging than the truth. This approach provides opportunities for the clinical team to better support the client and their family after diagnosis, identifying and working through unhelpful beliefs, and repairing ruptures in family relationships.

In order to thoughtfully, and sensitively communicate the diagnosis, it can help to frame FASD as a brain-based condition—for example, attributing the child's challenges to the idea that their brain developed and works differently than others. If the link to PAE is made, it is important to stress that their mother did not mean intentional harm (Todorow et al. 2012). Depending on the child's age and prenatal history, clinicians may speak about their mother's unawareness of the pregnancy and/or the harmful effects of alcohol use, and the impact of other factors (see Sect. 30.2). Additionally, the range of domains explored in FASD assessment affords plenty of opportunities to highlight specific strengths as well as weaknesses. Clinicians must also consider the child's age and neurodevelopmental profile, tailoring feedback to maximize their understanding. This may entail using concrete and developmentally appropriate terminology, using visual aids to convey key points, and sharing resources to reinforce communicated concepts.

## 30.5 Ethical Considerations in the Diagnosis and Support of FASD in Adults

A key ethical issue when working with adults with FASD is finding an adaptive balance between autonomy and support (Hardy et al. 2023). These adults frequently have deficits in executive functioning abilities such as planning, organizing, reasoning, attention, and impulse control. As well, high degrees of emotion dysregulation result in elevated levels of depression, anxiety, suicidal ideation and attempts, and addiction (Barr et al. 2006; Temple et al. 2019). These in turn lead to high rates of unemployment, residential instability, financial stress, challenges in relationships, justice system involvement, and victimization (McLachlan et al. 2020), all of which require significant support. An additional complication is that FASD is typically an invisible disability. Unfamiliar observers may not realize that the individual has a disability, instead attributing problems to willful misbehaviour. Many adults with FASD try to pass as neurotypical rather than disclose their disability, a phenomenon described as wearing a *cloak of competence*, which can lead to even greater misunderstanding and vulnerability as individuals refuse supports or deny the need for them.

When seeking to balance autonomy and beneficence for adults with FASD, it can help to involve family members or supportive professionals using the model of interdependence, which views the person as an adult, values and respects their wishes, but assumes ongoing vulnerability and support needs. The adult with FASD is an equal partner with supportive others in attaining a sustainable level of well-being for all parties. Interdependence accepts that substantial support needs are permanent and creates mechanisms for monitoring and providing this assistance. Many families have found promoting interdependence, starting in young adulthood, to be helpful in preparing for later-life challenges.

One of the most substantial later-life challenges is a very high level of interaction with the criminal justice system. Community-based sampling of adolescents and adults with FASD has found rates of thirty to sixty percent for criminal justice involvement, while studies in prison populations have found that adolescents with FASD are nineteen times more likely to be incarcerated than their peers (McLachlan et al. 2020; Popova et al. 2011; Streissguth et al. 1996). Many individuals with a history of PAE who interact with the criminal justice system have not been assessed for FASD; consequently, court-ordered or facilitated evaluations have become a common pathway to diagnosis in adulthood (Anderson et al. 2019). Ethical challenges can arise, however, around this type of assessment. Veracity is needed to explain in plain language and without bias the advantages, disadvantages, and possible adverse outcomes associated with the diagnostic process, laying out how these might affect the individual now and in the future. While completing court-ordered assessments can be beneficial, the individual is coerced to some extent into participating in an assessment that they may not want. Some individuals find that having an alternative explanation for their challenges allows for a sense of relief and less personal guilt for past difficulties (Temple et al. 2020). However, a diagnosis of FASD also means accepting the existence of a permanent disorder. It is important to spend sufficient time ensuring that the individual is aware of the implications of the diagnosis, that they know who may access the information, and to offer the opportunity to decline services if they so choose. Should they proceed, follow-up support should be offered to help process what it means to have FASD and what they have learned about themselves.

Adults with FASD also face challenges with navigating the legal system. Ninety percent of a sample of young offenders with FASD showed impaired understanding of their rights and basic legal concepts; a rate significantly higher than in those without FASD (McLachlan et al. 2014). Another issue is confabulation, which occurs with different types of brain injuries, wherein a person fills gaps in their memory with fabricated information (Brown et al. 2020). This can be related to increased susceptibility to suggestion during questioning or reconstructing memories using faulty information in order to create a coherent narrative. Given the possible repercussions of presenting or relying on incorrect information in court, along with the invisible nature of this disability, it is easy to see how very poor legal outcomes might occur. Therefore, professionals ought to thoroughly corroborate information, using as many sources as possible to improve the reliability of forensic conclusions and opinions (Brown et al. 2020).

## 30.6 Summary of Recommendations

This chapter examined some of the ethical and practical challenges associated with the prevention and diagnosis of FASD and support for affected individuals and their families. Going forward, service delivery and policy development should endeavour to integrate an ethics-based lens and focus on ensuring that prevention messaging is positive, supportive, and inclusive, recognizing the important role played by the social circle surrounding a woman in ensuring a healthy pregnancy. The development of FASD-informed services for professionals in prenatal care settings, primary care, child protection, and legal contexts, with training on frameworks and guidelines for addressing specific ethical challenges commonly encountered with this disability, is also needed. Finally, early intervention, diagnostic, and support services for individuals and families living with FASD should be accessible across all regions, guided by ethical considerations, and delivered with a welcoming and empathetic atmosphere to reduce stigma and foster positive outcomes for children, adults, and families.

## References

American Psychiatric Association (APA). 2022. *Diagnostic and statistical manual of mental disorders* (5th ed., text rev.). https://doi.org/10.1176/appi.books.9780890425787.

Anderson, Tara, Mansfield Mela, Thomas Rotter, and Nancy Poole. 2019. A qualitative investigation into barriers and enablers for the development of a clinical pathway for individuals living with

FASD and mental disorder/addictions. *Canadian Journal of Community Mental Health* 38 (3): 43–60. https://doi.org/10.7870/cjcmh-2019-009.

Armstrong, Elizabeth M. 2003. *Conceiving risk, bearing responsibility: Fetal alcohol syndrome and the diagnosis of moral disorder*. Baltimore: Johns Hopkins University Press.

Armstrong, Elizabeth M., and Ernest L. Abel. 2000. Fetal alcohol syndrome: The origins of a moral panic. *Alcohol and Alcoholism* 35: 276–282. https://doi.org/10.1093/alcalc/35.3.276.

Barr, Helen M., Fred L. Bookstein, Kieran O'Malley, Paul D. O'Connor, Janet E. Huggins, and Ann P. Streissguth. 2006. Binge drinking during pregnancy as a predictor of psychiatric disorders on the structured Clinical Interview for DSM-IV in young adult offspring. *American Journal of Psychiatry* 163: 1061–1065.

Bell, Emily, Natalie Zizzo, and Eric Racine. 2015. Caution! warning labels about alcohol and pregnancy: Unintended consequences and questionable effectiveness. *The American Journal of Bioethics* 15: 18–20. https://doi.org/10.1080/15265161.2014.998376.

Bell, Emily, Gail Andrew, Nina Di Pietro, Albert E. Chudley, James F. Reynolds, and Eric Racine. 2016. It's a shame! stigma against fetal alcohol spectrum disorder: Examining the ethical implications for public health practices and policies. *Public Health Ethics* 9: 65–77. https://doi.org/10.1093/phe/phv012.

Brown, Jerrod, Erik Asp, Megan N. Carter, Vanessa Spiller, and Deanna Bishop-Deaton. 2020. Suggestibility and confabulation among individuals with fetal alcohol spectrum disorder: A review for criminal justice, forensic mental health, and legal interviewers. *International Journal of Law and Psychiatry* 73: 101646. https://doi.org/10.1016/j.ijlp.2020.101646.

Burgoyne, Wendy. 2006. *What we have learned: Key Canadian FASD awareness campaigns*. Public Health Agency of Canada. https://www.canada.ca/content/dam/phac-aspc/migration/phac-aspc/publicat/fasd-ac-etcaf-cs/pdf/fasd-ac-etcaf-cs_e.pdf. Accessed 9 Apr 2024.

Cook, Jocelynn L., Courtney R. Green, Christine M. Lilley, Sally M. Anderson, Mary E. Baldwin, Albert E. Chudley, Julianne L. Conry, Nicole LeBlanc, Christine A. Loock, Jan Lutke, Bernadene F. Mallon, Audrey A. McFarlane, Valerie K. Temple, and Ted Rosales. 2016. Fetal alcohol spectrum disorder: A guideline for diagnosis across the lifespan. *Canadian Medical Association Journal* 188: 191–197. https://doi.org/10.1503/cmaj.141593.

Corrigan, Patrick W., Juana L. Lara, Binoy B. Shah, Kathleen T. Mitchell, Diana Simmes, and Kenneth L. Jones. 2017. The public stigma of birth mothers of children with fetal alcohol spectrum disorders. *Alcoholism: Clinical and Experimental Research* 41: 1166–1173. https://doi.org/10.1111/acer.13381.

Drabble, Laurie, Sue Thomas, Lisa O'Connor, and Sarah C. M. Roberts. 2014. State responses to alcohol use and pregnancy: Findings from the alcohol policy information system. *Journal of Social Work Practice in the Addictions* 14: 191–206. https://doi.org/10.1080/1533256X.2014.900409.

Eggertson, Laura. 2013. Stigma a major barrier to treatment for pregnant women with addictions. *Canadian Medical Association Journal* 18: 1562. https://doi.org/10.1503/cmaj.109-4653.

Flannigan, Katherine, Kelly D. Coons-Harding, Tara Anderson, Lindsay Wolfson, Alanna Campbell, Mansfield Mela, and Jacqueline Pei. 2020. A systematic review of interventions to improve mental health and substance use outcomes for individuals with prenatal alcohol exposure and fetal alcohol spectrum disorder. *Alcoholism: Clinical and Experimental Research* 44: 2401–2430. https://doi.org/10.1111/acer.14490.

Ganong, Lawrence H., and Marilyn Coleman. 1995. The content of mother stereotypes. *Sex Roles* 32: 495–512. https://doi.org/10.1007/BF01544185.

Gonzales, Kelly L., Michelle M. Jacob, Amanda Mercier, Heather Heater, Lindsay N. Goes, Jillene Joseph Behind, and Suzie Kuerschner. 2021. An indigenous framework of the cycle of fetal alcohol spectrum disorder risk and prevention across the generations: Historical trauma, harm and healing. *Ethnicity & Health* 26: 280–298. https://doi.org/10.1080/13557858.2018.1495320.

Government of Canada. 2021. *Canadian alcohol and drugs survey CADS: Summary of results for 2019*. https://www.canada.ca/en/health-canada/services/canadian-alcohol-drugs-survey/2019-summary.html. Accessed 9 Apr 2024.

Hardy, Kelly, Jacqueline Pei, and Lauren Richardson. 2023. *Neurodiversity and FASD*. Canada FASD Research Network. https://canfasd.ca/wp-content/uploads/publications/Neurodiversity-and-FASD.pdf. Accessed 9 Apr 2024.

Health Canada. 2022. *Your guide to a healthy pregnancy*. https://www.canada.ca/content/dam/phac-aspc/documents/services/health-promotion/healthy-pregnancy/healthy-pregnancy-guide.pdf. Accessed 9 Apr 2024.

Helgesson, Gert, Göran. Bertilsson, Helena Domeij, Gunilla Fahlström, Emelie Heintz, Anders Hjern, Christina N. Gordh, Viviann Nordin, Jenny Rangmar, Ann-Margret. Rydell, Viveka S. Wahlsten, and Monica Hultcrantz. 2018. Ethical aspects of diagnosis and interventions for children with fetal alcohol spectrum disorder FASD and their families. *BMC Medical Ethics* 19: 1. https://doi.org/10.1186/s12910-017-0242-5.

Husain, Kazim, Rais A. Ansari, and Leon Ferder. 2014. Alcohol-induced hypertension: Mechanism and prevention. *World Journal of Cardiology* 6: 245–252. https://doi.org/10.4330/wjc.v6.i5.245.

Jacobson, Joseph L., Sandra W. Jacobson, Robert J. Sokol, Susan S. Martier, Joel W. Ager, and Melissa G. Kaplan-Estrin. 1993. Teratogenic effects of alcohol on infant development. *Alcohol Clinical & Experimental Research* 17: 174–183. https://doi.org/10.1111/j.1530-0277.1993.tb00744.x.

Joly, Vannesa, Kelly Harding, and Jacqueline Pei. 2022. Towards healthy outcomes for individuals with FASD: A philosophy and approach to intervention. *Journal of Fetal Alcohol Spectrum Disorder* 4: e171–e185. https://doi.org/10.22374/jfasd.v4iSP1.20.

McFarlane, Audrey. 2011. Fetal alcohol spectrum disorder in adults: Diagnosis and assessment by a multidisciplinary team in a rural area. *Canadian Journal of Rural Medicine* 16: 25–30.

McLachlan, Kaitlyn, Ronald Roesch, Jodi L. Viljoen, and Kevin S. Douglas. 2014. Evaluating the psycholegal abilities of young offenders with fetal alcohol spectrum disorder. *Law and Human Behaviour* 38: 10–22. https://doi.org/10.1037/lhb0000037.

McLachlan, Kaitlyn, Katherine Flannigan, Valerie K. Temple, K. Kathy Unsworth, and Jocelynn L. Cook. 2020. Difficulties in daily living experienced by adolescents, transition-aged youth, and adults with fetal alcohol spectrum disorder. *Alcohol Clinical & Experimental Research* 44: 1609–1624. https://doi.org/10.1111/acer.14385.

O'Leary, Colleen M., Natasha Nassar, Stephen R. Zubrick, Jennifer J. Kurinczuk, Fiona Stanley, and Carol Bower. 2010. Evidence of a complex association between dose, pattern and timing of prenatal alcohol exposure and child behaviour problems. *Addiction* 105: 74–86. https://doi.org/10.1111/j.1360-0443.2009.02756.x.

Osna, Natalia A., Terence M. Donohue, and Kusum K. Kharbanda. 2017. Alcoholic liver disease: Pathogenesis and current management. *Alcohol Research* 38: 147–161.

Paley, Blair, and Mary J. O'Connor. 2009. Intervention for individuals with fetal alcohol spectrum disorders: Treatment approaches and case management. *Developmental Disabilities Research Reviews* 15: 258–267. https://doi.org/10.1002/ddrr.67.

Pei, Jacqueline, Melissa Tremblay, Allison McNeil, Nancy Poole, and Audrey McFarlane. 2017. Neuropsychological aspects of prevention and intervention for FASD in Canada. *Journal of Pediatric Neuropsychology* 3: 25–37. https://doi.org/10.1007/s40817-016-0020-1.

Popova, Svetlana, Shannon Lange, Dennis Bekmuradov, Alanna Mihic, and Jürgen. Rehm. 2011. Fetal alcohol spectrum disorder prevalence estimates in correctional systems: A systematic literature review. *Canadian Journal of Public Health* 102: 336–340. https://doi.org/10.1007/BF03404172.

Popova, Svetlana, Shannon Lange, Vladimir Poznyak, Albert E. Chudley, Kevin D. Shield, James N. Reynolds, Margaret Murray, and Jürgen. Rehm. 2019. Population-based prevalence of fetal alcohol spectrum disorder in Canada. *BMC Public Health* 19: 845. https://doi.org/10.1186/s12889-019-7213-3.

Popova, Svetlana, Danijela Dozet, Shahela A. Laboni, Krista Brower, and Valerie K. Temple. 2022. Why do women consume alcohol during pregnancy or while breastfeeding? *Drug and Alcohol Review* 41: 759–777. https://doi.org/10.1111/dar.13425.

Streissguth, Ann P., Helen M. Barr, Julia Kogan and Fred L. Bookstein. 1996. Understanding the occurrence of secondary disabilities in clients with fetal alcohol syndrome and fetal alcohol effects. Centers for Disease Control and Prevention. http://lib.adai.uw.edu/pubs/bk2698.pdf. Accessed 14 Apr 2023.

Temple, Valerie K., Jocelynn L. Cook, Kathy Unsworth, Hasu Rajani, and Mansfield Mela. 2019. Mental health and affect regulation impairment in fetal alcohol spectrum disorder FASD: Results from the Canadian national FASD database. *Alcohol and Alcoholism* 54: 545–550. https://doi.org/10.1093/alcalc/agz049.

Temple, Valerie K., Sonal Prasad, Svetlana Popova, and Ann Lindsay. 2020. Long-term outcomes following fetal alcohol spectrum disorder FASD diagnosis in adulthood. *Journal of Intellectual & Developmental Disability* 46: 272–280. https://doi.org/10.3109/13668250.2020.1824612.

Todorow, Michelle, Karrela Paris, and Ellen Fantus. 2012. Ethical considerations when communicating a diagnosis of a fetal alcohol spectrum disorder to a child. *Journal of Population Therapeutics and Clinical Pharmacology* 19: e361-368.

Varin, Mélanie, Elia Palladino, Kate H. MacEachern, Lisa Belzak, and Melissa M. Baker. 2021. Prevalence of alcohol use among women of reproductive age in Canada. *Health Promotion and Chronic Disease Prevention in Canada* 41: 267–271. https://doi.org/10.24095/hpcdp.41.9.04.

Wolfson, Lindsay, Nancy Poole, Kelly D. Harding, and Julie Stinson. 2022. At a juncture: Exploring patterns and trends in FASD prevention research from 2015–2021 using the four-part model of prevention. *Journal of Fetal Alcohol Spectrum Disorder* 4: e36–e59. https://doi.org/10.22374/jfasd.v4iSP1.13.

Yan, Allison, Emily Bell, and Eric Racine. 2014. Ethical and social challenges in newborn screening for prenatal alcohol exposure. *Canadian Journal of Neurological Sciences* 41: 115–118. https://doi.org/10.1017/s0317167100016413.

Zizzo, Natalie, and Eric Racine. 2017. Ethical challenges in FASD prevention: Scientific uncertainty, stigma, and respect for women's autonomy. *Canadian Journal of Public Health* 108: 414–417. https://doi.org/10.17269/cjph.108.6048.

# Chapter 31
# A Proposal for Training ABA Practitioners in Ethical Deliberation

**Nicole Luke and Janet A. Vogt**

**Abstract** The practice of applied behavior analysis (ABA) can involve situations which cause moral tension for the practitioner. As professionals in the health and human service sectors, ABA practitioners (such as behavior analysts) must develop sophisticated repertoires to successfully connect complex concepts such as ethics with contextual and circumstantial phenomena in their daily practice. In this chapter, we discuss the process of ethical deliberation and some of the key principles that come into play and argue for the importance of developing skills in reflective practice. Ultimately, we propose using Rhodes and Alfandre's eight-step process, in the context of the Behavioral Skills Training model, and informed by the principles of reflective practice, to encourage a structured approach to training ABA practitioners in ethical deliberation. A composite case study is used to illustrate a common situation that may create moral tension for the ABA practitioner and to model the application of a decision protocol to a specific case. Although the case study is specific to ABA practice with autistic children and youth, this approach can be used in any clinical context where professional behavior analysts collaborate to provide services to others.

**Keywords** ABA · Ethical deliberation · Reflective practice · Behavior analysis · BST · Behavior skills training · Professional ethics

---

N. Luke (✉)
Brock University, St. Catharines, ON, Canada
e-mail: nl.behavioranalyst@gmail.com

J. A. Vogt
Surrey Place, Toronto, ON, Canada
e-mail: j.vogt@alumni.utoronto.ca; janet.vogt@surreyplace.ca



A. Bianchi and J. A. Vogt (eds.), *Intellectual Disabilities and Autism: Ethics and Practice*, The International Library of Bioethics 108,
https://doi.org/10.1007/978-3-031-61565-8_31

## 31.1 Introduction

This chapter assumes an audience of ethicists, behavior analysts, trainees in human services, researchers, and other stakeholders with an interest in intellectual disabilities and autism. This assumption leads us to focus on the practical and actionable elements of ethical decision making in professional human services settings and, in particular, how to train staff in ethical deliberation. As an example of how to train a clinical team, we have chosen to focus on applied behavior analysis (ABA) practitioners.

## 31.2 Background

Behavior analysts generally consider themselves a separate and distinct community of professionals, but in practice their profession overlaps significantly with those of education and psychology, along with other human service professions. Historically, behavior analysts were considered to be under the ethical umbrella of psychology, and for this reason the development of their own ethics guidelines was delayed. The Behavior Analyst Certification Board® (BACB) was formed in 1989 and the first Professional Disciplinary Standards were not published until 1999 (BACB 2023). At this point of growth as a profession, they were mostly concerned with practical matters, and it is this practical foundation that has driven the ethical guidelines that are accepted today by the profession. Certification as a behavior analyst did not require specific ethics training until 2015, when a 45-hour ethics course was introduced as a condition of certification or recertification. Guidelines for Responsible Conduct was renamed as Professional and Ethical Compliance Code in 2014, and eventually became an Ethics Code (BACB 2020). The BACB identified four core principles in that code, listed as: (1) benefit others; (2) treat others with compassion, dignity, and respect; (3) behave with integrity; and (4) ensure the competence of behavior analysts. According to the BACB, "The core principles are foundational concepts that should guide all aspects of a behavior analyst's work" (BACB 2020, 3).

## 31.3 Principles Commonly Invoked in Ethical Deliberation in the Clinical Context

Ethical theorists have proposed a number of ways to organize and talk about ethical principles or values, but all have roughly the same concerns. Beauchamp and Childress identified four core ethical principles and these are common across health services sectors: (1) beneficence, (2) non-maleficence, (3) respect for autonomy, and (4) justice (Beauchamp and Childress 2013). *Beneficence* refers to the moral obligation to act for the benefit of others. *Non-maleficence* requires that we abstain from causing harm or risk of harm to others. *Autonomy* entails two conditions: liberty (the absence of controlling influences) and agency (self-initiated action). In the very least, a person's autonomy encompasses self-rule that is free from the kinds of limitations that may prevent meaningful choice. *Justice* takes a number of forms in the context of ethical deliberation. *Procedural justice* is concerned with fair processes, whereas *distributive justice* is concerned with "fair, equitable, and appropriate distribution of benefits and burdens determined by norms that structure the terms of social cooperation" (Beauchamp and Childress 2013, 250). *Legal justice* is concerned with respect for morally acceptable laws, while *rights-based justice* involves respect for a person's rights (Gillon 1994). These principles are invoked in settings dedicated to care, such as health and human services or educational settings where behavior analysts often work. They are also reflected amongst the core ethical principles of many other professional associations, just as we see with the BACB Ethics Code.

Cox (2012) integrated several professional disciplines which intersect in the treatment of autistic individuals by linking their unique ethics guidelines. He proposed "common ethical principles and obligations encountered in the interdisciplinary setting" (Cox 2012, 2730) and listed them as: (1) informed consent, (2) confidentiality, (3) beneficence, (4) non-maleficence, (5) respect for autonomy, (6) clinical justice, (7) professional competence, (8) fiduciary responsibility, (9) peer communication, and (10) non-judgmental regard. From those, he proposed the following principles as a "code of ethics and conduct for the interdisciplinary autism therapist" (Cox 2012, 2735): (1) beneficence, (2) nonmaleficence, (3) respect for persons, and (4) professional commitments. Cox believed that creating a common ethical code could unite different professionals as they collaborated to treat their clients. He felt this was critical to providing appropriate care.

## 31.4 The Need for an Ethical Decision-Making Process

Many, if not all, clinicians will presumably encounter ethically complex clinical dilemmas at some point in their practice. For behavior analysts, tension may arise in their day-to-day practice when they are confronted with situations in which values (and associated ethical principles) come into conflict and the best course of action is not readily apparent. For instance, a behavior analyst may wish to act for the benefit of her client but in doing so might infringe on the client's liberty or find herself in conflict with the values and beliefs of another professional. Rhodes and Alfandre (2007) propose an eight-step process for working through ethical dilemmas. Although their process was developed with medical staff and trainees in mind, it is transferable across disciplines, thus making it suitable for the interdisciplinary settings in which most behavior analysts practice. We recommend its use in clinical and educational settings where behavior analysts are working, are being supervised, and have opportunities for learning at their place of work. Section 31.4.1 explains each step in the process.

### 31.4.1 Rhodes and Alfandre's Eight-Step Approach for Ethical Deliberation in Clinical Contexts

Step One: "Collect all relevant data that could help to answer your question" (Rhodes and Alfandre 2007, 67). Gather all the data that can inform a deep understanding of the context. If the case is presently occurring, communicate clearly and openly with the client and/or family in order to understand their needs and perspective(s).

Step Two: "Identify principles, duties and concepts involved, and explain how they relate to the case" (Ibid.). As a team, reflect on principles, duties, and concepts that are recognized as important to the case (e.g., duty to protect one's client from experiencing harm). One must also consider relevant laws, organizational policies, and professional guidelines. Professional behavior analysts should refer to the BACB Ethics Code (BACB 2020).

Step Three: "Consider whether principles conflict in this situation or whether there is uncertainty about what a particular principle (e.g., beneficence, respect for autonomy) directs you to do" (Rhodes and Alfandre 2007, 67). It is common for the principles of beneficence and respect for autonomy to come into conflict in clinical situations (Varkey 2021). In situations where only one principle is salient, the difficulty may lie in determining the best course of action.

Step Four: "Formulate a question that reflects the conflict" (Rhodes and Alfandre 2007, 67). The question should capture the primary dilemma and be phrased in such a way as to guide the team toward a decision and an action, or actions, for next steps.

Step Five: "Decide which principle should have priority in this case and support that choice with factors relevant to the case or find an alternative that avoids the dilemma" (Ibid.). Any suggestion offered should be accompanied by a clear argument or rationale to support its adoption.

Step Six: "When uncertainty persists, note whether there is some missing information that would help you resolve the dilemma. What information? How will it help resolve the dilemma?" (Ibid.). In any ethical deliberation, one should be prepared for a more iterative or cyclic process rather than assuming a strictly linear one. Remain open to revisiting and refining your steps, thereby gaining a better understanding of the case at hand.

Step Seven: "Evaluate your decision by asking if it is what a consensus of exemplary … [professional peers][1] … would agree to do" (Ibid.). Reflect on whether the team has reached their decision by adhering to the basic ethics principles and thereby cleared the bar for an ethical standard of care.

Step Eight: "Plan the practical steps that you should take, focusing on the details of the case and the future issues that you foresee" (Ibid.). The thoughtful implementation of the decision, considering its potential consequences (both intended and unintended), is achieved through clear and open communication with all key stakeholders. Depending on the case at hand, a simple cost–benefit analysis may figure into this step of the process. The last step also points to the circular nature of the

[1] The term *professional peers* replaces *doctors* (the term used by Rhodes and Alfandre).

entire process where the opportunity to revisit the case allows the professional to reflect on their decisions and the outcomes of the actions they take, as well as to evaluate whether a change to the course of action is warranted.

## 31.5 Reflective Practice

### *31.5.1 What is Reflective Practice?*

Reflection involves consideration of the broader context, meaning, and implications of an experience or action (Branch and Paranjape 2002). It does not have to be an independent pursuit but can be performed in the context of collaboration and dialogue (Richard et al. 2019). Reflective practice can be understood as a process wherein the practitioner, having experienced doubt regarding a clinical situation, reflects on that experience with the goal of identifying potential actions to remediate the situation and ameliorate/mitigate feelings of uncertainty (Richard et al. 2019; Schön 1983). According to Driessen and colleagues, (2008, 827) "… reflection means letting future behavior be guided by a systematic and critical analysis of past actions and their consequences." Both definitions represent what Schön referred to as *reflection-on-action*, meaning reflecting after the fact (Schön 1983; Branch and George 2017). Schön also articulated a *reflection-in-action*, meaning reflecting in the moment (Schön 1983; Branch and George 2017). This involves reflecting on your thoughts, feelings, and behaviors as you experience them in the clinical context (Davies 2012). We recommend professionals engage in both *reflection-on* and *reflection-in* activities across their practices. Furthermore, we suggest that reflective practice may enable practitioners to engage in more ethically defensible decision making. We provide an example of *reflection-on* in Sect. 31.8, where we model the multi-step process described above.

### *31.5.2 What Are the Benefits of Reflective Practice for the Learner?*

There are demonstrated benefits of developing the skill of reflective practice and thereby becoming self-directed (Davies 2012) and lifelong learners (Branch and George 2017). Richard et al. (2019) reviewed the use of reflective practice in interprofessional education and practice and reported acquisition of the following skills amongst learners: communication and dialogue, conflict resolution, understanding of roles, and critical thinking. According to Davies (2012), the benefits include promotion of deep learning; a better understanding of one's own beliefs, attitudes, and values; and improved personal and clinical confidence. Branch and George

(2017) have argued that guided group reflection using narratives can contribute to the learner's moral development and formation of professional identity.

### 31.5.3 Incorporation of Reflective Practice into Clinical Teaching

Assuming that behavior analysts will encounter ethically complex clinical dilemmas, (i.e., involving moral, ethical, social and/or professional considerations), clinical teaching must prepare them to reflect effectively on these dimensions. According to Branch and Paranjape (2002), successful use of reflection in clinical teaching entails three elements. First, the facilitator must be a good role model, exhibiting enthusiasm for learning and high levels of skills and knowledge, while emphasizing the psychological and social aspects of care provision. Second, the facilitator must establish a reflective atmosphere with learners by gaining their trust and displaying an appropriate interest in, and concern for, the learning professional and their clients. The reflective atmosphere requires a safe and open learning environment, free of assessment and judgement, where learners feel their contributions are met with understanding (Davies 2012; Branch and George 2017). Finally, the facilitator must demonstrate the following skills: listen carefully to learners, pick up on their thoughts or feelings, reflect these back to them, and use open-ended questions to encourage deeper reflection (Davies 2012). Such reflection should include the economic and power relationships involved. The facilitator should also reflect on their own assumptions and behaviors with respect to these interactive dynamics and one's own social role (Branch and George 2017). Thus, the facilitator must provide support and empathy to the learners while challenging them to consider the moral, ethical, social and/or professional dimensions of the situation being discussed (Davies 2012; Driessen et al. 2008).

When a learner raises a salient point that challenges the opinion of the facilitator, it is important that the facilitator models reflexivity (i.e., self-reflection); that is, the examination of one's own assumptions and behaviors (Branch and George 2017). This can be done by acknowledging the value of such a statement and articulating any resulting change in one's own understanding of the situation. The deliberate practice of reflexivity can contribute to a heightened awareness of the underlying ideologies, politics, and cultures of clinicians (McEldowney and Connor 2011).

There are various sources of clinical situations for reflection: for example, case studies, personal practice journals, or accounts of learners' own lived experiences (Richard et al. 2019). Branch and George (2017) use guided group reflection to consider critical incidents and appreciative inquiry narratives. The critical incident describes a lived experience considered important by the learner, whether they perceived it as positive or negative. Appreciative inquiry narratives are success stories describing a situation wherein the learner was at their best (Cooperrider and Whitney 2005). These have been used to increase both the learner's awareness of, and their motivation to meet, the emotional needs of their patients (Branch and Frankel 2016). We recommend that supervising behavior analysts consider incorporating these elements of reflective practice into their ethical training.

## 31.6 Behavioral Skills Training

One of the most common strategies for training professional behavior analysts is called Behavioral Skills Training (Miltenberger 2008; Sun 2022). This strategy has several components including instruction, modeling, practice, and feedback (Miltenberger 2008). Behavioral Skills Training typically requires the instructor to break down a skill into component parts and this component analysis allows the instructor to quickly identify areas where the learner may be having challenges (Weston et al. 2020). The model-practice-feedback loop can be repeated until the learner has achieved mastery of the target skill. Behavioral Skills Training (BST) can be used to teach a variety of complex skills (daily living skills and job skills) and can be used with many learners (e.g., people with an intellectual disability, caregivers, professionals). Behavior analysts have been taught job skills as complex as functional behavior assessment using the BST strategy (Nylen and King 2023). The complexity and need for repeated opportunities to practice ethical decision making align with the BST teaching strategy structure and goals.

## 31.7 Using Case Studies to Develop Ethical Deliberation Skills

We recommend that the supervising professional use case studies to model an ethical deliberation process, highlighting the *reflection-on* strategies, and employing the BST strategy to break down the component steps, and offer instruction, practice, and feedback. The illustrative case study outlined below is based on real world observations in professional practice. It has been designed to create ethical tension and the client, family, and their circumstances have been described in such a way as to help the reader appreciate some of the factors that *may* contribute to making the case particularly challenging. Not all of the details provided are relevant to the decision at hand. Part of ethical decision making is deciding which factors are creating the tension you seek to resolve. Which are the facts that will push the deliberative process? Your primary objective should be to arrive at an ethically sound and well-reasoned answer or "solution" and to articulate a plan for its implementation. Many times, a challenging case does not necessarily pose an ethical dilemma. Even so, a structured approach to deliberation can be very helpful. You can use this approach whether you are trying to determine the best action in a particularly challenging emergent situation or whether you are reflecting on a decision you made in the past with which you are now uncomfortable or wish to revisit. Bailey and Burch (2019) and Sush and Najdowski (2021) both provide excellent case studies set in the field of behavior analysis.

In the following section, a training plan is proposed that consists of four phases; each one providing opportunities for the supervisor to model Rhodes and Alfandre's eight-step process and opportunities for the learners to practice ethical deliberation and receive feedback. Applying the BST strategy, we outline a series of graduated opportunities for practice and suggest a possible route for advancing reflective, ethically sensitive professional behavior.

## 31.8 Proposed Training Protocol

When a supervisor first commences ethical reflection training with their team, they will need to allow time to introduce the deliberation process and key ethical principles. Although some staff may have encountered this content, it is important to ensure that the entire team is familiar with it. During the ethical deliberation and reflection training process, the supervisor must pay attention to which components of the skill (i.e., the eight-step process) are particularly challenging for the team. To take the team through an ethical deliberation process for the first time, we propose scheduling four weekly meetings of thirty to sixty minutes each.

In the first meeting, the supervisor presents a case study and introduces Rhodes and Alfandre's eight-step process and the ethics principles covered in Sect. 31.3. As homework, the team studies the case and completes *Step One*, requiring them to determine which elements are relevant to the ethical issue at hand.

**Case Study 31.1 Alternative Dietary Approaches in the Intensive Behavioral Intervention Setting**

Julia works for an intensive behavioral intervention (IBI) program that serves pre-school aged children with autism and emphasizes the importance of evidence-based practice. The program is delivered by a community-based service agency that stresses the importance of family-centred care. The agency does not provide a dietetics or nutrition service.

Each instructor therapist is assigned a primary client and Julia recently received a new child, Robyn. Robyn is non-verbal and has a very small number of requests he can make using a picture exchange communication system (PECS). When his attempts to communicate are unsuccessful, his behaviors are often challenging, including self-injurious ones (e.g., biting his hands, hitting his temple repeatedly with the heel of his hand).

Robyn is an only child. His parents are highly educated, motivated, and open to trying different therapeutic approaches to help him. His mother is a trained healthcare professional. After Robyn's birth, she quit her job and now devotes all her time to her son. Despite the lack of evidence to support the gluten-free casein-free (GFCF) diet as a treatment and cure for autism, Robyn's mother insists that he be on the diet and is convinced that he is showing improvement as a result of the dietary therapy. Even before his mother put him on the GFCF diet, like many children with autism, Robyn displayed very restricted dietary choices. His diet was limited to chicken nuggets, ketchup, Goldfish crackers and broccoli.

Robyn attends the IBI program Monday through Friday, from 9:00 a.m. to 3:00 p.m. The lunch hour is a structured activity during which the instructor therapists and supervising therapist monitor the children, encouraging them to eat the food provided by their parents, master the use of eating utensils, accept the introduction of new food items, and sit at the table without interfering with the other children. Robyn struggles during the lunch breaks; consistently refusing the food his mother sends and often grabbing at other children's food, especially chicken nuggets. When Julia and other IBI staff intervene, Robyn usually engages in one of his self-injurious behaviors.

Julia has exhausted the non-intrusive interventions available to her. Unfortunately, nothing has worked, and behavioral outbursts continue.

The outline of the second meeting includes debriefing on the *Step One* homework, working collaboratively through *Step Two* and *Step Three* of the deliberation process, and assigning *Step Four* as homework. The debrief regarding *Step One* should produce an articulation of the relevant data from the case. For example, the team discussion might produce the following:

- *The context for Robyn's case is a treatment centre for autistic children and more specifically, the structured lunch period, where there are clear behavioral expectations (e.g., children will eat the food provided by their families and not interfere with other children). These expectations are considered important in preparing the child for entry to school.*
- *The centre promotes evidence-based care which is family-centred.*
- *Julia is Robyn's primary therapist and is responsible for his daily care and interactions at the centre. She reports to a supervisor.*
- *Robyn is a young autistic boy with limited communication abilities, limited dietary interests, and some history of self-injurious behavior, especially when frustrated.*
- *During the lunch period, Robyn refuses to eat his food. He takes food from other children and emits self-injurious behavior if staff intervene.*
- *Continued efforts to implement the GFCF diet are causing upset and injury to Robyn, disruption of the lunch period for other children, and distress for the team.*

Next, the supervisor facilitates discussion of *Step Two* (identification of relevant principles, duties, and concepts) and *Step Three* (determination of conflicting principles). It is during discussions such as these that the facilitator needs to model reflexivity and create a safe and open learning environment conducive to reflection (see Sect. 31.5.3). Team reflection and discussion of *Step Two* might identify the following as being important principles and concepts:

- *Beneficence is related to (1) Julia's interest in teaching Robyn to eat; (2) Robyn's mother's insistence that the GFCF diet is an important component of his treatment and is helping him; and, conversely, (3) the option of allowing Robyn to have the desired foods because at least he would be eating something (rather than refusing the foods currently provided).*
- *Non-maleficence is relevant because of the therapist's interest in protecting Robyn and the other children from harm. In Robyn's case, the harms include not eating; the potential use of more intrusive measures; and the likelihood of escalating self-injurious behaviors.*
- *Respect for autonomy/persons is related to Robyn's clear preference to eat foods of his own choice, but also (under Cox's interpretation) to Robyn's mother's interest in having him follow the GFCF diet.*
- *Family-centred care, which aims for a mutually beneficial partnership among service providers and families, wherein the latter are empowered to determine how they will participate in care and decision making, could also support Robyn's mother's position.*
- *Evidence-based practice is relevant because the clinical team (including the program's pediatrician) believes there is insufficient evidence to support the GFCF diet.*
- *Professional commitments could relate to the therapist's interest in following her supervisor's guidance to lead the lunchtime circle, as well as the supervisor's interest in ensuring that a therapeutic diet is evidence-based.*
- *Relevant BACB Codes:1.10-Awareness of personal biases and challenges. Julia does not think that the GFCF diet is good for Robyn; 2.01-Providing effective treatment. Julia is not trained in nutrition and is unaware of scientific evidence for GFCF. She believes that sitting and eating lunch in the group without incident will improve Robyn's quality of life; 2.09-Involving clients and stakeholders. Julia wants to include Robyn's mother in treatment decisions but is conflicted about implementing the GFCF diet that the mother wants; 2.12-Considering medical needs. Julia is worried about Robyn's self-injury and about his health if he doesn't eat; 3.01-Responsibility to clients. Julia has a responsibility to do no harm to Robyn; 3.02-Identifying stakeholders. Julia has obligations to Robyn's mother; 3.12-Advocating for appropriate services. Julia has an obligation to advocate for evidence-based treatment for Robyn. 3.01 is likely the priority code to consider in this case. This relates to non-maleficence.*

Team reflection and discussion of *Step Three* may identify the following conflicts between principles (or important concepts):

- *Tension between beneficence (given the mother's belief about the GFCF diet and the therapist's interest in teaching Robyn to eat) and both non-maleficence and*

*respect for Robyn's autonomy. Trying to teach Robyn to eat using the GFCF-compliant foods provided, when Robyn will not eat them, has led to self-injury and grabbing food from other children, which are harmful to Robyn and others.*
- *Tension between family-centred care, which suggests that the mother should have a say in what her child eats, and professional commitments (e.g., the program's claim to provide evidence-based care), especially as the mother is advocating for a dietary plan that is controversial.*

At this point, the supervisor has led the team through some of the most challenging steps for learners who are relatively new to ethical analysis. The results of the reflection so far are shared with the team and the supervisor assigns *Step Four* as homework. This step asks the learner to consider the reflection to this point and articulate a question that captures the moral tension at the heart of the case.

The outline of the third meeting includes: debriefing on the *Step Four* homework; working collaboratively on *Step Five* through *Step Seven* of the deliberation process; and assigning *Step Eight* as homework. The debrief on *Step Four* should produce a question that highlights the moral tension. This can be a challenging skill to master. Team discussion might produce the following question:

- *Should the IBI team continue to implement the GFCF diet during the lunch period or should they ask Robyn's mother to provide something that Robyn will eat willingly?*

Next, the supervisor facilitates a discussion of *Step Five* (identifying the principle that should have priority), *Step Six* (encouraging reflection on whether additional information is needed), and *Step Seven* (considering whether a consensus of exemplary professional peers would agree with their decisions). For example, a team discussion of *Step Five* might identify the following principle:

- *The principle of non-maleficence trumps the others, as there are multiple opportunities for harm to both Robyn and the other children.*

Team reflection on *Step Six* might produce the following suggestions for further investigation:

- *The supervisor and therapist could review the data to determine if there is any relationship between an increase in self-injury and/or grabbing food from others and the introduction of the dietary plan. As the GFCF diet is not within the scope of applied behavior analysis, the team ought to consult with the medical team or a dietitian to ensure their understanding of the effectiveness of the GFCF diet is correct. As the organization claims to provide evidence-based services, it is very important that the team identifies the appropriate healthcare profession to consult regarding the implementation of this diet.*

Team reflection on *Step Seven* might produce the following suggestions:

- *The supervisor can confer with colleagues, look for clinical examples or experiences that resemble the current case, and engage with interprofessional peers in medicine and dietetics (regarding the diet).*

The fourth meeting may commence with a debrief on *Step Eight* and produce the following plan of action:

- *Depending on the outcomes of the previous two steps, the supervisor will arrange to meet with the mother, review Robyn's lunchtime data, explain the impact of his outbursts on himself and others, share findings from the literature regarding the GFCF diet, and recruit the mother to work with the supervisor to plan for Robyn's lunchtime learning going forward.*
- *The supervisor will explain that they consulted healthcare experts in arriving at their decision and, if these experts are part of their centre, the mother could be offered a consultation with them. Furthermore, the supervisor will seek the ongoing support of medical and dietetic colleagues to monitor the research and ensure the team adopts best practices based on the most up-to-date information.*

Once the ethical deliberation has been completed, the supervisor leads a debrief reinforcing three key messages. First, finding answers to ethical issues is not always straightforward but requires careful reflection and sometimes revisiting earlier steps in the deliberation. Second, working through one ethically complex case in this systematic manner is important, but does not serve as sufficient training. It is only through repeated exposure, employing a clear and reflective process, and receiving feedback from a facilitator skilled in ethical deliberation, that you develop competence. Third, the deliberation process may result in more than one equally defensible option, which may cause moral distress. Having used a defensible process to reach a decision should provide some degree of comfort. Once the debrief is finished, the supervisor should allow time for the team to provide feedback on the training as a whole. Such discussions should inform decisions on next steps in training. To ensure the team is comfortable providing feedback, opportunities for individual debriefing and anonymous feedback should be provided. Within two weeks of the fourth meeting, the supervisor should follow up with the team and review the process before introducing a new case in the second phase of training.

The remaining three training phases are structured following the BST model-practice-feedback loop. The number and frequency of meetings in subsequent training phases will depend on the availability of training time and the readiness of the team to take on more independent study (i.e., homework). For example, in the second training phase, the supervisor may be able to accomplish the deliberation process in three meetings. For instance, they might present the case and discuss *Step One* through *Step Three* in the first meeting; assign *Step Four* as homework; discuss *Step Four* through *Step Seven* in a second meeting; assign *Step Eight* as homework; and review the entire process in a third meeting. In general, the supervisor should aim to shift the focus of these meetings to creating a safe space and encouraging reflection and collaborative deliberation. Once a clinical team has achieved a certain proficiency in ethical deliberation, we recommend regular practice on ethical decision making be built into semi-annual training for all members of the team. Additionally, we recommend that clinical teams go through this training process before applying an ethical decision-making protocol to the day-to-day activities of the program.

## 31.9 Conclusion

Professionals in the health and human services sectors must often justify their recommendations, and such justifications must be based on logic and reason rather than personal feelings or emotions. When ethical dilemmas arise in clinical practice and the issue involves moral tension or distress, it is even more important that the professional has the skills to engage in an ethical deliberation process. Engaging in a transparent deliberative process should allow clinicians to explain their decision(s) in a manner that is justifiable and defensible. Traditionally, applied programs are demanding when it comes to learning practical skills, but expectations are quickly rising that ethics and ethical behavior are being prioritized as part of the daily operating habits of programs in which behavior analysts work, which means that learning in the workplace should include ethical decision making.

## References

BACB (Behavior Analyst Certification Board). 2020. "Ethics code for behavior analysts." *BACB.* https://bacb.com/wp-content/ethics-code-for-behavior-analysts/ Accessed 9 Apr 2024.

BACB (Behavior Analyst Certification Board). 2023. Behavior Analyst Certification Board Historical Ethics Timeline. *BACB.* https://www.bacb.com/ethics-information/ethics-resources/ Accessed 9 Apr 2024.

Bailey, Jon S., and Mary R. Burch. 2019. *Analyzing Ethics Questions from Behavior Analysts: A Student Workbook.* New York, NY: Routledge. https://doi.org/10.4324/9781351117784-1.

Beauchamp, Tom L., and James F. Childress. 2013. *Principles of Biomedical ethics*, 7th ed. Oxford: Oxford University Press.

Branch Jr., William T., and Richard Frankel. 2016. Not all stories of professional identity formation are equal: An analysis of formation narratives of highly humanistic physicians. *Patient education and counseling* 99 (8): 1394–1399.

Branch Jr., William T., and Maura George. 2017. Reflection-based learning for professional ethical formation. *AMA Journal of Ethics*19(4): 349–356.

Branch Jr., William T., and Anuradha Paranjape. 2002. Feedback and reflection: teaching methods for clinical settings. *Academic Medicine* 77 (12 Part 1): 1185–1188.

Cooperrider, David L., and Diana Whitney. 2005. *Appreciative Inquiry: A Positive Revolution in Change.* San Francisco: Berrett-Koehler Publishers Inc.

Cox, David J. 2012. From interdisciplinary to integrated care of the child with autism: The essential role for a code of ethics. *Journal of Autism and Developmental Disorders* 42: 2729–2738.

Davies, Samantha. 2012. Embracing reflective practice. *Education for Primary Care* 23 (1): 9–12.

Driessen, Erik, Jan van Tartwijk, and Tim Dornan. 2008. The self critical doctor: Helping students become more reflective. *BMJ* 336 (7648): 827–830.

Gillon, Raanan. 1994. Medical ethics: Four principles plus attention to scope. *BMJ* 309 (6948): 184–188.

McEldowney, Rose, and Margaret J. Connor. 2011. Cultural safety as an ethic of care: A praxiological process. *Journal of Transcultural Nursing* 22 (4): 342–349.

Miltenberger, Raymond G. 2008. Behavioral skills training procedures. In *Behavior modification: Principles and Procedures,* ed. by Raymond G. Miltenberger. 4th edn. Belmont: Thomson Wadsworth.

Nylen, Brendon, and Seth King. 2023. Training practitioners to conduct a functional behavior assessment via telehealth-delivered behavioral skills training: A systematic literature review. *Journal of Behavioral Education.* https://doi.org/10.1007/s10864-023-09514-8.

Rhodes, Rosamond, and David Alfandre. 2007. A systematic approach to clinical moral reasoning. *Clinical Ethics* 2 (2): 66–70.

Richard, Amélie., Mathieu Gagnon, and Emmanuelle Careau. 2019. Using reflective practice in interprofessional education and practice: A realist review of its characteristics and effectiveness. *Journal of Interprofessional Care* 33 (5): 424–436.

Schön, Donald A. 1983. *The Reflective Practitioner: How Professionals Think in Action.* New York: Basic Books.

Sun, Xiaohui. 2022. Behavior skills training for family caregivers of people with intellectual or developmental disabilities: A systematic review of literature. *International Journal of Developmental Disabilities* 68 (3): 247–273. https://doi.org/10.1080/20473869.2020.1793650.

Sush, Darren, and Adel C. Najdowski. 2021. *A Workbook of Ethical Case Scenarios in Applied Behavior Analysis*, 2nd ed. Amsterdam: Academic Press.

Varkey, Basil. 2021. Principles of clinical ethics and their application to practice. *Medical Principles and Practice* 30 (1): 17–28.

Weston, Regan, Tonya N. Davis, Supriya Radhakrishnan, Nicole O'Guinn, and Gabby Rivera. 2020. Comparing performance feedback and video self-monitoring within a BST package to train pre-service behavior analysts to conduct preference assessments. *Journal of Behavioral Education* 29 (3): 571–583. https://doi.org/10.1007/s10864-019-09333-w.

www.ingramcontent.com/pod-product-compliance
Lightning Source LLC
Chambersburg PA
CBHW072153120325
23430CB00003B/42
*9783031615641*